Annals of Mathematics Studies
Number 52

ADVANCES IN GAME THEORY

R. J. AUMANN
L. D. BERKOVITZ
M. DAVIS
W. H. FLEMING
A. R. GALMARINO
O. GROSS
J. C. HARSANYI
M. HEBERT
J. R. ISBELL
G. JENTZSCH
S. M. JOHNSON
M. MASCHLER
K. MIYASAWA
J. MYCIELSKI
E. D. NERING
H. NIKAIDO
G. OWEN
B. PELEG
H. RADSTRÖM
R. A. RESTREPO
J. L. ROSENFELD
C. RYLL-NARDZEWSKI
R. SELTEN
L. S. SHAPLEY
R. E. STEARNS
L. E. ZACHRISSON

EDITED BY

M. Dresher, L. S. Shapley, A. W. Tucker

PRINCETON, NEW JERSEY
PRINCETON UNIVERSITY PRESS
1964

L. C. Card 63-9985
ISBN 0-691-07902-1

Second Printing, 1971

Printed in the United States of America

PREFACE

This Study carries on the tradition of the four volumes entitled _Contributions to the Theory of Games_;* it is similar in format, but has a different reason for existence. The field of game theory is now well established and widely diffused through the mathematical world, thanks in part to the success of the earlier volumes; papers on game theory regularly appear in many scientific journals. But with no single, specialized journal as focal point, there remains a serious problem of communication within the field. It is often difficult, for isolated workers and especially for newcomers, to get abreast of developments, find out where the significant new problems lie, and avoid needless duplications of effort.

Accordingly, this Study was conceived as a one-volume cross section of current activity in the theory of games. Contributions were invited from a widely distributed and, it is hoped, representative group of investigators, and twenty-nine papers were finally accepted. Most of these might otherwise have appeared in one or another of the regular mathematics, economics, or operations research journals. It is hoped that the authors will agree that publication of their work in the present volume is a step toward improving the accessibility of the expanding frontiers of the theory.

The scope of this collection was arbitrarily drawn to exclude theoretical essays of primarily nonmathematical content, as well as mathematical applications of primarily nontheoretical interest. Also, no purely expository articles were solicited. Within these limits, however, the editors welcomed variety, not only in subject matter, but also in the level of the mathematical techniques invoked and in the level of abstractness or particularity in the approach. As the browsing reader will quickly discover, the contents of this book are far from homogeneous either in style, in

---

* Annals of Mathematics Studies 24 (1950), 28 (1953), 39 (1957), 40 (1959).

prerequisites, or in pertinence to any one area of interest.

As for subject matter, the editors can point to a good balance, achieved without premeditation, on many different accounts: two-person *vs.* n-person games; finite *vs.* infinite games; extensive *vs.* normal form; side-payment *vs.* no-side-payment games; cooperative *vs.* noncooperative games; existing theory *vs.* new departures; etc. This evidence of activity on all fronts speaks well for the vitality of the field as a whole, as does the large international authorship of this volume.

The contributions are arranged not by order of receipt, but by subject. It no longer seems appropriate, however, to organize the book into sections, as in earlier volumes, nor to attempt in the Preface to survey the contents from a unified viewpoint. Nevertheless, a brief description of the subjects that are covered will be given here to provide a bird's-eye view of the book as a whole and to supplement the bare titles in the Table of Contents.

The first thirteen papers deal with two-person games. In the first, SHAPLEY explores several topics growing out of the theory of matrix games, particularly their ordinal properties. Papers 2 and 3, by RESTREPO and JOHNSON, concern the solving of particular two-person games; the first is an abstract poker model, the second is a mathematical form of "hide and seek". Papers 4 and 5, by GROSS and NIKAIDO, apply the minimax theorem to point-set topology and mathematical economics, respectively. In Paper 6, ROSENFELD investigates the repeated play of a game matrix in which the values of some of the entries are not known to the players in advance. In Papers 7 and 8, DAVIS and MYCIELSKI carry on the study of infinite games of perfect information begun by Gale and Stewart in Study 28; the second paper makes an application of that discrete-time model to what could be called positional games with continuous time. This type of game is treated systematically in Paper 9 by RYLL-NARDZEWSKI, using general topological and set theoretical notions, and in Papers 10, 11, and 12 by BERKOVITZ and FLEMING, using techniques from the calculus of variations. Finally, ZACHRISSON, in Paper 13, considers games in which the strategies of the two players generate a Markov process; again information is essentially perfect and time may be a continuous parameter.

PREFACE

The last sixteen papers are in the field of n-person games. In Paper 14, ISBELL determines the existence, for various values of n, of a particular kind of n-person simple game. Papers 15 and 16, by SHAPLEY and OWEN, study the solutions of a class of games, mostly simple, that exhibit a certain compound structure. The next two papers, by GALMARINO and HEBERT, study two aspects of the solutions of all four-person constant-sum games. STEARNS, in Paper 19, obtains all solutions of all three-person games in the no-side-payment theory. In Paper 20, JENTZSCH considers the general theory of cooperative n-person games, with and without side payments. Papers 21, 22, and 23, by AUMANN and MASCHLER, MASCHLER, and PELEG, develop and apply a new solution concept of "bargaining set" for cooperative games in characteristic-function form. RADSTRÖM, in Paper 24, introduces a solution concept leading to certain "basic" imputations, each being associated with a hierarchical coalition structure of a certain kind. NERING, in Paper 25, advances still another solution concept for characteristic function games, based on the merging of players into successively larger coalitions. A common feature of these new solution concepts is that they all generalize the celebrated three-point classical solution to the zero-sum three-person game. MIYASAWA, in Paper 26, formulates a bargaining model for general cooperative games in normal form that leads to a one-point solution, or "value". SELTEN, in Paper 27, axiomatizes the valuation of n-person games in the classical theory, basing his considerations primarily on the extensive form. AUMANN, in Paper 28, considers infinite games in extensive form, in which continua of choices at each move as well as plays of infinite length are permitted; his results on behavior strategies would apply most directly to the noncooperative theory. Finally, HARSANYI, in the concluding paper, undertakes to analyze and refine the equilibrium-point concept with the goal of providing a unique solution to noncooperative games.

* * *

The editing of this volume was done primarily at The RAND Corporation, which provided the services of two of the editors, as well as secretarial help. The Office of Naval Research supported the work of the third editor through its Logistics project at Princeton University. Thanks are also due that project for providing the English translation of Paper 18, and to The

RAND Corporation for making available its translation of Paper 20. The burden assumed by several other authors in supplying texts in English, not their native language, should not be overlooked. A host of anonymous referees (including a majority of the contributors as well as several others) gave willing assistance; many of them performed beyond the call of duty. The typing of the master copy has been the painstaking and efficient work of Mrs. Margaret Wray. Finally, the Princeton University Press, through its Science Editor Jay Wilson, has been constantly interested and helpful throughout the preparation of this volume. To all those who have given their assistance, the editors offer their sincere thanks.

NOTE

A few of the papers in this Study were among the 39 papers presented at the Princeton University Conference on Recent Advances in Game Theory, October 4-6, 1961, directed by Oskar Morgenstern and A. W. Tucker. Proceedings of this Conference were prepared by Michael Maschler for the Econometric Research Program, Department of Economics, Princeton University, and were privately printed for members of The Princeton University Conference.

M. Dresher
L. S. Shapley
A. W. Tucker

December 1963

CONTENTS

Preface v

Paper 1. Some Topics in Two-Person Games 1
By L. S. Shapley

2. Games With a Random Move 29
By Rodrigo A. Restrepo

3. A Search Game 39
By Selmer M. Johnson

4. The Rendezvous Value of a Metric Space 49
By O. Gross

5. Generalized Gross Substitutability and Extremization 55
by Hukukane Nikaido

6. Adaptive Competitive Decision 69
By Jack L. Rosenfeld

7. Infinite Games of Perfect Information 85
By Morton Davis

8. Continuous Games of Perfect Information 103
By Jan Mycielski

9. A Theory of Pursuit and Evasion 113
By C. Ryll-Nardzewski

10. A Variational Approach to Differential Games 127
By Leonard D. Berkovitz

11. A Differential Game Without Pure Strategy Solutions on an Open Set 175
By Leonard D. Berkovitz

12. The Convergence Problem for Differential Games, II 195
By Wendell H. Fleming

13. Markov Games 211
By Lars Erik Zachrisson

14. Homogeneous Games, III 255
By J. R. Isbell

15. Solutions of Compound Simple Games 267
By L. S. Shapley

16. The Tensor Composition of Nonnegative Games 307
By Guillermo Owen

CONTENTS

Paper 17. On the Cardinality of Solutions of Four-Person Constant-Sum Games 327
By Alberto Raul Galmarino

18. The Doubly Discriminatory Solutions of the Four-Person Constant-Sum Game 345
By Michael H. Hebert

19. Three-Person Cooperative Games Without Side Payments 377
By R. E. Stearns

20. Some Thoughts on the Theory of Cooperative Games 407
By Gerd Jentzsch

21. The Bargaining Set for Cooperative Games 443
By Robert J. Aumann and Michael Maschler

22. Stable Payoff Configurations for Quota Games 477
By Michael Maschler

23. On the Bargaining Set $\mathfrak{M}_0$ of m-Quota Games 501
By Bezalel Peleg

24. A Property of Stability Possessed by Certain Imputations 513
By Hans Radström

25. Coalition Bargaining in n-Person Games 531
By Evar D. Nering

26. The n-Person Bargaining Game 547
By Koichi Miyasawa

27. Valuation of n-Person Games 577
By Reinhard Selten

28. Mixed and Behavior Strategies in Infinite Extensive Games 627
By Robert J. Aumann

29. A General Solution for Finite Noncooperative Games Based on Risk-Dominance 651
By John C. Harsanyi

ADVANCES IN GAME THEORY

SOME TOPICS IN TWO-PERSON GAMES

L. S. Shapley

INTRODUCTION

This note reports on half-a-dozen loosely related excursions into the theory of finite, two-person games, both zero-sum and non-zero-sum. The connecting thread is a general predilection for results that do not depend upon the full linear structure of the real numbers. Thus, most of our theorems and examples are invariant under order-preserving transformations applied to the payoff spaces, while a few (in § 1) are invariant under the group of transformations that commute with multiplication by -1.

We should make it clear that we are not interested in "ordinal" utility, as such, but rather the ordinal properties of "cardinal" utility. The former would require a conceptual reorientation which we do not wish to undertake here. Nevertheless, the ordinalist may find useful ideas in this paper.

Rather than summarize the whole paper here, we shall merely take a sample; for a more synoptic view the reader is invited to scan the section headings. Consider first the matrix game shown in the margin. The solution is easily found as soon as we recognize that the game is symmetric in the players. Indeed, if we map each player's i^{th} strategy into the other's $i{+}1^{st}$ (mod 4), we merely reverse the signs of the payoffs. It follows that the value is 0 and that there is a solution of the form (a,b,a,b), (b,a,b,a). (See § 1.)

0	1	2	−1
−1	0	1	−2
2	−1	0	1
1	−2	−1	0

Next, consider the class of matrix games in which the payoffs are ordered like those in the matrix at left (next page). In all these games, player I's third strategy is never playable, although it is not dominated in the usual sense. To verify this, observe that if the value of the 2-by-2

10	0	8
1	9	7
2	3	6
3	4	5

subgame in the upper left corner is greater than "3" the third row is dominated by a linear combination of the first two rows, while if it is less than "7" the third column can be dropped, and then the third row. (See § 3.)

Finally, consider the non-zero-sum game with outcome matrix as shown at right. Player I rates the outcomes $A > B > C$; player II rates them $B > A > C$. If we apply the algorithm of "fictitious play" to this game, a strange thing happens. Rather than converging to the unique equilibrium point (at which all probabilities are equal), the sequence of mixed-strategy pairs generated by the algorithm oscillates around it, keeping a finite distance away. (See § 5.)

A	C	B
B	A	C
C	B	A

The five main sections of this paper are essentially independent, both logically and topically. Our reason for combining them into a single paper is the hope that they will appeal to a single audience. Much of this work has already appeared in short RAND Memoranda [15], and some of it has been cited in the published literature [5], [11]. In reworking this material, however, we have added many new results.

§ 1. SYMMETRIC GAMES

1.1. Discussion

It is easy to see that a two-person zero-sum game can be symmetric in the players without having a skew-symmetric payoff matrix. "Matching Pennies" is a simple example; another is shown at the right; and another is given in the Introduction. The point is, of course, that an automorphism of the game that permutes the players can simultaneously shuffle the labels of the pure strategies. It would be interesting to know something about the abstract structure of such automorphisms.*

1	1	−1
−1	0	−1
−1	1	1

As a first step, we shall show that the matrix of a symmetric game can be decomposed into an array of square blocks in such a way that (a) each block has constant diagonals (in one direction), (b) the array as a whole is skew-symmetric in a certain sense, and (c) the size of each block is a power

* The narrow "skew-symmetric" definition is most often seen in the literature (e.g., [19], [10], [7]). But Nash uses the more general form in [13]; see also [19], p. 166.

	1	3	2
1	1	−1	1
3	−1	1	1
2	−1	−1	0

of 2. This is illustrated at left for the 3-by-3 example given above.

The "power of 2" property, (c), is quite interesting. It tells us, for example, that the 6-by-6 game illustrated below, which is obviously symmetric and has constant diagonals, can nevertheless be decomposed into smaller blocks. The 4-by-4 and 8-by-8 analogues of this matrix, on the other hand, do not decompose.

	1	2	3	4	5	6
1	1	2	3	−3	−2	−1
2	−1	1	2	3	−3	−2
3	−2	−1	1	2	3	−3
4	−3	−2	−1	1	2	3
5	3	−3	−2	−1	1	2
6	2	3	−3	−2	−1	1

$\longrightarrow$

	2	5	3	6	4	1
1	2	−2	3	−1	−3	1
4	−2	2	−1	3	1	−3
2	1	−3	2	−2	3	−1
5	−3	1	−2	2	−1	3
3	−1	3	1	−3	2	−2
6	3	−1	−3	1	−2	2

1.2. The Main Theorem

Let $A, B, \ldots$ denote n-by-n game matrices (n fixed throughout), and let $P, Q, R, \ldots$ denote permutation matrices of the same size. Primes will denote transposition, which is equivalent to inversion for permutation matrices. We define the following matrix properties:

<u>equivalence</u>: $A \equiv B \longleftrightarrow A = PBQ'$ for some P, Q.

<u>symmetry</u>: $A \in \Sigma \longleftrightarrow A \equiv -A'$.

<u>conjugates</u>: $P \approx Q \longleftrightarrow P = RQR'$ for some R.

Certain subclasses of Σ will also be of interest:

$$A \in \Sigma(P,Q) \longleftrightarrow B = -PB'Q' \text{ for some } B \equiv A.$$

These subclasses exhaust Σ, but are not disjoint. Indeed, we have

LEMMA 1. If $PQ \approx RS$ then $\Sigma(P,Q) = \Sigma(R,S)$.

PROOF. Given $PQ = TRST'$ and $A = -PA'Q'$, we require $B \equiv A$ such that $B = -RB'S'$. As it happens, $B = T'AP'TR$ serves the purpose. Indeed, since $P' = QTS'R'T'$, we have

$$\begin{aligned} B &= T'(-PA'Q')(QTS'R'T')TR \\ &= -T'PA'TS' \\ &= -(T'AP'T)'S' \\ &= -RB'S'. \end{aligned}$$

Note that $\Sigma(P,Q)$ depends only on PQ. If we define $\Sigma(P) = \Sigma(P,I)$, where I is the identity, then Lemma 1 may be restated:

LEMMA 1'. If $P \approx Q$ then $\Sigma(P) = \Sigma(Q)$.

It can be shown that the converse is valid. Since two permutations are conjugate if and only if their cyclic factors have matching periods, there is thus a one-to-one correspondence between the classes $\Sigma(P)$ and the partitions of n. This stronger result is not used in what follows.

LEMMA 2. If Q is an odd power of P then $\Sigma(Q) \supseteq \Sigma(P)$.

PROOF. Suppose $A = -PA'$. We must show that $A \in \Sigma(P^{2k+1})$ for all k. But $A = -P(-PA')' = PAP'$; hence $A = P^kAP'^k = -P^{k+1}A'P'^k$. Therefore $A \in \Sigma(P^{k+1}, P^k) = \Sigma(P^{2k+1})$.

THEOREM 1.1. Every symmetric game $A \in \Sigma$ is equivalent to a game B satisfying $B = -RB'$ for some permutation R, the order of which is a power of 2.

PROOF. Let $A \in \Sigma(P)$. The order of P can be represented in the form $c2^k$, with c odd. Then the order of P^c will be 2^k. By Lemma 2, $A \in \Sigma(P^c)$. Thus P^c will serve as the R of the theorem. Q.E.D.

1.3. Block Decomposition

The decomposition into blocks can now be described. By proper choice of B, we can give R the form:

$$R = (1\ 2\ 3\ \ldots\ \lambda_1)(\lambda_1+1\ \ldots\ \lambda_1+\lambda_2)\ \ldots\ (\ldots\ n),$$

where the periods λ_μ are powers of 2. B now breaks up into a square array of λ_μ-by-λ_ν blocks $B_{\mu\nu}$, and the equation $B = -RB'$ implies (and is implied by) the structure set forth in the following theorem. The proof is straightforward.

THEOREM 1.2. Let $\lambda = \min(\lambda_\mu, \lambda_\nu)$. The block $B_{\mu\nu}$ has constant diagonals, in the sense that numbers $\beta_1, \ldots, \beta_\lambda$ exist such that for $\overline{ij}$ in that block,*

$$b_{ij} = \beta_h \quad \text{if} \quad i - j \equiv h \pmod{\lambda}.$$

(If $\lambda_\mu \neq \lambda_\nu$, then $B_{\mu\nu}$ breaks up into identical square sub-blocks of size λ.) In the symmetrically located block $B_{\nu\mu}$ the same numbers appear; we have

$$b_{ij} = -\beta_{\lambda-h+1} \quad \text{if} \quad i - j \equiv h \pmod{\lambda}.$$

In particular, along the main diagonal of the array ($\mu = \nu$), we have $\beta_h = -\beta_{\lambda-h+1}$ for $h = 1, \ldots, \lambda$.

COROLLARY 1. Indecomposable symmetric n-by-n games exist only for n a power of 2.

COROLLARY 2. Every symmetric game of odd size has a zero in its payoff matrix.

1.4. Solutions

To solve a symmetric game we may (a) replace the $B_{\mu\nu}$ by their

---

* We write $\overline{ij}$ for the ordered pair (i, j).

average values, (b) solve the resulting skew-symmetric matrix,* and (c) distribute the mixed-strategy probabilities for each block equally among its constituent pure strategies.** Not every solution of the original can be obtained in this way, however. In the game at the right, for example, (0, 2/3, 1/3, 0) is a basic (extreme) optimal strategy of each player.

1	−1	2	−2
−1	1	−2	2
2	−2	4	−4
−2	2	−4	4

1.5. Symmetric Nonzero-sum Games

There is a direct extension to nonzero-sum games. Let us call the matrix pairs (A_1, A_2) and (B_1, B_2) equivalent if, for some P, Q, both $A_1 = PB_1Q'$ and $A_2 = PB_2Q'$. Let us call (A_1, A_2) symmetric if (A_1, A_2) and (A_2', A_1') are equivalent. Then the following counterpart to Theorem 1.1 can be established by essentially the same proof:

THEOREM 1.3. Every symmetric nonzero-sum games (A_1, A_2) is equivalent to a game of the form $(B_1, B_2) = (RB_2', RB_1')$ for some permutation R of order a power of 2.

The description of the block decomposition remains much as before, though the "main diagonal" loses some of its special significance. Corollary 1 remains valid, but not Corollary 2.

§ 2. SOME THEOREMS ABOUT SADDLEPOINTS

2.1. A Condition for the Existence of a Saddlepoint

THEOREM 2.1. If A is the matrix of a zero-sum two-person game, and if every 2-by-2 submatrix of A has a saddlepoint, then A has a saddlepoint.

PROOF. Let val[A] = v. Let j be the index of a column having the minimum number, π, of entries greater than v. Suppose $\pi > 0$; then, for

* See Kaplansky [10] and Gale, Kuhn, and Tucker [7].

** See for example, Gale, Kuhn, and Tucker [8], application (e).

some i we have $a_{ij} > v$. Since the value of the game is only v, we must have $a_{ij'} \leq v$ for some j'. But the column indexed by j' has at least π entries greater than v, too many to be paired off against the $\pi-1$ remaining entries $> v$ of the other column. Thus, for some i' we have $a_{i'j'} > v \geq a_{i'j}$. Since the 2-by-2 submatrix:

	j	j'
i	$> v$	$\leq v$
i'	$\leq v$	$> v$

has no saddlepoint, the assumption $\pi > 0$ was incorrect. Hence there is a column with <u>no</u> entries greater than v. Similarly there is a row with no entries less than v. Q.E.D.

2.2. Detached Rows and Columns

The hypothesis of Theorem 2.1 actually imposes a very special structure on the matrix A. Let us say that the pth row of A is *detached* if

$$\max_j a_{pj} \leq \max_{i \neq p} \min_j a_{ij}.$$

Similarly, the qth column is *detached* if

$$\min_i a_{iq} \geq \min_{j \neq q} \max_i a_{ij}.$$

Detachment obviously implies domination. For 2-by-2 matrices, the existence of a detached row or column is equivalent to the existence of a saddlepoint.

THEOREM 2.2. If every 2-by-2 submatrix of A has a saddlepoint, then A has a detached row or column.

PROOF. By Theorem 2.1, both A and A' (its transpose) have saddlepoints. Hence there is a column of A with no entries greater than val[A], as well as a column of A (row of A') with no entries less than val[A']. If val[A] $<$ val[A'], then these columns are distinct, and the latter is a detached column. Similarly, if val[A] $>$ val[A'], there is a detached row. If val[A] = val[A'] = v, then there is either a detached row or column, or a saddlepoint $\overline{pq}$ common to both A and A'. In the latter case, $a_{pj} \equiv a_{iq} \equiv v$, all i,j, and we can use the fact that the submatrix obtained by deleting row p and column q has a saddlepoint to show that either p or q is detached, depending on whether the value of the submatrix is $\geq v$ or $\leq v$. Q.E.D.

2.3. A Generalization

Theorem 2.1 can be generalized, after a fashion. Let us say that a matrix is "in general position" if no two collinear entries are equal.

THEOREM 2.3. Let A be an m-by-n matrix in general position, and let $2 \leq r \leq m$, $2 \leq s \leq n$. If every r-by-s submatrix of A has a saddlepoint, then A has a saddlepoint.

PROOF. It suffices to prove the case where $r = m \geq 2$ and $s = n-1 \geq 2$; the rest will follow by induction and symmetry. Let A^q denote A with the qth column deleted. Let $\overline{i_q j_q}$ be the location of the saddlepoint of A^q (which is unique, since A is in general position). If all of the j_q's are distinct, for q = 1, ..., n, then every column of A will contain one of the points $\overline{i_q j_q}$. Since each $a_{i_q j_q}$ is a column maximum, one of them will be the maximum of the whole matrix. On the other hand, it must also be the (strict) minimum of its row in A^q, which contains at least two entries. This impossibility implies that the j_q's are <u>not</u> all distinct. Let $j_q = j_t$, $q \neq t$. Then it is apparent that the point $\overline{i_q j_q} = \overline{i_t j_t}$ is a saddlepoint of A. Q.E.D.

At right we illustrate what can happen if two collinear entries are equal. Every 3-by-2 submatrix has a saddlepoint, but not the full matrix.

1	0	−1
−2	0	2
2	−1	3

§ 3. ORDER MATRICES

3.1. Definitions. The Saddle

By a <u>line</u> of a matrix we shall mean either a row or a column. Two numerical matrices of the same size will be called <u>order equivalent</u> if the elements of corresponding lines are ordered alike. An <u>order matrix</u>, $\mathcal{A}$, is an equivalence class of order-equivalent numerical matrices. Abstractly, it may be regarded as a partial ordering $\underset{\mathcal{A}}{<}$ on the set $I(\mathcal{A})$ of all index pairs $\overline{ij}$, with the property that collinear points are always comparable, while noncollinear points are never comparable, except as a result of transitivity.

If $K \subseteq I(\mathcal{A})$ is a set of index pairs, we shall write K_1 for the set of first members, K_2 for the set of second members, and $\overline{K}$ for $K_1 \times K_2$. In other words, K_1 is the smallest set of rows, K_2 the smallest set of columns, and $\overline{K}$ the smallest submatrix, that covers K. If $K = \overline{K}$ then K is <u>rectangular</u>.

A <u>generalized saddle point</u> (<u>GSP</u>) of an order matrix $\mathcal{A}$ is a rectangular set $K \subseteq I(\mathcal{A})$ such that (1) for each $i \notin K_1$ there is a $p \in K_1$ with $\overline{pj} \underset{\mathcal{A}}{>} \overline{ij}$ for all $j \in K_2$, and (2) for each $j \notin K_2$ there is a $q \in K_2$ with $\overline{iq} \underset{\mathcal{A}}{<} \overline{ij}$ for all $i \in K_1$.* (Note the strict inequalities.) A GSP that contains no other GSP is called a <u>saddle</u>.

THEOREM 3.1. Every order matrix has a unique saddle.

The proof consists in showing that the intersection of two GSP's is a GSP. Let K and L be two GSPs of $\mathcal{A}$. Then certainly $K \cap L \neq \phi$, since K and L must both carry the optimal mixed strategies of any numerical matrix belonging to $\mathcal{A}$.** It will suffice to show that for each $i \notin K_1$ there is a p in $K_1 \cap L_1$ with $\overline{pj} > \overline{ij}$ for all $j \in K_2 \cap L_2$. But we can find a $p' \in K_1$ with that property, since K is a GSP. If $p' \notin L_1$, then we can find a $p'' \in L_1$ with $\overline{p''j} \underset{\mathcal{A}}{>} \overline{p'j}$ for all $j \in K_2 \cap L_2$. If $p'' \notin K_1$, then $p''' \in K_1$ can be found bearing the same relation to p'', etc. The strict inequalities ensure that the sequence terminates. That is, there is a row in $K_1 \cap L_1$ that majorizes row i on $K_2 \cap L_2$. Q.E.D.

The saddle of $\mathcal{A}$ will be denoted by $S(\mathcal{A})$. It contains everything

---

* Compare [5], pp. 41-42.

** Bass [1] found a way to prove $K \cap L \neq \phi$ directly, without reference to optimal strategies.

relevant to the solutions of the numerical games A belonging to $\mathfrak{A}$. In particular, it contains Bohnenblust's "essential" submatrix* and the Shapley-Snow "kernels."**

We may define a weak GSP by using nonstrict inequalities $\underset{\mathfrak{A}}{\geq}$, $\underset{\mathfrak{A}}{\leq}$. The intersection of two weak GSP's need not be a weak GSP. However, as Bass has shown, the intersection of a weak GSP with the saddle is a weak GSP.

3.2. Residuals

A one-element saddle is an ordinary (strict) saddlepoint, and is easy to find, even in a very large matrix. Identification of the saddle in general may be more tedious. The next theorem, based on an ingenious idea of Harlan Mills, provides us a rapid method of generating points in the saddle, and thereby substantially reduces the search. (See also § 3.5.)

LEMMA 1. (Mills) Let $\mathfrak{A}$ be an order matrix with at least two columns, let $\overline{pq}$ be maximal with respect to $\underset{\mathfrak{A}}{<}$, and let $\mathfrak{A}^q$ be obtained from $\mathfrak{A}$ by deleting column q.*** Then $S(\mathfrak{A}^q) \subseteq S(\mathfrak{A})$.

PROOF. Write S for $S(\mathfrak{A})$ and S^q for $S_1 \times (S_2 - \{q\})$. If $q \notin S_2$, then S^q is at once seen to be a GSP of $\mathfrak{A}^q$. If $q \in S_2$, then $p \in S_1$ by maximality. To show that S^q is a GSP of $\mathfrak{A}^q$, note that the row condition is satisfied for S^q, just as for S, while the column condition fails only if column q was used (in $\mathfrak{A}$) to minorize some column outside S_2. But this is impossible, by the maximality of $\overline{pq}$. Thus S^q is always a GSP of $\mathfrak{A}^q$. Hence $S(\mathfrak{A}^q) \subseteq S^q \subseteq S$.

A similar lemma holds concerning the deletion of rows containing minimal elements.

The following lemma is easily established; we omit the proof.

LEMMA 2. If $\mathfrak{B}$ is obtained from $\mathfrak{A}$ by deletion of a strictly majorized row or a strictly minorized column,

* See [2], p. 52; also [9], p. 44.

** See [17], p. 32.

*** I.e., delete column q from any $A \in \mathfrak{A}$, then use order equivalence.

then $S(\mathcal{B}) \subseteq S(\mathcal{A})$.

Now consider a sequence of nonvoid order matrices $\mathcal{A}, \mathcal{A}^{(1)}, \mathcal{A}^{(2)}, \ldots,$ each obtained from the preceding by deleting a column containing a maximal point, <u>or</u> a row containing a minimal point, <u>or</u> a strictly majorized row, <u>or</u> a strictly minorized column. If $\mathcal{A}$ is an m-by-n matrix, then $\mathcal{A}^{(m+n-2)}$ is 1-by-1. The point so defined is called a <u>residual</u> of $\mathcal{A}$. Let $R(\mathcal{A})$ denote the set of all residuals of $\mathcal{A}$. Our lemmas, together with the fact that the saddle is rectangular, give us the following result:

THEOREM 3.2. $R(\mathcal{A}) \subseteq \overline{R(\mathcal{A})} \subseteq S(\mathcal{A})$.

3.3. Examples

Let us illustrate the residual concept.

(a) In the order matrix of "Matching Pennies" (at left), every entry is either maximal or minimal. By the maximality of $\overline{11}$ we can strike out the first column, then the first row goes by the minimality of $\overline{12}$, and $\overline{22}$ is a residual. The three other sequences give us $\overline{11}$, $\overline{12}$, and $\overline{21}$ as residuals. Thus $R(\mathcal{A}) = \overline{R(\mathcal{A})} = S(\mathcal{A}) = I(\mathcal{A})$.

1	−1
−1	1

(b) More generally, let $\mathcal{A}$ be a square matrix with a "detached" diagonal; for example let $\overline{ii}$ be maximal with respect to $\underset{\mathcal{A}}{<}$, for all i. Then for each i, we can delete all columns except i by Lemma 1, and then by row deletions obtain $\overline{ii}$ as a residual. Hence $\overline{R(\mathcal{A})} = S(\mathcal{A}) = I(\mathcal{A})$.

(c) We first conjectured [15b] that deletions of the first type (Lemma 1) would always suffice to determine the saddle. This was disproved by Bass [1] with the 4-by-4 example at right, which has no residuals "of the first type" in the top row. But if, for example, we strike out column 3, then row 3 goes by domination (Lemma 2), and the resulting 3-by-3 game has a detached diagonal, making $\overline{11}$ a residual by (b) above. (All residuals are circled.)

①	⑪	2	⑬
⑨	⑥	③	14
2	⑤	⑮	4
⑧	⑫	⑩	⑦

(d) Thus, a new counterexample is needed to show that $\overline{R(\mathcal{A})} \neq S(\mathcal{A})$ is possible. It is provided by the example described in the Introduction, which we reproduce here with residuals circled. As we shall see later, it is no coincidence that the residuals avoid the "unplayable" third row.

⑩	⓪	⑧
1	⑨	⑦
2	3	6
③	④	5

3.4. Minimax and Maximin Points

A minimax point of a numerical matrix is a column maximum of smallest value. A maximin point is a row minimum of largest value. Since noncollinear comparisons are usually involved, these terms are not well defined for order matrices. Nevertheless, we have the following:

THEOREM 3.3. The minimax and maximin points of any numerical matrix $A \in \mathfrak{A}$ are residuals of $\mathfrak{A}$.

PROOF. Let $\overline{pq}$ be a minimax point of some $A \in \mathfrak{A}$. If A has more than one column, then it will have a maximum in some column other than q, and the corresponding location in $\mathfrak{A}$ will be maximal. Delete this column and $\overline{pq}$ is still minimax. In this way we may strip the matrix down to just the single column q, and then by row deletions obtain $\overline{pq}$ as a residual. The proof for maximin points is similar.

Bass [1] has shown that every weak GSP contains a minimax point and a maximin point.

3.5. An Algorithm

The following method for determining the saddle was suggested by Mills (see [1], p. 3): Start with any submatrix K^0 known to be contained in the saddle (for example, a single residual). For $h = 0, 1, \ldots,$ given $K^h \subseteq S(\mathfrak{A})$, construct K^{h+1} by adjoining either (a) a new row that is maximal (i.e., not strictly majorized) in $I_1(\mathfrak{A}) \times K_2^h$, or (b) a new column that is minimal in $K_1^h \times I_2(\mathfrak{A})$. Then $K^{h+1} \subseteq S(\mathfrak{A})$. If neither (a) nor (b) is possible, then $K^h = S(\mathfrak{A})$.

3.6. The Center

Let $C_1(A)$ denote the set of "active" strategies for the first player in A, i.e., those that appear with positive probability in some solution. Similarly, $C_2(A)$ for the second player. Let $C(A) = C_1(A) \times C_2(A)$. Then the set

$$C(\mathfrak{A}) = \bigcup_{A \in \mathfrak{A}} C(A)$$

will be called the <u>center</u> of $\mathfrak{A}$.* The center is contained in the saddle. It is not necessarily rectangular, as may be seen from the example at the right (center shaded).

0	4	2
4	0	3
1	2	5

Example (d) of § 3.3 suggests that a point which never occurs in optimal play cannot be a residual, i.e., that $R(\mathfrak{A}) \subseteq C(\mathfrak{A})$. This is not always true. For example, a matrix not "in general position" can cause trouble, as at left. Another difficulty is more basic. In the matrix at right, columns 4 and 5 serve to establish that "10" < "13" and "15" < "18". These relations make row 1 unplayable, despite the presence of residuals.*

(0)	(1)	(1)
3	(0)	(2)

1	(10)	(4)	11	5
(15)	7	(6)	8	16
(2)	(18)	(3)	19	17
(13)	(9)	(14)	12	20

Accordingly, we modify the order matrix concept for the sequences $\{\mathfrak{A}^{(k)}\}$ that define $R(\mathfrak{A})$ (see § 3.2), retaining noncollinear comparisons from $\mathfrak{A}$ whenever applicable. The new relations can destroy maximal or minimal elements, but not create them; hence a <u>restricted</u> <u>residual</u> <u>set</u> $R^{\#}(\mathfrak{A}) \subseteq R(\mathfrak{A})$ results. Our aim is to prove that $R^{\#}(\mathfrak{A}) \subseteq C(\mathfrak{A})$, for $\mathfrak{A}$ in general position.

LEMMA 3. If $\overline{pq} \in C(A)$ then every neighborhood of A contains a B having a <u>unique</u> solution and such that $\overline{pq} \in C(B)$.

PROOF. If $\overline{pq} \in C(A)$ then a "basic" (extremal) solution (x^o, y^o) of A exists with $x_p^o > 0$, $y_q^o > 0$. We may assume that $\text{val}[A] = v \neq 0$. By the Shapley-Snow theorem on basic solutions [17], a nonsingular submatrix $\dot{A}$ can be found for which $(\dot{x}^o, \dot{y}^o)$ is an "equalizer":

$$\dot{x}^o\dot{A} = \dot{y}^o\dot{A}' = v\dot{1}, \text{ whence } \dot{x}^o = v\dot{1}\dot{A}^{-1},\ \dot{y}^o = v\dot{1}\dot{A}'^{-1}.$$

(Dots indicate suppression of indices not appearing in $\dot{A}$; $\dot{1}$ is a vector of 1's.) Given $\epsilon > 0$, define $\dot{B} = \dot{A} - (\epsilon/v)\dot{A}^2$. If ϵ is sufficiently small, we can invert $\dot{B}$, as follows:

$$\dot{B}^{-1} = \dot{A}^{-1} + (\epsilon/v)\dot{I} + O(\epsilon^2).$$

---

* Proof: If row 1 is active, then so is row 2, by an easy domination argument. Let $d = (7-15, 1-10, 10-7, 15-1)$. Then $\Sigma_i d_i = 0$, and we verify that $\Sigma_i a_{ij} d_i > 0$ for each j. Hence, any x with $x_1, x_2 > 0$ can be improved.

Then $\dot{B}$ has an equalizer solution proportional to $(\dot{x}^o, \dot{y}^o) + \epsilon(\dot{1}, \dot{1}) + 0(\epsilon^2)$. This is all-positive for small ϵ, and hence unique.*

We now replace $\dot{A}$ by $\dot{B}$ in the original matrix. Since the value and solution of $\dot{B}$ differ only slightly from those of $\dot{A}$, it is easy to perturb the rows and columns outside the submatrix so as to make them irrelevant. The full matrix now has a unique solution, and p and q are still active. Q.E.D.

LEMMA 4. Let A have at least two columns, let A^q result from A by deleting column q, and let $A(\omega)$ result from A by adding ω to the entry at $\overline{pq}$. Then $\lim_{\omega\to\infty} \mathrm{val}[A(\omega)] = \mathrm{val}[A^q]$.

PROOF. Let x^o be optimal in A^q. Given $\epsilon > 0$, choose x such that $\|x - x^o\| \leq \epsilon$ and $x_p > 0$. Then

$$xA(\omega) \geq (v-\epsilon a,\ v-\epsilon a,\ \ldots,\ \underbrace{-a+\omega x_p}_{(q^{th}\text{ component})},\ \ldots,\ v-\epsilon a),$$

where $v = \mathrm{val}[A^q]$ and $a = \max_{ij}|a_{ij}|$. Thus, for ω large, x guarantees the first player at least $v - \epsilon a$ in the game $A(\omega)$. Hence we have

$$v - \epsilon a \leq \mathrm{val}[A(\omega)] \leq v. \qquad \text{Q.E.D.}$$

The following lemmas involve the "modified" order matrix concept, in that $\underset{\mathfrak{A}}{>}$ and $\underset{\mathfrak{B}}{>}$ may include relations not derivable by collinear comparison.

LEMMA 5. Let $\mathfrak{A}$ be in general position; let $\overline{pq}$ be maximal in $\underset{\mathfrak{A}}{>}$; and let the elements of $\mathfrak{B}$ be the elements of $\mathfrak{A}$ with column q deleted.** Then $C(\mathfrak{B}) \subseteq C(\mathfrak{A})$.

---

* Another $\dot{B}$, suggested by O. Gross, is given by

$$b_{ij} = a_{ij} - \Sigma_i\dot{x}_i a_{ij} - \Sigma_j\dot{y}_j a_{ij} + 2\ \Sigma_{ij}\dot{x}_i\dot{y}_j a_{ij},$$

with $(\dot{x}, \dot{y})$ all-positive but close to $(\dot{x}^o, \dot{y}^o)$.

** Contrast $\mathfrak{B}$ with the $\mathfrak{A}^q$ of Lemma 1.

PROOF. Take $\overline{rs} \in C(\mathcal{B})$. Using Lemma 3 and the fact that $\mathcal{A}$ is in general position, find $A \in \mathcal{A}$ such that the associated $A^q \in \mathcal{B}$ has a unique solution (x^*, y^*) with $x_r^* > 0$ and $y_s^* > 0$. Define the functions

$$f(x) = \min_{j \neq q} \sum_i a_{ij}x_i, \qquad g(y) = \max_i \sum_{j \neq q} a_{ij}y_j.$$

Since $f(x)$ is polyhedral (piecewise linear) and has a unique maximum at x^*, there exists $\alpha > 0$ such that

$$f(x) \leq \text{val}[A^q] - \alpha\|x^* - x\| \tag{3.1}$$

for all mixed-strategy vectors x. Similarly, there exists $\beta > 0$ such that

$$g(y) \geq \text{val}[A^q] + \beta\|y^* - y\| \tag{3.2}$$

for all mixed-strategy vectors y with $y_q = 0$.

Now let $A(\omega)$ be obtained from A by adding $\omega > 0$ to the element a_{pq}, and let $(x(\omega), y(\omega))$ be any solution of $A(\omega)$. Then certainly $\text{val}[A(\omega)] \leq f(x(\omega))$, and we have by (3.1)

$$\alpha\|x^* - x(\omega)\| \leq \text{val}[A^q] - \text{val}[A(\omega)].$$

By Lemma 4, $x(\omega) \to x^*$ as $\omega \to \infty$; hence $x_r(\omega)$ is positive for large ω.

The corresponding statement for $y_s(\omega)$ is a bit harder to prove. First we observe that

$$\sum_j a_{pj}y_j(\omega) + \omega y_q(\omega) \leq \text{val } A(\omega) \leq a,$$

where $a = \max_{ij}|a_{ij}|$. Hence

$$y_q(\omega) \leq 2a/\omega.$$

Then we have

$$\begin{aligned}\mathrm{val}[A(\omega)] &= \max_i \sum_j a_{ij}(\omega)y_j(\omega) \geq \max_i \sum_j a_{ij}y_j(\omega) \\ &= \max_i \left[\sum_{j\neq q} a_{ij}y_j(\omega) + a_{is}y_q(\omega) + (a_{iq} - a_{is})y_q(\omega)\right] \\ &\geq \max_i \left[\sum_{j\neq q} a_{ij}y_j(\omega) + a_{is}y_q(\omega)\right] - 2ay_q(\omega) \\ &\geq g(y'(\omega)) - 4a^2/\omega,\end{aligned}$$

where $y'(\omega)$ is like $y(\omega)$ except for $y'_q(\omega) = 0$ and $y'_s(\omega) = y_s(\omega) + y_q(\omega)$. Hence, by (3.2)

$$\mathrm{val}[A(\omega)] \geq \mathrm{val}[A^q] + \beta\|y^* - y'(\omega)\| - 4a^2/\omega,$$

and $y'(\omega)$ converges to y^*, by Lemma 4. Since $y_q(\omega) \to 0$, we have $y_s(\omega) \to y_s^* > 0$. Thus both r and s are active in $A(\omega)$ for large ω. But $A(\omega) \in \mathcal{A}$ for all positive ω. Hence $\overline{rs} \in C(\mathcal{A})$. Q.E.D.

LEMMA 6. If $\mathcal{B}$ is obtained from $\mathcal{A}$ by the deletion (as in Lemma 5) of a strictly majorized row or a strictly minorized column, then $C(\mathcal{B}) \subseteq C(\mathcal{A})$.

PROOF. A strictly dominated strategy can never be active.

THEOREM 3.4. If $\mathcal{A}$ is in general position, then $R^{\#}(\mathcal{A}) \subseteq C(\mathcal{A})$.

PROOF. The theorem follows directly from Lemma 5 (and its "row" counterpart), Lemma 6, and the definition of restricted residual.

3.7. The Antisaddle

We close this part with a simple result concerning antisaddlepoints, i.e., saddlepoints of the negative matrix.

THEOREM 3.5. A strict antisaddlepoint of $\mathcal{A}$ cannot be in the center, unless $\mathcal{A}$ is 1-by-1.

PROOF. Suppose $S(-\mathfrak{A}) = \{\overline{pq}\} \subseteq C(\mathfrak{A})$. Take $A \in \mathfrak{A}$ such that $\overline{pq} \in C(A)$, and let $\dot{A}$ be the submatrix of A defined by $C(A)$. Then $\dot{A}$ has an equalizer solution, which solves $-\dot{A}$ as well. But $-\dot{A}$ has a strict saddlepoint at $\overline{pq}$, and hence a unique, pure-strategy solution. Therefore $\dot{A}$ is 1-by-1 and $\overline{pq}$ is a saddlepoint of A as well as of $-A$, strict in both cases. This is possible only if A is 1-by-1. Q.E.D.

Weak antisaddlepoints are sometimes found in the center, as shown (starred) in the accompanying examples. The smaller matrix has a saddlepoint; the larger one (which is symmetric in the sense of § 1) does not.

1*	1
0	2

0*	0	0	−1
0	1	−1	1
0	−1	1	1
1	−1	−1	0

Similar reasoning can be used to establish the following more general result, which applies however only to numerical matrices:

THEOREM 3.6. If $S(A) \subseteq C(-A)$ then $S(A) = C(-A)$.

§ 4. INSTANTANEOUS GAMES

4.1. Games of Almost-perfect Information

There is a class of games of timing, typified by the so-called "noisy duels" ([5], p. 128-134; also [11], Vol. II), in which the payoff depends discontinuously upon the order in which certain actions occur, but is a continuous function of the occurrence times, given the sequence of events. The (finite) set of actions available to a player at any moment may depend on what has gone before; in any case the players are informed at all times of all all previous history. Except for the possibility of simultaneous action, a solution in pure strategies would be indicated. This possibility, however, not only makes mixed strategies necessary in general, but even affects the existence of a value. Indeed, there are some surprisingly simple indeterminate (valueless) games of this type, perhaps worthy of a place beside the examples of Ville ([5], p. 115) and of Sion and Wolfe ([18]). (See below, Fig. 1.)

Simultaneity is not a serious problem in the middle of the game, since either player can avoid it at negligible cost by altering his timing by a small random amount. But this tactic may not work at the beginning or end of the game, since the random displacement would be in a fixed direction.

In order to focus on the problem of determinateness, in games of timing with almost-perfect information, we shall assume that there is a "critical instant" at the beginning of the game. The players will be motivated (by the payoffs) to take some action either at $t = 0$ or immediately thereafter. We thereby reduce the infinite game in extensive form to an essentially finite, "instantaneous" game, whose payoffs are defined by the values of the subgames that result after the first action, or pair of simultaneous actions.

One of our results (Theorem 4.4) states that an instantaneous game has a value provided that the payoff in the event of simultaneous actions lies between the payoffs for the same actions performed singly. Another result (Theorem 4.5) shows how to assign a formal value to the game when the true value does not exist.

4.2. Some Indeterminate Noisy Duels

Two gunfighters are crouched behind barriers, where they can neither see nor be seen. If they stand up at the same instant, we may assume that their chances are equal if they both shoot at once, or if they both hold their fire. If only one fires, that one wins. If one stands up without shooting, and finds the other still down, then he can take control of the situation, and eventually win. But if the one under cover hears the other stand up and fire, then *he* can stand up and take control before his opponent has time to reload.

This (with apologies to the real world!) is an example of an indeterminate noisy duel. The critical moment is at the beginning of the time interval, since to wait an appreciable amount of time is definitely bad strategy. The situation at $t = 0$ can be represented by the following matrix:

	Stand up shooting	Stand up quiet	Wait
Stand up shooting	0	+1	−1
Stand up quiet	−1	0	+1
Wait	+1	−1	Ⓡ

Here "Ⓡ" is meant to suggest "repeat" or "replay," as in a recursive game (see § 4.4 below).

Let us test for a value. In the <u>minorant</u> game, in which player I openly commits himself to a particular (mixed) strategy, the outcome Ⓡ is worth no more than -1, because player II, knowing the probable duration of the other's "wait," will almost always succeed in acting first. Setting Ⓡ $= -1$ and solving, we find that the value is $-1/9$. By symmetry, the value of the <u>majorant</u> game is $+1/9$. Hence the duel is indeterminate.

An even simpler example is shown at right. The minorant value is zero; the majorant value is $1/3$. Any matrix order-equivalent to this one (see § 3) also represents a game without a value, if we interpret "Ⓡ" as a number between 0 and 1 in determining order equivalence.

1	−1
−1	1
0	Ⓡ

The above can be viewed as a game on two unit squares, with discontinuities along the diagonals (Fig. 1).

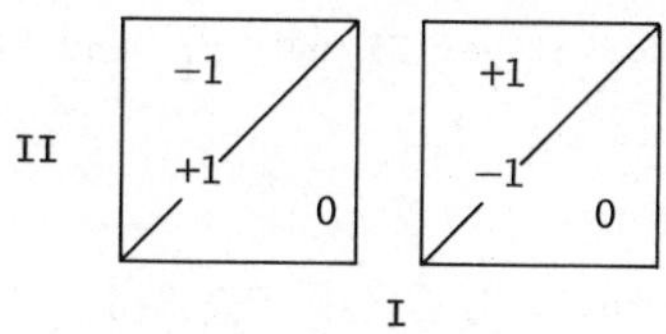

Figure 1.

4.3. The General Case

Let $M(\varphi)$ denote the following $m+1$-by-$n+1$ array:

	"act"			"wait"
"act"	a_{11}	$\cdots$	a_{1n}	a_{10}
	$\vdots$		$\vdots$	$\vdots$
	a_{m1}		a_{mn}	a_{m0}
"wait"	a_{01}		a_{0n}	φ

Here φ is either a real number or the symbol Ⓡ. The game represented by M(Ⓡ) is said to be *instantaneous* if

(4.1) $$\min_{j\neq 0} a_{0j} < \max_{i\neq 0} a_{i0}.$$

This condition puts a premium on immediate action. Without it, at least one player would be willing to delay, and other elements of the original game of timing (from which we assume the data in M(Ⓡ) to have been drawn) would become significant in the selection and timing of the first action. We shall be concerned henceforth only with the instantaneous case.

The minorant and majorant games have the matrices

$$M_1 = M(\min_{j\neq 0} a_{0j}) \quad \text{and} \quad M_2 = M(\max_{i\neq 0} a_{i0}),$$

respectively. Let their values be v_1 and v_2, and their sets of optimal strategies X_1 and X_2 (for player I) and Y_1 and Y_2 (for player II). Obviously,

(4.2) $$\min_{j\neq 0} a_{0j} \leq v_1 \leq v_2 \leq \max_{i\neq 0} a_{i0}.$$

If $v_1 < v_2$ the game is indeterminate; if $v_1 = v_2$ the game is determinate and the optimal strategies for the two players correspond exactly to the elements of X_1 and Y_2, respectively.

THEOREM 4.1. *A necessary and sufficient condition for the instantaneous game* M(Ⓡ) *to be determinate is that* $X_1 \subseteq X_2$ *and* $Y_2 \subseteq Y_1$.

PROOF. Necessity. Any strategy guaranteeing v_1 to player I in M_1 will guarantee him at least as much in M_2. Therefore, $v_1 = v_2$ implies $X_1 \subseteq X_2$. Similarly, $Y_2 \subseteq Y_1$.

Sufficiency. Let $X_1 \subseteq X_2$, $Y_2 \subseteq Y_1$, and suppose $v_1 < v_2$. Take $x \in X_1$. Then, since x is optimal in M_2, we have

$$\sum_{i=0}^{m} x_i a_{ij} \geq v_2 > v_1, \quad j = 1, \ldots, n.$$

Hence no $j \neq 0$ can be active in M_1. Hence Y_1 consists of the single strategy "wait." The same holds for Y_2, by inclusion, and for X_2 and X_1, by symmetry. Hence M_1 and M_2 have saddlepoints at "wait-wait," contradicting (4.1). Q.E.D.

THEOREM 4.2. A necessary condition for the instantaneous game $M(\textcircled{R})$ to be determinate is that "wait" is unplayable (i.e., not active) for at least one player in at least one of M_1, M_2.

PROOF. Let $v_1 = v_2$, and take $x \in X_1$, $y \in Y_2$. Then $x \in X_2$ and $y \in Y_1$, by Theorem 4.1, and we have

$$0 = v_2 - v_1 = xM_2y - xM_1y = x_0y_0(\max_{i\neq 0} a_{i0} - \min_{j\neq 0} a_{0j}).$$

By (4.1), $x_0y_0 = 0$. Since x and y may be chosen independently, it follows that either x_0 vanishes identically for $x \in X_1$, or y_0 vanishes identically for $y \in Y_2$. Q.E.D.

THEOREM 4.3. A sufficient condition for the instantaneous game $M(\textcircled{R})$ to be determinate is that "wait" is unplayable for both players in at least one of M_1, M_2.

PROOF. Under the hypothesis, val[M(a)] is independent of a.

The conditions given so far are somewhat impractical, since the values of the games M_1, M_2 are at least as easy to find as the optimal strategies. The condition in the next theorem is free from this drawback, and it also has a simple heuristic interpretation, namely: if two actions occur simultaneously, the result is intermediate in value between the results of either action performed separately.

THEOREM 4.4. If (4.1) holds, and, if, for all $i \neq 0$, $j \neq 0$.

$$\min(a_{i0}, a_{0j}) \leq a_{ij} \leq \max(a_{i0}, a_{0j}), \tag{4.3}$$

then M(®) is determinate.

PROOF. Let i^*, j^* be such that $a_{i^*0} = \max_{i \neq 0} a_{i0}$, $a_{0j^*} = \min_{j \neq 0} a_{0j}$. Let I be the set of $i \neq 0$ such that $a_{i0} \geq a_{0j^*}$, and J the set of $j \neq 0$ such that $a_{0j} \leq a_{i^*0}$. Then by (4.1), $i^* \in I$ and $j^* \in J$, so that neither I nor J nor their complements are empty. Take $j \in J$ and $i \in I$. If $i \neq 0$ then we have

$$a_{i0} < a_{0j^*} \leq a_{0j} \leq a_{i^*0},$$

by the definitions of I, j^*, and J, respectively. Applying the "betweenness" condition (4.3) twice, we obtain

$$a_{ij} \leq a_{0j} \leq a_{i^*j}.$$

Thus, for every $i \notin I$ (including $i = 0$), $a_{ij} \leq a_{i^*j}$ holds for all $j \in J$. Similarly, for every $j \in J$, $a_{ij} \geq a_{ij^*}$ holds for all $i \in I$. This means that $I \times J$ is a weak generalized saddlepoint of M(a) (see § 3.1). Hence val[M(a)] is independent of a. Q.E.D.

4.4. A Way of Resolving the Indeterminacy

One may wish to assign a formal value to the instantaneous game M(®) when no true value exists. A way to do this — perhaps the only "fair" way — is to pursue the analogy with stochastic or recursive games [16], [6], and attempt to solve the equation

$$a = \text{val}[M(a)]. \tag{4.4}$$

In effect, we replace the time continuum by a discrete, well-ordered set of points, making simultaneity <u>after</u> t = 0 a significant possibility. We shall see that condition (4.1), which characterizes "instantaneous" games, is intimately related to the existence of a unique solution to (4.4).

THEOREM 4.5. If $M(\circledR)$ is an instantaneous game, then (4.4) has a unique solution $a = \bar{v}$, and we have $v_1 \leq \bar{v} \leq v_2$.

PROOF. Uniqueness. Let (4.4) have two solutions $\bar{v} < \bar{\bar{v}}$. Let x be optimal in $M(\bar{\bar{v}})$ for player I and let y be optimal in $M(\bar{v})$ for player II. Then $xM(\bar{\bar{v}})y \geq \bar{\bar{v}}$ and $xM(\bar{v})y \leq \bar{v}$. Subtracting, we obtain

$$\bar{\bar{v}} - \bar{v} \leq x(M(\bar{\bar{v}}) - M(\bar{v}))y = x_0 y_0 (\bar{\bar{v}} - \bar{v}).$$

Hence $x_0 = y_0 = 1$, and x and y are pure. Thus, for all $i \neq 0$, $j \neq 0$,

$$a_{0j} \geq \bar{\bar{v}} > \bar{v} \geq a_{i0}.$$

This contradicts (4.1).

Existence. The function $F(a) = \mathrm{val}[M(a)]$ is monotonic nondecreasing. Hence, by (4.2), we have

$$F(\min_{j \neq 0} a_{0j}) \leq F(v_1);$$

in other words $v_1 - F(v_1) \leq 0$. Similarly, $v_2 - F(v_2) \geq 0$. But $a - F(a)$ is continuous; hence it has a zero in $[v_1, v_2]$. Q.E.D.

COROLLARY. In the determinate case, $\bar{v}$ is equal to the true value.

We note that if (4.1) is not satisfied, then all values in the interval $[\max a_{i0}, \min a_{0j}]$ are solutions of (4.4), and no others.

From the fact that $F(a)$ and $a - F(a)$ are both monotonic nondecreasing functions, it follows that the sequence $b, F(b), F(F(b)), \ldots$ (b arbitrary) converges monotonically to a solution of (4.4). This can be useful in making sharper numerical estimates; e.g., we have $F(v_1) \leq \bar{v} \leq F(v_2)$, etc.

In the first indeterminate game of § 4.2 we have $\bar{v} = 0$, by symmetry. In the second example we also have $\bar{v} = 0$; thus, the first player's advantage in this game, such as it is, vanishes if time is made discrete. This feature, also present in the order-equivalent variants, shows that strict inequality need not hold in Theorem 4.5 for indeterminate games.

§ 5. FICTITIOUS PLAY IN NON-ZERO-SUM GAMES

5.1. Discussion

The method of fictitious play (FP) resembles a multistage learning process. At each stage, it is assumed that the players choose a strategy that would yield the optimum result if employed against all past choices of their opponents. Various conventions can be adopted with regard to the first move, indifferent alternatives, simultaneous vs. alternating moves, and weighting of past choices. The method can meaningfully be applied to any finite game, and to many infinite games as well. (See [3], [4], [5] pp. 82-85, and [14].)

It was once conjectured that the mixed strategies defined by the accumulated choices of the players would always converge to the equilibrium point of the game, or, in the event of nonuniqueness, to a set of mutually compatible equilibrium points. This is the natural generalization of Robinson's theorem [14] for the zero-sum two-person case; it was recently verified by Miyasawa [12] for the special case of two players with two pure strategies apiece.

The trouble begins, as we shall see, as soon as we add a third strategy for each player. It appears, intuitively, that this size is necessary to produce enough variety; if FP is to fail, the game must contain elements of both coordination and competition. Our counterexamples include a whole class of order-equivalent games, and thus do not depend on numerical quirks in the payoff matrices; nor are they sensitive to the minor technicalities of the FP algorithm. It is clear that games with more players, or with more strategies per player, can exhibit the same kind of misbehavior.

5.2. A Class of Nonconvergent Examples

We shall elaborate slightly on the game described in the Introduction, to eliminate any question of "degeneracy." The payoff matrices are shown at right. We assume that $a_i > b_i > c_i$ and $\alpha_i > \beta_i > \gamma_i$, for $i = 1, 2, 3$. It follows that the game is not constant-sum. It is easily shown that the equilibrium point must be completely mixed (all strategies active), and hence unique.

$$\underset{\text{I}}{\begin{array}{|ccc|}\hline a_1 & c_2 & b_3 \\ b_1 & a_2 & c_3 \\ c_1 & b_2 & a_3 \\ \hline\end{array}} \quad \underset{\text{II}}{\begin{array}{|ccc|}\hline \beta_1 & \gamma_1 & \alpha_1 \\ \alpha_2 & \beta_2 & \gamma_2 \\ \gamma_3 & \alpha_3 & \beta_3 \\ \hline\end{array}}$$

For simplicity, we shall assume that the FP choices are made simultaneously, and that the first choice pair is $\overline{11}$. Consider any occurrence of $\overline{11}$ in the FP sequence. The next choice of player I will certainly be 1 again, since that strategy will have become more desirable. Player II will either stay with 1, or shift to 3. Eventually, in fact, he must shift to 3, since $\alpha_1 > \beta_1$. Thus, after each run of $\overline{11}$ we will find a run of $\overline{13}$. By a similar argument, this will be followed by a run of $\overline{33}$, and then runs of $\overline{32}$, $\overline{22}$, $\overline{21}$, $\overline{11}$, ..., in a never-ending cycle.

Suppose that a run of $\overline{11}$ is just about to begin. Let X represent the current "accumulated-choices" vector for player I (thus, X_i is the number of times he has chosen row i), and let Y be the same for player II. Let H denote the current "comparative-payoffs" vector for player I (thus, $H_1 = Y_1a_1 + Y_2c_2 + Y_3b_3$, etc.). Since he is about to choose 1, we have

$$H_1 = \max(H_1, H_2, H_3).$$

Let there occur r_{11} choices of $\overline{11}$, followed by r_{13} choices of $\overline{13}$. Then I's new comparative payoffs are

$$\begin{cases} H_1' = H_1 + r_{11}a_1 + r_{13}b_3, \\ H_2' = H_2 + r_{11}b_1 + r_{13}c_3, \\ H_3' = H_3 + r_{11}c_1 + r_{13}a_3. \end{cases}$$

Since he shifts to 3 at this point, we must have $H_3' \geq H_1'$. But $H_3 \leq H_1$; hence $H_3' - H_3 \geq H_1' - H_1$, and we have

$$r_{13} \geq \frac{a_1 - c_1}{a_3 - b_3} r_{11}.$$

Let r_{33} be the length of the $\overline{33}$ run that follows. By analogous reasoning, we have

$$r_{33} \geq \frac{\alpha_1 - \gamma_1}{\alpha_3 - \beta_3} r_{13}.$$

Repeating the argument four times more, we obtain

$$r'_{11} \geq \left(\prod_{i=1}^{3} \frac{(a_i - c_i)(\alpha_i - \gamma_i)}{(a_i - b_i)(\alpha_i - \beta_i)} \right) r_{11}, \tag{5.1}$$

where r'_{11} denotes the length of the next $\overline{11}$ run. Since the constant in (5.1) is greater than 1, the runs of $\overline{11}$ increase in length exponentially (or faster). The same rate of increase occurs for the other choice pairs. Hence the ratios X_i/X_j, Y_i/Y_j, $i \neq j$, do not converge. In particular, the normalized strategy vectors $x = X/\Sigma X_i$, $y = Y/\Sigma Y_i$ do not converge to the unique equilibirum point strategies.

It can be shown that x and y, in their respective strategy spaces, approach limit cycles which do not contain the equilibrium point strategies. Hence there is no convergence to an equilibrium point even through subsequences.

The argument we have given is independent of the tie-breaking rule. With minor modifications it can also handle the case of alternating moves, as well as the case of nonintegral run lengths. The latter implies that the differential-equation version of FP (see [3]) will also fail to converge to the solution.

5.3. An Example

We have computed the limit cycles for the numerical example shown at right. The payoffs do not satisfy the strict inequalities assumed in the preceding proof, and the constant in the estimate (5.1) is exactly 1. A more refined analysis, however, shows that the run lengths do increase exponentially; in fact, the ratio of r'_{11} to r_{11} tends in the limit to

$$\text{I} = \begin{bmatrix} 1 & 0 & 0 \\ 0 & 1 & 0 \\ 0 & 0 & 1 \end{bmatrix} \qquad \text{II} = \begin{bmatrix} 0 & 0 & 1 \\ 1 & 0 & 0 \\ 0 & 1 & 0 \end{bmatrix}$$

θ^6, where $\theta \doteq 1.466$ is a root of $\theta^3 - \theta^2 = 1$.

The limit cycles in the separate strategy spaces are illustrated in Fig. 2. In the product space they form a hexagon; typical vertices are

$$(\bullet) \quad (1, \theta^4, \theta^2; \theta^3, \theta^3, \theta)/C,$$

$$(\blacksquare) \quad (\theta^4, \theta^4, \theta^2; \theta^5, \theta^3, \theta)/C\theta, \text{ etc.},$$

where C is a normalizing constant. The unique equilibrium point is

$$(\odot) \quad (1, 1, 1; 1, 1, 1)/3.$$

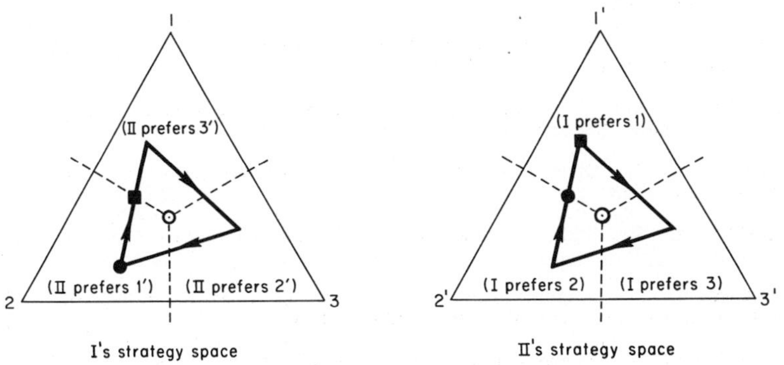

Figure 2

BIBLIOGRAPHY

[1] BASS, H., Order Matrices, Junior Paper, Princeton University, 1954.

[2] BOHNENBLUST, H. F., KARLIN, S. and SHAPLEY, L. S., "Solutions of discrete, two-person games," Annals of Mathematics Study No. 24 (Princeton, 1950), pp. 51-72.

[3] BROWN, G. W., "Iterative solution of games by fictitious play," Activity Analysis of Production and Allocation, John Wiley and Sons, 1951, pp. 374-376.

[4] DANSKIN, J. M., "Fictitious play for continuous games," Naval Research Logistics Quarterly, Vol. 1 (1954), pp. 313-320.

[5] DRESHER, M., Games of Strategy: Theory and Applications, Prentice Hall, 1961.

[6] EVERETT, H., "Recursive games," Annals of Mathematics Study No. 39, (Princeton, 1957), pp. 47-48.

[7] GALE, D., KUHN, H. W. and TUCKER, A. W., "On symmetric games," Annals of Mathematics Study No. 24 (Princeton, 1950), pp. 81-87.

[8] GALE, D., KUHN, H. W. and TUCKER, A. W., "Reductions of game matrices," Annals of Mathematics Study No. 24 (Princeton, 1950), pp. 89-96.

[9] GALE, D. and SHERMAN, S., "Solutions of finite two-person games," Annals of Mathematics Study No. 24 (Princeton, 1950), pp. 37-49.

[10) KAPLANSKY, I., "A contribution to von Neumann's theory of games," Annals of Mathematics, Vol. 46 (1945), pp. 474-479.

[11] KARLIN, S., Mathematical Methods and Theory in Games, Programming and Economics, Addison-Wesley, 1959.

[12] MIYASAWA, K., On the Convergence of the Learning Process in a 2 × 2 Non-zero-sum Two-person Game, Economic Research Program, Princeton University, Research Memorandum No. 33 (1961).

[13] NASH, J. F., "Non-cooperative games," Annals of Mathematics, Vol. 54 (1951), pp. 286-295.

[14] ROBINSON, J., "An iterative method of solving a game," Annals of Mathematics, Vol. 54 (1951), pp. 296-301.

[15a] SHAPLEY, L. S., The Noisy Duel: Existence of a Value in the Singular Case, The RAND Corporation, Rm-641, July, 1951.

[15b] SHAPLEY, L. S., Order Matrices. I, The RAND Corporation, RM-1142, September, 1953.

[15c] SHAPLEY, L. S., Order Matrices. II, The RAND Corporation, RM-1145, September, 1953.

[15d] SHAPLEY, L. S., A Condition for the Existence of Saddle-points, The RAND Corporation, RM-1598, September, 1955.

[15e] SHAPLEY, L. S., Symmetric Games, The RAND Corporation, RM-2476, June, 1960.

[15f] SHAPLEY, L. S., On the Nonconvergence of Fictitious Play, The RAND Corporation, RM-3026, March, 1962.

[16] SHAPLEY, L. S., "Stochastic games," Proceedings of the National Academy of Sciences, Vol. 39 (1953), pp. 1095-1100.

[17] SHAPLEY, L. S. and SNOW, R. N., "Basic solutions of discrete games," Annals of Mathematics Study No. 24 (Princeton, 1950), pp. 27-35.

[18] SION, M. and WOLFE, P., "On a game without a value," Annals of Mathematics Study No. 39 (Princeton, 1957), pp. 299-306.

[19] VON NEUMANN, J. and MORGENSTERN, O., Theory of Games and Economic Behavior, Princeton, 1944.

L. S. Shapley

The RAND Corporation

GAMES WITH A RANDOM MOVE

Rodrigo A. Restrepo

§ 1. INTRODUCTION

In some models of poker investigated by von Neumann [1], [2], Gillies [2], Mayberry [2], Karlin [3] and others, the strategies are functions of the outcome of a random move. This move corresponds to the dealing of a hand from a deck of cards, each possible hand being identified with some point in the real line. It seems desirable to include some of these examples in a more general class of games to which a general method of attack is applicable.

This paper considers a class of games that includes the first model in [3]. Here the random move is described by two uniformly distributed random variables with values x and y in the unit interval. The maximizing and minimizing strategies are n-tuples $f(x) = (f_1(x), f_2(x), \ldots, f_n(x))$ any $g(y) = (g_1(y), g_2(y), \ldots, g_n(y))$ respectively. These strategies are designed to fit a situation where, having received some information about the outcome of the random move, each player may select one of several courses of action from a finite set of alternatives, the number of alternatives being the same for both players. The number $f_i(x)$ will be the probability that the first player, knowing only the value x, will select the ith action. Consequently f will satisfy the restriction $0 \leq f_i(x) \leq 1$, and if the different courses of action are mutually exclusive, additional restrictions must be imposed on the strategies. Similar remarks apply to the minimizing strategies. Thus, the following strategy spaces may be considered:

$$\mathfrak{F} = \{f \mid 0 \leq f_i(x) \leq 1, \quad \text{all} \quad x \quad \text{and} \quad i\}$$

$$\mathcal{G} = \{g \mid 0 \leq g_i(y) \leq 1, \quad \text{all} \quad y \quad \text{and} \quad i\}$$

$$\mathfrak{F}^o = \{f \in \mathfrak{F} \mid \sum_{i=1}^{n} f_i(x) \leq 1, \quad \text{all} \quad x\}$$

$$\mathcal{G}^o = \{g \in \mathcal{G} \mid \sum_{i=1}^{n} g_i(y) \leq 1, \quad \text{all} \quad y\} .$$

It is assumed that f_i and g_i are Lebesgue measurable, and no distinction is made between strategies that differ only on a set of measure zero. Thus the strategies are completely defined by functions F_i and G_i with

$$(1) \qquad F_i(x) = \int_0^x f_i(x)\, dx, \quad G_i(y) = \int_0^y g_i(y)\, dy.$$

The pay-off $P(f, g)$ shall be the expected value of a function $R(x, y; f(x), g(y))$ with respect to the distribution of x and y, that is,

$$P(f, g) = \int_0^1 \int_0^1 R(x, y; f(x), g(y))\, dx\, dy$$

where

$$(2) \qquad R(x, y; f(x), g(y))$$

$$= \sum_{i=1}^{n} \{a_i f_i(x) - b_i g_i(y) + [c_i + d_i \operatorname{sign}(x - y)]\, f_i(x)\, g_i(y)\} .$$

The choice of R is motivated by the examples mentioned above. The solution of the game defined by $(P, \mathfrak{F}, \mathcal{G})$, i.e., with pay-off P and strategy spaces $\mathfrak{F}$ and $\mathcal{G}$, is given explicitly below. The precise characterization of the solution of the game $(P, \mathfrak{F}^o, \mathcal{G})$ is, in general, a difficult problem that requires the consideration of a large number of inequalities. However, it is possible to give a qualitative description of its optimal strategies that simplifies considerably the solution of this game and in a number of cases reduces the problem of finding optimal strategies to the problem of solving a rather simple system of equations. The game $(P, \mathfrak{F}, \mathcal{G}^o)$ can be reduced to the previous one by interchanging the role of the players. The solution of $(P, \mathfrak{F}^o, \mathcal{G}^o)$ is not known.

In order to simplify the description of the optimal strategies it is convenient to introduce constants α_i and β_i defined by

$$\alpha_i = 1 - \frac{a_i}{d_i - c_i}, \quad \beta_i = 1 - \frac{b_i}{d_i + c_i} .$$

§ 2. A CHARACTERIZATION OF THE OPTIMAL STRATEGIES

Two strategies f^* and g^* are optimal if and only if $P(f^*, g^*) = \max_f P(f, g^*)$ and $P(f^*, g^*) = \min_g P(f^*, g)$. Explicitly,

$$P(f^*, g^*) = -\sum_{i=1}^{n} b_i G_i^*(1) + \max_f \sum_{i=1}^{n} \int_0^1 f_i(x)\varphi_i(x, g^*)dx \tag{3}$$

and

$$P(f^*, g^*) = \sum_{i=1}^{n} a_i F_i^*(1) + \min_g \sum_{i=1}^{n} \int_0^1 g_i(y)\Psi_i(y, f^*)dy \tag{4}$$

where

$$\varphi_i(x, g^*) = a_i + (c_i - d_i)G_i^*(1) + 2\, d_i G_i^*(x) \tag{5}$$

and

$$\Psi_i(y, f^*) = -b_i + (c_i + d_i)F_i^*(1) - 2\, d_i F_i^*(y) . \tag{6}$$

Thus, if the strategy space is $\mathfrak{F}$, then the strategy f^* satisfies the following conditions:

$$\begin{aligned} f_i^*(x) &= 0 & &\text{if } \varphi_i(x, g^*) < 0 \\ 0 \le f_i^*(x) &\le 1 & &\text{if } \varphi_i(x, g^*) = 0 \\ f_i^*(x) &= 1 & &\text{if } \varphi_i(x, g^*) > 0 . \end{aligned}$$

Furthermore, if the strategy space is $\mathfrak{F}^o$, then the conditions are as follows:

$$f_i^*(x) = 0 \quad \text{if} \quad \varphi_i(x, g^*) < 0 \quad \text{or} \quad \varphi_i(x, g^*) < \max_{j \neq i} \varphi_j(x, g^*)$$

$$0 \leq f_i^*(x) \leq 1 \quad \text{if} \quad \varphi_i(x, g^*) = \max_j \varphi_j(x, g^*) \geq 0$$

$$f_i^*(x) = 1 \quad \text{if} \quad \varphi_i(x, g^*) > 0 \quad \text{and} \quad \varphi_i(x, g^*) > \max_{j \neq i} \varphi_j(x, g^*) .$$

These conditions characterize f^*, and the minimizing strategy g^* can be characterized in a similar manner. With this characterization (due to S. Karlin), any knowledge about one of the optimal strategies yields information about the optimal strategies for the other player. The process is iterative, and the coefficients φ_i and Ψ_i play a fundamental role. Some of their properties are given in the next two lemmas; the proofs are simple computations which are omitted.

LEMMA 2.1. For each x,

$$\max_g \varphi_i(x,g) = (1-x) \max (a_i, a_i+c_i-d_i) + x \max (a_i, a_i+c_i+d_i)$$

$$\min_g \varphi_i(x,g) = (1-x) \min (a_i, a_i+c_i-d_i) + x \min (a_i, a_i+c_i+d_i)$$

$$\min_f \Psi_i(y,f) = (1-y) \min (-b_i, -b_i+c_i+d_i) + y \min (-b_i, -b_i+c_i-d_i)$$

$$\max_f \Psi_i(y,f) = (1-y) \max (-b_i, -b_i+c_i+d_i) + y \max (-b_i, -b_i+c_i-d_i) .$$

LEMMA 2.2. Let $\chi(E)$ denote the characteristic function of the set E. If $d_i > 0$, then

$$\varphi_i(x, \chi([y_i, 1])) = (d_i - c_i)(y_i - \alpha_i) + \max [0, 2d_i(x - y_i)]$$

$$\Psi_i(y, \chi([x_i, 1])) = -(d_i + c_i)(x_i - \beta_i) - \max [0, 2d_i(y - x_i)].$$

§ 3. GAMES DEFINED BY $(P, \mathfrak{F}, \mathfrak{G})$

In these games the components of each strategy are uncorrelated, and in order to solve the game it is sufficient to find the minmax strategies for each index i. For simplicity it will be assumed in this section that $a_i b_i \geq 0$, for each i. Furthermore, if a_i and b_i are negative, reversing the roles of the two players one obtains an equivalent game where the corresponding coefficients are positive. If $a_i \geq 0$, $b_i \geq 0$ but $d_i < 0$, then reversing the orientation of the unit interval (i.e., replacing sign $(x - y)$ by sign $(y - x)$) one obtains a game with $d_i \geq 0$; if $d_i = 0$, then φ_i and Ψ_i are constant functions and the game is trivial. Finally, if $a_i \geq 0$, $b_i \geq 0$, $d_i > 0$ and $|c_i| \geq d_i$, then either $-b_i + c_i + d_i \leq 0$ or $a_i + c_i - d_i \geq 0$. In the first case Lemma 2.1 shows that $\Psi_i(y, f) \geq 0$ for all f, and $g^*(y) \equiv 1$ is uniformly best against any f. Since g^* is known, the characterization of Section 2 yields the optimal f^*. The case where $a_i + c_i - d_i \geq 0$ is equally trivial.

One must consider now the games where $a_i \geq 0$, $b_i \geq 0$, $|c_i| < d_i$, $a_i + c_i - d_i < 0$ and $-b_i + c_i + d_i > 0$. These are games where α_i and β_i are in the unit interval; their solutions are given in the following theorem:

THEOREM 3.1. Under the given assumptions the game has solutions of the form

$$f_i^* = m_i\chi([0, x_i]) + \chi([x_i, 1])$$

$$g_i^* = n_i\chi([0, y_i]) + \chi([y_i, 1]) ,$$

where $\chi(E)$ denotes the characteristic function of the set E. Furthermore, $m_i n_i = 0$, and $x_i = \alpha_i$ or $y_i = \beta_i$.

PROOF. Consider any strategies f^* and g^* of the given form with $x_i = y_i = \alpha_i$, $n_i = 0$ and $0 \leq m_i \leq 1$. By Lemma 2.2 f^* maximizes $P(f, g^*)$, but g^* minimizes $P(f^*, g)$ if and only if m_i is chosen so that $\Psi_i(\alpha_i, f^*) = 0$. This equation is equivalent to

$$\alpha_i = \frac{d_i + c_i - b_i}{d_i + c_i + m_i(d_i - c_i)}. \tag{7}$$

As m_i varies over the unit interval, the right hand side of (7) varies between β_i and $\frac{d_i + c_i}{2d_i}\beta_i$. Thus the given strategies are optimal if α_i is in the given range. Considering also the possibility that $m_i > 0$ and $y_i > \alpha_i$, the results can be summarized as follows:

CASE 1. $\frac{d_i + c_i}{2d_i} \beta_i > \alpha_i$. The optimal strategies are f^*, g^* with $m_i = 1$, $n_i = 0$, $x_i = \alpha_i$ and $y_i = \frac{d_i + c_i}{2d_i} \beta_i$.

CASE 2. $\beta_i \geq \alpha_i \geq \frac{d_i + c_i}{2d_i} \beta_i$. The optimal strategies are f^*, g^* with $x_i = y_i = \alpha_i$, $n_i = 0$, and m_i satisfying equation (7). Using the symmetry of the problem, the remaining cases have the following solutions:

CASE 3. $\alpha_i \geq \beta_i \geq \frac{d_i - c_i}{2d_i} \alpha_i$. The optimal strategies are f^*, g^* with $x_i = y_i = \beta_i$, $m_i = 0$, and n_i satisfying the equation

$$\beta_i = \frac{d_i - c_i - \alpha_i}{d_i - c_i + n_i(d_i + c_i)} . \tag{8}$$

CASE 4. $\frac{d_i - c_i}{2d_i} \alpha_i > \beta_i$. The optimal strategies are f^*, g^* with $m_i = n_i = 0$, $y_i = \beta_i$ and $x_i = \frac{d_i - c_i}{2d_i} \alpha_i$.

A simple qualitative description clarifies the results obtained above. Indeed, if $\beta_i >> \alpha_i$, then $f_i^* = \chi([0, \alpha_i]) + \chi([\alpha_i, 1])$ and $g^* = \chi([y_i, 1])$ with $y_i > \alpha_i$. As β_i decreases relative to α_i, y_i decreases until $y_i = \alpha_i$. Then, as β_i decreases further, g_i^* remains fixed but $f^* = m_i\chi([0, \alpha_i]) + \chi([\alpha_i, 1])$, and m_i decreases until $m_i = 0$ when $\alpha_i = \beta_i$. As β_i decreases further the process reverses itself: g_i^* is changed to become $n_i\chi([0, \beta_i]) + \chi([\beta_i, 1])$ while f_i^* remains fixed. Finally, after $n_i = 1$, g_i^* remains fixed, but $f_i^* = \chi([x_i, 1])$ with x_i increasing beyond β_i.

UNIQUENESS. Let f^o and g^o be any strategies obtained from f^* and g^* by replacing the constants m_i and n_i by functions $m_i(t)$, $n_i(t)$ with values between 0 and 1 and such that $\int_0^{x_i} m_i(t)\, dt = m_i x_i$ and $\int_0^{y_i} n_i(t)\, dt = n_i y_i$. Then f^o and g^o are optimal, and all optimal strategies are of this type.

§ 4. GAMES DEFINED BY $(P, \mathfrak{F}^o, \mathfrak{G})$ or $(P, \mathfrak{F}, \mathfrak{G}^o)$

Since the game is essentially symmetric, it is sufficient to consider only the games defined by $(P, \mathfrak{F}^o, \mathfrak{G})$. The existence of optimal strategies for these games is an immediate consequence of the general minmax theorem [4].

THEOREM 4.1. If $d_i \geq 0$ for all i, then there exist numbers $y_1, \ldots, y_n$ such that the strategy $g = (\chi([y_1, 1]), \chi([y_2, 1]), \ldots, \chi([y_n, 1]))$ is optimal for the game defined by $(P, \mathfrak{F}^o, \mathfrak{G})$.

PROOF. Let f^* and g^* be any pair of optimal strategies; let g_i^* denote the i-th component of g^*, and let i be a fixed index such that g_i^* cannot be expressed (a. e.) as $\chi([y_i, 1])$ for any y_i. By choice of i there exist a largest number u_i and a smallest number v_i such that $u_i < v_i$, $g_i^*(y) \equiv 0$ for $y \leq u_i$, and $g_i^*(y) \equiv 1$ a.e. for $y \geq v_i$. Furthermore, the optimal character of g^* implies that $\Psi_i(y, f^*) \equiv 0$ for $u_i \leq y \leq v_i$, and then equation (6) implies that $f_i^*(x) \equiv 0$ in the same interval. This in turn implies that $\varphi_i(x, g^*) \leq 0$ or $\varphi_i(x, g^*) \leq \max_{j \neq i} \varphi_j(x, g^*)$, for $u_i \leq x \leq v_i$. Now let y_i be chosen so that $v_i - y_i = \int_{u_i}^{v_i} g_i^*(t)dt$, and let g^o be the strategy obtained from g^* by replacing the (fixed) i-th component g_i^* by $\chi([y_i, 1])$. To show that g^o is optimal it is sufficient to show that f^* and g^o satisfy equations (3) and (4). Clearly, $P(f^*, g^o) = P(f^*, g^*) = \min_g P(f^*, g)$ because $\Psi_i(y, f^*) = 0$ wherever $g^o \neq g^*$. Also, since $\varphi_i(x, g^o) \leq \varphi_i(x, g^*)$ with strict inequality only on sets where $f_i^* \equiv 0$, the fact that f^* maximizes $P(f, g^*)$ implies that f^* also maximizes $P(f, g^o)$. Iterating the process one obtains an optimal strategy of the given form.

The same argument can be used to show that if some d_i is negative, the corresponding component of g^* is of the form $\chi([0, y_i])$.

LEMMA 4.1. If $|c_i| < d_i$ and $b_i \leq 0$, then there exists an optimal minimizing strategy g^* and a constant λ^* such that for all i, $g_i^* = \chi([y_i, 1])$ and for each i either $y_i = \dfrac{\lambda^*}{d_i - c_i} + \alpha_i$ or $y_i = 1$.

PROOF. Let f^* and g^o be optimal strategies such that g^o is of the type described in Theorem 4.1, i.e., $g_i = \chi([y_i, 1])$ for $i = 1, \ldots, n$.

Let $\lambda_i = (y_i - \alpha_i)(d_i - c_i)$ and assume that there exist two indices i and j such that $\lambda_i > \lambda_j$. Also, if $y_j < 1$ let g^δ be the strategy obtained from g^o by replacing y_j by $y_j + \delta$ where $\delta < (2d_j)^{-1}(\lambda_i - \lambda_j)$, and $\delta < 1 - y_j$.

By Lemma 2.2, if $x < y_j + \delta$ then $\varphi_j(x, g^o) \leq \lambda_j + 2d_j\delta < \lambda_i \leq \varphi_i(x, g^o)$, implying that $f_j^*(x) \equiv 0$ a.e., in the interval $[0, y_j + \delta]$. Furthermore, if $x > y_j + \delta$ then $\varphi_j(x, g^\delta) < \varphi_j(x, g^o)$ implying that $f_j^*(x) \equiv 0$ a.e. for $x > y_j + \delta$; otherwise one would have $P(f^*, g^\delta) < P(f^*, g^o)$ (see Equation 3) contradicting the optimal character of f^* and g^o. Since f^* maximizes $P(f, g^o)$ and $f_j^* \equiv 0$, the inequalities $\varphi_j(x, g^\delta) < \varphi_i(x, g^o) = \varphi_i(x, g^\delta)$ for $x < y_j + \delta$, and $\varphi_j(x, g^\delta) < \varphi_j(x, g^o) < \max\{0, \max_{i \neq j} \varphi_i(x, g^o)\}$ imply that f^* maximizes also $P(f, g^\delta)$. Finally, since $f_j^* \equiv 0$, Equation 6 shows that $\Psi_j(y, f^*) \equiv -b_j \geq 0$. Thus, if g^o minimizes $P(f^*, g)$, then g^δ also minimizes $P(f^*, g)$. Thus f^* and g^δ form a pair of optimal strategies. Since the perturbation is a linear function of $\lambda_i - \lambda_j$, the process can be continued until $\lambda_j = \lambda_i$ or $y_j = 1$. Applying this result to all components one obtains a strategy g^* such that $y_i^* = 1$ or $\lambda_i^* = \max \lambda_i$.

REMARK. If $b_i \geq 0$ for all i, one can perform a similar perturbation, replacing y_i by $y_i - \delta$ provided that $\varphi_i(x, g^o) < \max_{j \neq i} \varphi_j(x, g^o)$ for $x > y_i$. However, the perturbation δ is not a linear function of $\lambda_i - \lambda_j$, and the method does not lead to any general theorem.

THEOREM 4.2. If $b_i \leq 0$ and $|c_i| < d_i$ for all i then there exists an optimal minimizing strategy $g^* = (\chi([y_1^*, 1]), \ldots, \chi([y_n^*, 1]))$ such that for each i either $y_i = 1$ or $y_i = \alpha_i + \dfrac{\lambda^*}{d_i - c_i}$, where λ^* is a nonnegative constant independent of i. Furthermore, the solution of the game is determined completely by the number λ^*.

PROOF. The fact that $\lambda^* \geq 0$ can be established by a perturbation argument similar to that of Lemma 4.1. To show that the solution is completely determined by λ^*, it is sufficient to observe that setting

$y_i^* = \alpha_i + \dfrac{\lambda^*}{d_i - c_i}$ one completely determines g^* and $\varphi_i(x, g^*)$. Then f^* can be determined by means of the characterization of Section 2. In the indeterminate case where $\varphi_i(x, g^*) = \max_j \varphi_j(x, g^*) \geq 0$, the indeterminacy of f can be eliminated by the conditions $\Psi_i(y_i^*, f^*) = 0$ for all i. Viewing λ^* as a parameter, these conditions constitute a system of equations whose unknowns are λ^* and the values of f_i^* in the indeterminate regions. There is no loss of generality in assuming that each f_i is a constant in each of these intervals.

The previous result can be improved slightly. Indeed, since $\lambda^* = \varphi_i(0, g^*)$, Lemma 2.1 implies that $a_i \geq \lambda^* \geq a_i + c_i - d_i$. For these values of λ^*, the relative order of $y_1, \ldots, y_n$ can be deduced from the following theorem:

THEOREM 4.3. Let $a_i \geq 0$, $|c_i| < d_i$, and let the indices be arranged so that $a_1 \geq a_2 \geq \cdots \geq a_n$. If $a_i \geq \lambda \geq a_i + c_i - d_i$ for all i, and if $y_i = \alpha_i + \dfrac{\lambda}{d_i - c_i}$, then $y_1 \leq y_2 \leq \cdots \leq y_n$ for all λ if and only if $a_i + c_i - d_i$ is a monotone decreasing function of i.

PROOF. By definition, $y_i = 1 - \dfrac{a_i - \lambda}{d_i - c_i}$, and the inequality $y_i \geq y_j$ is equivalent to

$$\frac{d_i - c_i}{d_j - c_j} \geq 1 - \frac{a_j - a_i}{a_j - \lambda} .$$

If $a_i \leq a_j$, the right hand side of (9) is a monotone decreasing function of λ; it achieves its minimum value (zero) when $\lambda = a_i$; it achieves its maximum value when $\lambda = a_j + c_j - d_j$ and in this case the inequality (9) becomes $a_i + c_i - d_i \leq a_j + c_j - d_j$.

EXAMPLE. The poker model considered in [3] has $a_i \equiv 2$, $b_i \equiv 0$, $c_i \equiv -1$ and $d_i = A_i + 1$ with $2 \leq A_1 \leq \cdots \leq A_n$. The assumptions of Theorem 4.3 are satisfied, $\lambda^* = 0$ and the optimal g^* has

$$y_i = \frac{A_i}{A_i + 2} .$$

Then $\varphi_i(x, g^*) = \lambda^* + 2d_i G^*(x)$, and the optimal f^* can be deduced immediately from the characterization given in Section 2.

BIBLIOGRAPHY

[1] von NEUMANN, J. and MORGENSTERN, O., Theory of Games and Economic Behavior, Princeton, 1944.

[2] GILLIES, D. B., MAYBERRY, J. P., and von NEUMANN, J., "Two variants of poker," Annals of Mathematics Study No. 28 (Princeton 1953), pp. 13-50.

[3] KARLIN, S., and RESTREPO, R., "Multi-stage poker models," Annals of Mathematics Study No. 39 (Princeton 1957), pp. 337-363.

[4] KARLIN, S., "Operator treatment of the minmax principle," Annals of Mathematics Study No. 24 (Princeton 1950), pp. 133-154.

Rodrigo A. Restrepo

The University of British Columbia

A SEARCH GAME

Selmer M. Johnson

§ 1. INTRODUCTION

The following search game was first suggested to the author by Melvin Dresher several years ago.

Blue chooses h, an integer, from the set of integers 1 to n (a region to hide). Red guesses an integer from 1 to n, is told whether he is too high or too low, and repeats until he guesses h. The payoff to Blue is one unit for each guess by Red (including the last guess h). This game is illustrated in [1], pages 32-35.

Progress was reported by the author in a 1958 internal RAND document which presented the solution for $n \leq 11$ using a special notational device to describe Red strategies.

Recently Gilbert [2] discussed the same game, along with related problems, and gave its solution for $n \leq 6$. He stated that the calculations, even for $n = 6$, were quite lengthy. Later, in [3] Konheim computed the number of "bisecting" strategies for Red. This does not pertain directly to solving the game.

We present both the improved notational device for describing Red strategies and some recent theorems concerning necessary conditions for optimality which greatly reduce the size of the game matrix. This makes the solutions for $n \leq 10$ quite simple. The case $n = 11$ is more complicated, and incidentally exhibits a qualitative feature of Blue's optimal strategy conjectured by Gilbert in [2], namely that Blue ought to avoid $n/4$ and $3n/4$ as well as $n/2$.

§ 2. NOTATION

At first glance it would appear that this is a multimove game for Red, since after each try he has new information and makes his next move accordingly. However, a better formulation is to describe a hunting strategy as a complete pattern of moves which takes care of all possible hiding places for Blue. Thus, if $1 \leq j \leq n$ are the n numbers Blue can choose from, each with probability p_j, then a strategy for Red is an ordered set of n integers, denoted by $S_i = \{S_{ij}\}$. The following example will illustrate the meaning of the S_{ij}:

j	1	2	3	4	5	6	7
p_j	p_1	p_2	p_3	p_4	p_5	p_6	p_7
S_{ij}	2	3	1	3	4	2	3

Here Red guesses 3 first. If too high, he tries 1 for his second guess; if too low, he tries 6 for his second guess, etc., S_{ij} is the number of the guess when j is tried.

Red will play a mixture of these strategies S_i, each with probability t_i. The payoff, if Red plays S_i and Blue plays $\{p_j\}$, is $\sum_{j=1}^{n} S_{ij}p_j$. The value of the game is

$$V = \min_{\{t_i\}} \max_{\{p_j\}} \sum_i \sum_j S_{ij}p_jt_i = \max_{\{p_j\}} \min_{\{t_i\}} \sum_j \sum_i S_{ij}p_jt_i .$$

§ 3. PROPERTIES OF OPTIMAL BLUE STRATEGIES

First it is clear from the definition of the game that Blue may play symmetrically about the center of the interval. Thus, we may assume

$$p_j = p_{n-j+1} . \tag{1}$$

Also it is clear that all p_j are positive in an optimal Blue strategy.

THEOREM 1.

$$p_1 \geq p_2. \tag{2}$$

PROOF. If $p_1 < p_2$, then Red would always play 2 before 1, so that these two column payoffs in the game matrix could not be equal. To see this consider any S_i where 1 is played before 2. Compare S_i with a related $\overline{S}_i$ where 2 is played at the kth guess rather than 1, and the rest of $\overline{S}_i$ matches S_i in the relative order of play:

	p_1	p_2				p_a				p_b	
S_i	k	k+m	.	t	.	k+m−1	.	r	.	k−1	.
$\overline{S}_i$	k+1	k	.	t−1	.	k+m−1	.	r	.	k−1	.

(Note that the values of S_{ij} will be reduced by 1 in the interval $2 < j < a$, where $S_{ia} = S_{i2} - 1$.) Then

$$\sum_j (S_{ij} - \overline{S}_{ij})p_j = -\,p_1 + mp_2 + \sum_{2<j<a} p_j > 0. \tag{3}$$

It can be shown that $p_1 > p_2$ for $n \geq 5$. Other properties of optimal Blue strategies can be conjectured from the list of solutions in Sec. 5, but seem to be difficult to prove in general. For instance, $p_1 = p_2 + p_3$ holds for $5 \leq n \leq 11$, and $p_1 = 2p_2$ holds for $7 \leq n \leq 11$. Also, one might conjecture that $p_1 > p_j$, $1 < j < n$, for $n > 4$.

§ 4. PROPERTIES OF OPTIMAL RED STRATEGIES

In this section we greatly reduce the list of possible optimal Red strategies against any trial Blue strategy.

THEOREM 2. Suppose at a given stage that Red, playing S_i, has located h on the interval $k \leq j \leq m$, and that S_i calls for next playing at a, left of the median of the hider's frequency distribution on this interval, and if a is too small, next playing at b to the right of a. Then a necessary condition for optimality of S_i against $\{p_j\}$ is that

$$\sum_{k\leq j\leq a} p_j \geqq \sum_{b\leq j\leq m} p_j . \tag{4}$$

PROOF. Assume that S_i does not satisfy (4). Then consider a related strategy $\overline{S}_i$ formed by interchanging play on numbers a and b while leaving the rest of the sequence of moves unchanged.

Thus

$$\begin{aligned} \overline{S}_{ij} &= S_{ij} + 1 & k \leq j \leq a, \\ \overline{S}_{ij} &= S_{ij} & a < j < b \\ \overline{S}_{ij} &= S_{ij} - 1 & b \leq j \leq m. \end{aligned} \tag{5}$$

The difference in payoff for playing $\overline{S}_i$ rather than S_i is

$$\sum_{j=k}^{m} \overline{S}_{ij}p_j - \sum_{j=k}^{m} S_{ij}p_j = \sum_{j=k}^{a} p_j - \sum_{j=b}^{m} p_j < 0. \tag{6}$$

Therefore S_i is dominated by $\overline{S}_i$.

A similar argument holds when a is at the right of the median of Blue's frequency distribution on $k \leq j \leq m$.

For example, if Red's first try at 3 is too small, his second try must be $\geq n - 2$ as a consequence of (1) and Theorem 2.

THEOREM 3. At each stage Red should make his guess inside the middle third of Blue's probability distribution on the current interval of uncertainty.

PROOF. We prove a somewhat sharper result. Suppose at a given stage Red has located Blue on the interval $k \leq j \leq m$, and S_i calls for next playing at a to the left of the median of Blue's probability distribution on this interval; and if a is too small, next playing at b where $b > a$, and if b is too large, next playing at c where $a < c < b$. Suppose that subsequent play is optimal. Compare this S_i with a related $\overline{S}_i$ formed by interchanging the order of play on a and c. That is, choose c first; then if c is too large, choose a next, and if c is too small choose b next. The order of the rest of the play in $\overline{S}_i$ matches subsequent play for S_i. The relative payoffs (each decreased by a constant) for S_i and $\overline{S}_i$ on the current interval are as follows:

	k				a				c				b				m
S_i		·	t	·	1	·	r	·	3	·	u	·	2	·	v	·	
$\overline{S}_i$		·	t+1	·	2	·	r−1	·	1	·	u−1	·	2	·	v	·	

Comparing payoffs we have

$$\sum_{j=1}^{n} (\overline{S}_{ij} - S_{ij})p_j = \sum_{j=k}^{m} (\overline{S}_{ij} - S_{ij})p_j = \sum_{j=k}^{a} p_j - \sum_{j=a+1}^{b-1} p_j - p_c .$$

Thus if S_i is optimal against $\{p_j\}$ then

$$\sum_{j=k}^{a} p_j \geq \sum_{j=a+1}^{b-1} p_j + p_c . \tag{7}$$

From Theorem 2, $\sum_{j=k}^{a} p_j \geq \sum_{j=b}^{m} p_j$, so that if $\sum_{j=k}^{a} p_j < \frac{1}{3} \sum_{j=k}^{m} p_j$ then $\sum_{k}^{a} p_j < \sum_{a+1}^{b-1} p_j$, violating (7). Thus Theorem 3 is proved.

§ 5. SOLUTIONS FOR $n \leq 11$

In this section we exhibit optimal strategies and the game value for $n \leq 11$. From (1) it follows that Red can play a given S_i and its symmetric counterpart S_i' equally often. Thus we list only the strategies S_i for which $S_{ij} = 1$ for $j \leq \frac{n+1}{2}$ and remember the probability associated with this strategy is really split equally between S_i and S_i'.

We shall list frequency distributions rather than probability distribution to further simplify the description of the solutions.

The general method of solution was as follows. For small values of n, the reduced list of Red strategies satisfying the theorems of the previous section gave a game matrix which was solved directly. As n grew longer an optimal Blue strategy was conjectured from the previous case, a list of Red strategies optimal against it was prepared, and a probability mixture was sought which gave the same expected payoff for each column for 1 to n in the game matrix.

A little trial and error work sufficed to find solutions for $n \leq 11$, although the last case was considerably harder. Further cases could be solved by linear programming on a computer.

The optimal mixture for Red is usually not unique. No attempt was made to find the mixture over the smallest number of Red pure strategies.

<u>Case $n = 2$ (trivial)</u>. There is just one S_1; $p_1 = p_2 = .5$; and $v(2) = 3/2$.

<u>Case $n = 3$</u>. Here only $S_1 = (2, 1, 2)$ and $S_2 = (1, 3, 2)$ with $S_2' = (2, 3, 1)$ are undominated. By the remarks concerning symmetry, the 3 by 3 game matrix can be reduced to an equivalent 2 by 2 matrix.

Red \ Blue	p_1	p_2	p_1
S_1	2	1	2
S_2	1	3	2

becomes

Red \ Blue	2	1
3	4	1
2	3	3

Thus $P = \{p_j\} = \{.4, .2, .4\}$, $t_1 = .6$, and $t_2 = .4$, the latter split equally between S_2 and S_2' of course. $V(3) - 9/5$.

<u>Case $n = 4$</u>. Here there are only 3 undominated strategies S_i. The reduced game matrix is 3 by 2.

Red \ Blue	p_1	p_2	p_2	p_1
S_1	2	1	2	3
S_2	2	1	3	2
S_3	1	3	4	2

Red \ Blue	1	1
0	5	3
1	4	4
0	3	7

A solution gives $p_j = 1/4$ for each j, while S_2 and S_2' are played with probability 1/2 each, and $V(4) = 2$.

Case $n = 5$. The list of undominated strategies S_i and the reduced game matrix are as follows:

Red \ Blue	p_1	p_2	p_3	p_2	p_1
S_1	2	3	1	3	2
S_2	3	2	1	2	3
S_3	2	1	3	2	3
S_4	2	1	4	3	2
S_5	2	1	3	4	2
S_6	1	4	3	4	2
S_7	1	3	5	4	2

Red \ Blue	5	3	2	
4	4	6	1	40
	6	4	1	40
4	5	3	3	40
1	4	4	4	40
	4	5	3	41
	3	8	3	45
	3	7	5	46
	40	40	20	

Thus Blue plays with frequency (5, 3, 2, 3, 5) and Red plays S_1 with probability 4/9, S_3 and S_3' each with 2/9 probability and S_4 and S_4' each with 1/18 probability. $V(5) = 20/9$.

Case $n = 6$.

	p_1	p_2	p_3	p_3	p_2	p_1
S_1	2	3	1	4	3	2
⋮	2	3	1	3	4	2
	2	3	1	3	2	3
	2	3	1	2	4	3
	2	1	4	3	4	2
	2	1	5	4	3	2
	2	1	4	3	2	3
	1	4	3	5	4	2
	1	4	3	4	5	2
S_{10}	1	3	5	6	4	2

	5	3	2	
2	4	6	5	48
	4	7	4	49
6	5	5	4	48
	5	7	3	52
	4	5	7	49
	4	4	9	50
2	5	3	7	48
	3	8	8	55
	3	9	7	56
	3	7	11	56
	48	48	48	

Blue plays a frequency distribution (5, 3, 2, 2, 3, 5), Red plays $t_1 = .2$, $t_3 = .6$, $t_7 = .2$, and $V(6) = 2.4$.

Case $n = 7$. The solution for $n = 7$, giving optimal S_i only, is

Red \ Blue	2	1	1	1	1	1	2
S_1	3	2	3	1	3	4	2
S_2	3	2	3	1	3	2	3
S_3	2	3	1	3	4	2	3
S_4	2	3	1	4	3	2	3

	2	1	1	1	
2	5	6	6	1	23
1	6	4	6	1	23
4	5	5	5	3	23
2	5	5	4	4	23
	46	46	46	23	

with frequency distributions as indicated for each player. $V(7) = 23/9$.

Case $n = 8$.

Red \ Blue	2	1	1	1	1	1	1	2
S_1	2	4	3	1	4	3	2	3
S_2	3	2	3	1	4	3	2	3
S_3	2	3	1	4	3	2	4	3

	2	1	1	1	
2	5	6	6	5	27
2	6	4	6	5	27
1	5	7	3	7	27
	27	27	27	27	

with $V(8) = 2.7$.

Case $n = 9$.

Red \ Blue	2	1	1	1	1	1	1	1	2
S_1	3	2	3	4	1	4	3	2	3
S_2	3	2	4	3	1	3	4	2	3
S_3	2	4	3	1	4	3	2	4	3
S_4	3	2	3	1	4	3	4	2	3
S_5	3	2	3	1	3	4	2	4	3
S_6	2	3	1	4	3	4	2	4	3

	2	1	1	1	1	
1	6	4	6	8	1	31
2	6	4	8	6	1	31
2	5	8	5	4	4	31
2	6	4	7	4	4	31
2	6	6	5	5	3	31
2	5	7	3	8	3	31
	62	62	62	62	31	

Here the number of S_i strategies is larger than necessary, but suffices. $V(9) = 31/11$.

<u>Case n = 10</u>. Here $V(10) = 35/12$.

Red \ Blue	2	1	1	1	1	1	1	1	1	2
S_1	3	2	3	4	1	4	3	4	2	3
S_2	3	2	3	4	1	4	3	2	4	3
S_3	3	2	4	3	1	4	3	2	4	3
S_4	2	3	4	1	4	3	4	2	4	3
S_5	3	2	3	1	4	3	4	2	4	3

	2	1	1	1	1	
2	6	4	7	7	5	35
1	6	6	5	7	5	35
4	6	6	6	6	5	35
2	5	7	6	5	7	35
3	6	6	5	5	7	35
	70	70	70	70	70	

<u>Case n = 11</u>. Here $V(11) = 3\,\frac{1}{62}$.

Red \ Blue	58	29	29	34	25	22	25	34	29	29	58
S_1	3	4	2	3	4	1	4	3	4	2	3
S_2	3	4	2	3	4	1	4	3	2	4	3
S_3	3	2	4	3	1	4	3	4	2	4	3
S_4	3	2	3	1	4	5	3	4	2	4	3
S_5	3	2	3	1	4	3	4	2	4	3	4
S_6	3	2	3	1	4	3	4	2	5	4	3

	58	29	29	34	25	22	
54	6	6	6	6	8	1	1122
6	6	8	4	6	8	1	1122
87	6	6	6	7	4	4	1122
18	6	6	5	5	7	5	1122
6	7	5	7	3	8	3	1122
15	6	6	8	3	8	3	1122
	1122	1122	1122	1122	1122	561	

The computation appears to get more difficult for $n = 12$ and beyond. Nevertheless, considerable reduction in the list of Red strategies is accomplished by the techniques of this paper. For instance, if $f(n)$ is the total number of Red pure strategies, the recursion relation

$$f(n) = \sum_{k=1}^{n} f(k-1)\, f(n-k) \tag{8}$$

with $f(0) = f(1) = 1$, etc., gives $f(11) = 58{,}786$ and $f(12) = 208{,}012$. One could also use a simple explicit formula ((2) of reference [2]) for computing f(n), namely,

$$f(n) = \frac{1}{n+1}\binom{2n}{n}. \tag{9}$$

BIBLIOGRAPHY

[1]. DRESHER, Melvin, Games of Strategy: Theory and Application, Prentice Hall (1961).

[2] GILBERT, E. N., "Games of identification or convergence," SIAM Review Vol. 4, No. 1, January 1962, pp. 16-24.

[3] KONHEIM, A. G., "The number of bisecting strategies for the game (N)," SIAM Review, Vol. 4, No. 4, October 1962, pp. 379-384.

Selmer M. Johnson

The RAND Corporation

THE RENDEZVOUS VALUE OF A METRIC SPACE

O. Gross

§ 1. INTRODUCTION AND THEOREMS

It is the purpose of this short note to prove a few theorems regarding average (arithmetic mean) distances of points in a finite collection from some point in a compact connected metric space. To each such space, in fact, corresponds a unique constant which we shall call its rendezvous value. We shall not use the term in the sequel, but introduce it solely for its connotative value in those instances in which the space is equipped with suitable geodesics. At any rate it adds spice to the title.

Perhaps the theorems are well known, but since they seem _not_ to be to the few people polled (expressions ranged from surprise to total disbelief), it would seem advisable to jot down short proofs of them.

A stronger motivation or raison d'être for this paper, however, is furnished by the not too well exploited fact that game theory can be used as a tool in other branches of mathematics and physics, including geometry, as in this instance. In fact, well known results in game theory render the proofs of the three theorems almost trivial.

The theorems we wish to prove can be phrased as follows:

THEOREM 1. Relative to a compact connected metric space there exists a unique constant K with the property that given any finite collection of points one can find a point P such that the average distance of the points in the collection from P is equal to K.

THEOREM 2. If K is the constant according to Theorem 1

of a compact connected metric space of positive diameter d then $\frac{d}{2} \leq K < d$.

THEOREM 3. Let d and K be numbers such that $\frac{d}{2} \leq K < d$, then there exists a compact connected metric space of diameter d such that its constant as defined by Theorem 1 is equal to K.

Before proceeding to the proofs, we remark that Theorem 3 asserts that the conclusion of Theorem 2 cannot be sharpened without restricting the hypothesis. For example, one might restrict the spaces to be subsets of the Euclidean plane, since this seems to be fashionable, and ask for a sharp upper bound for the ratio $\frac{K}{d}$. Computations seem to indicate that the bound is not much larger than 2/3. Perhaps this will be the subject of another paper. So without further digression we shall proceed with the proofs.

§ 2. PROOF OF THEOREM 1

We shall prove first the existence of a constant with the stated property and then show that there can be at most one such corresponding to the given space. To this end, consider the two-person zero sum game in which each person independently chooses a point in the space and the payoff is given by the distance between them. According to a theorem of Glicksberg [1], for example, the game has a value and optimal strategies. Let its value be V. We assert that V has the requisite property. To see this, let $\{x_1, \ldots, x_n\}$ be any finite collection of points, D the metric of the space, and consider the mixed strategy by the minimizer which assigns weight $\frac{1}{n}$ to each of these points. Then, whether this strategy is optimal or not, we have (by virtue of continuity, compactness, etc., and the definition of value of a game)

$$\max_P \frac{1}{n} \sum_{i=1}^{n} D(x_i, P) \geq V. \tag{1}$$

Of course the maximizer can use the same strategy, and upon exploiting the symmetry of the metric, we obtain

$$\text{(2)} \qquad \min \frac{1}{n} \sum_{i=1}^{n} D(x_i, P) \leq V.$$

At this point connectedness plays a role. Since the functions of P in (1) and (2) are really the same function and since the _range of a real-valued continuous function on a connected set is connected_, there exists a P where equality is achieved:

$$\frac{1}{n} \sum_{i=1}^{n} D(x_i, P) = V.$$

Thus, existence is proved. To see that no other constant (than V) works, assume we have another such K. Assume for the moment that $K > V$. Now, for a game on a compact space it is known that one can approximate to an optimal strategy by a finite mixture, provided the payoff is continuous. Thus in this instance, given $\epsilon > 0$ there exists a finite collection of points $\{x_1, \ldots, x_n\}$ and positive weights $\{\lambda_1, \ldots, \lambda_n\}$ summing to unity such that for all P

$$\sum_{i=1}^{n} \lambda_i D(x_i, P) \leq V + \epsilon;$$

the strategy being employed, say, by the minimizing player.

We remark that the theorem is true if the space consists of a single point. But if otherwise, since the space is connected, there are infinitely many points in every neighborhood, in particular we can obtain a cluster of such about each of the x_i above. Thus, if the λ_i were rational numbers with common denominator N and numerators n_i, we can select N different points x_i', $i = 1, \ldots, N$, with n_i of them clustered near x_i to obtain an arithmetic average approximating an optimal strategy. This follows by the continuity of the metric. Since, however, the rational numbers are dense in the reals, we see that this can be done in any event, so that given any $\epsilon > 0$ there exists a finite collection of points $\{x_1, \ldots, x_N\}$ such that for all P

$$\frac{1}{N} \sum_{i=1} D(x_i, P) \leq V + \epsilon.$$

But since $K > V$, we can choose ϵ sufficiently small so that the value K cannot be achieved by any point P in the collection. The case $K < V$ can obviously be treated in the same manner, and uniqueness follows. This concludes the proof of Theorem 1.

§ 3. PROOF OF THEOREM 2

By the arguments used to prove Theorem 1, the K involved in the statement of Theorem 2 is the value of a two person zero sum game in which each player independently picks a point in the space with the payoff being the distance between their choices. To show that the value of the game satisfies the lower inequality, we observe first that the compactness of the space guarantees the existence of a pair of diametrically distant points x_1 and x_2. Let the maximizer play a pair of such, each point with probability 1/2. Whether this strategy is optimal or not, we have

$$K \geq \min_P \left(\frac{1}{2} D(x_1, P) + \frac{1}{2} D(x_2, P)\right).$$

But by the triangle inequality, etc.

$$D(x_1, P) + D(x_2, P) \geq D(x_1, x_2) = d;$$

whence

$$K \geq \frac{d}{2}.$$

To see that $K < d$, we observe first that any pure strategy for the minimizer guarantees that $K \leq d$, since d is a bound on the payoff. By assumption, however, $d > 0$, and it remains to show that $K \neq d$. We can without loss in generality, therefore, take $d = 1$. However, the assumption that $K = d = 1$ leads to the existence of an infinite sequence of points in the space all one unit apart from each other (by repeated application of Theorem 1). But such a sequence cannot contain a convergent subsequence, and the space is therefore not compact, contrary to hypothesis. Thus $K < d$. This concludes the proof of Theorem 2.

§ 4. PROOF OF THEOREM 3

The hypothesis of Theorem 3 is not satisfied unless $d > 0$, so we can without loss in generality assume that $d = 1$. If we rephrase the theorem in game theoretic language, as we can, in view of the previous arguments, we are required to find a suitable compact connected metric space such that the game value is a given number K on the interval $(\frac{1}{2}, 1]$. In fact, the space we shall select will be homeomorphic to the closed unit interval. Consider the closed unit interval $[0, 1]$. One verifies that if we select

$$D_\lambda(x, y) = \frac{(\lambda+1)\ |x-y|}{\lambda|x-y| + 1}, \quad 0 \leq {x \atop y} \leq 1, \quad \lambda \geq 0$$

as the payoff, then D_λ as defined above is <u>metric</u> if $|x - y|$ evaluates the Euclidian metric. Since, moreover, D_λ is continuous etc., the new space is still connected and compact and has diameter 1, for any choice of λ. If $\lambda = 0$, the value of the game is $\frac{1}{2}$, and it is an easy exercise to verify that as $\lambda \to \infty$ the value of the game tends to 1. Since the value of the game is readily shown to be a continuous function of the game parameter λ, we can with the appropriate choice thereof obtain a game with the prescribed value. This concludes the proof of Theorem 3.

BIBLIOGRAPHY

[1] GLICKSBERG, I., "Minimax theorem for upper and lower semi-continuous payoffs," The RAND Corporation Research Memorandum, RM-478, October 1950.

O. Gross

The RAND Corporation
Santa Monica, California

GENERALIZED GROSS SUBSTITUTABILITY AND EXTREMIZATION

Hukukane Nikaidô

§ 1. INTRODUCTION

This paper is a sequel to the two previous papers by this writer [15, 16]. It may also be regarded as an answer to the question how far we can go without fixed point theorems when handling systems of inequalities which generalize those arising in game theory.

Ever since recent contributions [1, 6, 7, 8, 10, 13, 14], with the aid of fixed point theorems or their equivalent propositions, gave general conditions for the existence of competitive equilibrium, attention has shifted to the detailed observation of competitive economic systems under rather special conditions such as gross substitutability. Several interesting results have been quite recently obtained along this line (e.g., [2, 3, 4, 9, 11, 12]). The purpose of this paper, which is also in accordance with the line above, is to give an elaborated version of the existence theorem of competitive equilibrium in the case of generalized gross substitutability. A criterion, in the form of a boundary condition, for the existence of equilibrium will be given, so powerful that it can be still effective even when the known fixed point techniques break down. This criterion is also useful to locate equilibrium within a given, possibly small, open subset of the whole price set. A rearrangement of the argument also entails the unique determination of some system of functions by their boundary values, a situation similar to those observed in the theories of analytic functions and harmonic functions. Generalized gross substitutability is a unified generalization of skew symmetricity and gross substitutability. Finally, the dynamic implication of generalized gross substitutability will be made clear for the Brown-von Neumann differential equation, which was originally devised to solve zero-sum two-person games.

Our method is extremely elementary and consists of the extremization of the sum of the squared positive excess demands of goods. It takes advantage of a simple fact that some quadratic form is negative definite, which was observed by McKenzie [9]. This fact can be derived from the following well-known basic theorem on matrices with nonpositive off-diagonal elements, which reads as follows:

For any $n \times n$ matrix $B = [b_{ij}]$, $b_{ij} \leqq 0$ $(i \neq j)$, the following three conditions are equivalent:

(I) $Bx > 0$ for some nonnegative vector $x \geqq 0$.

(II) B has the nonnegative inverse $B^{-1} \geqq 0$.

(III) The Hawkins-Simon condition holds, i.e., all the principal minors are positive.

As is well-known, this basic theorem can be directly proved by bringing B into a triangular matrix, through elimination, without any reference to the Frobenius theorem on nonnegative matrices.

§ 2. EXISTENCE AND LOCATION OF EQUILIBRIUM

2.1 We shall be concerned with a competitive economy where all goods, labeled $i = 1, 2, \ldots, n$ $(n \geqq 2)$, are <u>generalized weak gross substitutes</u> in the sense as defined below. The price system will be denoted by a nonnegative n-vector $p = (p_1, p_2, \ldots, p_n) \geq 0$. We shall pay special attention to a set P of price vectors in which the excess demand functions are defined. In the sequel, all topological considerations will be done with regard to the relative topology of the nonnegative orthant R^n_+ of R^n, as induced by R^n, unless otherwise indicated.

The basic assumptions throughout this paper are as follows:

(α) <u>P is a nonempty subcone of the nonnegative orthant R^n_+, the origin 0 deleted, and it is open in R^n_+.</u>

(β) <u>The excess demand functions $E_i(p)$ $(i = 1, 2, \ldots, n)$ are single-valued continuous functions and have continuous partial derivatives $\frac{\partial E_i}{\partial p_j}$ in P, which will be denoted by $E_{ij}(p)$.</u>

(γ) <u>Positive homogeneity of degree m, $0 \leqq m \leqq 1$; that is, $E_i(\lambda p) = \lambda^m E_i(p)$ for $\lambda > 0$, $p \in P$.</u>

(δ) *The Walras law holds*: *i.e.*, $\sum_{i=1}^{n} p_i E_i(p) = 0$ *identically in* P.

(ε) *Generalized gross substitutability prevails*: *that is*, $E_{ij}(p) + E_{ji}(p) \geqq 0 \ (i \neq j)$ *everywhere in* P.

2.2 In the terminology of economists, gross substitutability prevails when $E_{ij}(p) \geqq 0 \ (i \neq j)$. On the other hand, $E_{ij}(p) + E_{ji}(p) = 0$ for all i, j when the Jacobian matrix of the system of functions $E_i(p)$ is skew symmetric. In the both cases condition (ε) is clearly true.

2.3 The central problem at issue here is the possibility and stability of a competitive equilibrium. $\hat{p} \in P$ is said to be an equilibrium price vector, if

$$E_i(\hat{p}) \leqq 0 \qquad (i = 1, 2, \ldots, n) \tag{2.1}$$

$$E_i(\hat{p}) = 0 \quad \text{for } i \text{ with } \hat{p}_i > 0. \tag{2.2}$$

The well-established general theorem on the existence of competitive equilibrium premises only the continuity of the excess demand functions and the Walras law, and not any special condition such as gross substitutability. But it heavily relies upon the simpler topological structure of the domain of the excess demand functions such as convexity or acyclicity. Compared with it, the novelty of our result lies in the fact that at the cost of assuming a type of substitutability the domain of $E_i(p)$ is allowed to be of more complicated nature than as is usually assumed. P need neither be convex nor even acyclic, so that our result seems to be more than an alternative proof. The figure on the right illustrates the intersecting portion of a P with the normalizing hyperplane $\sum_{i=1}^{n} p_i = 1$.

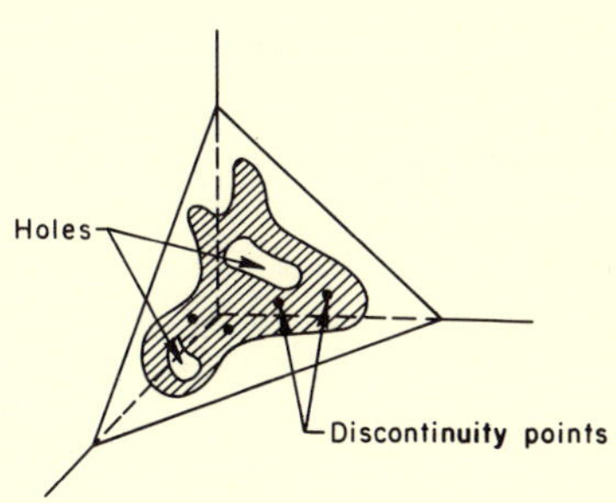

2.4 As a preliminary step, a lemma, which is observed by McKenzie [9] for the case of $m = 0$, will be given for a variety of values of m.

LEMMA. Under conditions (α), (β), (γ), (δ) and (ε) the following results are true:

[i] When $0 \leqq m < 1$, let $N(p) = \{j \mid E_j(p) > 0\}$ for any $p \in P$. Then, the quadratic form

$$\sum_{i,j \in N(p)} E_{ij}(p)\xi_i\xi_j \tag{2.3}$$

is negative definite, provided $N(p) \neq \phi$.

[ii] When $m = 1$, the quadratic form

$$\sum_{i,j=1} E_{ij}(p)\xi_i\xi_j \tag{2.4}$$

is negative semi-definite at every $p \in P$.

PROOF. The proof will be done by adapting that of McKenzie [9] to our case. Upon differentiation of equation $\sum p_i E_i(p) = 0$ given in (δ) we have

$$\sum_{i=1}^{n} p_i E_{ij}(p) + E_j(p) = 0 \qquad (j = 1, 2, \ldots, n).$$

On the other hand, (γ) together with (β) implies, by virtue of the Euler theorem on homogeneous functions,

$$\sum_{i=1}^{n} p_i E_{ji}(p) - mE_j(p) = 0 \qquad (j = 1, 2, \ldots, n).$$

Hence summing up these equations gives

$$\sum_{i=1}^{n} e_{ij}(p)p_i + (1-m)E_j(p) = 0 \qquad (j = 1, 2, \ldots, n). \tag{2.5}$$

where $e_{ij}(p) = E_{ij}(p) + E_{ji}(p)$. The subsequent portions of the argument will be separately developed for [i] and [ii].

The case of [i]. Relation (2.5) can be rearranged to

$$(2.6)\qquad \sum_{i\in N(p)} -e_{ij}(p)p_i = (1-m)E_j(p) + \sum_{i\notin N(p)} e_{ij}(p)p_i \quad (j\in N(p)).$$

By (ϵ), $e_{ij}(p) \geqq 0$ for $i \notin N(p)$, $j \in N(p)$, so that $\sum_{i\notin N(p)} e_{ij}(p)p_i \geqq 0$ for $j \in N(p)$ in view of p_i being nonnegative. Furthermore, $(1-m)E_j(p) > 0$ for $j \in N(p)$, since $1 > m$. Thus the right-hand side of (2.6) is positive for any $j \in N(p)$, that is

$$(2.7)\qquad \sum_{i\in N(p)} -e_{ij}(p)p_i > 0, \quad p_i \geqq 0 \qquad (i,\ j\in N(p)).$$

Since the off-diagonal elements of the matrix $[-e_{ij}(p)]$ $(i,\ j\in N(p))$ are non-positive by (ϵ), condition (I) in the foregoing section is satisfied by this matrix. Therefore, the matrix satisfies also the Hawkins-Simon condition (III), so that all the principal minors of $[-e_{ij}(p)]$ are positive. This proves that (2.3) is negative definite, because (2.3) $= \frac{1}{2} \sum_{i,j\in N(p)} e_{ij}(p)\xi_i\xi_j$.

The case of [ii]. Since $m = 1$, relation (2.5) reduces to

$$(2.8)\qquad \sum_{i=1}^{n} e_{ij}(p)p_i = 0 \qquad (j = 1,\ 2,\ \ldots,\ n).$$

Let $\epsilon > 0$ be an arbitrary positive number. We rearrange (2.8) to

$$(2.9)\qquad \sum_{i=1}^{n} (\epsilon\delta_{ij} - e_{ij}(p))p_i = \epsilon p_j \qquad (j = 1,\ 2,\ \ldots,\ n),$$

where δ_{ij} are Kronecker's deltas.

By (α), P is open in R^n_+. Therefore, P contains some vectors all of whose components are positive, and furthermore, every point of P can be regarded as a limit point of the former. Thus, once (2.4) is negative semi-definite at any $p \in P$ with $p > 0$, the continuity of $e_{ij}(p)$'s implied by (β) assures that the same is true everywhere in P.

Let us now suppose that $p_j > 0$ $(j = 1,\ 2,\ \ldots,\ n)$ in (2.9). Then, in the same way as in [i], it can be shown that the principal minors of the

$n \times n$ matrix $[\epsilon\delta_{ij} - e_{ij}(p)]$ are positive. Hence $\sum_{i,j=1} (e_{ij}(p) - \epsilon\delta_{ij})\xi_i\xi_j \leqq 0$. In view of the arbitrariness of $\epsilon > 0$, this proves (2.4) $= \frac{1}{2} \sum_{i,j=1}^{n} e_{ij}(p)\xi_i\xi_j \leqq 0$ at every $p > 0$, in P. Q.E.D.

2.5 Next we shall give a boundary condition for the existence of a competitive equilibrium in P.

Let

$$\Phi(p) = \sum_{i=1}^{n} \theta_i(p)^2, \tag{2.10}$$

where

$$\theta_i(p) = \max [E_i(p), 0] \qquad (i = 1, 2, \ldots, n). \tag{2.11}$$

Throughout this paper, as in [5, 15, 16], the function $\Phi(p)$ will play an important role.

Boundary Condition (*) on $\Phi(p)$. For any nonzero boundary point $\hat{p} \in \bar{P}$, $\notin P$, if any exists, and for any sequence $\{p^\nu\}$ in P converging to $\hat{p}$, there exists some $q \in P$, depending possibly on $\hat{p}$ and $\{p^\nu\}$ such that

$$\sum_{i=1}^{n} \hat{p}_i = \sum_{i=1}^{n} q_i \tag{2.12}$$

$$\Phi(q) < \limsup_{\nu \to \infty} \Phi(p^\nu). \tag{2.13}$$

THEOREM. If (α), (β), (γ), (δ), (ϵ) and the boundary condition (*) are fulfilled, then there is an equilibrium price system $\hat{p}$ in P which satisfies (2.1) and (2.2).

PROOF. Let S be the intersection of P with the normalizing hyperplane $\sum_{i=1}^{n} p_i = 1$. Clearly we have

$$\inf_{p \in S} \Phi(p) = \delta \geqq 0.$$

Take a sequence $\{p^\nu\}$ in S such that $\lim \Phi(p^\nu) = \delta$. Since $p^\nu \in S$, we may assume that $\{p^\nu\}$ itself converges to a limit $\hat{p}$. If $\hat{p} \notin S$, then naturally $\hat{p} \notin P$. On the other hand, $\hat{p} \in \overline{P}$. Thus, by the boundary condition (*), there is a $q \in S$ for these $\hat{p}$ and $\{p^\nu\}$ such that (2.13) is valid. But this yields a contradiction: $\delta \leqq \Phi(q) < \lim_{\nu\to\infty} \Phi(p^\nu) = \delta$. Therefore, $\hat{p} \in P$, and continuity implies $\Phi(\hat{p}) = \delta$. That is, $\Phi(p)$ is minimized at $p = \hat{p}$ subject to the condition $p \in S$. Since P is open in R^n_+, the following marginality conditions hold:

$$\text{(2.14)} \qquad \left[\frac{\partial\Phi}{\partial p_j}\right]_{p=\hat{p}} \geqq \sum_{k=1}^{n} \hat{p}_k \left[\frac{\partial\Phi}{\partial p_k}\right]_{p=\hat{p}} \qquad (j = 1, 2, \ldots, n).$$

Since $\Phi(p)$ is positively homogeneous of degree $2m$, the right-hand side of (2.14) equals $2m\Phi(\hat{p})$, by the Euler theorem on homogeneous functions. On the other hand, in view of (2.10) and (2.11), we have $\frac{\partial\Phi}{\partial p_j} = 2 \sum_{i=1}^{n} \theta_i(p)E_{ij}(p)$ $(j = 1, 2, \ldots, n)$, so that (2.14) reduces to

$$\text{(2.15)} \qquad \sum_{i=1}^{n} \theta_i(\hat{p})E_{ij}(\hat{p}) \geqq m\Phi(\hat{p}) \qquad (j = 1, 2, \ldots, n).$$

Multiplying the jth inequality of (2.15) by $\theta_j(\hat{p})$, we have

$$\text{(2.16)} \qquad \sum_{i,j=1}^{n} E_{ij}(\hat{p})\theta_i(\hat{p})\theta_j(\hat{p}) \geqq m\Phi(\hat{p}) \sum_{j=1}^{n} \theta_j(\hat{p}).$$

Now, if $1 > m \geqq 0$, supposing that the set $N(\hat{p}) = \{j \mid E_j(p) > 0\}$ is nonempty, and in view of [i] in the foregoing lemma, (2.16) leads to a contradiction

$$0 > \sum_{i,j\in N(\hat{p})} E_{ij}(\hat{p})\theta_i(\hat{p})\theta_j(\hat{p}) \geqq m\Phi(\hat{p}) \sum_{j=1}^{n} \theta_j(\hat{p}) \geqq 0.$$

Also, if $m = 1$, by [ii] in the lemma, the left-hand side of (2.16) is nonpositive, so that $\Phi(\hat{p}) \sum_{j=1}^{n} \theta_j(\hat{p}) \leqq 0$. Therefore, in all the cases we have (2.1). (2.2) is an immediate consequence of (2.1) and (δ). Q.E.D.

2.6 The following corollaries give some special but important cases on which the boundary condition (*) works well.

COROLLARY 1. Suppose that a system of functions $E_i(p)$ which satisfies (α), (β), (γ), (δ) and (ϵ) in P also fulfills the following boundary condition (**): For any nonzero boundary point $\hat{p} \in \overline{P}$, $\notin P$, there is an i such that $\lim_{p \to \hat{p}} E_i(p) = + \infty$. Then, there is an equilibrium price vector in P.

PROOF: The validity of (**) implies that $\lim_{p \to \hat{p}} \Phi(p) = + \infty$ for any nonzero boundary point $\hat{p} \in \overline{P}$, $\notin P$, if any. This surely means the validity of (*) so that the theorem applies to the present case.

COROLLARY 2. Suppose that P as well as its closure minus the origin is contained in a larger set where $E_i(p)$'s are still defined and continuous. Suppose, furthermore, that the system of functions $E_i(p)$ which satisfies (α), (β), (γ), (δ) and (ϵ) in P also fulfills the following boundary condition (***): For any nonzero boundary point $\hat{p} \in \overline{P}$, $\notin P$, there is some $q \in P$, depending on $\hat{p}$, such that $\sum_{i=1}^{n} \hat{p}_i = \sum_{i=1}^{n} q_i$ and $\Phi(q) < \Phi(\hat{p})$. Then, there is an equilibrium price vector in P.

PROOF. Since this time $E_i(p)$'s are defined and continuous even on the boundary (minus the origin) of P, (*) is equivalent to (***).

Corollary 1 includes the most familiar case in which P equals the totality of all positive vectors. Corollary 2 gives rise to the location of an equilibrium within a possibly thin neighborhood cone P. It is also noted that a version of the existence theorem as considered by Arrow and Hurwicz [4], in which $E_i(p)$'s are allowed to take an infinite value, can be easily reduced to Corollary 1, if generalized gross substitutability is assumed.

§ 3. DETERMINATION OF FUNCTION VALUES BY BOUNDARY VALUES

3.1 In this section it will be observed to what extent a system of functions $E_i(p)$ satisfying (α), (β) and (γ) is determined by their boundary values. As before, all the topological concepts should be understood in the relative topology of R^n_+. Denote by S and $\overline{S}$ the intersections of the normalizing hyperplane with P and $\overline{P}$, respectively.

THEOREM. Suppose that $E_i(p)$'s satisfy (α), (β), (γ), (δ) and (ϵ) and $0 \leqq m < 1$. Then, unless $E_i(p) = 0$ $(i = 1, 2, \ldots, n)$ everywhere in P, the function $\Phi(p)$ of (2.10) is maximized subject to $p \in S$ at no point of P.

PROOF. Suppose that a maximizing point $\hat{p}$ lies in P. Since (α), (β), (γ), (δ) and (ϵ) hold in P, we have the marginality condition

$$(3.1) \qquad \left[\frac{\partial \Phi}{\partial p_j}\right]_{p=\hat{p}} \leqq \sum_{k=1}^{n} \hat{p}_k \left[\frac{\partial \Phi}{\partial p_k}\right]_{p=\hat{p}} \qquad (j = 1, 2, \ldots, n),$$

which is the counterpart of (2.14) in the maximizing case. It is noted that equality holds in the jth relation in (3.1) unless $\hat{p}_j = 0$. We wish to show that $\Phi(\hat{p}) = 0$. In fact, if $\Phi(\hat{p}) > 0$, the set $N(\hat{p}) = \{j \mid E_j(\hat{p}) > 0\}$ is nonempty. Since (2.7) holds for $p = \hat{p}$ and $- e_{ij}(\hat{p}) \leqq 0$ $(i \neq j)$, it is immediate that $\hat{p}_j > 0$ for $j \in N(\hat{p})$, so that

$$\left[\frac{\partial \Phi}{\partial p_j}\right]_{p=\hat{p}} = \sum_{k=1}^{n} \hat{p}_k \left[\frac{\partial \Phi}{\partial p_k}\right]_{p=\hat{p}} \qquad (j \in N(\hat{p})).$$

Thus, similarly as in the proof of [i] in the theorem in Section 2, we arrive at a contradiction

$$0 > \sum_{i,j \in N(\hat{p})} E_{ij}(\hat{p})\theta_i(\hat{p})\theta_j(\hat{p}) = m\Phi(\hat{p}) \sum_{j=1}^{n} \theta_j(\hat{p}) \geqq 0.$$

Therefore, $0 = \Phi(\hat{p}) = \max \Phi(p)$ over all $p \in S$, which, combined with (α), (β), (γ) and (δ), implies that $E_i(p) = 0$ $(i = 1, 2, \ldots, n)$ everywhere in P.

3.2 The following corollaries are some of the immediate consequences of the theorem.

COROLLARY 1. Let $F_i(p)$ $(i = 1, 2, \ldots, n)$ and $G_i(p)$ $(i = 1, 2, \ldots, n)$ be two systems of functions satisfying (α), (β) and (γ) with $m < 1$ in a common P. Suppose furthermore that they are defined and continuous in $\overline{S}$, and

$$\sum_{i=1}^{n} p_i F_i(p) = \sum_{i=1}^{n} p_i G_i(p) \tag{3.2}$$

$$\frac{\partial F_i}{\partial p_j} + \frac{\partial F_j}{\partial p_i} \geqq \frac{\partial G_i}{\partial p_j} + \frac{\partial G_j}{\partial p_i} \qquad (i \neq j) \tag{3.3}$$

everywhere in P. Then, if $F_i(p) = G_i(p)$ $(i = 1, 2, \ldots, n)$ for $p \in \overline{S}, \notin S$, we have

$$F_i(p) = G_i(p) \qquad (i = 1, 2, \ldots, n)$$

everywhere in P.

PROOF. Let $E_i(p) = F_i(p) - G_i(p)$ $(i = 1, 2, \ldots, n)$. The corresponding $\Phi(p)$ is continuous on $\overline{S}$ which is compact, so that $\Phi(p)$ is maximized subject to $p \in \overline{S}$ at some $\hat{p}$. Unless $E_i(p) = 0$ $(i = 1, 2, \ldots, n)$ identically in P, clearly $\Phi(\hat{p}) > 0$. Since, by assumption, $E_i(p) = 0$ $(i = 1, 2, \ldots, n)$ for $p \in \overline{S}, \notin S$, this $\hat{p}$ must lie in P. But this result is in contradiction to the above theorem, because $E_i(p)$'s satisfy all the assumptions of the theorem in P. Therefore $F_i(p) - G_i(p) = E_i(p) = 0$ $(i = 1, 2, \ldots, n)$ identically in P. Q.E.D.

COROLLARY 2. Except for the system of identically vanishing functions, there is no system of functions

$E_i(p)$ satisfying (α), (β), (γ), (δ) and (ϵ), if $0 \leqq m < 1$ and P consists of all nonzero nonnegative vectors $p \geq 0$.

PROOF. Since this time $\overline{S} = S \subset P$, the corresponding $\Phi(p)$ really takes on a maximum subject to $p \in S$ at some point $\hat{p} \in P$. Therefore, by the above theorem, $E_i(p) = 0$ $(i = 1, 2, \ldots, n)$ identically in P. Q.E.D. It should be noted, however, that if P consists of all positive vectors $p > 0$, conditions (α), (β), (γ), (δ) and (ϵ), are met by systems of nontrivial functions. The following system of functions provides us with an example. $E_i(p) = \frac{1}{p_i} \sum_{j=1}^{n} a_{ij} p_j^{1+m} - p_i^m$ $(i = 1, 2, \ldots, n)$, where $a_{ij} \geqq 0$ $(i \neq j)$, $\sum_{i=1}^{n} a_{ij} = 1$ $(j = 1, 2, \ldots, n)$ and $1 \geqq m \geqq 0$. On the other hand, the most familiar example of nontrivial systems satisfying these conditions for $m = 1$ on the totality of all nonnegative vectors $p \geq 0$ is undoubtedly that arising from the skew symmetric payoff matrix $[a_{ij}]$ of a zero-sum two-person game: $E_i(p) = \sum_{j=1}^{n} a_{ij} p_j$ $(i = 1, 2, \ldots, n)$.

§ 4. $\Phi(p)$ AS A LIAPOUNOV FUNCTION

4.1 This final section will be devoted to discussions on the dynamic stability of equilibrium by the Brown-von Neumann differential equation for the case of generalized gross substitutability. Stability considerations by this type of differential equation originated with Brown and von Neumann [5], and the later contribution by this writer [16] put their results in a general form. In the following lines it will be shown that the lemma in Section 2 motivates an extension of their results in another direction.

In discussing stability, in addition to conditions (α), (β), (γ), (δ) and (ϵ), we assume on $E_i(p)$'s the following additional condition: (∞) *For any point* $q \in \overline{S}$ *but* $\notin S$, *if any exists*, $\lim_{p \to q} E_i(p) = +\infty$ *for some* i.

4.2 Consider the Brown-von Neumann differential equation

$$\frac{dp_i}{dt} = \theta_i(p) - p_i \sum_{k=1}^{n} \theta_k(p) \qquad (i = 1, 2, \ldots, n), \tag{4.1}$$

where $\theta_i(p)$'s are given by (2.11).

THEOREM. Under conditions (α), (β), (γ), (δ), (ϵ) and (∞), for any initial position $p(0) = p^o \in S$ equation (4.1) has a global solution $p(t) \in S$ $(t \geqq 0)$ starting from $p(0)$. Furthermore,

$$\lim_{t \to \infty} \Phi(p(t)) = 0$$

for any such solution $p(t)$.

PROOF. Let, for any p which need not be nonnegative,

$$|p| = (|p_1|, |p_2|, \ldots, |p_n|)$$

where $|p_i|$ is the absolute value of p_i. Since the mapping $p \to |p|$: $R^n \to R^n_+$ is continuous, and P is assumed to be open in R^n_+, the set $Q = \{p| \ |p| \in P\}$ is open (in R^n). Furthermore, in view of (α) and (β), the functions $E_i(|p|)$ satisfy the Lipschitz condition in any compact neighborhood of p^o lying entirely in Q. Therefore, by virtue of the Cauchy-Lipschitz existence theorem, the modified equation

$$\frac{dp_i}{dt} = \theta_i(|p|) - p_i \sum_{k=1}^{n} \theta_k(|p|) \qquad (i = 1, 2, \ldots, n) \tag{4.2}$$

has a local solution $p(t)$ $(\sigma \geqq t \geqq 0)$, with $p(0) = p^o$, where σ is a positive number. It is readily seen, similarly as in [16], that $p(t) \in S$ for $t \in [0, \sigma]$. This means that this $p(t)$ is a local solution of (4.1). Next, let $p(t)$ be any local solution of (4.1) defined in $[0, \tau]$, where $\tau > 0$. Then, upon differentiation and taking into account the relations $p(t) \in S$, (γ) and $\theta_i(p)^2 = \theta_i(p)E_i(p)$, we have

$$\frac{d}{dt} \Phi(p(t)) = 2 \sum_{i,j=1}^{n} E_{ij}(p)\theta_i(p)\theta_j(p) - 2m\Phi(p) \sum_{i=1}^{n} \theta_i(p) \tag{4.3}$$

in $[0, \tau]$. By the lemma in Section 2, the right-hand side is nonpositive, so that $\Phi(p(t))$ is nonincreasing in $[0, \tau]$. Let ω be the least upper bound of all τ for which the local solution $p(t)$ $(\sigma \geq t \geq 0)$ can be continued to $[0, \tau]$. Suppose that $\omega < +\infty$. Since $p(t)$ can be continued to $[0, \omega)$, the above result assures that $p(t) \in \Gamma = \{p \mid p \in S, \Phi(p) \leq \Phi(p(0))\}$ for $t \in [0, \omega)$. Let $p^{\nu} \in \Gamma$ and $\lim_{\nu\to\infty} p^{\nu} = p \in \overline{S}$. Then, assumption (∞) implies $p \in S$, that is, Γ is closed in $\overline{S}$. Since $\overline{S}$ is compact, Γ is also compact. Thus, in the light of the compactness of Γ, we can let $|\theta_i(p) - p_i \sum_{k=1}^{n} \theta_k(p)| \leq K$ on Γ for some $K > 0$. Then, for any t_1, $t_2 \in [0, \omega)$, we have

$$|p_i(t_2) - p_i(t_1)| \leq \int_{t_1}^{t_2} |\theta_i(p) - p_i \sum_{k=1}^{n} \theta_k(p)| dt \leq K|t_1 - t_2|,$$

the extreme right-hand side of which tends to zero as $t_1, t_2 \to \omega$. Hence $\lim_{t\to\omega} p(t)$ exists and the limit belongs to Γ. This implies the possibility to continue the solution beyond ω, so that ω must be $+\infty$.

Now take any global solution $p(t)$. Then, (4.3) holds over time, and we have

$$\frac{d}{dt} \Phi(p(t)) \leq \begin{cases} 2 \sum_{i,j\in N(p)} E_{ij}(p)\theta_i(p)\theta_j(p) & (\text{if } 1 > m \geq 0) \\ -2\Phi(p)^{3/2} & (\text{if } m = 1). \end{cases}$$

This proves, in the light of the lemma in Section 2, that $\Phi(p(t)) \to 0$ as $t \to +\infty$. Since $p(t) \in \Gamma$ over time, and Γ is compact, the distance from $p(t)$ to the set of all equilibrium vectors converges to zero as t tends to infinity. Q.E.D. It is finally noted that, if P consists of all nonzero nonnegative $p \geq 0$, assumption (∞) does not become effective and can be dispensed with, because $S = \overline{S}$.

BIBLIOGRAPHY

[1] ARROW, K. J. and DEBREU, G., "Existence of an equilibrium for a competitive economy," Econometrica, Vol. 22, No. 3 (1954), pp. 265-290.

[2] ARROW, K. J., BLOCK, H. D. and HURWICZ, L., "On the stability of the competitive equilibrium, II," Econometrica, Vol. 27, No. 1 (1959), pp. 82-109.

[3] ARROW, K. J. and HURWICZ, L., "Competitive stability under weak gross substitutability: The 'Euclidean distance' approach," International Economic Review, Vol. 1, No. 1 (1960) pp. 38-49.

[4] ARROW, K. J. and HURWICZ, L., "Some remarks on the equilibria of economic systems," Econometrica, Vol. 28, No. 3 (1960), pp. 640-646.

[5] BROWN, G. W. and von NEUMANN, J.,"Solution of games by a differential equation," Annals of Mathematics Studies, No. 24, (1950), pp. 73-79.

[6] DEBREU, G., Theory of Value, (John Wiley, 1959).

[7] GALE, D., "The law of supply and demand," Math. Scand. 3 (1955), pp. 155-169.

[8] MCKENZIE, L. W., "On equilibrium in Graham's model of world trade and other competitive systems," Econometrica, Vol. 22 (1954), pp. 147-161.

[9] MCKENZIE, L. W., "Matrices with Dominant Diagonals and Economic Theory," Proceedings of the First Stanford Symposium, (Stanford University Press, 1960), pp. 47-62.

[10] MCKENZIE, L. W., "On the existence of general equilibrium for a competitive market," Econometrica, Vol. 27 (1959), pp. 54-71.

[11] MCKENZIE, L. W., "Stability of equilibrium and the value of positive excess demand," Econometrica, Vol. 28, No. 3 (1960), pp. 606-617.

[12] MORISHIMA, M., "On the three Hicksian laws of comparative statics," The Review of Economic Studies, Vol. XXVII, 3 (1960), pp. 195-201.

[13] NIKAIDÔ, H., "On the classical multilateral exchange problem," Metroeconomica, Vol. 8, No. 2 (1956), pp. 135-145.

[14] NIKAIDÔ, H., "Coincidence and some systems of inequalities," Journal of the Mathematical Society of Japan, Vol. 11, No. 4 (1959) pp. 354-373.

[15] NIKAIDÔ, H., "On a method of proof for the minimax theorem," Proceedings of the American Mathematical Society, Vol. 10, No. 1 (1959), pp. 205-212.

[16] NIDAIDÔ, H., "Stability of equilibrium by the Brown-von Neumann differential equation," Econometrica, Vol. 27, No. 4 (1959), pp. 654-671.

Hukukane Nidaidô

The Institute of Social and Economic Research
Osaka University

ADAPTIVE COMPETITIVE DECISION*

Jack L. Rosenfeld

§ 1. INTRODUCTION

During recent years, much interest has been shown in a class of problems called by the author general adaptive processes. In these processes some measure of performance, or "return," is increased as the experimenter gathers more information about the process. The characteristic feature, however, is that the actions taken by the experimenter determine both the actual return to him and the type of information gathered by him. The two-armed bandit problem described by Bradt et al. [1] is a good example. The question posed is how to decide which of two slot machines to use at each of n trials in order to maximize the total return. Both machines pay either nothing or one dollar when played. The a priori probability distributions of the payoff odds of each machine are available to the experimenter. This problem is a finite length general adaptive process. Other general adaptive processes are described in [2] to [6].

An adaptive competitive decision process consists of an $m \times n$, two-person, zero-sum game that is to be played an infinite number of times. After each step (a single play of the $m \times n$ game), the payoff is made and each player is told what alternative (pure strategy of the $m \times n$ game) was selected by his opponent. The payoff matrix is not completely specified in advance. The nature of the uncertain specification of the matrix and the process by which the uncertainty can be resolved form the heart of the adaptive decision process. Some of the payoffs of the matrix are unknown to the players when the game is begun. Unknown payoffs are selected initially according to a priori probability distributions, and players are told only these probability distributions. If a_{ij} is one of the unknown payoffs, the players do not learn the true value of a_{ij} until player A (the maximizing

* Agencies supporting this research are listed at the end of the paper.

player) uses alternative i and player B (the minimizing player) uses alternative j at some step of the infinite process. Then A receives a_{ij}; B loses a_{ij}; and both players are told the true value of a_{ij}, so that it is no longer unknown. Of course, when all the unknown payoffs have been received, the process is reduced to the repeated play of a conventional $m \times n$, two-person, zero-sum game.

(One can visualize a stack of matrices, each with all its payoffs permanently recorded. Some of these payoffs are hidden by covers on all matrices. A probability is assigned to each matrix in the stack. The players know this probability distribution. One of the matrices is chosen according to the probability distribution by a neutral referee, and that matrix is shown to the players with the covers in place. The game is played repeatedly until a pair of alternatives corresponding to one of the covered payoffs is used. The cover is then removed, and the play resumes until the next cover must be removed, and so on. This process continues until no cover remains. The completely specified game is then repeated indefinitely.)

There are two subclasses of adaptive competitive decision: the equal information case, in which A and B are given the same a priori knowledge about the unknown payoffs; and the unequal information case, in which their a priori data are different. (The "stack of matrices" illustration is not valid in certain unequal information cases.) One example of the unequal information subclass is the situation in which B knows the true value of some payoff that A does not know.

Adaptive competitive decision processes belong to a class of problems involving the repeated play of certain game matrices; these processes are all infinite games that have supinf solutions [7, 8, 9].

§ 2. MEASURE OF PERFORMANCE

It appears that A should play so as to receive a large payoff, learn the payoffs that are unknown to him, and prevent B from learning the payoffs that are unknown to B. Also, A should extract information about the payoffs unknown to him by observing the alternatives that B has chosen during previous steps, while not divulging to B, by the alternatives A chooses, any information about the payoffs unknown to B. A quantitative measure of performance has been found, such that A can essentially satisfy these goals by selecting a strategy that minimizes the measure of performance. This measure is called the mean loss. Player B should select a strategy that maximizes the mean loss.

Before the mean loss is defined, it must be noted that the players of adaptive competitive decision processes lose no flexibility by restricting their strategies to "behavior strategies." A player is said to be using a behavior strategy if at each step he selects a probability distribution over his m (or n) alternatives and uses that distribution to select an alternative. The distribution he chooses may depend upon his knowledge of the history of the process, which consists of the alternatives selected by both players and payoffs received at all preceding steps. (The analysis used by Kuhn [10] is easily applied to adaptive competitive decision processes, which are infinite games of perfect recall, to reach this conclusion.) Because behavior strategies are completely general for adaptive competitive decision processes, the following discussion will assume that both players restrict themselves to the use of behavior strategies.

The probability distributions used by players A and B at the kth step are respectively denoted by

$$p^k \equiv (p_1^k, p_2^k, \ldots, p_m^k), \qquad q^k \equiv (q_1^k, q_2^k, \ldots, q_n^k) .$$

Here p_i^k (q_j^k) is the probability that A (B) selects alternative i (j) at the kth step. Hence,

$$\sum_{i=1}^{m} p_i^k = 1, \qquad \sum_{j=1}^{n} q_j^k = 1 .$$

Note that k is a superscript—not a power. In general, p^k and q^k depend upon the past history of the process.

The value of the payoff matrix is denoted by v; it represents the maximum expected return that A could guarantee himself at one step, if all the unknown payoffs were known. Of course, v is a function of the true values of the unknown payoffs. The expected return to player A at the kth step when A uses p^k and B uses q^k is denoted by r^k:

$$r^k \equiv \sum_{i=1}^{m} \sum_{j=1}^{n} p_i^k q_j^k a_{ij} .$$

Since some of the payoffs a_{ij} are unknown, r^k is a function of p^k, q^k, and the values of the unknown payoffs. The term

$$L^k \equiv v - r^k$$

is called the single-step loss at the kth step; it is the difference between the expected payoff that A could guarantee himself if the values of the unknown payoffs were known and the expected payoff A does receive. The single-step loss, L^k, represents the expected loss to A at the kth step of the process due to his lack of data about the unknown payoffs.

The loss for N steps of the process, called the N-truncated loss, L_N, is

$$L_N \equiv \sum_{k=1}^{N} L^k .$$

The N-truncated loss is a function of the true values of the unknown payoffs as well as the strategies used by both players. The mean value of the N-truncated loss, with respect to the probability distribution of the unknown payoffs is

$$\bar{L}_N \equiv \int L_N dP ,$$

where P denotes the cumulative distribution function of the unknown payoffs. $\bar{L}_N$ is a function only of the strategies of A and B. The mean N-truncated loss can also be written

$$\bar{L}_N = N\bar{v} - \sum_{k=1}^{N} \overline{r^k} , \tag{1}$$

where $\overline{r^k}$ and $\bar{v}$ are the mean values of the expected return to A at the kth step and the value of the payoff matrix, respectively. If the process were to continue for N steps and then terminate, an optimum strategy for player A would be the one that maximizes the mean value of the sum of his expected returns at each step. According to Equation 1, this is also the strategy that minimizes the mean N-truncated loss. Conversely, B's optimum strategy maximizes $\overline{L}_N$.

Because the adaptive competitive decision process involves an infinite repetition of the game, the mean N-truncated loss is useful only if it can be extended to an infinite number of plays. If the limit as N approaches infinity of $\overline{L}_N$ exists, then this quantity is the measure of performance that A should try to minimize and B to maximize. This measure is called the mean loss, $\overline{L}$:

$$\overline{L} \equiv \lim_{N\to\infty} \overline{L}_N .$$

It will be shown for the equal information case that a supinf solution exists:

$$\overline{L}_{opt} \equiv \inf_{S_A} \sup_{S_B} \overline{L} = \sup_{S_B} \inf_{S_A} \overline{L} , \tag{2}$$

where S_A and S_B represent all possible mixed strategies of A and B, respectively, for the infinite game of adaptive competitive decision. $\overline{L}_{opt}$ may be negative as well as positive or zero.

Measures of performance other than the mean loss (e.g., the mean sum of discounted expected returns) can be used. The strategies resulting from the optimization of these other measures will differ from those arrived at by optimizing the mean loss. However, the mean loss has been found to be a very useful measure.

§ 3. EQUAL INFORMATION — SINGLE UNKNOWN PAYOFF

The following intuitive argument presents the essence of the more formal proof that follows. This matrix is used as an illustration. The game is to be played repeatedly.

$$A \begin{matrix} \\ 1 \\ 2 \end{matrix} \overset{\begin{matrix} & B & \\ 1 & & 2 \end{matrix}}{\begin{bmatrix} a_{11} & 0 \\ 0 & -2 \end{bmatrix}} \qquad \begin{aligned} Pr(a_{11} = -4) &= \tfrac{1}{2} \\ Pr(a_{11} = -1) &= \tfrac{1}{2} \end{aligned} .$$

Thus

$$\bar{v} = -1, \qquad \bar{a}_{11} = -2.5 \ .$$

Consider the auxiliary game defined by the matrix

$$\begin{bmatrix} \bar{v} - \bar{a}_{11} & \bar{v} - a_{12} + \bar{L} \\ \bar{v} - a_{21} + \bar{L} & \bar{v} - a_{22} + \bar{L} \end{bmatrix} = \begin{bmatrix} 1.5 & \bar{L} - 1 \\ \bar{L} - 1 & \bar{L} + 1 \end{bmatrix} .$$

What relationship does this auxiliary matrix have to the original process? Let $\bar{L}$ be the mean loss associated with the competitive decision process, if a mean loss exists. When one step of the process is played, the mean loss for that step is $\bar{v} - \bar{a}_{11}$ if a_{11} is received, and $\bar{v} - a_{ij}$ if a_{11} is not received. In the former case, the process terminates with no further loss, because both players should use optimum strategies from the second step on. If a_{11} is not received, the mean loss from the second step on is $\bar{L}$, because the situation faced by the players is identical to the situation faced by them before the first step. This reasoning leads to the auxiliary matrix.

Players A and B should be as willing to make one play of the game specified by the auxiliary matrix as they are to participate in the original process. Furthermore, the minimax value of the auxiliary matrix should equal the mean loss for the adaptive competitive decision process upon which that matrix is based ($y(\bar{L})$ denotes the value of the auxiliary matrix):

$$y(\bar{L}) = \bar{L} \ . \tag{3}$$

Note that A is the minimizing player of the auxiliary matrix and B is the maximizing player, since the payoff represents a loss to A and a gain to B. The value of $\bar{L}$ that satisfies Equation 3 is called the optimum mean loss, $\bar{L}_{opt}$. The minimax value of the auxiliary matrix for the example is as follows:

$$y(\bar{L}) = \begin{cases} \dfrac{-(\bar{L})^2 + 3.5\bar{L} + 0.5}{-\bar{L} + 4.5} & \text{if } \bar{L} \leq 2.5 \\ \bar{L} - 1 & \text{if } \bar{L} \geq 2.5 \ . \end{cases}$$

Therefore, $\bar{L}_{opt} = \frac{1}{2}$.

Because neither player's information about the process changes from one step to the next, it seems reasonable that each player should repeatedly use one probability distribution for his alternatives before the unknown payoff is received, and afterwards repeatedly use the minimax distribution for the completely known matrix. Strategies that consist of the repeated use of one distribution until an unknown payoff is received, and then another repeated distribution until the next unknown payoff is received, and so on, are called "piecewise-stationary." The first segment of A's optimum piecewise-stationary strategy is the distribution that minimizes the expected payoff from the auxiliary matrix, and B's first distribution maximizes the expected payoff. These strategies, of course, are functions of $\bar{L}$. When $\frac{1}{2} = \bar{L}_{opt}$ is substituted for $\bar{L}$ in the auxiliary matrix for the example, the minimax strategies are found to be

$$p_o = (\tfrac{1}{2}, \tfrac{1}{2}) \qquad q_o = (\tfrac{1}{2}, \tfrac{1}{2}).$$

(It will be shown later that the strategy q_o is not truly optimum.)

The complete proof of the preceding solution is given in [11] and is briefly outlined here; it adheres closely to the proof given by Everett for "simple stochastic games" [9]. Consider an adaptive competitive decision process with a single unknown payoff a_{11}. The total loss for N steps of the process is

$$L_N = \sum_{k=1}^{N} (v - r^k) \prod_{t=0}^{k-1} (1 - p_1^t q_1^t) , \tag{4}$$

and the mean value of the total loss for N steps is

$$\bar{L}_N = \sum_{k=1}^{N} (\bar{v} - \overline{r^k}) \prod_{t=0}^{k-1} (1 - p_1^t q_1^t) , \tag{5}$$

where $p_1^0 q_1^0$ is defined equal to 0. The validity of these expressions is based upon the assumption that once a_{11} is received the players will repeatedly use optimum strategies for the remaining known game; thus the loss will equal zero from that point on. Equations 4 and 5 are valid only for "memoryless" sequences of distributions, $p^1, \ldots, p^N$ and $q^1, \ldots, q^N$.

Memoryless sequences are those in which p^k and q^k are independent of the actual alternatives used by the players at steps 1, 2, ..., k−1. However, the use of Equations 4 and 5 will be restricted to cases in which one may assume with complete generality that the sequences are memoryless.

Next an auxiliary matrix is constructed such that the (1, 1) entry is $\bar{v} - a_{11}$, and all other entries are $\bar{v} - a_{ij} + x$. If player A is the minimizing player and B is the maximizing player, this single-step game has a minimax solution with value denoted by $y(x)$ and optimum strategies p_x and q_x. The value and minimax strategies are functions of variable x. The auxiliary matrix can be used to divide the set of all real numbers into two subsets, S_1 and S_2:

$$x \in S_1 \quad \text{if} \begin{cases} y(x) < x & \text{and} \quad x < 0, \text{ or} \\ y(x) \leq x & \text{and} \quad x \geq 0; \end{cases}$$

(6)

$$x \in S_2 \quad \text{if} \begin{cases} y(x) > x & \text{and} \quad x > 0, \text{ or} \\ y(x) \geq x & \text{and} \quad x \leq 0 . \end{cases}$$

(The point $x = 0$ can be in both subsets.)

If $x \in S_1$, if N steps of the adaptive competitive process are played, and if A uses distribution p_x repeatedly until a_{11} is received or until the number of steps reaches N, then an upper bound to $\bar{L}_N$ can be found by substituting into Equation 5 inequalities based upon the definition of S_1 in Equation 6. (Equation 5 can be used despite the restriction to memoryless sequences, since it has been demonstrated [11] that if player A uses a given piecewise-stationary strategy, player B can do as well using a memoryless strategy as he can with any more general strategy.) The resultant bound to $\bar{L}_N$ is

$$\bar{L}_N \leq \begin{cases} x - x \prod_{k=0}^{N} (1 - p_{1x} q_1^k) - \sum_{k=1}^{N} c \prod_{t=0}^{k-1} (1 - p_{1x} q_1^t) & \text{if} \quad x < 0 \\ \\ x - x \prod_{k=0}^{N} (1 - p_{1x} q_1^k) & \text{if} \quad x \geq 0 \end{cases}$$

for any sequence $q^1, \ldots, q^N$. The term p_{1x} denotes the first component of p_x, and c is a constant dependent only upon x. It is easy to show from this expression that either

$$\lim_{N\to\infty} \bar{L}_N \leq x;$$

or there exists an integer N_o such that if N is greater than N_o, $\bar{L}_N \leq x$. In other words, by using distribution p_x repeatedly until a_{11} is received and then playing optimally, player A can guarantee that either a mean loss no greater than x exists ($\bar{L}_N$ converges), or after a sufficient number of plays the mean value of his total loss is no greater than x.

An analogous derivation leads to the conclusion that if $x \in S_2$ and if player B uses a piecewise-stationary strategy beginning with distribution q_x, then either

$$\lim_{N\to\infty} \bar{L}_N \geq x,$$

or there is an integer N_o such that $\bar{L}_N \geq x$ for all $N > N_o$, no matter what strategy player A uses.

The point $x = +\infty$ is in S_1, and the point $x = -\infty$ is in S_2; therefore, there is at least one point $x = \bar{L}_{opt}$, in any epsilon neighborhood of which there is a point in S_1 and a point in S_2. It is easy to show that the point $\bar{L}_{opt}$ is unique, that $y(\bar{L}_{opt}) = \bar{L}_{opt}$, and that Equation 2 is valid (L_{opt} is the supinf solution of the process).

In the numerical example used earlier in this section, $S_1 = [\frac{1}{2}, +\infty]$, and $S_2 = [-\infty, \frac{1}{2})$. Thus, A can guarantee that his mean loss is no greater than x for any x greater than or equal to $\frac{1}{2}$, and B can guarantee that A's loss is no less than x for any x less than $\frac{1}{2}$. In order to minimize his mean loss, A should choose the minimax distribution corresponding to $x = \frac{1}{2}(p_o = (\frac{1}{2}, \frac{1}{2}))$, and B should choose a minimax distribution for some x arbitrarily close to — but less than — $\frac{1}{2}$. Thus B can guarantee that A's mean loss is no less than $\frac{1}{2} - \epsilon$ by using

$$q = (\frac{2}{4+\epsilon}, \frac{2+\epsilon}{4+\epsilon}) .$$

(The strategy $q_o = (\frac{1}{2}, \frac{1}{2})$ that was given earlier will not guarantee that $\bar{L} \geq \frac{1}{2}$. This can be demonstrated by showing that $\bar{L} = 0$ when B uses q_o and A selects alternative 2 repeatedly.)

§ 4. EQUAL INFORMATION — MULTIPLE UNKNOWN PAYOFFS

In the case of two unknown payoffs, it is assumed that once one of them is received, both players play in an optimum fashion; therefore, the remaining mean loss to player A is $\bar{L}_{opt}$ for the resultant adaptive competitive decision problem with a single unknown payoff. The following auxiliary matrix is used to find the optimum mean loss, $\bar{L}_{opt}$, for the two unknown payoff case (in a 3×3 process with unknown payoffs a_{11} and a_{22}):

$$\begin{bmatrix} \bar{v} - \bar{a}_{11} + \overline{L_{opt}|a_{11}} & \bar{v} - a_{12} + \bar{L} & \bar{v} - a_{13} + \bar{L} \\ \bar{v} - a_{21} + \bar{L} & \bar{v} - \bar{a}_{22} + \overline{L_{opt}|a_{22}} & \bar{v} - a_{23} + \bar{L} \\ \bar{v} - a_{31} + \bar{L} & \bar{v} - a_{32} + \bar{L} & \bar{v} - a_{33} + \bar{L} \end{bmatrix} .$$

$\bar{L}_{opt}|a_{11}$ denotes the optimum mean loss for the case of a single unknown payoff, a_{22}, when a_{11} is known; $\overline{L_{opt}|a_{11}}$ denotes the mean value of $\bar{L}_{opt}|a_{11}$, with respect to the probability distribution of a_{11}:

$$\overline{L_{opt}|a_{11}} = \int \bar{L}_{opt}|a_{11} \, dP(a_{11}) .$$

$\overline{L_{opt}|a_{22}}$ is defined similarly. S_1 and S_2 can be defined by Equation 6. Then there is an optimum mean loss, $\bar{L}_{opt}$, which is the greatest lower bound of S_1 and the least upper bound of S_2. All the conclusions drawn about the single unknown payoff case are true for the two unknown payoff case. The derivation of these results is a very simple extension of the preceding analysis.

When r payoffs are unknown $(r > 1)$, an auxiliary matrix is used with entries $\bar{v} - \bar{a}_{ij} + \overline{L_{opt}|a_{ij}}$ corresponding to the r unknown payoffs and $\bar{v} - a_{ij} + \bar{L}$ as the remaining entries. $\bar{L}_{opt}|a_{ij}$, which is a function of a_{ij}, is the optimum mean loss for the process with a_{ij} known and the $r - 1$ other payoffs unknown, and $\overline{L_{opt}|a_{ij}}$ is the mean value of $\bar{L}_{opt}|a_{ij}$

with respect to the probability distribution for a_{ij}.

§ 5. UNEQUAL INFORMATION

The concept of adaptive competitive decision with unequal information requires some explanation. It seems obvious that situations can occur in which some payoffs are unknown to one player and other payoffs are unknown to the other player, and some payoffs are unknown to both players with the same probability distributions given to both. However, cases in which the players are given different probability distributions for the same unknown payoffs may appear unrealistic. The example that follows demonstrates how such a situation can logically arise.

Two biased coins are flipped by the referee. The first has probability p of landing heads up — the second, q. Unknown payoff a_{11} has value $+a$ if both coins land with the same side up and $-a$ if the coins land with different sides up. Player A is shown the outcome of the toss of the first coin, and player B, that of the second coin. The complete process may be tabulated as follows:

Outcome	Probability	a_{11}	$Pr_A(a_{11} = +a)$	$Pr_B(a_{11} = +a)$
HH	pq	$+a$	q	p
HT	$p(1 - q)$	$-a$	q	$1 - p$
TH	$(1 - p)q$	$-a$	$1 - q$	p
TT	$(1 - p)(1 - q)$	$+a$	$1 - q$	$1 - p$,

where $Pr_A(a_{11} = +a)$ denotes A's a posteriori probability that $a_{11} = +a$ after he is shown the outcome of the toss of the first coin. $Pr_B(a_{11} = +a)$ has a similar meaning. Notice that the quantities $Pr_A(a_{11} = +a)$ and $Pr_B(a_{11} = +a)$ can be different from each other for all four outcomes.

The general problem of adaptive competitive decision with unequal information is an infinite, two-person, zero-sum game, since an infinite game tree can always be constructed. Figure 1 illustrates the first part of the game tree for an adaptive competitive decision process with unequal information based upon the coin example. It is assumed, for convenience,

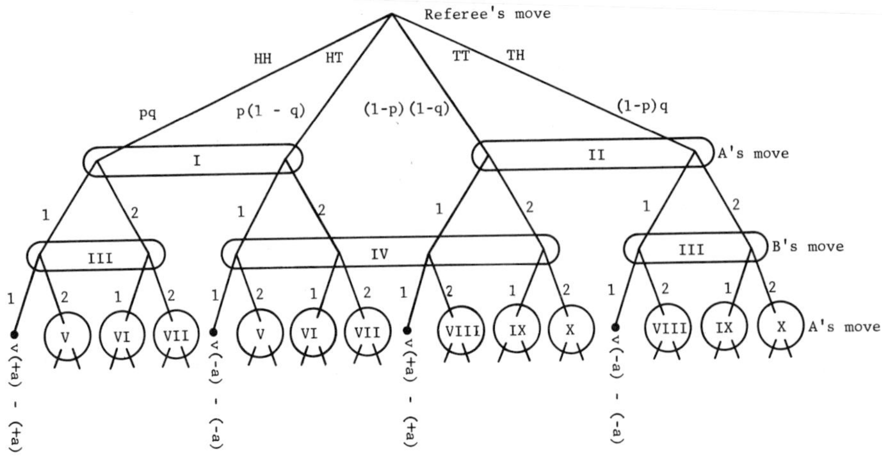

FIGURE 1

that the payoff matrix is 2×2 with a single unknown payoff a_{11}. Roman numerals denote information sets; all nodes indexed by the same Roman numeral are indistinguishable. The four terminal nodes represent cases in which the unknown payoff is received and no further loss is sustained. The quantities written beneath these nodes are the total losses sustained by A.

(In a more general unequal information situation, at the first move — made by the referee — the unknown payoffs are selected according to some initial probability distribution. If the number of possible outcomes for the referee's move is finite, then a finite number of branches leaves the first node. Let the second move be that of player A. (The choice is arbitrary.) If the payoff matrix is $m \times n$, player A has m alternatives from all information sets. There is one information set for every possible outcome of the referee's random move about which A is informed. The next move belongs to B, and his information sets correspond similarly to the possible outcomes about which he is told. The game has a terminating node after B's move if all the unknown payoffs have been received, and the payoff equals the total loss A sustains for the sequence of moves leading to that node. When the game does not terminate, both players have at least one more move. The information sets for A at the first move are subdivided to form the information sets for A's second move. Each subdivision corresponds to one of the mn possible pairs of alternatives used by A and B at the previous two moves. (One of these subdivisions may be a terminating node.) Player B's information sets at his first move are similarly subdivided. The construction of the game tree continues in the same manner indefinitely with moves by A and B alternated.)

Although infinite games do not necessarily possess supinf solutions, it was demonstrated in Sections 3 and 4 that adaptive competitive decision processes with equal information do have supinf solutions. Equal information situations are special cases of unequal information situations in which the a posteriori probability distributions of the unknown payoffs for both players are identical for all possible outcomes of the referee's move.

The solution of the following example represents the present extent of our knowledge about adaptive competitive decision with unequal information. Consider this matrix in which payoff a_{11} is unknown to player A but known to B:

$$
\begin{array}{cc} & \text{B} \\ & \begin{array}{cc} 1 & 2 \end{array} \\ \text{A} \begin{array}{c} 1 \\ 2 \end{array} & \begin{bmatrix} a_{11} & 0 \\ 0 & -1 \end{bmatrix} \end{array}
\qquad
\begin{array}{l} \Pr(a_{11} = +1) = \frac{1}{2} \\ \Pr(a_{11} = -1) = \frac{1}{2} . \end{array}
$$

The probability distribution function of a_{11} is known to both players, but B also knows the true value of payoff a_{11}.

Assume that A uses alternative 1 repeatedly until a_{11} is received. What strategy can B use that will maximize the loss to player A? If $a_{11} = +1$, then $v = 0$, so B's optimum strategy is to use alternative 2 repeatedly. If $a_{11} = -1$, then $v = -\frac{1}{2}$. In this case the single-step loss is $-\frac{1}{2}$ for each step at which player B uses alternative 2, and the single-step loss is $+\frac{1}{2}$ for the first step at which B uses alternative 1, and 0 thereafter. Therefore, B's optimum strategy is to use alternative 1 at the first step and $q = (\frac{1}{2}, \frac{1}{2})$ thereafter. If B uses the optimum strategies the mean loss is

$$\bar{L} = \lim_{N\to\infty} \bar{L}_N = \frac{1}{4} .$$

Hence, A can guarantee that his mean loss is no greater than $\frac{1}{4}$ by using the assumed strategy.

Assume that B uses the appropriate minimax distribution repeatedly from the second step on $(q = (0, 1)$ if $a_{11} = +1$; $q = (\frac{1}{2}, \frac{1}{2})$ if $a_{11} = -1)$, but at the first step uses alternative 2 if $a_{11} = +1$ and alternative 1 if $a_{11} = -1$. What strategy can A use to minimize his own mean loss? Player A need sustain no loss after the first step, because B's choice of alternative at the first step reveals the true value of a_{11}; A can use this information to play in an optimum fashion from the second step on. When A uses alternative 1 at step 1, his single-step loss is 0 if $a_{11} = +1$ and $+\frac{1}{2}$ if $a_{11} = -1$. When A uses alternative 2 at step 1, his single-step loss is 1 if $a_{11} = +1$ and $-\frac{1}{2}$ if $a_{11} = -1$. Therefore, no matter what A does at the first step, $\bar{L} = \frac{1}{4}$. Hence, B can guarantee that the mean loss to A is at least $\frac{1}{4}$ by using the assumed strategy.

Thus, a minimax solution exists for this unequal information process, with $\bar{L}_{opt} = \frac{1}{4}$. The minimax strategy given for player B is not piecewise-stationary. It can be demonstrated [11] that there exists no piecewise-stationary strategy for B, the use of which guarantees that the mean loss is at least as large as $\frac{1}{4}$. The significant conclusions that can be drawn from this example are (a) there exist examples of unequal information processes for which minimax solutions exist, and (b) piecewise-stationary strategies may not be optimum when minimax solutions exist. Nothing more has been learned about the general existence of supinf strategies.

ACKNOWLEDGEMENT

The research reported in this paper was made possible in part by support extended to the Research Laboratory of Electronics, the Massachusetts Institute of Technology, jointly by the United States Army (Signal Corps), the United States Navy (Office of Naval Research), and the United States Air Force (Office of Scientific Research, Air Research and Development Command) under Signal Corps Contract DA 36-039-sc-78108, Department of the Army Task 3-99-20-001, and Project 3-99-00-000, and was also supported in part by the Navy Bureau of Ships Contract Nos. 72716 and 77603.

BIBLIOGRAPHY

[1] BRADT, R. N., JOHNSON, S. M., and KARLIN, S., "On sequential designs maximizing the sum of n observations," Annals of Mathematical Statistics, Vol. 27, December 1956, pp. 1060-1074.

[2] FLOOD, Merrill M., "On game-learning theory and some decision-making experiments," Decision Processes, ed. by R. M. Thrall, C. H. Coombs, and R. L. Davis, John Wiley and Sons, Inc., New York, 1954, pp. 139-158.

[3] BUSH, Robert R., and MOSTELLER, Frederick, Stochastic Models for Learning, John Wiley and Sons, Inc., New York, 1955.

[4] ROBBINS, Herbert, "A sequential decision problem with a finite memory," Proceedings of the National Academy of Sciences, Vol. 42, December 1956, pp. 920-923.

[5] FRIEDBERG, R. M., "A learning machine: Part I," IBM Journal of Research and Development, Vol. 2, January 1958, pp. 2-13; FRIEDBERG, R. M., DUNHAM, B., and NORTH, J. H., "A learning machine: Part II," IBM Journal of Research and Development, Vol. 3, July 1959, pp. 282-287.

[6] KOCHEN, Manfred, and GALANTER, Eugene H., "The acquisition and utilization of information in problem solving and thinking," Information and Control, Vol. 1, September 1958, pp. 267-288.

[7] SHAPLEY, L. S., "Stochastic games," Proceedings of the National Academy of Sciences, Vol. 39, October 1953, pp. 1095-1100.

[8] GILLETTE, Dean, "Stochastic games with zero stop probabilities," Contributions to the Theory of Games, Vol. III, ed. by M. Dresher, A. W. Tucker, and P. Wolfe, Annals of Mathematics Study No. 39, Princeton University Press, Princeton, N. J., 1957, pp. 179-187.

[9] EVERETT, H., "Recursive games," Contributions to the Theory of Games, Vol. III, ed. by M. Dresher, A. W. Tucker, and P. Wolfe, Annals of Mathematics Study No. 39, Princeton University Press, Princeton, N. J., 1957, pp. 47-78.

[10] KUHN, H. W., "Extensive games and the problem of information," Contributions to the Theory of Games, Vol. II, ed. by H. W. Kuhn and A. W. Tucker, Annals of Mathematics Study No. 28, Princeton University Press, Princeton, N. J., 1953, pp. 193-216.

[11] ROSENFELD, Jack Lee, Adaptive Decision Processes, Doctoral Thesis, Department of Electrical Engineering, Massachusetts Institute of Technology, Cambridge, Mass., February 1961; also Technical Report 403, Research Laboratory of Electronics, Massachusetts Institute of Technology, Cambridge, Mass., September 1962.

Jack L. Rosenfeld

International Business Machines Corporation
Thomas J. Watson Research Center

INFINITE GAMES OF PERFECT INFORMATION

Morton Davis

§ 1. INTRODUCTION

It is well-known that finite, two-person, zero-sum games with perfect information are strictly determined [1]. There have been attempts to remove from this result each of these restrictions. It is the first of these with which we will be concerned, i.e., we consider infinite games.

In a paper by Gale and Stewart [2], zero-sum, two-person, <u>infinite</u> games with perfect information are defined. Familiarity with this paper will be assumed.

The notation of this paper will lean heavily on the above paper, but some additions and modifications will be made. In referring to the game Γ we will mean the $(x_0, X_I, X_{II}, X, f, S, S_I, S_{II})$ of Gale and Stewart, where each element is understood as given in the game. We will also write $\Gamma(S_I = A)$ to stand for the game $(x_0, X_I, X_{II}, X_m, f, S, A, A^c)$ where $S = A \cup A^c$. We use here and elsewhere in the paper the superscript c to denote complement. We assume $f^{-1}(x)$ is always a finite set.

In this paper we extend the results of Gale and Stewart [2] and Wolfe [3], answer Questions 1 and 2 of [2] (assuming the Continuum Hypothesis), and finally, characterize the winning sets of a game suggested to me by Professor L. E. Dubins.

I would like to express my appreciation to Professor David Blackwell, who introduced me to the problem and who made several valuable suggestions, to Professor L. E. Dubins, who suggested the game in Section 4 and discussed this problem with me, and finally to Professor Jan Mycielski, who gave several alternative proofs which were shorter and clearer, and suggested the extensions of Section 2, which was more general than my original form.

§ 2. THE DETERMINATENESS IN THE CASE $G_{\delta\sigma} \cup F_{\sigma\delta}$.

Gale and Stewart [2] have shown that if the winning sets (S_I or S_{II}) are either open or closed sets, then the game is determined. Wolfe has extended this result to G_δ and F_σ sets. In this section, this is still further extended to $G_{\delta\sigma}$ and $F_{\sigma\delta}$ sets.

DEFINITION 2.1. Player I(II) has a loss at $x \in X$ in the game Γ if and only if player I(II) has a loss for the subgame Γ_x.

DEFINITION 2.2. x is a position of s if and only if $s(i) = x$ for some i (or alternately, s is said to pass through x).

DEFINITION 2.3. A subgame Γ_C of Γ is I-imposed if and only if for all $x \in X_{II}^C$, $f^{C^{-1}}(x) = f^{-1}(x)$; (a similar definition is given for II-imposed) (see [2], Def. 8, Th. 4, 5 and Notation). Intuitively, this states that a I-imposed subgame is a subgame that may "be imposed" by player I, i.e., one in which player II has no restriction of his alternatives.

DEFINITION 2.4. For $A \subseteq X$, $O(A) = \bigcup_{x \in A} u(x)$ (see [2], p. 252, Notation).

LEMMA 2.1. If player II does not have a loss at $\bar{x} \in X_I$ then player II does not have a loss for all $x \in f^{-1}(\bar{x})$.

PROOF. The proof is trivial. If player II had a loss at $y \in f^{-1}(\bar{x})$ then at $\bar{x}$ player I could choose y so player II would then have had to have had a loss at $\bar{x}$ as well, contrary to the hypothesis.

LEMMA 2.2. If player II does not have a loss at $\bar{x} \in X_{II}$, then player II does not have a loss for some $x \in f^{-1}(\bar{x})$.

PROOF. Trivial.

LEMMA 2.3. If Γ is not a loss for player II, $P \subseteq S_I$, P is an open set, then there exists a II-imposed subgame Γ_C of Γ such that Γ_C is not a loss for player II and $C \cap P = \emptyset$.

PROOF. Let C be all the plays in S that have the following property: if x is any position of s (for $s \in C$), then Γ_x is not a loss for player II. Since C is clearly a closed set, Γ_C is a subgame, and, in particular, Γ_C is a II-imposed subgame by Lemmas 2.1 and 2.2.

If player I had a winning strategy σ in Γ_C he would have had one in Γ as well, for if player I plays consistently with σ as long as he can in Γ_C (i.e., as long as player II does not reach a lost position) and conforms to a winning strategy σ_y when player II reaches his first lost position, y, player I will clearly win. Hence Γ_C is not a loss for player II.

Also $C \cap P = \emptyset$ for if $s \in P$ there is some position of s, x, where $u(x) \in P$ and hence player II has a loss at x and $s \notin C$, and Lemma 2.3 follows.

LEMMA 2.4. If Γ is not a loss for player II, $H \subseteq S_I$, H is a G_δ set, then there is a II-imposed subgame Γ_C of Γ such that $C \cap H = \emptyset$ and Γ_C is not a loss for player II.

PROOF. Let B be the set of all $x \in X$ such that in Γ_x (see [2], p. 254, Notation) there is a II-imposed subgame $[\Gamma_x]_{D_x}$ which is not a loss for player II and such that $D_x \cap H = \emptyset$. Let $H = \bigcap_{i=0}^{\infty} 0(C_i)$ where $C_i \subseteq X$ and $X = C_0 \supseteq C_1 \supseteq C_2 \cdots$.

We first prove the following sequence of statements:

(A_i) If $x \in B^c \cap C_i$, then $\Gamma'_x = \Gamma_x(S'_I = S_I \cup 0\{B^c \cap C_{i+1}\})$

is a loss for player II.

If this were not true there would exist by Lemma 2.3 a II-imposed subgame $[\Gamma'_x]_E$ of Γ'_x which is no loss for player II and such that $E \cap 0(B^c \cap C_{i+1}) = \emptyset$. We may further modify E so that the first time a position y in $B \cap C_{i+1}$ is reached, player II plays in the II-imposed subgame D_y (the existence of which follows from y being in B) which is no loss for him, and where $D_y \cap H = \emptyset$.

Clearly, the resulting play will either not pass through a point of C_{i+1} (and therefore not be in H) <u>or</u> it will pass through a point of $B \cap C_{i+1}$ and also not be in H. Since this modified II-imposed subgame is not a loss for player II, $x \in B$ contrary to the hypothesis, and (A_i) follows.

Suppose that the initial position x_0 of the game belongs to B^c, i.e., $x_0 \in B^c \cap C_0$ contrary to the conclusion of Lemma 2.4. Then by (A_0) player I has a winning strategy in $\Gamma'(S'_I = S_I \cup 0\{B^c \cap C_1\})$. Player I can use that strategy until a position in $B_c \cap C_1$ is reached. He then can shift to the winning strategy assured by (A_1) in $\Gamma''(S'_I = S_I \cup 0\{B^c \cap C_2\})$ until (and if) a position in $B^c \cap C_2$ is reached, when he shifts to the strategy guaranteed by (A_2) etc. The resulting play must pass through x_0, $x_1, x_2, \ldots$ where $x_i \in C_i$ and hence $s \in \bigcap_{i=0}^{\infty} 0(C_i) = H \subseteq S_I$ or $s \notin 0(C_i)$ for some i and so $s \in S_I$. So player I has a win contrary to the hypothesis. Hence $x_0 \in B$ and Lemma 2.4 follows.

THEOREM 2.1. If $S_I \in G_{\delta\sigma} \cup F_{\sigma\delta}$, then Γ is determined.

PROOF. Since the two players play essentially dual roles in this game, we may assume $S_I \in G_{\delta\sigma}$. Let $S_I = \bigcup_{i=1}^{\infty} H_i$ where $H_i \in G_\delta$.

Assume Γ is not a loss for player II. Then by Lemma 2.2 there is a II-imposed subgame Γ_{D_1} of Γ such that $D_1 \cap H_1 = \emptyset$ and Γ_{D_1} is not a loss for player II.

In general, if Γ_{D_i} is a II-imposed subgame of Γ such that $D_i \cap H_i = \emptyset$ and Γ_{D_i} is not a loss for player II, there is another II-imposed subgame of this subgame (and hence of the original game), $\Gamma_{D_{i+1}}$, such that $D_{i+1} \cap H_{i+1} = \emptyset$, and $\Gamma_{D_{i+1}}$ is no loss for player II. Now $D = \bigcap_{i=1}^{\infty} D_i$ is a non-null closed set, and determines a II-imposed subgame Γ_D where $D \cap S_I = \emptyset$, so player II has a win. Q.E.D.

§ 3. EXAMPLES OF NON-DETERMINED GAMES

In the Gale-Stewart paper [2], the definition of absolutely-determined games is made and certain questions are raised concerning it. In particular, the question is asked if the property of absolute determinacy is invariant under unions and intersections of the winning sets. (If this were true for countably many unions and intersections of winning sets, this would imply games that had Borel sets as winning sets would be absolutely determined since it is shown that closed and open winning sets form absolutely determined games.) Unfortunately, this is not the case; in the next section this will be shown and an answer to another question raised by Gale and Stewart will be given.

DEFINITION 3.1. $\underline{P_\sigma}$ is the set of all plays that may occur if player I uses strategy $\underline{\sigma}$.

DEFINITION 3.2. $\underline{P^\tau}$ is the set of all plays that may occur if player II uses the strategy $\underline{\tau}$.

DEFINITION 3.3. A vertex $x \in X$ is <u>a win (loss) for player I (II)</u> if and only if the subgame Γ_x is <u>a win (loss) for player I (II)</u>.

DEFINITION 3.4. A vertex $x \in X$ is <u>undetermined</u> if and only if it is not a win for either player I or II.

DEFINITION 3.5. A play $s \in S$ is <u>plausible</u> if and only if one of these three statements is true:

(i) $\Gamma_{s(i)}$ is a win for player I for all i,

(ii) $\Gamma_{s(i)}$ is a win for player II for all i,

(iii) $\Gamma_{s(i)}$ is undetermined for all i.

REMARK. The set of all plausible plays, $\tilde{P}$, is a closed set.

DEFINITION 3.6. $\Gamma_{\widetilde{P}}$ the subgame determined by the closed set $\widetilde{P}$, will be called the <u>plausible subgame</u> of Γ.

DEFINITION 3.7. A game is <u>absolutely undetermined</u> if and only if x is undetermined for all $x \in X$.

DEFINITION 3.8. <u>Γ^*</u> will denote the following game (mentioned in Gale-Stewart):

X_I = {binary sequences with an even number of bits}

X_{II} = {binary sequences with an odd number of bits}

S = {set of infinite binary sequences}

$f(x)$ = carries finite binary sequences into the same finite binary sequence with the last bit removed. (A bit is a binary digit, i.e., 0 or 1.)

REMARK. We note the following obvious facts:

If $x \in X_I$: x a loss for player I implies for all $y \in f^{-1}(x)$, y a loss for player I; x undetermined implies there is no $y \in f^{-1}(x)$ which is a win for player I <u>and</u> there exists some $y \in f^{-1}(x)$ which is not a loss for player I; x a win for player I implies there exists a $y \in f^{-1}(x)$ which is also a win for player I; σ a winning strategy for player I, $s \in P_\sigma$, implies $s(i)$ is a win for player I for all i.

DEFINITION 3.9. $s \in S$ is <u>consistent</u> with $\sigma(\tau)$ if and only if $s \in P_\sigma$ $(s \in P^\tau)$.

DEFINITION 3.10. $x \in X$ is <u>consistent</u> with $s \in S$ if and only if $s(i) = x$ for some i.

DEFINITION 3.11. $x \in X$ is <u>consistent</u> with $A \subseteq S$ if and only if x is consistent with s for some $s \in A$.

DEFINITION 3.12. $\sigma(\tau)$ is consistent with Γ_C if and only if for all $x \in X_I$ $(x \in X_{II})$ such that x is consistent with C, $\sigma(x)$ $(\tau(x))$ is also consistent with C.

DEFINITION 3.13. A mesh of a strategy σ is a subset Q of P_σ such that there is a $1-1$ correspondence between each x consistent with P_σ and some s, consistent with x, in Q. $s(\sigma, x, Q)$ will denote that $s \in Q$ is associated with x.

DEFINITION 3.14. A game is semi-absolutely determined if and only if every subgame is determined.

DEFINITION 3.15. The negative game of a game is the identical game with S_I, S_{II} replacing S_{II}, S_I, respectively.

DEFINITION 3.16. A set $M \subseteq S$ has a countable $\sigma(\tau)$ span if and only if there exists a finite or countable number of σ's (τ's) such that

$$M \subseteq \bigcup_{i=1}^{\infty} P_{\sigma_i} \quad \left(M \subseteq \bigcup_{i=1}^{\infty} P^{\tau_i}\right).$$

LEMMA 3.1. If (1) $\Gamma = \Gamma^{x}$, (2) $P_\sigma \subseteq \left(\bigcup_{i=1}^{\infty} s_i\right) \cup \left(\bigcup_{i=1}^{\infty} P_{\sigma_i} \cap C_i\right)$, and (3) Q_i = (a fixed mesh of σ_i) $\cap C_i$, then there exists $\underline{s}$ such that $\underline{s} \in P_\sigma \cap Q_i$ for some i.

PROOF. Assume the contrary; i.e., $P_\sigma \cap Q_i = \emptyset$ (for all i). Construct $\underline{s}$ as follows:

Step 1: Let $\underline{s}(0) = x_0$; let $n_1 = \min_k$ such that $\sigma(x_0) = \sigma_k(x_0)$; let $s^1 = s(\sigma_{n_1}, x_0, Q_{n_1})$.

By hypothesis, $s^1 \notin P_\sigma$; therefore there exists $x \in X_I$ consistent with s^1 and such that $\sigma(x) \neq \sigma_{n_1}(x)$; define $\underline{s}(i) = s^1(i)$ for all i such that $i \leq i_1$ where $s^1(i_1) = x$; take $\underline{s}(i_1 + 1) = \sigma(x_1)$, $s_1(i_1 + 2) \neq \underline{s}(i_1 + 2) = x_1$.

Step m: Assume $\underline{s}(i)$ has been defined for all $i \leq i_{m-1} + 2$; define $n_m = \min_k \{k \mid x_{m-1}$ is consistent with σ_k and $\sigma_k(x_{k-1})\}$. (If there is not such a k, assumption (2) is false, since it cannot happen that

$$P_\sigma \cap u(x) \subseteq \bigcup_{i=1}^{\infty} s_i$$

for any x consistent with σ.)

Since $s^m = s(\sigma_{n_m}, x_{k-1}, Q_{n_m}) \notin P_\sigma$, by hypothesis, there exists $x \in X_I$ consistent with s^m such that $\sigma(x) \neq \sigma_{n_m}(x)$. Define $\underline{s}(i) = s^m(i)$ for all $1 \leq i_m$ where $s^m(i_m) = x$; take $\underline{s}(i_m + 1) = \sigma(x)$ and $x_m = \underline{s}(i_m + 2) \neq s_m(i_m + 2)$.

The resulting play $\underline{s}$ is inconsistent with each s_i, σ_i, which contradicts hypothesis (2). Q.E.D.

LEMMA 3.2. If Γ is any Gale-Stewart game where S is compact and if there exists a countable set $\bigcup_{i=1}^{\infty} s_i \subseteq S_{II}$ such that

$$P_\sigma \cap \left[\bigcup_{i=1}^{\infty} s_i\right] \neq \emptyset$$

for all strategies σ for player I, then player II has a win.

PROOF. Player I has a win for neither the above game nor the game Γ' where $S_{II} = \bigcup_{i=1}^{\infty} s_i$. But the above S_{II} set is a F_σ set and therefore Γ' is strictly determined and player II must have a win for the game Γ', and a fortiori, for the original game Γ. Q.E.D.

LEMMA 3.3. If x_0 is an undetermined vertex of Γ, then $\Gamma_{\tilde{p}}$, the plausible subgame of Γ, is absolutely undetermined.

PROOF. Assume the contrary, i.e., that there exists $\sigma_{\tilde{p}}$, a winning strategy for $\Gamma_{\tilde{p}}$. For each $x \in X_I$ where player I has a win, we associate σ_x : a winning strategy for player I in Γ_x.

For each $x \in X_I$, x an undetermined vertex, x consistent with

$\sigma_{\tilde{P}}$, let $\sigma(x) = \sigma_{\tilde{P}}(x)$.

For each $x \in X_I$, x a winning vertex for player I, we will take $\sigma(x) = \sigma_z(x)$ where $z = f^n(x)$ and $n = \max_n \{m \mid \Gamma_{(f^m(x))}$ is a win for player I$\}$.

For all other $x \in X_I$, define $\sigma(x)$ arbitrarily.

Let $s \in P_\sigma$: If (1) $s \in \tilde{P}$ then $\sigma = \sigma_{\tilde{P}}$ and $s \in S_I$; (2) $s \notin \tilde{P}$, take first i such that $x = s(i)$ is determined. x is in X_{II} since $\sigma_{\tilde{P}}$ carries $\{x \mid x$ consistent with $\tilde{P}\}$ into itself. Also, it is clear that x must be a win for player I. Therefore, $P_\sigma \subseteq S_I$, σ is a winning strategy, which contradicts the hypothesis. Q.E.D.

LEMMA 3.4. (a) If $S_I \subseteq \bigcup_{i=1}^{\infty} P_{\sigma_i}$, then the game Γ is determined.

(b) If $S_I \subseteq \bigcup_{i=1}^{\infty} P^{\tau_i}$, then the game Γ is determined.

PROOF of (a). Assume the contrary. Then Γ is undetermined and so is $\Gamma_{\tilde{P}}$; in the game $\Gamma_{\tilde{P}}$, $(S_I) \subseteq \bigcup_{i=1}^{\infty} P_{\sigma_{\tilde{P}_i}}$. Now, given $x \in (X_{II})_{\tilde{P}}$ there exists $j = \min_i$ such that x is consistent with $\sigma_{\tilde{P}_i}$. Since Γ_x is undetermined for all $x \in X$, there exists $\tau_{\tilde{P}_j}$ such that $\langle \sigma_{\tilde{P}_j}, \tau_{\tilde{P}_j} \rangle \in S_{II}$. Let $\tau_{\tilde{P}}(x) = \tau_{\tilde{P}_j}(x)$. It is clear that for $s \in P^{\tau_{\tilde{P}}}$, <u>either</u> $s \notin \bigcup_{i=1}^{\infty} P_{\sigma_{\tilde{P}_j}}$, and therefore $s \in S_{II}$, <u>or</u> there exists $\min j$ such that $s \in P_{\sigma_{\tilde{P}_j}}$, and [illegible] $s = \langle \sigma_{\tilde{P}_j}, \tau_{\tilde{P}_j} \rangle \in S_{II}$. Hence player II has a win which contradicts the hypothesis by Lemma 3.3.

PROOF of (b). Assume the contrary, i.e., that Γ, and therefore $\Gamma_{\tilde{P}}$, is undetermined. Also, it is clear that $(S_I \cap P \subseteq \bigcup_{i=1}^{\infty} P_{\sigma_{\tilde{P}_i}}$. With each $x \in X_{II}$ where player II has two choices, take $j = \min_i$ where x is consistent with τ_i, and take $\tau(x) \neq \tau_i(x)$. Since $(\Gamma_{\tilde{P}})_x$ is undetermined for all x it follows that for all $s \in P$, there exist an infinite number of $x \in X_{II}$ such that player II has two choices. Hence, $s \in S_{II}$, which contradicts the hypothesis.

REMARK. It is clear that the roles of σ and τ may be interchanged and a similar theorem obtained.

LEMMA 3.5. P_σ is a closed set.

PROOF. $P_\sigma = S - \bigcup_{x \in A} u(x)$; $A = \{x \mid x$ not consistent with $\sigma\}$.

LEMMA 3.6. If Γ^* is a win for player I, $S_I = \bigcup_{x \in A} u(x)$, then there exists a finite subset of A, B where $S_I = \bigcup_{x \in B} u(x)$ still is a win for player I.

PROOF. This follows directly from: $u(x)$ is open (a base set), S is compact, P_σ is a closed subset of a compact set, and the Heine-Borel theorem.

Answer to Question 2 of Gale and Stewart.

We will consider three games of the form Γ^*, differing only in the sets S_I, as follows:

(1) Γ' where S_I' is the winning set for player I,

(2) Γ'' where S_I'' is the winning set for player I,

(3) Γ where $S_I' \cap S_I''$ is the winning set for player I,

We will define S_I', S_I'', so that Γ' and Γ'' are absolutely determined games but Γ is undetermined, thus answering the question raised by Gale and Stewart (by symmetry the same could be obtained for union).

It was shown by Gale and Stewart that the set of strategies for player I or II is of the power of the continuum. We note that the cardinal number of the set of all closed sets is also of the power of the continuum since the space is Hausdorff, has a countable basis, and has an infinite number of points.

The first thing we do is associate the σ's, the τ's, and the C's (the closed sets) with the set of ordinal numbers less than Ω (the first uncountable ordinal). We then proceed by induction over these ordinals, executing the following steps:

Step 1:

(i) With σ_1, the first strategy for player I, take an arbitrary strategy for player II, τ_{σ_1}, and set $< \sigma_1, \tau_{\sigma_1} > = s_{1a}$ into S'_{II}.

(ii) With τ_1, take $< \sigma_{\tau_1}, \tau_1 > = s_{1b}$ $(s_{1b} \neq s_{1a})$ and assign s_{1b} to $S'_I \cap S''_I$.

(iii) Consider C_1, the first closed set, and the associated subgame Γ_{C_1}. (In (iii) to (vi) we will consider all notation as pertaining to the subgame Γ_{C_1}.)

We ask whether there exists a strategy for player I, σ_1^*, such that $P_{\sigma_1^*}$ has no points thus far assigned to S'_{II}, and that for each x consistent with σ_1^* there exists an infinite number of points in $\mathcal{U}(x) \cap P_{\sigma_1^*}$ not yet assigned to S''_1. If the answer is yes to the above question, choose one such σ_1^*, assign $P_{\sigma_1^*}$ to S'_I and assign Q_i to S''_{II} where Q_i is some mesh of σ_1^*; then go to (iv). If the answer is no, leave out (iii) and go directly to (iv).

(iv) If there exists σ_1^{**} such that no point of $P_{\sigma_1^{**}}$ has yet been assigned to S''_{II}, assign $P_{\sigma_1^{**}}$ to S''_I and go to (v); if not, go to (v) directly.

(v) If there exists τ_1^* such that no point of $P^{\tau_1^*}$ has yet been assigned to S'_{II}, assign $P^{\tau_1^*}$ to S'_I and go on to (vi); if not, go to (vi) directly.

(vi) If there exists τ_1^{**} such that no point of $P^{\tau_1^{**}}$ has yet been assigned to S''_{II}, assign $P^{\tau_1^{**}}$ and go to Step α (i) directly.

<u>Step α (the inductive step)</u>:

(i) If $P_{\sigma_\alpha} \leq \left[\bigcup_{K<\alpha} s_{Kb}\right] \cup \left[\bigcup_{K<\alpha} < \sigma_\alpha, \tau_K^* >\right] \cup \left[\bigcup_{K<\alpha} (P_{\sigma_K} \cap C_1)\right]$ take $s_{\alpha a} = < \sigma_\alpha, \tau_{\alpha_a} >$ which is not in the above union and assign it to S'_{II}. If the inclusion does hold, we note that the above union consists of a countable number of points and $\bigcup_{K<\alpha} (P_{\sigma_K^*} \cap C_1)$.

By Lemma 3.1 there exists a point in Q_j for some j, which also lies in P_{σ_α}; hence there exists $s \in P_{\sigma_\alpha}$ already assigned to S''_{II}.

(ii) At each step there are only a countable number of points assigned to $S'_{II} \cup S''_{II}$, so we may assign $s_{ab} = < \sigma_{\tau_\alpha}, \tau_\alpha >$ to $S'_I \cap S''_I$.

(iii), (iv), (v), and (vi) are conditional, so we may do these without diffifulty, substituting the subscript α for the subscript 1 (used in Step 1), where appropriate.

It is clear from the previous construction that the game Γ, with winning set for player I, $S_I = S'_I \cap S''_I$, is undetermined. It remains only to show the absolute determinacy of the two games Γ' and Γ''.

Consider C_α: Consider the negative of the subgame Γ'_{C_α}, $\overline{\Gamma}$. By construction, there existed at the αth step a τ^*_α such that $P^{\tau^*_\alpha}$ was assigned to S'_I, or, since there had been only a countable number of points (s_{1a}) assigned to S'_{II}, by Lemma 3.2 there must have been a sufficient number of points assigned to S'_{II} to insure a win for player I. In any case the negative subgame is determined.

We may show in precisely the same way that the negative of subgame Γ'' is also semi-absolutely determined.

Again consider C_α. Assume Γ'_{C_α} is an undetermined subgame. Let $C_\beta = P$ and $\Gamma_{C_\beta} = \Gamma_P$, the plausible subgame of Γ'_{C_β}. Now, at the βth step there must have existed σ^*_β, such that no element of $P_{\sigma^*_\beta}$ had yet been assigned to S'_{II}, and with each x consistent with σ^*_β and C_β, there existed an infinite number of plays (also consistent with σ^*_β and C_β) not yet assigned to S''_I. This follows from Lemmas 3.3 and 3.4 (since $[\Gamma_P]_x$ is undetermined for all x), and hence, σ^*_β is a win for Γ_{C_β}. But this contradicts Lemma 3.3, which states the plausible subgame of an undetermined game is undetermined. So Γ' is semi-absolutely determined.

By a method analogous to the proof of the negative determinacy of Γ', we may show that the game Γ'' is also semi-absolutely determined. Therefore, both games, Γ' and Γ'', are absolutely determined, and the proof is complete.

REMARK. Those points not assigned to any set (or assigned to a set in the game $\Gamma'(\Gamma'')$, but not assigned in the game $\Gamma''(\Gamma')$) may be assigned arbitrarily without affecting the construction.

Answer to Question 1 of Gale and Stewart:

Again we use the game Γ^*. We first well-order the σ's, τ's, and the closed sets C, putting them, as we did in the first construction, into $1-1$ correspondence with the set of ordinal numbers less than Ω, the first uncountable ordinal number. We then construct a game whose negative is semi-absolutely determined, but which itself is undetermined.

Step 1:

(i) With σ_1 we associate the play $s_1' = < \sigma_1, \tau_{\sigma_1} >$ and assign it to S_{II}.

(ii) With τ_1 we associate the play $s_1'' = < \sigma_{\tau_1}, \tau_1 >$ and assign it to S_I.

(iii) Consider C_1. If there exists a σ_1' consistent with C_1 such that no point of P'_{σ_1} has yet been assigned to S_I, choose one such σ_1', and assign P'_{σ_1} to S_{II} and go on to Step α; if there does not exist such a σ_1', go to Step α directly.

Step α:

(i) We note that any countable step there has been only a countable number of points assigned to S_I; therefore, there must exist $s_\alpha' = < \sigma_\alpha, \tau_{\sigma_\alpha} >$ not yet assigned to S_I (since P_{σ_α} is uncountable). Assign s_α' to S_{II}.

(ii) We note first there have been only a countable number of points consistent with τ_α, assigned thus far to S_{II}:

$$\left[\bigcup_{K<\alpha} < \sigma_K', \tau_\alpha >\right] \cup \left[\bigcup_{K<\alpha} s_K'\right] .$$

Since there are an uncountable number of points in P^{τ_α}, it follows that there must exist $s_\alpha'' = < \sigma_{\tau_\alpha}, \tau_\alpha >$ not yet assigned to S_{II}; assign s_α'' to S_I.

(iii) This step (where α replaces 1 in the appropriate places of Step I (iii)) is conditional so there is no difficulty here.

We assign all points in S not assigned by the induction, arbitrarily, and note the game is undetermined. We show the negative game is semi-absolutely determined as follows:

Let C_α be an arbitrarily closed set. At Step α (iii), either we assigned P'_{σ_α} to S_{II}, assuring a win for player I, or (since only a countable number of points, $\bigcup_{K<\alpha} s_K''$, have thus far been assigned to S_I) the hypotheses of Lemma 3.2 are met, and the subgame is determined. Therefore, the negative of the original game is semi-absolutely determined and the construction is complete.

§ 4. THE UNSYMMETRIC GAME

Consider the following game, suggested by L. E. Dubins:

Let S be the set of all infinite binary sequences, and divide S into two complementary sets S_I and S_{II}, the winning sets for players I and II, respectively, as in the earlier sections. Also, as before, X is the set of all finite binary sequences, and $u(x) = \{s \in S \mid s(i) = x$ for some $i\}$. (Here $s(i)$ denotes the ith initial segment of the play s.) P^{τ} will again designate the set of plays that may occur if player II uses strategy τ, play being in the sense of sequence s, without regard to history.

Player I plays first and chooses any binary sequence of length n_1, where n_1 is a finite integer chosen by player I. Player II may then choose a 0 or a 1 to be the next bit (binary digit) of the sequence. Player I may now choose a finite sequence (say of length n_2) of bits and add it on to the $n_1 + 1$ bits already formed. The players alternate in this manner, and the final infinite sequence will lie in either set S_I or set S_{II}, in which case player I or II will win, respectively.

In the earlier games considered, if we were given a sequence, the way in which it was formed (i.e., which player chose which bits) was determined. Now a sequence may be formed in many ways, and, in general, a strategy may dictate at a position x different choices depending on how x was formed.

DEFINITION 4.1. A <u>history</u> $h = (x \; ; \; f)$ is an ordered pair pair consisting of x, a finite sequence of bits, and f, the way in which x was formed (x is said to be formed by f).

DEFINITION 4.2. A strategy τ is said to be <u>independent</u> at x if $\tau(x \; ; \; f)$ is invariant with respect to f, i.e., $\tau(x \; ; \; f_1) = \tau(x \; ; \; f_2)$ for all f_1, f_2 that form x. The sequence x_1 followed by x_2 will be denoted by (x_1, x_2). If x_1 is a finite sequence formed by f_1 followed by x_2 formed by f_2, we may write the total history $h = (x_1, x_2 \; ; \; f_1, f_2)$. (We assume the move of one of the players actually starts at the beginning of x_2.)

DEFINITION 4.3. A strategy, τ, is said to be a <u>positional</u> strategy if τ is independent at x for all x in X.

We now show that in the above defined game, player I has a win if and only if S_I contains a perfect set and player II has a win if and only if S_I has at most a countably infinite number of points. This characterizes the winning sets for both players, and from the existence of sets which have a non-countable number of points and contain no perfect sets one infers the existence of undetermined games.

THEOREM 4.1. Player I has a win if and only if S_I contains a perfect set.

PROOF. Let S_I contain a perfect set P.

Let $M = \{x \mid \mathcal{U}(x, 1) \cap P$ and $\mathcal{U}(x, 0) \cap P$ are both perfect$\}$.

We note $\mathcal{U}(x) \cap P$ is null or perfect for any x in X since $\mathcal{U}(x)$ is both open and closed. Also M is not null, for if we take 2 points in P and take x to be the initial portion they have in common, we find this x is in M. Therefore, if player I chooses an x in M initially, $\mathcal{U}(x, y) \cap P$ is perfect (where y is 0 or 1 depending upon player II's choice), and as before player I may choose another x in M, etc. Continuing thus, we obtain a point in closure $(P) = P$, and therefore in S_I.

On the other hand, if player I has a winning strategy σ, it is trivial to verify that P_σ (Def. 3.1) is perfect.

LEMMA 4.1. Let τ_1 be any strategy for player II and let x_1 be a fixed finite sequence. Then player II has a strategy τ_2 such that $P^{\tau_1} \supseteq P^{\tau_2}$ (Def. 3.2), where τ_2 is independent at x. We will call τ_2 a refinement of τ_1 at x_1.

PROOF. Assume there are f_1, f_2 such that $\tau_1(x_1 ; f_1) \neq \tau_1(x_1 ; f_2)$.

(If this is not the case, τ_1 is its own refinement with respect to x_1.) Pick a fixed f_1 which form x_1.

Define τ_2 as follows:

$$\tau_2(x_1 \; ; \; f) = \tau_1(x_1 \; ; \; f_1),$$

$$\tau_2(x_1, x_2 \; ; \; f, f_2) = \tau_1(x_1, x_2 \; ; \; f_1, f_2),$$

if player II chose the first bit of x_2;

otherwise, define τ_2 as

$$\tau_2(x \; ; \; f) = \tau_1(x \; ; \; f).$$

In effect, τ_2 dictates the same plays as τ_1, <u>except</u> when a history arises where the initial sequence is x_1 and player II made the following move, in which case player II "forgets" how x_1 was actually formed and plays as though the initial sequence x_1 was formed by f_1. It is easy to see that τ_2 is a new strategy meeting the requirements of the lemma.

LEMMA 4.2. If player II has a winning strategy, then he has a winning positional strategy.

PROOF. Assume τ_0 is a winning historical strategy and let x_1, x_2, ... be the positions at which τ is not independent and such that if $i > j$ then the length of $x_i \geq$ the length of x_j. Let τ_i be defined inductively as a refinement of τ_{i-1} at x_i. If we define $\tau(x) = \tau_i(x \; ; \; f)$, where i is so large that length $x_i >$ length x, then $P^\tau \subseteq P^{\tau_0}$ (by construction) and $P^{\tau_0} \subseteq S_{II}$ (by hypothesis), so $P^\tau \subseteq S_{II}$, and τ is a winning positional strategy for player II.

THEOREM 4.2. Player II has a win if and only if S_I is at most countably infinite.

PROOF. If S_I is countably infinite, i.e., $S_I = \{s_1, s_2, \ldots\}$, player II may make at his k^{th} turn to move, a choice that precludes the formation of s_k, which assures the resulting play being in S_{II}.

If player II has a winning historical strategy he has a winning positional strategy, say τ, by Lemma 4.2.

Each s in S_I has the property that for some N (dependent on s), $i \geq N$ implies $\tau(s(i)) \neq s(i + 1)$. Define $N(s)$ to be the least such N. (If this were not the case there would exist an s in S and an infinite number of n_i, $i = 1, 2, 3, \ldots$, such that $\tau(s(n_i)) = s(n_i + 1)$ for all i. If player I chooses his move so that it terminates in one such $s(n_i)$, the resulting play would be s in S_I, which contradicts the hypothesis that τ is a winning strategy for player II.)

The set $B_K = \{s \mid s \text{ in } S_I \text{ and } N(s) \leq K\}$ is finite since for any x of length K there exists at most one s in B_K whose initial sequence is x (for $i > K$, $s(i) \neq \tau[s(i - 1)]$, and there exist at most 2^K different sequences of length K. $S_I \subseteq \bigcup_{K=1}^{\infty} B_K$, so S_I is at most countable and Theorem 4.2 follows.

BIBLIOGRAPHY

[1] VON NEUMANN, John and MORGENSTERN, Oskar, The Theory of Games and Economic Behavior, Princeton 1944, 3rd ed., 1953.

[2] GALE, David and STEWART, F. M., "Infinite games with perfect information," Annals of Mathematics Study No. (Princeton 1953), pp. 245-266.

[3] WOLFE, Philip, "The strict determinateness of certain infinite games," Pacific J. Math., Vol. 5, Supplement I, 1955.

Morton Davis

Princeton University
Rutgers University

CONTINUOUS GAMES WITH PERFECT INFORMATION*

Jan Mycielski

§ 1. INTRODUCTION

This paper contains several remarks and results on the application of general theorems on infinite games with perfect information to some special games of that kind and also the games of pursuit and evasion.

The theory of infinite games with perfect information was started by Gale and Stewart [2]. Some work was also done independently in Poland and preliminary announcements were published [8], [7].** A proof and a natural generalization of the result announced in Sec. 2 of [7] was published by Oxtoby [9]. The result announced in Sec. 1 of [7] was a rediscovery of a theorem of Philip Wolfe [11]. This theorem has been recently refined by M. D. Davis [1] and this will be exploited in the present paper.

By a <u>0-sum</u> <u>2-person</u> <u>game</u> we mean a triple $\langle A, B, v\rangle$ where A and B are nonempty sets, called <u>sets</u> <u>of</u> <u>strategies</u>, and $v = v(a, b)$ is a real valued function,*** called <u>pay-off</u> <u>function</u>, defined on $A \times B$. The game is called <u>determined</u>, if

$$\sup_{a\in A} \inf_{b\in B} v(a, b) = \inf_{b\in B} \sup_{a\in A} v(a, b).$$

All the games studied in that paper would have a more special structure. A mapping $\pi : A \times B \to S$ would be given (S in some set) and we will consider a class of games $\Gamma(\varphi) = \langle A, B, v_\varphi\rangle$, where $v_\varphi(a, b) = \varphi(\pi(a, b))$, where φ is any real valued function on S. All the theorems of this paper

* This paper was written when the author was working on a project in foundations of mathematics supported by the National Science Foundation (Grant No. G-14006).

** Secs. 1 and 3 of [7] contain some mistakes.

*** Throughout this paper the real line is always completed with $+\infty$ and $-\infty$.

would have the form "for such and such a class of functions φ the game $\Gamma(\varphi)$ is determined." S will always be a topological space and most of these classes of φ's will have the following form

(1.1) $\varphi \in K_n(S)$ if and only if there exists such a set $R(\varphi)$ of real numbers which is dense in the real line and such that for every $r \in R(\varphi)$

$$\{s : \varphi(S) < r\} \in B_n \quad \text{or} \quad \{s : \varphi(S) \leq r\} \in B_n,$$

where $B_0 = F \cup G$, $B_1 = F_\sigma \cup G_\delta$, $B = F_{\sigma\delta} \cup G_{\delta\sigma}$, ...*

Let us mention without proof a general theorem of Ryll-Nardzewski on such structures:

If A, B and S are compact topological Hausdorff spaces and π is continuous then the following conditions are equivalent

(i) $\Gamma(\varphi)$ is determined for every continuous function φ.

(ii) $\Gamma(\varphi)$ is determined for every φ which is the characteristic function of an open subset of S.

(iii) $\Gamma(\varphi)$ is determined for every φ which is the characteristic function of a closed subset of S.

(iv) $\Gamma(\varphi)$ is determined for every $\varphi \in K_0(S)$.

Of course K_0 includes upper and lower semicontinuous functions. We do not know whether (i) implies that $\Gamma(\varphi)$ is determined for a class of functions larger than $K_0(S)$, e.g., the first class of Baire (even in the case of compact metric spaces).

The most natural approach for defining games of pursuit and evasion is to take for A and B some classes of vector valued functions such that for every $a \in A$ and $b \in B$ the system of differential equations

$$\dot{p}(t) = a(p(t), q(t), t), \quad \dot{q}(t) = b(p(t), q(t), t)$$

---

* As usual F denotes the class of closed sets and G the class of open sets of the space. The subscript σ (δ) denotes the closure of the class with respect to denumerable union (intersection).

with the initial condition

$$p(0) = p_0, \quad q(0) = q_0$$

has a unique solution (p, q). Here $\pi(a, b) = (p, q)$ and $\varphi(p, q)$ may be defined in many ways, e. g., $\varphi(p, q) = \|p(1) - q(1)\|$ or $\varphi(p, q) = \inf\{t : p(t) = q(t)\}$.* But this approach presents some essential difficulties discovered by Zięba [12] (such games are not determined in general and therefore are not adequate for pursuit and evasion).

Another approach is presented by Ryll-Nardzewski [10]. His theory is more strictly dealing with pursuit and evasion (since φ is always of the type $\inf\{t : p(t) = q(t)\}$) but it provides some tools for effective "analytic" solutions of games — while the considerations of the present paper have purely existential character.

In the present paper another approach is given. It is the method of approximation by positional games.** We give also a short exposition of the theory of positional games with perfect information.

§ 2. POSITIONAL GAMES

2.1 The General Theory

We introduce the following definitions:

M is a nonempty set, which we treat often as a discrete Hausdorff space.

S is a nonempty closed subset of the product space $M \times M \times \ldots$.

For every $s = (s_0, s_1, \ldots) \in S$, $s|n$ $(n = 0, 1, \ldots)$ denotes the finite (or empty) sequence $(s_0, s_1, \ldots, s_{n-1})$.

$J(s|n)$ is a function defined on all $s|n$, taking the values I or II.

$\Phi(s)$ is a real valued function on S.

The system (M, S, J, Φ) is denoted by $\Gamma(\Phi)$.

$\Gamma(\Phi)$ represents a 0-sum 2-person game which is played as follows: The players I and II are constructing a sequence $s \in S$ choosing consecutively

* We put $\inf \emptyset = +\infty$.

** Some discussion concerning such approaches is given in [12].

$s_0, s_1, \ldots,$ the nth choice being performed by I if $J(s|n) = I$ and by II if $J(s|n) = II$ (where $s|n$ denotes the segment already constructed). Each of the players is supposed to know (M, S, J, Φ) and $s|n$ when he is choosing s_n. Clearly the nth choice s_n is supposed to be done in such a way that there exists such an $s' \in S$, that $s|n = s'|n$ and $s_n = s_n'$ and hence, since S is closed, they get $s \in S$. Then player II pays to player I the value $\Phi(s)$. A definition of the corresponding triple $\langle A, B, v\rangle$ does not present any difficulties.

The above definition is a slight modification of that of Gale and Steward [2]. But this modification is essential for the formulation of our results (e.g., Theorem 2 below). In fact, it is obvious how to transform the collection $(x_0, X_I, X_{II}, X, f, S, \Phi)$ of [2] into our (M, S, J, Φ) and vice versa. E.g., the S of [2] corresponds to $\{(s|0, s|1, \ldots): s \in S\}$ in our terminology. It is important that this correspondence induces a homeomorphism of the S of [2] onto our S. This permits a simple translation of the results proved for the formalism of [2] into our terminology (a translation of the proofs would be also easy).

By these remarks a fundamental theorem of Morton Davis [1] can be stated as follows:

(A) <u>If Φ is a characteristic function of a set $X \subseteq S$ such that $X \in B_2$ (see (1.1)) then $\Gamma(\Phi)$ is determined</u>.*

The following assertion is a consequence of the axiom of determinateness of Mycielski and Steinhaus [6] (for the proof see [5]):

(B) <u>If M is denumerable and Φ is the characteristic function of any set $X \subseteq S$ then $\Gamma(\Phi)$ is determined</u>.

(Let us recall that this statement is inconsistent with the axiom of choice, since using this axiom one can prove the existence of such an $X \subseteq S$ for which the corresponding $\Gamma(\Phi)$ is not determined see [2], [8]).

The following statements are easy consequences of (A) and (B).

(A*) <u>If $\Phi \in K_2(S)$ then $\Gamma(\Phi)$ is determined</u>.

(B*) <u>The axiom of determinateness implies that</u>

---

* In [1] there is an additional supposition that S is compact. But it is easy to avoid it by a slight modification of the proof.

$\Gamma(\Phi)$ is determined whenever M is denumerable.

$K_2(S)$ is actually the largest class of functions for which (A*) is known to be valid (see [2], [11], [1]). Any improvement of this result would give improvements of most of the theorems of this paper. It is important for applications that if $\Phi_{ij}(s)$ is a double sequence of continuous functions on S, then the functions

$$\underline{\lim}_{i=\infty}\ \overline{\lim}_{j=\infty} \Phi_{ij}, \qquad \overline{\lim}_{i=\infty}\ \underline{\lim}_{j=\infty} \Phi_{ij},$$

belong to $K_2(S)$.

By (B*) all the results of Secs. 2 and 3 admit obvious generalizations on the assumption of the axiom of determinateness and the denumerability of M.

2.2 Special Games

Let M be a metric space and let every sequence $s \in S$ be convergent in M (we do not suppose that S is closed in $M \times M \times \ldots$ with respect to the nondiscrete topology in M). Let φ be a real valued function over M, and $\Phi(s) = \varphi(\lim_{n=\infty} s_n)$. The game $\Gamma(\Phi)$ is denoted by Γ_φ.

THEOREM 1. If $\varphi \in K_1(M)$ then Γ_φ is determined.

PROOF. The mappings $\ell_n : S \to M$ defined by $\ell_n(s) = s_n$ are continuous (even if M is discrete). Therefore the mapping $\ell(s) = \lim_{n=\infty} s_n$ is of the first class of Baire. We have

$$\{s : \Phi(s) < (\leq)\ r\} = \{s : \varphi\ell(s) < (\leq)\ r\} = \ell^{-1}\{x : \varphi(x) < (\leq)\ r\}.$$

Since $\varphi \in K_1(M)$ then for every $r \in R(\varphi)$ (see (1.1)) one of the sets $\{x : \varphi(x) < (\leq)\ r\}$ belongs to B_1. Therefore, since S is a metrizable space, for every $r \in R(\varphi)$ one of the sets $\ell^{-1}\{x : \varphi(x) < (\leq)\ r\}$ belongs to B_2. Hence $\Phi \in K_2(S)$ and the theorem follows by (A*).

COROLLARY 1. If φ is a characteristic function of

a set $X \subseteq M$ such that $X \in B_1$ then Γ_φ is determined.

Now we formulate an additional condition on the space S.

(C) For every $s \in S$ there exists a sequence of real positive numbers $a_0, a_1, \ldots$ such that $\lim_{n=\infty} a_n = 0$ and for every m and every $s' \in S$ such that $s'|m = s|m$ there is $\rho(\lim_{n=\infty} s_n, \lim_{n=\infty} s'_n) \leq a_m$, where ρ is the distance in M.

This condition implies a slightly weaker one.

(C') The mapping $\ell : S \to M$, where $\ell(s) = \lim_{n=\infty} s_n$, is continuous.

THEOREM 2. If (C') holds and $\varphi \in K_2(M)$, then Γ_φ is determined.

The proof is analogous to that of Theorem 1 (using the continuity of ℓ).

COROLLARY 2. If (C') holds and φ is the characteristic function of a set $X \subseteq M$ such that $X \in B_2$ then Γ_φ is determined.

2.3 Examples

1. A subset X of the real line is given. The players I and II choose consecutive real numbers $x_0, x_1, \ldots$; the even choices are done by player I and the odd choices by player II. The choices are supposed to satisfy the condition

$$(*) \qquad x_n < x_{n+1} \text{ and } x_{n+1} - x_n \leq x_n - x_{n-1}$$

Player I wins if $\lim_{n=\infty} x_n$ exists and $\lim_{n=\infty} x_n \in X$; otherwise II wins.

<u>This game is determined if</u> $X \in B_1$. (This follows from Corollary 1; here the space M is the real line completed with the point $+\infty$).

The same result holds for various modifications of (*) e.g., those considered by Hanani [3].

2. Replace (*) in Example 1 by

(**) $$|x_{n+1} - x_n| \leq 1/2^n.$$

<u>This game is determined if</u> $X \in B_2$. (This follows from Corollary 2; The essential point of (**) is that the series $\Sigma\ 1/2^n$ is convergent, which implies (C)).

3. Replace (*) in Example 1 by

$$x_n < x_{n+1} < x_{n-1} \quad \text{or} \quad x_{n-1} < x_{n+1} < x_n \quad (n = 1, 2, \ldots)$$

and let player I win if $\lim_{n=\infty} x_{2n} \in X$ and player II win if $\lim_{n=\infty} x_{2n} \notin X$.

<u>This game is determined if</u> $X \in B_2$. To prove this let us remark that player II can force the condition

$$|x_n - x_{n+1}| < 1/2^n$$

without weakening his possibilities, i.e., this condition gives a game which is determined if and only if the former game was. But this condition implies (C), hence the result holds by Corollary 2; (we have to take for M the real plane and the consecutive choices are to be interpreted as modifications of one of the coordinates (depending on the parity of the choice) of a point in M).

2.4 Some Win-Lose Games

Let M be a complete metric space and S any subset of $M \times M \times \ldots$ closed with respect to the topology induced by the discrete topology in M. Let $S^{(c)} = \{s : s \in S$ and s is convergent in the metric of $M\}$. A subset $X \subseteq M$ is given. Player I wins if $\lim_{n=\infty} s_n$ exists and $\lim_{n=\infty} s_n \in X$; otherwise II wins. Let Γ_X denote this game.

THEOREM 3. If $X \in G_\delta$ then Γ_X is determined.

THEOREM 4. If $S^{(c)}$ satisfies (C') (i.e., the mapping $\ell : S^{(c)} \to M$ is continuous) and $X \in F_{\sigma\delta}$, then Γ_X is determined.

PROOFS. Player I wins if and only if $s \in \ell^{-1}(X)$, where ℓ is the limit mapping confined to $S^{(c)}$. Since $S^{(c)} \in F_{\sigma\delta}$ (since M is complete this follows by a standard analysis of the Cauchy definition of convergence) it follows in both cases that $\ell^{-1}(X) \in F_{\sigma\delta} \subseteq B_2$. Hence the theorems follow from (A).

§ 3. APPROXIMATION OF CONTINUOUS GAMES BY POSITIONAL GAMES

A game of pursuit and evasion or any other continuous game with perfect information (we suppose only an intuitive knowledge of such games) can be in some sense represented by the games considered in the previous sections. Here follows a brief description of such a presentation of a special game of pursuit and evasion.*

P is a complete metric space with the metric $\rho(\cdot, \cdot)$. $p_0, q_0 \in P$ are two points (the initial positions). V is a positive real number (the ratio of velocities). Player I (the pursuer) chooses any point $p_1 \in P$. Then player II (the evader) chooses any point $q_1 \in P$ such that $\rho(q_0, q_1) \leq V \cdot \rho(p_0, p_1)$. Then I chooses any $p_2 \in P$ and II chooses any $q_2 \in P$ with $\rho(q_1, q_2) \leq V \cdot (p_1, p_2)$ etc.

Player I wins if and only if both limits $\lim_{n=\infty} p_n$, $\lim_{n=\infty} q_n$ exist and are equal. Otherwise II wins. Let Γ denote that game.

COROLLARY 3. Γ is determined.

PROOF. We put $M = P \times P$ and we interprete the consecutive choices in the same way as in Example 3 (Sec. 2.3). Let X be the diagonal of $P \times P$. We have $X \in G_\delta$. Therefore the corollary follows from Theorem 3.

Let us consider another game which differs from Γ in that way that a pay-off function is introduced

$$\Phi(s) = \sum_{n=0}^{\infty} \rho(p_n, p_{n+1}) + \lim_{m=\infty} \overline{\lim_{n=\infty}}\, m \cdot \rho(p_n, q_n).$$

Φ could be interpreted as a length of the play (which is infinite if p_n

* One approach of this kind was formulated also by G. Choquet (unpublished).

does not approach q_n) and we suppose that the pursuer wants to make it small while the evader has an opposite tendency.* Since $\Phi \in K_2(S)$, then by (A*) such a game is determined. Here the condition of completeness of P is non-essential.

Given the space P, the value of this game, i.e., the number $\sup \inf v(a, b)$, could be considered as a function of the initial positions $p_0, q_0 \in P$ and V. Let us denote this function by $f(p_0, q_0, V)$. Of course $f(p, p, V) = 0$.

Let P be the unit circle $|z| \leq 1$ and suppose $p_0 \neq q_0$. It is clear that

$$f(p_0, q_0, V) \begin{cases} < \infty & \text{if} \quad V < 1 \\ = \infty & \text{if} \quad V > 1 \end{cases}$$

but no reasonable way of calculating f (for $V < 1$) is known. It has been proved** that $f(p_0, q_0, 1) = \infty$.

We could generalize the above games supposing that the two players are picking their choices in two different spaces P and Q and taking some pay-off functions Φ. Also various modifications of the condition $\rho(q_n, q_{n+1}) \leq V \cdot \rho(p_n, p_{n+1})$ could be introduced together with some conditions on the choices of player I. In fact the sets of possible choices may be any nonempty subsets of the spaces depending on the previous choices. The only essential condition for the determinateness is $\Phi \in K_2$.

---

* We have accepted (Sec. 1) a definition of games according to which player I wants to get large values while player II small values. To get this we should take $-\Phi$ rather than Φ.

** The argument of Besicovitch described by Littlewood [4] can be modified to get a proof of this fact.

BIBLIOGRAPHY

[1] DAVIS, Morton D., "Infinite games of perfect information." (This Study.)

[2] GALE, David, and STEWART, F. M., "Infinite games with perfect information," Contributions to the Theory of Games, Vol. II, Annals of Math. Studies 28, 1953, pp. 245-266.

[3] HANANI, Haim, "A generalization of the Banach and Mazur game," Trans. Amer. Math. Soc. 94 (1960), pp. 86-102.

[4] LITTLEWOOD, J. E., "A Mathematician's Miscellany. Lion and Man," London, 1953, pp. 135-136.

[5] MYCIELSKI, Jan, "On the axiom of determinateness," Fund. Math. (To appear.)

[6] MYCIELSKI, Jan, and STEINHAUS, H., "A mathematical axiom contradicting the axiom of choice," Bull. Acad. Polon. Sci. Serie Math. Astr. Phys. 10, (1962), pp. 1-3.

[7] MYCIELSKI, Jan, SWIERCZKOWSKI, S., and ZIĘBA, A., "On infinite positional games," ibid. 4, (1956), pp. 485-488.

[8] MYCIELSKI, Jan and ZIĘBA, A., "On infinite games," ibid. 3, (1955), pp. 133-136.

[9] OXTOBY, J. C., "The Banach Mazur game and Banach category theorem," Contributions to the Theory of Games, Vol. III, Annals of Math. Studies 39, 1957, pp. 159-163.

[10] RYLL-NARDZEWSKI, C., "A theory of pursuit and evasion." (This Study)

[11] WOLFE, Philip, "The strict determinateness of certain infinite games," Pacific J. of Math. 5, (1955), pp. 891-897.

[12] ZIĘBA, A., "An example in pursuit theory," Studia Math. 22, (1962), pp. 1-6.

Jan Mycielski

Polish Academy of Sciences
Institute of Mathematics

A THEORY OF PURSUIT AND EVASION

C. Ryll-Nardzewski*

The study of problems of the pursuit and evasion are connected with difficulties of two kinds. The first is the fact that "effective solution" of quite simple games is very complicated and hard to achieve. The second, which seems interesting for future work, is the fact that the foundations of this theory are not yet definitely settled. As a main problem arises the correct choice of the definition of the strategy. This trouble is of course connected with the continuity of time and it seems that the tools of classical analysis (differential equations) are not sufficient although there are several interesting results in this field (see e.g., [1]).

The purpose of this paper is to present an approach to the theory of pursuit and evasion with perfect information based only on general topological and set theoretical notions.

The author's definition of the strategy is rather strange because multivalued strategies are allowed, such a strategy prescribes to a given motion of the enemy a whole class of possible motions. There are serious reasons for this. Namely, A. Zieba [2] has proved, under very general conditions, that a continuous, univalent strategy (which guarantees the value of the game) does not exist (the domain of the pursuit is the circle and the mentioned fact is a consequence of Brouwer fixed point theorem). On the other hand, it seems that the "value of the game" given by the theory presented in this paper (this value can be also, in some typical cases, described in a more direct way) corresponds very well to our intuitions.

Another, quite different approach to the theory of pursuit has been given by J. Mycielski [3]. It is based on a method of approximation of

---

* This paper was written when the author was working in the University of California at Berkeley on a research project on the foundation of mathematics supported by the U.S. National Science Foundation (Grant 19673).

continuous games by positional games and treats also games of a more general character than pursuit and evasion.

The schema adopted in this paper is general enough to enclose the pursuit and evasion with some restrictions on velocities or accelerations (which may depend on the positions and directions of motions — for this reason asymmetric distance functions are introduced (Sec. 1)). Our payoff function is the amount of time used by the pursuer and the evader to reach a configuration from a given set of final configurations. The main result is given in Sec. 4 (Theorem 9), it gives some general condition for the determinateness of our games.

In this paper some easy but laborious proofs are omitted.

§ 1. DESCRIPTION OF THE PURSUIT AND EVASION

We will consider two spaces P and E. Points p of P will represent positions of the first player (the pursuer) and $e \in E$ positions of the second one (the evader). Rules of admissible motions of the pursuer are given by a <u>distance function</u> $d(p_1, p_2)$, $(p_1, p_2 \in P)$ satisfying the following conditions:

1. $d(p_1, p_2) > 0$ for $p_1 \neq p_2$ and $d(p_1, p_2) = 0$ for $p_1 = p_2$,
2. $d(p_1, p_2) \leq d(p_1, p_3) + d(p_3, p_2)$,
3. for every pair $p_1, p_2 \in P$ there exists at least one point $p_3 \in P$ such that $d(p_1, p_3) = d(p_3, p_2) = 1/2\, d(p_1, p_2)$.
4. If $\lim_n d(p_1^{(n)}, p_2^{(n)}) = 0$ then $\lim_n d(p_2^{(n)}, p_1^{(n)}) = 0$,
5. from every infinite sequence $\{p_n\}$ we can select a "convergent" subsequence $\{p_{n_i}\}$, i.e., there exists a point $p_\infty \in P$ such that $\lim_i d(p_{n_i}, p_\infty) = 0$.

Such a pair $< P, d >$ will be called a <u>compact space with asymmetric, convex distance function</u> (CSACD). It is very easy to see that P is a complete, metric and compact space if we will introduce the symmetric distance ρ by the formula:

$$\rho(p_1, p_2) \overset{\text{df}}{=} d(p_1, p_2) + d(p_2, p_1).$$

Moreover it follows from the condition 3. that every pair of points $p_1, p_2 \in P$ can be connected by a <u>segment</u>, i.e., there is a function $p(t)$ $(0 \leq t \leq d(p_1, p_2))$ such that $d(p_1, p(t)) = t$ and $d(p(t), p_2) = d(p_1, p_2) - t$ for all nonnegative $t \leq d(p_1, p_2)$.

We assume that E is also a CSACD. The distance function in E will be denoted by the same letter d. Let us emphasize that in many typical cases the spaces P and E are quite different. For example, in the pursuit of n ships by m ships P and E are $2m$-dimensional and $2n$-dimensional Euclidean domains respectively.

Intuitively the distance $d(p_1, p_2)(d(e_1, e_2))$ is the shortest time needed by the pursuer (the evader) to pass from the position p_1 to the position $p_2(e_1$ and e_2, respectively).

To complete the description of our game it is necessary to distinguish a <u>set of final positions</u> F, it is an arbitrary closed subset of the product $P \times E$.

The <u>admissible motions</u> of the pursuer and the evader are all functions $p(t)$ and $e(t)$, defined for $t \geq 0$, and satisfying the conditions:

$$d(p(t_1), p(t_2)) \leq t_2 - t_1 \quad \text{and} \quad d(e(t_1), e(t_2)) \leq t_2 - t_1$$
$$\text{for all } t_2 \geq t_1 \geq 0.$$

It is easy to see the spaces of all admissible motions of the pursuer (AMP) and admissible motions of the evader (AME) are compact, metric spaces with respect to the topology of the uniform convergence in each finite time interval.

§ 2. STRATEGIES

In this section we consider admissible motions satisfying some fixed <u>initial conditions</u> $p(0) = p_0$ and $e(0) = e_0$.

Every subclass $A(e(\cdot))$, depending on $e(\cdot) \in \text{AME}$, will be called a <u>strategy of the pursuer</u> (SP) if the following conditions hold:

(i) for every $e(\cdot)$ the set $A(e(\cdot))$ is nonempty,

(ii) if $e(t) = e^*(t)$ for $t \leq t_0$, then the corresponding classes $A(e(\cdot))$ and $A(e^*(\cdot))$

are the same in this time interval $0 \leq t \leq t_0$.

The condition (ii) intuitively means that the pursuer has no information about the future motion of the evader.

Sometimes we will identify a strategy A (it is an abbreviation of $A(\cdot)$) with its <u>diagram</u> $\{< p(\cdot), e(\cdot) > : p(\cdot) \in A(e(\cdot))\}$ which is a subset of the product $AMP \times AME$. We will say that a strategy is <u>closed</u> if its diagram is closed. Every strategy A has a closure $\bar{A}$, whose diagram is obtained by taking the closure of the diagram of A. It is not difficult to verify that (i) and (ii) remain valid. The class of all closed strategies will be denoted by $\overline{SP}$.

If for every $e(\cdot) \in AME$ the set $A(e(\cdot))$ contains exactly one motion $p(\cdot)$ of the pursuer (in general it is not this case) then we will say that this strategy A is a <u>univalent strategy</u>. Every univalent strategy can be treated as a mapping from AME into AMP, and let us observe that this mapping is continuous if our strategy is a closed one.

Let us denote by $T(A)$ <u>the time of the pursuit guaranteed for the pursuer by the strategy</u> A, or more precisely the smallest number T such that for every $e(\cdot)$ and every $p(\cdot) \in A(e(\cdot))$ there exists a moment $t \leq T$ such that $< p(t), e(t) > \in F$. The number $T(A)$ can be infinite. If $T(A)$ is finite then A is called a <u>successful strategy of the pursuit</u>. Let us remark that $T(A) = T(\bar{A})$ always holds.

Minimizing $T(A)$ with respect to all strategies A, the initial positions p_0 and e_0 are fixed, we get the quantity

$$\bar{V}(p_0, e_0) \stackrel{df}{=} \inf_A T(A),$$

and it will be called the <u>upper value of the game</u>.

In a quite similar way we define the notion of the <u>strategy of the evader</u> (SE), and the notion of <u>the time of the escape</u> $t(B)$ <u>guaranteed by a given strategy</u> B.

Nevertheless there is one essential difference between strategies of the pursuit and the evasion, namely, in many typical examples we have $\sup_{B \in SE} t(B) = \infty$. For this reason only closed strategies will be considered and <u>the lower value of the game</u> $\underline{V}(p_0, e_0)$ is defined in an analogous way

$$\underline{V}(p_0, e_0) \stackrel{df}{=} \underline{\sup}_{B \in SE} t(B).$$

We have the following theorem:

THEOREM 1. For arbitrary strategies $A \in \overline{SP}$ and $B \in \overline{SE}$, corresponding to the same initial conditions, the system of the functional relations

(1) $$p(\cdot) \in A(e(\cdot)) \quad \text{and} \quad e(\cdot) \in B(p(\cdot))$$

has at least one solution.

PROOF. Let h be a positive number. We introduce an auxiliary strategy A^* defined as follows:

$p(\cdot) \in A^*(e(\cdot))$ if $p(t) = p(0)$ for $t \leq h$, and $p^*(\cdot) \in A(e(\cdot))$, where $p^*(t) = p(t + h)$ for all $t \geq 0$.

We replace the given system (1) by the new one

(1*) $$p(\cdot) \in A^*(e(\cdot)) \quad \text{and} \quad e(\cdot) \in B(p(\cdot)),$$

which can be easily solved (step by step in time intervals $\langle nh, (n + 1)h \rangle$, $n = 0, 1, 2, \ldots$). When h goes to zero, we can select, by means of some compactness arguments, from those solutions, a subsequence convergent to a solution of (1).

REMARK 1. The above theorem is a generalization of some existence theorems of the theory of differential and other equations.

REMARK 2. Theorem 1 justifies our idea of strategy: every pair A,B of closed strategies is consistent, i.e., both players can move simultaneously according to them.

As a corollary we get the fundamental inequality:

THEOREM 2. For all initial positions p_0, e_0 we have

(2) $$\underline{V}(p_0, e_0) \leq \overline{V}(p_0, e_0).$$

PROOF. If arbitrary closed strategies A and B are given, there exists, in view of Theorem 1, a solution of the system (1), hence it is easy to see that $t(B) \leq T(A)$ and this inequality yields (2).

The game is called _determined_ if its upper and lower values are equal and finite, i.e., for all initial positions p_0, e_0 we have $\underline{V}(p_0, e_0) = \overline{V}(p_0, e_0) < \infty$. In this case their common value will be called the _value of the game_ and denoted by $V(p_0, e_0)$. It represents the optimal time of the duration of the pursuit and can be forced by each of the players.

§ 3. MAJOR AND MINOR FUNCTIONS

Now we introduce some notions useful for the formulation of a condition under which our games are determined. A continuous, real function, $M(p, e)$, positive outside of the set F of final positions and vanishing on F, will be called a _major function_ (MF) if (α) for every p_0, and every $e(\cdot) \in AME$ there exists a motion $p(\cdot) \in AMP$ starting from the point p_0 such that the inequality

(3) $$M(p(t_2), e(t_2)) - M(p(t_1), e(t_1)) \leq -(t_2 - t_1)$$

holds for all $t_1 \leq t_2$ before some configuration from F was reached.

In an analogous way we will say that a function $m(p, e)$ (positive outside of F, vanishing on F and continuous) is a minor function (mF) if (β) for every e_0, and every $p(\cdot) \in AMP$ there exists an $e(\cdot) \in AME$ with $e(0) = e_0$ and satisfying the inequality

(4) $$m(p(t_2), e(t_2)) - m(p(t_1), e(t_1)) \geq -(t_2 - t_1)$$

under the same conditions on t_1, t_2.

We have the following theorem:

THEOREM 3. The inequality

(5) $$m(p, e) \leq M(p, e)$$

for all $p \in P$ and $e \in E$ holds for an arbitrary pair of minor and major functions $m(p, e)$ and $M(p, e)$.

PROOF. Given m and M, we define two strategies A_M and B_m as follows:

$$p(\cdot) \in A_M(e(\cdot)) \quad \text{if} \quad (3) \quad \text{holds,}$$

and

$$e(\cdot) \in B_m(p(\cdot)) \quad \text{if} \quad (4) \quad \text{holds.}$$

It is easy to check that A_M and B_m are closed strategies for the pursuer and the evader, respectively.

REMARK 3. The definitions of A_M and B_m are universal in the sense that these strategies are well defined for all initial positions.

Let $p(\cdot) \in AMP$ and $e(\cdot) \in AME$ be an arbitrary solution of the system (1) (it exists by Theorem 1) and let us denote by τ the first moment for which $< p(\tau), e(\tau) > \in F$, it might be that $\tau = \infty$. If we put $t_1 = 0$ and $t_2 < \tau$ in (3) and (4) we obtain the following inequalities,

$$M(p(t_2), e(t_2)) - M(p(0), e(0)) \leq - t_2 \tag{6}$$

and

$$m(p(t_2), e(t_2)) - m(p(0), e(0)) \geq - t_2 . \tag{7}$$

The function M is bounded (by our assumptions it is continuous on a compact space $P \times E$), hence (6) implies

$$\tau < \infty . \tag{8}$$

Taking $t_2 \uparrow \tau$ and subtracting (7) from (6) we get

$$m(p(0), e(0) - M(p(0), e(0) \leq M(p(\tau), e(\tau) - m(p(\tau), e(\tau)).$$

Moreover $M(p(\tau), e(\tau)) = m(p(\tau), e(\tau) = 0$ because $< p(\tau), e(\tau) > \epsilon F$ and both functions M and m vanish on F. The start points $p(0)$ and $e(0)$ were arbitrary which yields (5).

REMARK 4. The strategy A_M induced by the major function M is always successful. This is equivalent to (8).

We will give now another characterization of major and minor functions which is simpler and will be useful. The sets $\{p:d(p_0, p) \leq h\}$ and $\{e:d(e_0, e) \leq h\}$, where $h > 0$, will be denoted by $D(p_0, h)$ and $D(e_0, h)$ respectively. We assume that

(9) the cartesian product $D(p_0, h) \times D(e_0, h)$ and the set F of final positions are disjoint.

THEOREM 4. A $M(p, e)$ is a major function if

$$(\alpha') \qquad \max_{p \epsilon D(p_0,h)} \ \min_{e \epsilon D(e_0,h)} M(p, e) \leq M(p_0, e_0) - h,$$

for all $p_0 \epsilon P$ and $e_0 \epsilon E$ and positive numbers h satisfying (9). And $m(p, e)$ is a minor function if

$$(B') \qquad \min_{e \epsilon D(e_0,h)} \ \max_{p \epsilon D(p_0,h)} m(p, e) \geq m(p_0, e_0) - h,$$

under the same conditions.

PROOF. The necessity of (α') and (β') is quite obvious. Let us assume that (α') holds, a motion $e(\cdot) \epsilon AME$ is given, and some initial point p_0 for the pursuer is fixed. We choose a small positive number h and, step by step, construct a function $p_h(\cdot) \epsilon AMP$ satisfying the conditions:

$$(10) \qquad p_h(0) = p_0 \text{ and } M(p_h(t_{k+1}), e(t_{k+1})) \leq M(p_h(t_k), e_h(t_k) - h$$
$$\text{for } k = 0, 1, 2, \ldots, \text{ where } t_k = kh.$$

Further, as h goes to zero, by compactness arguments, we can select from the set of functions $p_h(\cdot)$ a sequence convergent to a motion $p(\cdot)$ satisfying the condition (α). In the same way we can prove that (β') is a sufficient condition. In the above reasoning h is supposed to be large enough but diminishing so that (9) is always satisfied.

Relations between minor and major functions and lower and upper values of the game are given by

THEOREM 5. The inequalities

$$m(p, e) \leq \underline{V}(p, e) \leq \overline{V}(p, e) \leq M(p, e) \quad \text{for} \quad p \in P,\ e \in E$$

hold for all minor and major functions m and M (if they exist at all).

PROOF. It is enough to take the special strategies A_M and B_m induced by M and m respectively (introduced in the proof of Theorem 3) and apply the definition of $\underline{V}$ and $\overline{V}$.

From Theorems 3 and 5 follows immediately:

THEROEM 6. There exists at most one function being simultaneously a minor and major function. If it exists, then the game is determined and this function is the value of the game.

§ 4. A CRITERION FOR DETERMINATENESS

The proof of the main result of this paper, Theorem 9, still requires some auxiliary propositions.

THEOREM 7. If for an initial position $< p_0, e_0 >$ the upper value $\overline{V}(p_0, e_0)$ is finite (that means that there exists at least one successful strategy for the pursuer), then there exists also an optimal successful strategy forcing this value $\overline{V}$, i.e.,

(10) $$T(A^{Opt}) = \overline{V}(p_0, e_0).$$

PROOF. We take a minimizing sequence of strategies $\{A_n\}$, $\lim T(A_n) = \overline{V}(p_0, e_0)$, and choose as A^{Opt} an arbitrary limit-strategy of this sequence. More precisely, the family of the diagrams of all strategies (and also the family of all strategies corresponding to a given initial position $< p_0, e_0 >$) can be considered as a subset of the topological space of all closed sets contained in AMP × AME. It is well known that this topological space is a compact one, and it is easy to verify that the families in question are closed. Moreover let us observe that $T(A)$, treated as a function on strategies A, is upper semicontinuous. The last remark implies that indeed our limit strategy A^{Opt} satisfies (10).

THEOREM 8. If $\overline{V}(p, e) < \infty$ for all $p \in P$ and $e \in E$, and $\overline{V}(p, s)$ tends to zero as $< p, s >$ tends to F, then $\overline{V}(p, s)$ is a continuous function on the whole product $P \times E$.

PROOF. Let $\varepsilon > 0$ and $p_0, p_1 \in P$, $e_0, e_1 \in E$ and we put $h_1 \stackrel{df}{=} d(p_0, p_1)$ and $h_2 \stackrel{df}{=} d(e_1, e_0)$. Let, further, A^{Opt} denote an optimal strategy corresponding to the initial position $< p_1, e_1 >$ (it exists in view of Theorem 7). We will define a strategy A corresponding to the initial conditions $< p_0, e_0 >$ in the following way: if a motion $e(\cdot) \in \text{AME}$ $(e(0) = e_0)$ is given then

$$e^*(t) \stackrel{df}{=} \begin{cases} u(t) & \text{for } t \leq h_2 \\ e(t - h_2) & \text{for } t > h_2 \end{cases},$$

where $u(\cdot)$ and $v(\cdot)$ denote segments connecting the pairs e_1, e_0 and p_1, p_0 respectively. Let $p^*(\cdot) \in A^{Opt}(e^*(\cdot))$, and let τ denote the first moment for which $< p^*(\tau), e^*(\tau) \in F$. We put now

$$p(t) \stackrel{df}{=} \begin{cases} v(t) & \text{for } t \leq \min(h_1, \tau) \\ p^*(t - h_1) & \text{for } \min(h_1, \tau) < t \leq h_1. \end{cases}$$

It is easy to check that $d(p(\tau), p^*(\tau)) \leq h_1$ and $d(e^*(\tau), e(\tau)) \leq h_2$.

We complete now the definition of $p(\cdot)$ for all times $t > \tau$, taking an arbitrary optimal strategy for the pursuer corresponding to the new initial position $< p(\tau), e(\tau) >$. The class of all motions $p(\cdot)$ obtained in this way will be denoted by A, and it is not hard to verify that it is a strategy in our sense. If h_1 and h_2 are sufficiently small, say less than a positive δ, then by the second supposition of Theorem $\overline{V}(p(\tau), e(\tau)) < \varepsilon$ and consequently

$$\overline{V}(p_0, e_0) \leq T(A) \leq \tau + \varepsilon \leq V(p_1, e_1) + \varepsilon.$$

Hence we have proved the inequality

$$\overline{V}(p_0, e_0) - \overline{V}(p_1, p_2) < \varepsilon$$

for all pairs of initial positions $< p_0, e_0 >$ and $< p_1, e_1 >$ which are sufficiently close, one to the other. Then $\overline{V}$ is uniformly continuous.

THEOREM 9. If for all initial positions $< p_0, e_0 >$ the pursuer has at least one successful strategy and the upper value $\overline{V}(p, s)$ tends to zero as $< p, s >$ tends to the set of final positions F (i.e., in an appropriate neighborhood of the set F the evader can be caught in a short time), then the game is determined and its upper value $\overline{V}$ (in this case $\overline{V} = V$) is simultaneously a minor and major function.

PROOF. In view of Theorem's 4, 6, and 8 it is enough to verify that the upper value $\overline{V}(p, e)$ satisfies both conditions (α') and (β').

Let us assume that $D(p_0, h) \times D(e_0, h)$ and F are disjoint and $e \in D(e_0, h)$. There exists a motion $e(\cdot) \in AME$ such that $e(0) = e_0$ and $e(h) = e$. Let us take an arbitrary motion $p(\cdot) \in A^{Opt}$ (corresponding to initial values $< p_0, e_0 >$). Of course $p(h) \in D(p_0, h)$ and consequently

$$h + \min_{p \in D(p_0, h)} \overline{V}(p, e) \leq \overline{V}(p(h), e) + h \leq \overline{V}(p_0, e_0).$$

Hence

$$\min_{p\in D(p_0,h)} \overline{V}(p, e) \leq \overline{V}(p_0, h_0) - h \quad \text{for all} \quad e\in D(e_0, h).$$

Obviously the last inequality is equivalent to (α').

Now let us fix an arbitrary point $p\in D(p_0, h)$, and define a strategy A as follows: in the time interval $< 0, h >$ the pursuer moves from p_0 to p along an arbitrary segment independently from the behavior of the evader. At the moment h some position $< p, e(h) >$ is obtained, where $e(h)\in D(e_0, h)$, then the pursuer applies one of his optimal strategies. It is easy to see that the described strategy A guarantees the time $T(A) \leq h + \max_{e\in D(e_0,h)} \overline{V}(p, e)$. Hence, by the definition of $\overline{V}$, we have for all $p\in D(p_0,h)$:

$$\overline{V}(p_0, h_0) \leq h + \max_{e\in D(e_0,h)} \overline{V}(p, e).$$

This concludes the proof.

It is obvious that, if at least one major function exists, then both assumptions of Theorem 9 are satisfied. Hence we have

THEOREM 10. If there exists some major function, then the game is determined.

REMARK 5. This theorem implies for example the determinateness of a game of pursuit and evasion of two points in a circle since the function

$$\frac{|p - e|}{V_p - V_e}$$

is a major function (where V_p, V_e are maximal velocities of the players and $V_p > V_e$). However we do not have an efficient way to compute the value of this quite simple game.

The following example shows that a game can be determined but its value is a discontinuous function, consequently some condition of continuity in Theorem 9 is really necessary.

EXAMPLE. $P \overset{df}{=} \langle 0, 1 \rangle \overset{df}{=} E$, $d(p_1, p_2) \overset{df}{=} |p_1 - p_2|$ and $d(e_1, e_2) \overset{df}{=} |e_1 - e_2|$ (i.e., the maximal velocities of both players are limited by 1). It is easy to see that this game is determined and its value is given by the formula

$$V(p, e) = \begin{cases} 0 & \text{if } p = e \\ p & \text{if } e < p \\ 1-p & \text{if } e > p \end{cases},$$

and V is not continuous.

§ 5. GENERALIZATIONS AND PROBLEMS

Let us mention some possibilities of generalization of the above theory which are for some applications, e.g., to replace compactness by local compactness or to introduce more general payoff functions of the form

$$\int_0^{\tau} \varphi(p(t), e(t))dt,$$

where $\tau = \min \{t: \langle p(t), e(t) \rangle \in F\}$ and φ is a given continuous real valued function. Functionals of this type are used in optimal control theory.

PROBLEM. Let us assume that a game is determined and its value is a continuous function. A strategy can be called nice if it is univalent, optimal, defined for all initial positions, and continuous with respect to all admissible motions of the enemy. It seems that such nice strategies of evasion exist only in exceptional cases when the space E is without "barriers." On the other hand we suspect that they exist more often for the pursuer, e.g., in the pursuit of one point by another in a simply connected domain with a sufficiently smooth boundary, under some natural restrictions on velocities.

BIBLIOGRAPHY

[1] KELENDZERIDZE, D. L., "Theory of optimal pursuit strategy," Steklov Mathematical Institute, January 13, 1961.

[2] ZIEBA, A,, "An example in pursuit theory," Studia Math. 22 (1962), pp. 1-6.

[3] MYCIELSKI, J., "Continuous games with perfect information," The RAND Corporation, P-2591, June, 1962.

C. Ryll-Nardzewski

University of Wroclaw
Wroclaw, Poland

A VARIATIONAL APPROACH TO DIFFERENTIAL GAMES

Leonard D. Berkovitz

§ 1. INTRODUCTION

A differential game is a two-person zero-sum game that can be described roughly as follows. Let t denote time, let $x = (x^1, \ldots, x^n)$ be a vector in real Euclidean n space, and let $\mathcal{R}$ be a fixed region of (t, x) space. The position, or state, $x(t)$ of the game is determined by a system of first-order differential equations

$$\frac{dx^i}{dt} = G^i(t;x,y,z), \quad i = 1, \ldots, n, \tag{1.1}$$

where $y = (y^1, \ldots, y^\sigma)$ is a vector chosen at each instant of time by player I and $z = (z^1, \ldots, z^s)$ is a vector similarly chosen by player II. The choice of y is governed by a vector-valued function of position and time,

$$Y(t, x) = (Y^1(t, x), \ldots, Y^\sigma(t, x)),$$

defined on $\mathcal{R}$, similarly, the choice of z is governed by a vector-valued function of position and time,

$$Z(t, x) = (Z^1(t, x), \ldots, Z^s(t, x)),$$

defined on $\mathcal{R}$. The functions $Y(t, x)$ and $Z(t, x)$ are called <u>pure strategies</u>, and the variables y and z are called <u>strategic variables</u>. Each player selects his strategy from a class of permissible functions before the start of play. The game is one of perfect information: Both players know the past history; they know how the game proceeds [the system of differential equations (1.1)]; and at time t they know the state $x(t)$ of the game. Therefore, in selecting a pure strategy, a player selects a set of instructions for choosing his strategic variable in all possible situations. Play begins at some initial time and position (t_0, x_0) in $\mathfrak{R}$, and terminates whenever t and the position vector $x(t)$ are such that $(t, x(t))$ is a point of a previously specified surface $\mathfrak{J}$ contained in the boundary of $\mathfrak{R}$. If (T, x_T) denotes the point of termination of the play of the game starting at (t_0, x_0), then the payoff to player I is given by

$$g(T, x_T) + \int_{t_0}^{T} f(t, x(t), y(t), z(t))dt,$$

where g is a real-valued function on $\mathfrak{J}$, f is a real-valued function on (t, x, y, z) space, $y(t) = Y(t, x(t))$, and $z(t) = Z(t, x(t))$. The objective of player I is to select a strategy $Y(t, x)$ that maximizes the payoff; that of player II is to select a strategy $Z(t, x)$ that minimizes the payoff.

The study of differential games was initiated by Rufus Isaacs [6]. Isaac's treatment of the problem was quite formal and heuristic. Nevertheless, very little additional work has been done on the problem. Fleming [4], [5], and Scarf [7] have considered discrete approximations to certain types of differential games. In [1], Fleming and the present author studied a special class of differential games in the plane by means of techniques and results from the calculus of variations.

In this paper, we shall study a much wider class of differential games than those considered in [1], again using techniques and results from the calculus of variations. The present investigation, however, is more complete, even for the special case corresponding to [1]. First, we shall describe the class of games to be studied. Necessary conditions that must hold along a path resulting from the employment of optimal pure strategies by the two players will then be deduced for the present class of games. This will be done by relating the problem of determining these necessary conditions to a problem of Bolza with differential inequalities as added side conditions. The theory of such Bolza problems will then be used to obtain the desired necessary conditions. The continuity and differentiability properties of the value will be studied next, and we shall show that the value satisfies an analogue of the Hamilton-Jacobi equation. In the last part of the paper, we shall develop a sufficiency theory that, in principle, enables us to construct optimal pure strategies. In some examples, this construction can actually be carried out. Some of our results were obtained by Isaacs [6] in a formal manner or under more restrictive assumptions.

We conclude this introductory section by listing certain definitions and conventions that we shall use throughout the paper. Vector matrix notation will generally be used. Vectors and matrices will be denoted by single letters. Superscripts will be used to denote the components of a vector; subscripts will be used to distinguish vectors. Vectors will be written as matrices consisting of either one row or one column. We shall not use a transpose symbol to distinguish between the two usages, as it will be clear from the context how the vector is to be considered. Thus, the product of two n-dimensional vectors λ and G, say, will be written as λG. All scalars that occur will be real and all vectors will have real components. A vector will be called <u>positive</u> if each of its components is positive. Negative, nonpositive and nonnegative vectors are defined similarly. The length

$$\left(\sum_{i=1}^{n} (x^i)^2 \right)^{1/2}$$

of a vector x will be denoted by $\|x\|$.

Let

$$\chi(t,x,y,z) = (\chi^1(t,x,y,z), \ldots, \chi^m(t,x,y,z))$$

be a vector-valued function defined and differentiable on a region of (t, x, y, z) space. If z is an s-dimensional vector, we can form a matrix of partial derivatives in which the (i,j)-th element is

$$\frac{\partial \chi^i}{\partial \chi^j}, \quad i = 1, \ldots, m, \text{ and } j = 1, \ldots, s.$$

The symbol χ_z will denote this matrix. Note that when $m = 1$, $s = 1$, this reduces to the usual notation for a partial derivative. The symbols χ_x and χ_y will have similar meanings. The determinant of a square matrix M will be denoted by $\det M$.

The term region will mean, as usual, an open connected set. The closure of a region $\mathfrak{D}$ will be denoted by $\bar{\mathfrak{D}}$, and $\bar{\mathfrak{D}}$ will be called a closed region. A real-valued function will be said to be of class $C^{(k)}$ on $\bar{\mathfrak{D}}$ if it is $C^{(k)}$ on $\mathfrak{D}$ and all of its derivatives up to and including those of order k have continuous extensions to $\bar{\mathfrak{D}}$. A vector-valued function will be said to be of class $C^{(k)}$ on $\bar{\mathfrak{D}}$ if each of its components is $C^{(k)}$ on $\bar{\mathfrak{D}}$. A region will be said to have a piecewise smooth boundary if its boundary consists of a finite number of manifolds with boundary.

The letter t will denote time, and the operator (d/dt) will be denoted by a prime.

§ 2. FORMULATION OF THE GAME

2.1 The Functions f and G

Let x be an n-dimensional vector, let y be an $\bar{s}$-dimensional vector, and let z be an s-dimensional vector. We shall be concerned with a bounded region $\mathcal{G}$ of (t, x) space and a bounded region $\mathcal{S}$ of (t, x, y, z) space. We assume that $\mathcal{G}$ is contained in the projection of $\mathcal{S}$ into (t, x) space. As we indicated in Section 1, we shall be concerned with a real-valued function $f(t, x, y, z)$ and a vector-valued function

$$G(t,x,y,z) = (G^1(t,x,y,z), \ldots, G^n(t,x,y,z))$$

with range contained in Euclidean n space. We assume that f and G are of class $C^{(2)}$ on $\mathcal{S}$. We may write the system (1.1) in vector notation as

$$x' = G(t,x,y,z), \tag{2.1}$$

where the prime denotes differentiation with respect to time.

2.2 The Constraints

We now discuss functions $K(t, x, z)$ that will constrain the choice of strategic variable z, and functions $\bar{K}(t, x, y)$ that will constrain the choice of strategic variable y. Let $\mathcal{S}^{(z)}$ denote the projection of $\mathcal{S}$ into (t, x, z) space. Let $K(t, x, z)$ be a vector-valued function of class

$C^{(2)}$ on $\mathcal{S}^{(z)}$, having range on p-dimensional space and satisfying the following <u>constraint conditions</u>:

(2.2) (i) The set of points (t, x, z) in $\mathcal{S}^{(z)}$ such that $K(t, x, z) \geq 0$ is nonvoid.

(ii) If $p > s$, then at each (t, x, z) in $\mathcal{S}$, at most s of the components of K can vanish. The matrix with elements $\partial K^i/\partial z^j$ formed from those components K^i of K that vanish at (t, x, z) has maximum rank at (t, x, z).

A third constraint condition will be stated presently.

Let ω be a naturally ordered subset of the integers $1, \ldots, p$ containing at most s elements. For each such ω consider the system of equations

$$K^{\omega}(t, x, z) = 0, \tag{2.3}$$

where the symbol K^{ω} denotes a vector whose components are those components of K having indices that belong to ω, and the order of the components is the natural order. Let $\mathcal{S}^{(z)}_{\omega}$ denote the subset of $\mathcal{S}^{(z)}$ on which (2.3) holds. It follows from (2.2) (ii) that $\mathcal{S}^{(z)}_{\omega}$ is either the null set or a differentiable manifold, and the solutions of (2.3) can be used to obtain a system of local coordinates for $\mathcal{S}^{(z)}_{\omega}$. We can now state our third constraint condition:

(iii) Each manifold $\mathcal{S}^{(z)}_{\omega}$ can be represented by a finite number of overlapping coordinate systems obtained by using solutions of (2.3) that always include (t, x) in the set of independent variables.

These solutions of (2.3) will be referred to as the <u>coordinate solutions</u>.

The constraints on the choice of y are furnished by vector-valued function $\overline{K}(t, x, y)$ of class $C^{(2)}$ on $\mathcal{S}^{(y)}$, having range in a $\overline{p}$-dimensional space, and satisfying (2.2) and (2.3) with appropriate changes in notation.

An important special case, which also serves to illustrate (2.2) and (2.3), is that of <u>separable constraints</u>. Let S_1 and S_2 be two subsets, not necessarily disjoint, of the set $1, \ldots, s$. Let $A^i(t, x)$ be of class $C^{(2)}$ on $\mathcal{G}$ for i in S_1, and let $B^j(t, x)$ be of class $C^{(2)}$ on $\mathcal{G}$ for j in S_2. If k is an integer in both S_1 and S_2, we suppose that $A^k(t, x) > B^k(t, x)$ on $\mathcal{G}$. For each (t, x) in $\mathcal{G}$, let the set of z such that (t, x, z) is in $\mathcal{S}^{(z)}$ and such that the inequalities

$$A^i(t, x) - z^i \geq 0, \qquad i \text{ in } S_1,$$

$$z^j - B^j(t, x) \geq 0, \qquad j \text{ in } S_2,$$

hold, be nonvoid. Then we can form a vector-valued function $K(t, x, z)$ that satisfies (2.2) and (2.3) by taking the functions $A^i(t, x) - z^i$ and $z^j - B^j(t, x)$ as the components of K and, if necessary, relabeling the superscripts.

2.3 Terminal Manifolds

For each $i = 1, \ldots, \alpha$, let $\mathfrak{J}_i'$ be an n-dimensional manifold of class $C^{(2)}$ that lies in $\mathcal{G}$ and that is given parametrically by equations

$$t = t_i(\sigma) \qquad x = x_i(\sigma), \tag{2.4}$$

where $\sigma = (\sigma^1, \ldots, \sigma^n)$ ranges over an open cube $\mathcal{K}_i$ in n-dimensional space. We select a connected submanifold $\mathfrak{J}_i$ of each $\mathfrak{J}_i'$. Let

$$\mathfrak{J} = \bigcup_{i=1}^{\alpha} \mathfrak{J}_i.$$

We shall call $\mathfrak{J}$ the <u>terminal</u> surface of the game. Points (t, x) that belong to $\mathfrak{J}$ will be denoted by (T, x_T).

Let $\mathcal{G}_1$ be a subregion of $\mathcal{G}$ such that $\mathfrak{J}$ is contained in $\mathcal{G}_1$. Let $g(t, x)$ be a real-valued function of class $C^{(2)}$ on $\mathcal{G}_1$. The function $g(T, x_T)$ is therefore defined on $\mathfrak{J}$ and is of class $C^{(2)}$ on each $\mathfrak{J}_i$. We shall call $g(T, x_T)$ the <u>terminal payoff</u> function.

We remark that the assumption that all of $\mathfrak{J}_i$ can be coordinatized by parameters σ in $\mathcal{K}_i$ is made to simplify the subsequent exposition. The essential assumption is that $\mathfrak{J}_i$ is a $C^{(2)}$ manifold.

2.4 Strategies

A notion that will be used in our discussion of strategies and other subsequent work is that of a decomposition of a region. A finite collection of subregions $\mathfrak{D}_1, \ldots, \mathfrak{D}_\alpha$ of a region $\mathfrak{D}$ will be said to constitute a decomposition of $\mathfrak{D}$ whenever the following conditions hold: (i) Each $\mathfrak{D}_i$, $i = 1, \ldots, \alpha$, is connected and has a piecewise-smooth boundary, (ii) $\mathfrak{D}_i \cap \mathfrak{D}_j = 0$ if $i \neq j$, and (iii) $\bar{\mathfrak{D}} = \cup_{i=1}^{\alpha} \bar{\mathfrak{D}}_i$. A real-valued function defined on $\bar{\mathfrak{D}}$ will be said to be piecewise $C^{(k)}$ on $\bar{\mathfrak{D}}$ if there is a decomposition of $\mathfrak{D}$ such that on each $\mathfrak{D}_i$ the function agrees with a function that is $C^{(k)}$ on $\bar{\mathfrak{D}}_i$. A vector-valued function will be said to be piecewise $C^{(k)}$ on $\mathfrak{D}$ if each component is piecewise $C^{(k)}$ on $\mathfrak{D}$.

Let $\mathfrak{R}$ be a region with closure $\bar{\mathfrak{R}}$ contained in $\mathcal{G}$, and such that the terminal surface $\mathfrak{J}$ forms a part of the boundary of $\mathfrak{R}$. Further, assume that $\mathfrak{J}$ can be imbedded in an n-dimensional manifold that separates $\mathcal{G}$, and that, for any given $i = 1, \ldots, \alpha$, the region $\mathfrak{R}_i$ is always on the same side of $\mathfrak{J}_i$. The region $\mathfrak{R}$ will be the region of (t, x) space in which the play of the game takes place. Note that since $\bar{\mathfrak{R}} \subset \mathcal{G}$, and $\mathcal{G}$ is bounded, it follows that $\bar{\mathfrak{R}}$ is compact.

Let $\mathcal{Y}$ denote the class of functions Y that are piecewise $C^{(2)}$ on $\bar{\mathfrak{R}}$ and satisfy the following conditions: (i) The values $y = Y(t, x)$ lie in an $\bar{s}$-dimensional space. (ii) For every (t, x) in $\bar{\mathfrak{R}}$, the point $(t, x, Y(t, x))$ lies in $\mathcal{S}^{(y)}$, the projection of $\mathcal{S}$ into (t, x, y) space, and satisfies $\bar{K}(t, x, Y(t, x)) \geq 0$; that is, the functional values satisfy the y-constraints. Similarly, let $\mathcal{Z}$ denote the class of functions Z that are piecewise $C^{(2)}$ on $\bar{\mathfrak{R}}$ and satisfy the following conditions: (i) The values $z = Z(t, x)$ lie in an s-dimensional space. (ii) For every (t, x) in $\bar{\mathfrak{R}}$, the point $(t, x, Z(t, x))$ lies in $\mathcal{S}^{(z)}$ and satisfies $K(t, x, Z(t, x)) \geq 0$. We assume that $\mathcal{Y}$ and $\mathcal{Z}$ are nonvoid.

Since a function Y in $\mathcal{Y}$ is piecewise $C^{(2)}$ on $\bar{\mathfrak{R}}$, there is associated with it a decomposition of $\bar{\mathfrak{R}}$, and the same is true of a function

Z in $\mathcal{Z}$. Hence a pair of functions (Y, Z) with Y in $\mathcal{Y}$ and Z in $\mathcal{Z}$ also has associated with it a decomposition of $\bar{\mathcal{R}}$, say $\mathcal{R}_1, \ldots, \mathcal{R}_q$. It follows from standard existence theorems for ordinary differential equations and the hypotheses concerning the functions G, Y, and Z, that corresponding to a given point (t_0, x_0), interior to some $\mathcal{R}_i$, there exists a unique solution

$$x(t) = x(t;\, t_0, x_0) \tag{2.5}$$

of the system of differential equations

$$x' = G(t, x, Y(t, x), Z(t, x)), \tag{2.6}$$

defined on some interval about t_0 and satisfying the initial condition

$$x(t_0) = x(t_0;\, t_0, x_0) = x_0. \tag{2.7}$$

What we shall need, however, is the assurance that either (i) $x(t;\, t_0, x_0)$ can be continued across the boundary of $\mathcal{R}_i$ and through other regions $\mathcal{R}_j$ so as to reach $\mathfrak{J}$, or (ii) $x(t;\, t_0, x_0)$ extends to a point on the boundary of $\mathcal{R}_i$ that is also a point of $\mathfrak{J}$. In terms of the game, we need to be assured that a play starting at (t_0, x_0) will terminate. Further, if (t_0, x_0) is both an interior point of $\mathcal{R}$ and a boundary point of $\mathcal{R}_i$ and $\mathcal{R}_j$ for some pair of indices i, j, then several possibilities exist for solutions of (2.6) satisfying the initial conditions (2.7). Thus, to ensure that play can take place, we need to impose restrictions on the functions Y and Z chosen as strategies by the players. The restrictions presently to be made are imposed partly for this purpose, and partly to enable us to carry out our analysis.

We define a <u>regular decomposition</u> of a region $\mathcal{R}$ to be a decomposition in which the constituent subregions can be designated

$$\mathcal{R}_{11}\ \mathcal{R}_{12}, \ldots, \mathcal{R}_{1j_1}, \mathcal{R}_{21}, \ldots, \mathcal{R}_{2j_2}, \ldots, \mathcal{R}_{\alpha 1}, \ldots, \mathcal{R}_{\alpha j_\alpha},$$

in such a way (see Fig. 1) that the following conditions are satisfied:

(i) The regions $\mathcal{R}_i$, defined for each $i = 1, \ldots, \alpha$ by the formula

$$\mathcal{R}_i = \left(\bigcup_{j=1}^{j_i} \mathcal{R}_{ij} \right) \cap \mathcal{R} ,$$

constitute a decomposition of $\mathcal{R}$.

(ii) For each $i = 1, \ldots, \alpha$, $\mathcal{R}_i$ always lies on the same side of $\mathfrak{J}_i$, and $\mathcal{R}_i \cap \mathfrak{J}_k = 0$ whenever $i \neq k$.

(iii) For each $i = 1, \ldots, \alpha$, $\overline{\mathcal{R}}_{ij_i} \cap \mathfrak{J}_i \neq 0$, and $\overline{\mathcal{R}}_{ij} \cap \mathfrak{J}_i = 0$ whenever $j \neq j_i$.

(iv) For each $i = 1, \ldots, \alpha$, and $j = 1, \ldots, j_i - 1$, the set

$$\mathfrak{M}_{ij} = (\overline{\mathcal{R}}_{ij} \cap \overline{\mathcal{R}}_{i,j+1}) \cap \mathcal{R}_i$$

is a connected and oriented manifold of dimension n and class $C^{(2)}$. We suppose that $\mathfrak{M}_{ij}$ can be described by equations

$$t = t_{ij}(\sigma) \qquad x = x_{ij}(\sigma), \tag{2.8}$$

where $\sigma = (\sigma^1, \ldots, \sigma^n)$ ranges over a cube $\mathcal{K}_{ij}$ in Euclidean n-space.

(v) Each manifold $\mathfrak{M}_{ij}$ divides $\mathcal{R}_i$ into two disjoint regions such that each region lies entirely on one side of $\mathfrak{M}_{ij}$. Furthermore, $\overline{\mathfrak{M}}_{ij} \cap \overline{\mathfrak{M}}_{ik} = 0$ for $i = 1, \ldots, \alpha$, and $j, k = 1, \ldots, j_i - 1$.

(vi) For each subset $i_1, \ldots, i_k$ of the integers $1, \ldots, \alpha$, the set $\mathfrak{N}_{i_1, \ldots, i_k}$, defined by the formula

$$\mathfrak{N}_{i_1,\ldots,i_k} = \left(\overline{\mathcal{R}}_{i_1} \cap \overline{\mathcal{R}}_{i_2} \cap \ldots \cap \overline{\mathcal{R}}_{i_k}\right) \cap \mathcal{R},$$

is either the null set or a connected, nonsingular oriented differentiable manifold.

We remark here that, as in the definition of $\mathfrak{T}_i$, we assume that all of $\mathfrak{M}_{ij}$ can be coordinatized by parameters in a single $\mathcal{K}_{ij}$. This is made for the purpose of simplifying the exposition.

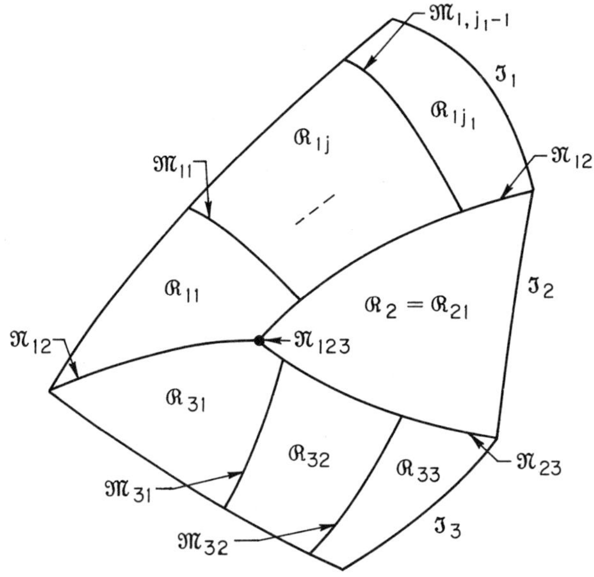

FIGURE 1. Regular decomposition of a region.

We now introduce additional notation that will be used in the sequel. In order to simplify the exposition, we shall sometimes denote the terminal manifolds $\mathfrak{T}_i$ as $\mathfrak{M}_{ij_i}$. Also, we define $\mathfrak{M}_{i0}$ to be the union of all manifolds $\mathfrak{N}_{i_1, \ldots, i_k}$ such that i belongs to the set $i_1, \ldots, i_k$. We define

$$\tilde{\mathcal{R}}^{+}_{ij} = \mathcal{R}_{ij} \cup \mathfrak{M}_{ij},$$

(2.9) $$\tilde{\mathcal{R}}^{-}_{ij} = \mathcal{R}_{ij} \cup \mathfrak{M}_{i,j-1},$$

$$\tilde{\mathcal{R}}_{ij} = \mathcal{R}_{ij} \cup \mathfrak{M}_{i,j-1} \cup \mathfrak{M}_{ij},$$

where $i = 1, \ldots, \alpha$, and $j = 1, \ldots, j_i$. Points belonging to $\mathfrak{M}_{ij}$ will be written as (t_{ij}, x_{ij}).

Let (Y, Z), with Y in $\mathcal{Y}$ and Z in $\mathcal{Z}$, be a pair of functions such that the decomposition associated with (Y, Z) is regular. Let Y_{ij} denote the function of class $C^{(2)}$ on $\bar{\mathcal{R}}_{ij}$ that agrees with Y on $\mathcal{R}_{ij}$, and let Z_{ij} denote the function of class $C^{(2)}$ on $\bar{\mathcal{R}}_{ij}$ that agrees with Z on $\mathcal{R}_{ij}$. When we consider the differential equation (2.6) with such functions Y and Z as arguments of G, we shall mean the forward differential equation; that is, the following equation on each $\tilde{\mathcal{R}}^{-}_{ij}$:

(2.10) $$x' = G(t, x, Y_{ij}(t, x), Z_{ij}(t, x)), \quad (t, x) \in \tilde{\mathcal{R}}^{-}_{ij}.$$

For $j = 0, \ldots, j_i - 1$ we shall use the notation $Y^{+}(t_{ij}, x_{ij})$ for $Y_{i,j+1}(t_{ij}, x_{ij})$ and the notation $Y^{-}(t_{ij}, x_{ij})$ for $Y_{i,j-1}(t_{ij}, x_{ij})$. The symbols Z^{+} and Z^{-} are defined in similar fashion.

A pair of functions (Y, Z), with Y in $\mathcal{Y}$ and Z in $\mathcal{Z}$, will be called a playable pair if the following conditions hold: (i) The decomposition of $\mathcal{R}$ associated with (Y, Z) is regular. (ii) For every (t_0, x_0) in $\tilde{\mathcal{R}}^{-}_{ij}$, $i = 1, \ldots, \alpha$, and $j = 1, \ldots, j_i$, the unique solution $x(t) = x(t; t_0, x_0)$ of (2.10) satisfying the initial condition (2.7) is defined on an interval $t_0 \leq t < t_{ij}$, such that, for $t_0 < t < t_{ij}$, the point $(t, x(t))$ is interior to $\mathcal{R}_{ij}$. (iii) If x_{ij} denotes $\lim x(t)$ as $t \to t^{-}_{ij}$, then (t_{ij}, x_{ij}) belongs to $\mathfrak{M}_{ij}$. Stated somewhat less precisely, conditions (ii) and (iii) impose the following restrictions on Y and Z: First, if (t_0, x_0) is a point of $\mathfrak{M}_{ij}$, $i = 1, \ldots, \alpha$, and $j = 0, \ldots, j_i - 1$, then the unique solution of (2.10) starting at (t_0, x_0) (the existence of this solution follows from the properties of Y, Z, and G, and from standard theorems) proceeds into the interior of $\mathcal{R}_{i,j+1}$. Second, the solution starting from any point (t_0, x_0) in $\tilde{\mathfrak{R}}_{ij}^-$ is interior to $\mathfrak{R}$ for $t > t_0$ and extends as far as $\mathfrak{M}_{ij}$, $j = 1, \ldots, j_i$. The existence of $\lim x(t)$ as $t \to t_{ij}^-$ follows from the properties of Y, Z, and G, and from standard theorems.

Let (t_0, x_0) be a point of $\bar{\mathfrak{R}}_i \cap \mathfrak{R}$, or, equivalently, a point of $\mathfrak{R}_i \cup \mathfrak{M}_{i0}$. It follows from the definition of a playable pair that we may piece together solutions of equations (2.10) in the obvious way to obtain a unique continuous solution $x(t) = x(t; t_0, x_0)$ of (2.6), satisfying the initial conditions (2.7) and possessing the following additional properties: (i) The solution is defined on an interval $t_0 \leq t \leq T$, where T may depend on (t_0, x_0) and $(T, x(T; t_0, x_0))$ is a point of $\mathfrak{I}_i$. (ii) The solution is interior to $\mathfrak{R}_i$ for $t_0 < t < T$. Such a solution will be called a <u>path resulting from</u> (Y, Z), or simply a <u>path</u>. The point (t_0, x_0) will be called the <u>initial point</u> of the path and $(T, x_T) \equiv (T, x(T; t_0, x_0))$ will be called the <u>terminal point</u> of the path. Although we interpret (2.6) as a forward equation, it is clear that at points where a path intersects a manifold $\mathfrak{M}_{ij}$, $j = 1, \ldots, j_i - 1$, $x'(t; t_0, x_0)$ has both right- and left-hand limits and that (2.6) holds from the left as well as from the right. At $\mathfrak{M}_{ij_i} = \mathfrak{I}_i$, of course, only left-hand limits exist, and we shall merely write $x'(t)$ and $x(t)$.

Since G is of class $C^{(2)}$ and Y and Z are piecewise $C^{(1)}$, it follows from (2.6) and standard theorems that $x'(t)$ has continuous derivatives between corners; i.e., $x(t)$ is of class $C^{(2)}$ between corners.

If (t_0, x_0) belongs to $\mathfrak{N}_{i_1, \ldots, i_k}$, it follows that, for each i in the set $i_1, \ldots, i_k$, there will be a path having (t_0, x_0) as initial point and lying in $\mathfrak{R}_i$, except for the initial and terminal points. Thus, if κ denotes the number of elements in the set $i_1, \ldots, i_k$, there will be κ paths emanating from (t_0, x_0).

Let $\mathcal{Y}_1$ be a subclass of $\mathcal{Y}$ and $\mathcal{Z}_1$ a subclass of $\mathcal{Z}$ such that, for every Y in $\mathcal{Y}_1$ and Z in $\mathcal{Z}_1$, the pair (Y, Z) is playable. In this event, we shall call the functions Y in $\mathcal{Y}_1$ and Z in $\mathcal{Z}_1$ <u>pure strategies.</u> Since we shall be dealing only with pure strategies in this paper, we shall henceforth drop the adjective <u>pure</u>. Player I chooses a strategy Y and player II chooses a strategy Z. There is a defect in this definition in that if we replace the subclass $\mathcal{Z}_1$ by a subclass $\mathcal{Z}_2$, $\mathcal{Z}_2 \subseteq \mathcal{Z}$, and consider the classes $\mathcal{Y}_1$, $\mathcal{Z}_2$, the functions Y in $\mathcal{Y}_1$ need no longer be playable against every Z in $\mathcal{Z}_2$.

2.5 Payoff

Let (Y, Z) be a playable pair, let (t_0, x_0) be a point of $\mathcal{R}$, and let $x_{(k)}(t)$, $k = 1, \ldots, \kappa$, where k depends on (t_0, x_0), denote the paths starting at (t_0, x_0) and terminating at (T, x_T), where T depends on k and (t_0, x_0). Then for each path $x_{(k)}(t)$ we obtain two additional functions,

$$(2.11) \qquad \begin{aligned} y_{(k)}(t) &= y_{(k)}(t;\, t_0, x_0) = Y(t, x_{(k)}(t;\, t_0, x_0)) \\ z_{(k)}(t) &= z_{(k)}(t;\, t_0, x_0) = Z(t, x_{(k)}(t;\, t_0, x_0)), \\ & \qquad\qquad k = 1, \ldots, \kappa, \end{aligned}$$

that are continuous for $t_0 \leq t \leq T$, with the possible exception of jumps at a finite number of points. The functions $y_{(k)}(t)$ and $z_{(k)}(t)$ defined in (2.11) will be referred to as the associated functions of the path $x_{(k)}(t)$, or simply, associated functions.

If we substitute a path $x_{(k)}(t)$ and its associated functions (2.11) into the function f discussed in Sec. 2.1, we obtain a function $f[t, x_{(k)}(t), y_{(k)}(t), z_{(k)}(t)]$ that is integrable on the interval $t_0 \leq t \leq T$. Thus, if g is the function discussed in Sec. 2.3, then

$$(2.12) \qquad \begin{aligned} P_{(k)}(t_0, x_0;\, Y, Z) &= g(T, x_{(k)}(T)) \\ &\quad + \int_{t_0}^{T} f[t, x_{(k)}(t), y_k(t), z_k(t)]dt \end{aligned}$$

is defined for $k = 1, \ldots, \kappa$. We call (2.12) the payoff resulting from the playable pair (Y, Z) and from the path $x_{(k)}(t)$ starting at (t_0, x_0). Since (t_0, x_0) is an arbitrary point of $\mathcal{R}$, we get in this fashion a multivalued function $P_{(k)}(t, x;\, Y, Z)$, $k = 1, \ldots, \kappa(t, x)$, defined on $\mathcal{R}$ for each playable pair (Y, Z). We call this function the payoff function for the pair (Y, Z).

2.6 Value and Optimal Strategies

Let (Y^*, Z^*) be a playable pair such that the payoff $P_{(k)}(t, x; Y^*, Z^*)$ is single valued on $\mathcal{R}$. In other words, if there is more than one path starting at a point (t, x), then all the values $P_{(k)}(t, x; Y^*, Z^*)$ are equal. Denote this common value as $W(t, x)$. Let $\mathcal{Y}_1$ and $\mathcal{Z}_1$ denote the classes of strategies being used. Suppose further that, for all Y in $\mathcal{Y}_1$ and Z in $\mathcal{Z}_1$, all indices σ that may arise from the pairs (Y, Z^*) at (t, x), and all indices ν that may arise from the pairs (Y^*, Z) at (t, x), the inequalities

$$(2.13)\qquad P_{(\sigma)}(t, x; Y, Z^*) \leq W(t, x) \leq P_{(\nu)}(t, x; Y^*, Z)$$

hold for all (t, x) in $\mathcal{R}$. In this event, we say that (Y^*, Z^*) is a saddle point relative to the classes $\mathcal{Y}_1$ and $\mathcal{Z}_1$. The strategies Y^* and Z^* are said to be optimal. The function $W(t, x)$ is called the value of the game. It follows from (2.12) and the definition of W that W is continuous on $\mathcal{R} \cup \mathfrak{T}$. Furthermore, we have $W(T, x_T) = g(T, x_T)$.

The notation $x^*_{(k)}(t)$ or $x^*_{(k)}(t; t_0, x_0)$ will be used for paths resulting from (Y^*, Z^*) and starting at (t_0, x_0); such paths will be called optimal paths. Optimal paths will sometimes also be denoted by $E^*_{(k)}$. The notation $y^*_{(k)}(t)$, $z^*_{(k)}(t)$ will be used for the functions associated with optimal paths. If we restrict our attention to one of the regions $\mathcal{R}_i$, as we shall often do in the sequel, we may drop the subscript (k) when discussing optimal paths and speak of the optimal path starting at (t_0, x_0) and lying in $\mathcal{R}_i$, where (t_0, x_0) is a point of $\mathcal{R}_i \cup \mathfrak{M}_{i0}$. We shall denote this optimal path simply by $x^*(t)$, or by E^*. An optimal path E^* satisfies the differential equation

$$(2.14)\qquad x'(t) = G(t, x, Y^*(t, x), Z^*(t, x)),$$

and has associated functions (2.11), which can be written as

$$(2.15)\qquad y^*(t) = Y^*(t, x^*(t)), \quad z^*(t) = Z^*(t, x^*(t)).$$

2.7 Saddle-Point Restrictions

One of the objectives of this paper is to deduce necessary conditions that must hold along an optimal path. In order to carry out the analysis, we further restrict the class of saddle-points (Y^*, Z^*) and classes of strategies $\mathcal{Y}_1$, $\mathcal{Z}_1$ as follows:

(2.16) (i) An optimal path $x^*_{(i)}(t)$ resulting from (Y^*, Z^*) is never tangent to a manifold $\mathfrak{M}_{ij}$, $j = 1, \ldots, j_i - 1$. More precisely, we have

$$dt_{ij} + (x'_{(i)})^+ \; dx_{ij} \neq 0,$$
$$dt_{ij} + (x'_{(i)})^- \; dx_{ij} \neq 0,$$

where dt_{ij} and dx_{ij} are determined from (2.8).

(ii) An optimal path $x^*_{(i)}(t)$ resulting from (Y^*, Z^*) and terminating at (T, x^*_T) is not tangent to $\mathfrak{T}_i$; in other words,

$$dt_i + x^{*\prime}_{(i)} \; dx_i \neq 0$$

at (T, x^*_T), where dt_i and dx_i are determined from (2.4).

(iii) An optimal path $x^*_{(i)}(t)$ resulting from (Y^*, Z^*) and starting at (t_0, x_0) in $\mathfrak{N}_{i_1,\ldots,i_k}$ is not tangent to $\mathfrak{N}_{i_1,\ldots,i_k}$.

(iv) There is a constant $b > 0$, independent of (t, x) in $\mathfrak{R}$ and of Y in $\mathcal{Y}_1$, such that if (t, x) is a point of a manifold $\mathfrak{N}_{i_1,\ldots,i_k}$ or of $\mathfrak{M}_{ij}$, $i = 1, \ldots, \alpha$, and $j = 1, \ldots, j_i - 1$, in the regular decomposition of $\mathfrak{R}$ associated with (Y, Z^*), then

$$|dt + G(t, x, Y^+, Z^{*+})dx| \geq b.$$

for all differentials dt, dx on the manifold. A similar condition holds for (Y^*, Z), where Z is any strategy in $\mathcal{Z}_1$.

§ 3. THE RELATED BOLZA PROBLEM

3.1 Statement of the Problem

Let (Y^*, Z^*) be a saddle-point relative to the classes $\mathcal{Y}_1$ and $\mathcal{Z}_1$, and let (t_0, x_0) be a point of $\tilde{\mathcal{R}}_{ij}^-$, where the region $\mathcal{R}_{ij}$ belongs to the regular decomposition of $\mathcal{R}$ associated with (Y^*, Z^*). Let $\mathcal{P}_0$ denote the class of associated functions $z(t)$ of (2.11), that result from pairs (Y^*, Z), where Z is in $\mathcal{Z}_1$. It follows from the definition of saddle-point that the function $z^*(t)$, associated with (Y^*, Z^*) furnishes a solution for the following problem:

PROBLEM A. Determine a function $z(t)$ in $\mathcal{P}_0$ that minimizes

$$g(t, x_T) + \int_{t_0}^{T} f(t, x, Y^*(t, x), z)dt$$

subject to the side condition

$$x' = G(t, x, Y^*(t, x), z) \qquad x(t_0) = x_0.$$

Consider the class of all functions $u(t)$ that are piecewise C'' on the closure of the projection of $\mathcal{G}$ on the t-axis and that have a range such that $(t, x, Y^*(t, x), u(t))$ belongs to $\mathcal{S}$. For each such u we can obtain a continuous solution of

$$x' = G(t, x, Y^*(t, x), u) \qquad x(t_0) = x_0$$

that defines a curve E, with possible corners in $\mathcal{G}$. Let $\mathcal{A}_0$ be the subclass of this class of functions u with the following properties:

(i) The curve E is defined and is interior to $\mathcal{G}$ for $t_0 \leq t \leq T$, where $(T, x(T))$ is a point of $\mathfrak{J}$ and E does not intersect $\mathfrak{J}$ for any $t_0 \leq t < T$.

(ii) Along E, the constraints (2.2)-(i) are satisfied; i.e., $K(t, x(t), u(t)) \geq 0$.

Clearly, $\mathcal{P}_0 \subset \mathcal{A}_0$.

The following problem, which is a problem of optimal control, differs from Problem A only in that the class $\mathcal{P}_0$ is replaced by the class $\mathcal{A}_0$.

PROBLEM B. Determine a function u^* in $\mathcal{A}_0$ that minimizes the functional

$$g(T, x_T) + \int_{t_0}^{T} f(t, x, Y^*(t, x), u)dt$$

over all u in $\mathcal{A}_0$, where the integral is taken along the curve E correspondent to u, and (T, x_T) is the point at which E intersects $\mathfrak{J}$.

In [2] it is shown how Problem B can be transformed into a Problem of Bolza in the calculus of variations, and how the necessary conditions for this Bolza problem can be translated back into necessary conditions that hold along the path corresponding to the optimal control u^*. We have seen that $z(t)$ furnishes a minimum for Problem B in the class $\mathcal{P}_0$. If $z^*(t)$ were also to furnish a minimum in the class $\mathcal{A}_0$, then the necessary conditions for optimal control would hold along the optimal path E^* resulting from (Y^*, Z^*) and starting at (t_0, x_0). All that is known, however, about E^* is that it furnishes a minimum for Problem A. The assumption that z^* furnishes a minimum for Problem B would impose another restriction on the paths obtained from saddle-points (Y^*, Z^*), and thus would be undesirable. Fortunately, it turns out that the class $\mathcal{P}_0$ is large enough for the necessary conditions associated with the solution of Problem B to hold along E^*. This will be shown in Sec. 3.3 below. In the next section we shall state the necessary conditions; they are transcriptions for Problem B of Theorem 2 of [2] and the discussion in Sec. 9 of [2].

3.2 The Necessary Conditions

THEOREM 1. Let

$$H(t,x,z,\lambda_0,\lambda) = \lambda_0 f(t,x,Y^*(t,x,),z) + \lambda G(t,x,Y^*(t,x),z). \tag{3.1}$$

Let $z^*(t) = z^*(t;\ t_0,\ x_0)$ be the function (2.11) associated with the path E^* resulting from $(Y^*,\ Z^*)$ and starting at $(t_0,\ x_0)$. Then there exists a constant $\lambda_0 \geq 0$, an n-dimensional vector $\lambda(t)$, and a p-dimensional vector $\mu(t)$ with the following properties. The vector $\lambda(t)$ is defined and continuous on $[t_0,\ T]$, except perhaps at values of t for which E^* intersects a manifold of discontinuity of Y^*. At such points, however, $\lambda(t)$ possesses unique right- and left-hand limits. The vector $\mu(t)$ is defined and continuous in $[t_0,\ T]$, except perhaps at values of t corresponding to corners of E^*. At such points, however, $\mu(t)$ possesses unique right- and left-hand limits. The vector $(\lambda_0,\ \lambda(t))$ never vanishes, and $\mu(t) \leq 0$ in $[t_0,\ T]$. Along E^* the following conditions are fulfilled:

Condition I.

$$x'(t) = H_\lambda \tag{3.2}$$

$$\lambda'(t) = -(H_x + \mu K_x) \tag{3.3}$$

$$H_z + \mu K_z = 0 \tag{3.4}$$

$$\mu^i K^i = 0 \qquad i = 1, \ldots, p\ . \tag{3.5}$$

At the end point $(T,\ x_T^*)$ of E^*,

$$\lambda_0(g_T t_{i\sigma} + g_x x_{i\sigma}) + H t_{i\sigma} - \lambda x_{i\sigma} = 0 \tag{3.6}$$

where $t_i(\sigma)$ and $x_i(\sigma)$ are as in (2.4). Along E^*, H is continuous, except at points of intersection with manifolds $\mathfrak{M}$ of discontinuity of Y^*, where the following holds

$$(H^+ - H^-)dt - (\lambda^+ - \lambda^-)dx = 0 \tag{3.7}$$

for all differentials dt, dx, along $\mathfrak{M}$.

Condition II. Between corners of E^*,

$$H(t, x, z, \lambda_0, \lambda) \geq H(t, x, z^*, \lambda_0, \lambda) \tag{3.8}$$

for every z such that $z = Z(t, x)$ for some Z in $\mathcal{Z}_I$. At corners (3.8) holds for the appropriate one-sided limits.

Condition III. At each point of E^*, let $\hat{K}$ denote the vector formed from K by taking those components of K that vanish at that point. Let $e = (e^1, \ldots, e^s)$ be a nonzero solution vector of the linear system $\hat{K}_z e = 0$ at this point. Then $e((H + \mu K)_{zz})e \geq 0$ at this point.

Since the terminal manifold $\mathfrak{I}_i$, which forms part of the boundary of $\mathfrak{R}_i$, is n-dimensional and since we assumed in (2.16)-(ii) that E^* is not tangent to $\mathfrak{I}_i$, it follows by the Corollary of Theorem 3 of [2] that E^* is normal. Hence, we may take $\lambda_0 = 1$ in Theorem 1, and with this choice of λ_0 the multipliers (λ, μ) are unique.

Let $\mathcal{P}_0'$ denote the class of associated functions y(t) of (2.11) that result from pairs (Y, Z^*), where Y is in $\mathcal{Y}_1$. Two maximization problems analogous to Problems A and B can now be posed and a theorem similar to Theorem 1 can be obtained for the maximum problem. Here we again suppose for the moment that the class $\mathcal{P}_0'$ is sufficiently large. The theorem for the maximum problem will assert the existence of a vector $(\bar{\lambda}_0, \bar{\lambda}, \bar{\mu})$ with properties similar to $(\lambda_0, \lambda, \mu)$, except that now $\bar{\mu} \geq 0$ and the inequalities are reversed in Conditions II and III.

3.3 Adequacy of $\mathcal{P}_0$ and $\mathcal{P}_0'$

In this section we shall sketch the arguments showing that $\mathcal{P}_0$ is sufficiently large for Theorem 1 to hold along E^*. Analogous arguments apply to $\mathcal{P}_0'$. The proof of Theorem 1 for Problem B is effected by transforming Problem B into the following equivalent Bolza problem in (t, x, ξ, w)-space and then using the necessary conditions for the Bolza problem (see [2], [8]).

PROBLEM C. Minimize

$$g(T, x_T) + \int_{t_0}^{T} f(t, x, Y^*(t, x), w')dt$$

in the class of $C^{(2)}$ arcs that satisfy the differential equations

$$\begin{aligned} x' &= G(t, x, Y^*(t, x), w') \\ (\xi')^2 &= K(t, x, w') \end{aligned} \tag{3.9}$$

and end conditions

$$\begin{aligned} &x(t_0) = x_0 \qquad x(T) = x_1(\sigma), \qquad T = t_1(\sigma) \\ &\xi(t_0) = 0 \qquad \xi(T) \text{ arbitrary} \\ &w(t_0) = 0 \qquad w(T) \text{ arbitrary.} \end{aligned} \tag{3.10}$$

Here, $(\xi')^2 = ((\xi^{1'})^2, \ldots, (\xi^{p'})^2)$ and w' is merely the variable z relabeled.

Condition I of Theorem 1 is a transcription of the multiplier rule and some of its corollaries for Problem C. An examination of the proof of the multiplier rule as given in [3], for example, shows that to carry out the proof we need only the following information. The minimizing arc E^*, given by $(x^*(t), \xi^*(t), w^*(t))$ furnishes a minimum in a class of arcs E_δ, defined by $(x(t, \delta), \xi(t, \delta), w(t, \delta))$, in which E^* is embedded in a certain way. The functions defining the arcs E_δ are solutions of (3.9), with end conditions (3.10), the parameter δ can be taken to lie in an interval $[-\delta_0, \delta_0]$, and for $\delta = 0$ we get the arc E^* itself. For definiteness in the following discussion, let us suppose that (t_0, x_0) belongs to $\tilde{\mathcal{R}}_{11}^-$. The arcs E_δ can only have corners at points at which $x(t, \delta)$ intersects the manifold $\mathfrak{M}_{ij}$, $j = 1, \ldots, j_1 - 1$. Furthermore, the arcs of this family satisfy the following conditions:

$$\|x(t, \delta) - x^*(\bar{t})\| \leq \delta$$

$$(3.11) \qquad \|x_t(t, \delta) - x^{*\prime}(\bar{t})\| \leq \delta$$

$$\|w_t(t, \delta) - w_t(\bar{t})\| \leq \delta ,$$

uniformly in t, whenever $|t - \bar{t}|$ is sufficiently small, provided that $(t, x(t, \delta))$ and $(\bar{t}, x^*(\bar{t}))$ are in the same $\mathfrak{M}_{ij}$, $j = 1, \ldots, j_1$.

To show that $\mathcal{P}_0$ is sufficiently large for Condition I to hold, we need only show that every arc E_δ can be obtained from a function $Z(t, x, \delta)$ in $\mathcal{Z}_1$. From the preceding discussion it is clear that this will be done if we can establish the following Lemma:

LEMMA 1. For every $-\delta_0 \leq \delta \leq \delta_0$, there exists a function $Z(t, x, \delta)$ in $\mathcal{Z}_1$ such that $Z(t, x(t, \delta), \delta) = w_t(t, \delta)$.

Recall that we assume that (t_0, x_0) is in $\tilde{\mathcal{R}}_{11}^-$. If (t_0, x_0) is in $\mathcal{R}_{11}$ we let the symbol $\mathfrak{M}_{10}$ denote the manifold $t = t_0$; otherwise it has its usual meaning. We sketch the construction of $Z(t, x, \delta)$ in $\mathcal{R}_{11}$; the definition in $\mathcal{R}_{1j}$, $j \geq 1$, is similar. For (t, x) not in $\mathcal{R}_1$ we take $Z(t, x, \delta) = Z^*(t, x)$.

Let

$$(3.12) \qquad z^*(t, \delta) = Z^*(t, x(t, \delta)).$$

Then, since $z^*(\bar{t}) = w_t^*(\bar{t})$,

$$\|w_t(t, \delta) - z^*(t, \delta)\| \leq \|w_t(t, \delta) - w_t^*(\bar{t})\| + \|z^*(\bar{t}) - z^*(t, \delta)\| .$$

It follows from the last inequality of (3.11), from (2.15), and from the uniform continuity of Z^* on $\bar{\mathcal{R}}_{11}$ that for arbitrary $\varepsilon > 0$ and $\delta = \delta(\varepsilon)$ sufficiently small, we have

$$\|w_t(t, \delta) - z^*(t, \delta)\| < \varepsilon, \tag{3.13}$$

uniformly in t for $t_{10}(\delta) \leq t \leq t_{11}(\delta)$, where $t_{1j}(\delta)$ denotes the value of t at which $x(t, \delta)$ intersects $\mathfrak{M}_{1j}$, $j = 0, 1$. We henceforth denote the curve defined by $x(t, \delta)$ in $\mathcal{R}_{11}$ by E_δ.

Since E^* is not tangent to $\mathfrak{M}_{10}$ or $\mathfrak{M}_{11}$ it follows that, for δ sufficiently small, the arcs E_δ also will not be tangent to these manifolds. Since the arc length along E_δ is given by

$$\int_{t_0}^{t} [\|x_t(t, \delta)\|^2 + 1]dt,$$

and $x(t, \delta)$ is $C^{(2)}$ on $t_{10}(\delta) \leq t \leq t_{11}(\delta)$, it follows that E_δ has radius of curvature bounded away from zero in this interval. Hence there exists an arbitrarily small neighborhood $\mathfrak{U}$ of E_δ that lies in $\mathcal{R}_{11}$ and a homeomorphism Φ of $\bar{\mathfrak{U}}$ onto a tube

$$t_0 \leq t \leq t_1 \qquad \|\rho\| \leq a$$

in $(n + 1)$-dimensional Euclidean (t, ρ)-space such that Φ is $C^{(2)}$ in $\bar{\mathfrak{U}}$, Φ maps E_δ onto the axis of the tube (i.e., onto $\rho = 0$), $\Phi(\mathfrak{M}_{10} \cap \bar{\mathfrak{U}}) = t_0$, and $\Phi(\mathfrak{M}_{11} \cap \bar{\mathfrak{U}}) = t_1$. It is now easy to show that there exists a real valued function φ that is $C^{(2)}$ in $\mathcal{R}_{11}$, that satisfies

$$0 \leq \varphi \leq 1, \tag{3.14}$$

and that takes on the value 1 in E_δ and the value 0 in $\tilde{\mathcal{R}}_{11} - \mathfrak{U}$.

We now define a function $\zeta^{(0)}(t, x, \delta)$ on $\tilde{\mathcal{R}}_{11}$ as follows:

$$\zeta^{(0)}(t, x, \delta) = \varphi w_t(t, \delta) + (1 - \varphi)Z^*(t, x). \tag{3.15}$$

The function $\zeta^{(0)}$ clearly is $C^{(2)}$ on $\tilde{\mathcal{R}}_{11}$ and satisfies the following conditions:

$$\zeta^{(0)}(t, x(t, \delta), \delta) = w_t(t, \delta), \tag{3.16}$$

$$\zeta^{(0)}(t, x, \delta) = Z^*(t, x), \quad \text{for} \quad (t, x) \in \widetilde{\mathfrak{R}}_{11} - \mathfrak{U}. \tag{3.17}$$

It follows from (3.14), (3.15), (3.13), (3.12), and the uniform continuity of Z^*_{11} on $\bar{\mathfrak{R}}_{11}$, that for arbitrary $\epsilon > 0$ and (t, x) in $\mathfrak{U}$, we have

$$\|\zeta^{(0)}(t, x, \delta) - Z^*(t, x)\| < \epsilon \tag{3.18}$$

whenever δ and $\mathfrak{U}$ are sufficiently small. Finally, from (3.16), (3.17), and $K(t, x(t, \delta), w_t(t, \delta)) \geq 0$ we obtain

$$K(t, x, \zeta^{(0)}(t, x, \delta)) \geq 0 \quad \text{for} \quad (t, x) \in E_\delta, \ (t, x) \in \widetilde{\mathfrak{R}}_{11} - \mathfrak{U}. \tag{3.19}$$

We note, however, that $\zeta^{(0)}$ need not satisfy the condition

$$K(t, x, \zeta^{(0)}(t, x, \delta)) \geq 0, \qquad (t, x) \in \mathfrak{U}. \tag{3.20}$$

We now sketch the construction of a function that has the requisite differentiability and continuity properties and that satisfies conditions (3.16) to (3.20).

Let $\mathcal{K}_j$, $j = 1, \ldots, p$, denote the set of points (t, x) such that

$$\begin{aligned} &K^i(t, x, \zeta^{(0)}(t, x)) \geq 0 \quad \text{if} \quad i \neq j, \\ &K^j(t, x, \zeta^{(0)}(t, x)) < 0. \end{aligned} \tag{3.21}$$

The sets $\mathcal{K}_j$ clearly are disjoint. From (3.20), it follows that if $\mathcal{K}_j$ is not void then $\mathcal{K}_j \subset \mathfrak{U}$. From continuity considerations, it follows that if $\mathcal{K}_j$ is not void it is open.

Let the set ω defined in Sec. 2.2 consist of the single element j. For typographical convenience we shall let subscripts denote components in this and the next paragraph. By assumption (2.3), the manifold $\mathcal{S}^{(z)}_j$ can be coordinatized by a finite number of overlapping coordinate systems, each of the form

$$
\begin{aligned}
z_1 &= z_1, \\
&\cdots \\
z_{k-1} &= z_{k-1}, \\
z_k &= \Omega(t, x, z_1, \ldots, z_{k-1}, z_{k+1}, \ldots, z_p), \qquad (3.22) \\
z_{k+1} &= z_{k+1}, \\
&\cdots \\
z_p &= z_p.
\end{aligned}
$$

Define a function $\zeta^{(1)}$ of $\widetilde{\mathcal{R}}_{11}$ in the following manner. For (t, x) not in any $\mathcal{K}_j$, let $\zeta^{(1)}(t, x) = \zeta^{(0)}(t, x)$. On each nonvoid $\mathcal{K}_j$, $j = 1, \ldots, p$, let $\zeta^{(1)}$ be given by a finite number of overlapping representations, each representation of $\zeta^{(1)}$ corresponding to a coordinate system (3.22) on $\mathcal{S}_j^{(z)}$ as follows:

$$
\begin{aligned}
\zeta_1^{(1)}(t, x) &= \zeta_1^{(0)}(t, x), \\
&\cdots \\
\zeta_{k-1}^{(1)}(t, x) &= \zeta_{k-1}^{(0)}(t, x), \\
\zeta_k^{(1)}(t, x) &= \Omega_k(t, x, \zeta_1^{(0)}, \ldots, \zeta_{k-1}^{(0)}, \zeta_{k+1}^{(0)}, \ldots, \zeta_p^{(0)}), \\
\zeta_{k+1}^{(1)}(t, x) &= \zeta_{k+1}^{(0)}(t, x), \\
&\cdots \\
\zeta_p^{(1)}(t, x) &= \zeta_p^{(0)}(t, x).
\end{aligned}
$$

The region of validity in $\mathcal{K}_j$ of each representation is that of the corresponding coordinatization (3.22) of $\mathcal{S}_j^{(z)}$.

It follows from the definition of $\zeta^{(1)}$ that $\zeta^{(1)}$ is piecewise $C^{(2)}$ in $\widetilde{\mathcal{R}}_{11}$, and that the various derivatives are uniformly bounded. Discontinuities of $\zeta^{(1)}$ occur at those boundary points of sets $\mathcal{K}_j$ at which $K^j(t, x, \zeta^{(0)}(t, x)) < 0$; discontinuities in the derivatives of $\zeta^{(1)}$ can occur only at boundary points of $\mathcal{K}_j$. All relations involving $\zeta^{(0)}$, except the second inequality in (3.21) hold with $\zeta^{(0)}$ replaced by $\zeta^{(1)}$.

The validity of (3.18) with $\zeta^{(0)}$ replaced by $\zeta^{(1)}$ follows from the properties of $\mathcal{S}_j^{(z)}$ and from the ability to make the distance between any point of $\mathcal{K}_j$ and that portion of the boundary of $\mathcal{K}_j$ in which $K^j(t, x, \zeta^{(0)}(t, x)) = 0$ arbitrarily small, uniformly for all components of $\mathcal{K}_j$, by choosing $\mathcal{U}$ sufficiently small to start with. If (t, x) is in $\mathcal{K}_j$, then $K(t, x, \zeta^{(1)}) \geq 0$ and $K^j(t, x, \zeta^{(0)}) = 0$. Hence there exist no points at which precisely one of the inequalities $K^j(t, x, \zeta^{(1)}(t, x)) < 0$ holds.

We next consider subsets $\mathcal{K}_{jk}$ of $\mathcal{R}_{11}$ in which precisely two components of K fail to satisfy the relationship $K^i(t, x, \zeta^{(1)}(t, x)) \geq 0$, $i = 1, \ldots, p$. We define a function $\zeta^{(2)}$, which is a modification of $\zeta^{(1)}$, by means of the coordinate solutions of (2.3), where corresponding to each $\mathcal{K}_{j,k}$, ω is now the set (j, k). The function $\zeta^{(2)}$ has continuity and differentiability properties similar to $\zeta^{(1)}$, and all the pertinent relationships involving $\zeta^{(1)}$ hold with $\zeta^{(1)}$ replaced by $\zeta^{(2)}$. There will now exist no points at which precisely one or precisely two inequalities $K^j(t, x, \zeta^{(2)}(t, x)) < 0$ hold.

Proceeding in this fashion, we obtain a sequence of functions $\zeta^{(0)}, \ldots, \zeta^{(p)}$. The function $\zeta^{(p)}$ is continuous on $\bar{\mathcal{R}}_{11}$, satisfies (3.19) and (3.20), and is piecewise $C^{(2)}$ on $\bar{\mathcal{R}}_{11}$. Furthermore, conditions (3.16), (3.17), and (3.18) are satisfied with $\zeta^{(0)}$ replaced by $\zeta^{(p)}$. The derivatives of $\zeta^{(p)}$ are bounded on $\mathcal{R}_{11}$. It can be shown as a consequence of (2.2)-(ii) that the function $\zeta^{(p)}$ can be modified in neighborhoods of the points at which $\zeta^{(p)}$ has discontinuous derivatives so as to obtain a function $\zeta_{11}(t, x, \delta)$ that is $C^{(2)}$ on $\bar{\mathcal{R}}_{11}$ and satisfied conditions (3.17) to (3.20).

We take

$$Z(t, x, \delta) = \zeta_{11}(t, x, \delta) \quad \text{on } \mathcal{R}_{11}.$$

From the preceding discussion it follows that, for (t, x) in $\mathcal{R}_1$,

$$\begin{aligned} &Z(t, x(t, \delta), \delta) = z^*(t, \delta), \\ (3.23) \qquad &\|Z(t, x, \delta) - Z^*(t, x)\| < \varepsilon, \\ &K(t, x, Z(t, x, \delta)) \geq 0; \end{aligned}$$

the second line holds for δ sufficiently small.

To complete the proof we must show that $(Y, Z(t, x, \delta))$ is a playable pair for all Y in $\mathcal{Y}_1$. Since $Z(t, x, \delta)$ and Y are piecewise $C^{(2)}$, this pair gives rise to a decomposition of $\mathcal{R}$ which, by the construction of $Z(t, x, \delta)$, is the same as that associated with the pair (Y, Z^*). Hence this decomposition is regular. Requirements (ii) and (iii) in the definition of playable pair hold as consequences of standard existence theorems concerning differential equations, the properties of G, Y, $Z(t, x, \delta)$, (2.16)-(i), and the second inequality in (3.23).

Condition II of Lemma 1 follows from the Weierstrass condition for Problem C; Condition III and the condition $\mu \leq 0$ follow from the Clebsch condition for Problem C (see [2]). To establish the Weierstrass condition, and consequently the Clebsch condition for the Bolza problem it is required only that E^* furnish a minimum in a certain subclass of piecewise $C^{(2)}$ arcs satisfying (3.9) and (3.10) (see [3]). By methods similar to those used above, it can be shown that every arc in this subclass can be obtained as a path in $\mathcal{P}$. Hence Conditions II and III also hold along E^*.

§ 4. NECESSARY CONDITIONS

4.1 Introduction

In this chapter we shall give necessary conditions that must hold along an optimal path E^* resulting from (Y^*, Z^*). Our discussion will be simplified by introducing a function $F(t, x, y, z, \lambda)$, defined as follows:

$$F(t, x, y, z, \lambda) = f(t, x, y, z) + \lambda G(t, x, y, z), \tag{4.1}$$

where λ is an n-dimensional vector. When we speak of considering F, or any other function, along an optimal path E^* resulting from (Y^*, Z^*), we mean that the arguments (t, x, y, z, λ) of the functions are to be evaluated at $(t, x^*(t), y^*(t), z^*(t), \lambda(t))$, where $y^*(t)$ and $z^*(t)$ are defined by (2.15).

4.2 The First Necessary Condition

THEOREM 2. Let E^* be an optimal path resulting from

(Y^*, Z^*), starting at (t_0, x_0) in $\mathcal{R}_i$, and terminating at $(T, x_T) = (T, x^*(T))$. Then there exist unique functions $\lambda(t)$, $\mu(t)$, $\bar{\mu}(t)$, that depend on E^*, that have values in n space, p space and $\bar{p}$ space, respectively, and that possess the following properties: (i) The functions λ, μ, and $\bar{\mu}$ are continuous on the interval $t_0 \leq t \leq T$, except perhaps at values of $t = t_{ij}$, $j = 1, \ldots, j_i$, corresponding to points at which E^* intersects the manifolds $\mathfrak{M}_{ij}$; at these values of t, the vectors λ, μ, and $\bar{\mu}$ have unique right- and left-hand limits. (ii) Along E^* the following hold.

(4.2) $$x' = F_\lambda$$

(4.3) $$\lambda' = -(F_x + F_y Y_x^* + F_z Z_x^*)$$

(4.4) $$F_z + \mu K_z = 0 \qquad F_y + \bar{\mu} \bar{K}_y = 0$$

(4.5) $$\mu^i K^i = 0 \qquad \bar{\mu}^j \bar{K}^j = 0 \qquad i = 1, \ldots, p$$

(4.6) $$\mu \leq 0 \qquad \bar{\mu} \geq 0$$

At points at which E^* intersects the manifolds $\mathfrak{M}_{ij}$, $j = 1, \ldots, j_i - 1$, the relations (4.2)-(4.6) hold for the one-sided limits. At the terminal point $(T, x^*(T))$ of E^*,

(4.7) $$(F + g_t)dt_i - (\lambda - g_x)dx_i = 0$$

for all differentials dt_i, dx_i along $\mathcal{J}_i$. If $\mathfrak{M}_{ij}$, $j = 1, \ldots, j_i - 1$, is a manifold of discontinuity of at most one of the functions Y^*, Z^*, then λ and F are continuous along E^* at $t = t_{ij}$. If $\mathfrak{M}_{ij}$, $j = 1, \ldots, j_i - 1$, is a manifold of discontinuity of both Y^* and Z^* then at $t = t_{ij}$

(4.8) $$(F^+ - F^-)dt_{ij} - (\lambda^+ - \lambda^-)dx_{ij} = 0,$$

where the one-sided limits are taken along E^*.

To establish the theorem we first consider Y^* to be fixed and look upon Z^* as furnishing a minimum against Y^* in the class $\mathcal{Z}_1$. Theorem 1 is therefore applicable to this minimization problem. Moreover, we may take $\lambda_0 = 1$ and get unique (λ, μ). We now transcribe the statement of Theorem 1 as follows: From (3.1) and (4.1) we get

$$(4.9) \qquad H_x = F_x + F_y Y_x^* \qquad H_\lambda = F_\lambda \qquad H_z = F_z$$

Let (t^*, x^*) be a point of E^*, lying in some $\mathcal{R}_{ij}$, such that $K^i(t^*, x^*, Z^*) = 0$ for some component K^i, $i = 1, \ldots, p$, of K. Since $K^i(t, x, Z^*(t, x)) \geq 0$ in $\mathcal{R}_{ij}$, we have that K^i has a minimum at the interior point (t^*, x^*) of $\mathcal{R}_{ij}$. Hence

$$\frac{\partial K^i}{\partial x} = K_x^i + K_z^i Z_x^* = 0.$$

From this relation it follows that along E^*, $\mu(K_x + K_z Z_x^*) = 0$. From (3.4) and the relation $H_z = F_z$, it follows that $F_z Z_x^* = -\mu K_z Z_x^*$; hence

$$(4.10) \qquad \mu K_x = F_z Z_x^*.$$

With the help of (4.9) and (4.10) we see that all of the statements of Theorem 2 follow from Theorem 1, with the exception of the relations involving μ in (4.4)-(4.6) and the statements concerning the continuity of λ and F.

Let us now consider Z^* to be fixed and let us look upon Y^* as furnishing a maximum against Z^* in the class $\mathcal{Y}_1$. We then obtain from the version of Theorem 1 for the maximum problem that the statements of Theorem 2 hold with λ replaced by a vector $\bar{\lambda}$, with the exception of the relations involving μ in (4.4)-(4.6) and the statements concerning the continuity of λ and F.

We now show that $\bar{\lambda} = \lambda$. Let M denote the $n \times n$ matrix with typical (j, k)-th element

$$(4.11) \qquad G^j \frac{\partial t_i}{\partial \sigma^k} - \frac{\partial x_i^j}{\partial \sigma^k}, \qquad \begin{array}{l} j = 1, \ldots, n \\ k = 1, \ldots, n \end{array}$$

where the elements are evaluated at the terminal point of E^*. From (4.7) we get that both $\lambda(T)$ and $\overline{\lambda}(T)$ must satisfy the following system of linear equations

$$(4.12) \qquad \lambda M = -(f + g_T)t_{i\sigma} - g_x x_{i\sigma}.$$

Since, by assumption (2.16)-(ii), E^* is not tangent to $\mathfrak{J}_i$, the matrix

$$\begin{pmatrix} 1 & t_{i\sigma} \\ G & x_{i\sigma} \end{pmatrix}$$

has rank $n + 1$. If for each $k = 1, \ldots, n$, we multiply the first column of this matrix by $-\partial t_i/\partial\sigma^k$ and add the result to the k-th column, we get the matrix

$$\begin{pmatrix} 1 & 0 \\ G & -M \end{pmatrix}.$$

Hence M has rank n, and the system (4.12) has a unique solution. Hence $\lambda(T) = \overline{\lambda}(T)$. Furthermore, both λ and $\overline{\lambda}$ are solutions of (4.3), where the arguments on the right-hand side are evaluated along E^*. Hence $\lambda(t) = \overline{\lambda}(t)$ on the interval $t_{i,j_i-1} < t \le T$, and $\lambda^+ = \overline{\lambda}^+$ at $t = t_{i,j_i-1}$.

Since $\lambda^+ = \overline{\lambda}^+$ at $t = t_{i,j_i-1}$, it follows from (4.8) and the non tangency assumption (2.16)-(i), by an argument similar to that used to show the equality of $\lambda(T)$ and $\overline{\lambda}(T)$, that $\lambda^- = \overline{\lambda}^-$ at $t = t_{i,j_i-1}$. Hence, since λ and $\overline{\lambda}$ both satisfy (4.3) along E^* on the interval $t_{i,j_i-2} \le t \le t_{i,j_i-1}$, it follows that $\lambda = \overline{\lambda}$ in this interval. We may proceed inductively in this manner, working our way backwards across the manifolds $\mathfrak{M}_{ij}$, and show that λ and $\overline{\lambda}$ are identical for $t_0 \le t \le T$.

To complete the proof we note that if $\mathfrak{M}_{ij}$ is a manifold of discontinuity of at most one of the functions Y^* and Z^* — say, for definiteness, Y^* — then by Theorem 1 λ is continuous at t_{i,j_i-1}. Hence $\bar{\lambda}$ is also continuous and so is F.

4.3 Special Cases

In the important special case of separable constraints, defined in Sec. 2.2, the statements (4.4) to (4.6) can be combined into a useful and easily applicable criterion.

COROLLARY 1. Let the constraints be separable. If $i(i = 1, \ldots, s)$ belongs to S_1 and S_2, then at every point (t, x) interior to some $\mathcal{R}_{jk}$,

$$(4.13)\qquad F_{z^i} \begin{cases} \geq 0 & \text{if } z^{*i}(t, x) = B^i(t, x), \\ = 0 & \text{if } B^i(t, x) < z^{*i}(t, x) < A^i(t, x), \\ \leq 0 & \text{if } z^{*i}(t, x) = A^i(t, x). \end{cases}$$

If i does not belong to $S_1(S_2)$, then (4.13) holds with statements involving $A^i(B^i)$ deleted. If i is in neither S_1 nor S_2, then $F_{z^i} = 0$. Similarly, if $i(i = 1, \ldots, \bar{s})$ belongs to $\bar{S}_1$ and $\bar{S}_2$, then at every point (t, x) interior to some $\mathcal{R}_{jk}$,

$$(4.14)\qquad F_{y^i} \begin{cases} \geq 0 & \text{if } Y^{*i}(t, x) = \bar{A}^i(t, x), \\ = 0 & \text{if } \bar{B}^i(t, x) < Y^{*i}(t, x) < \bar{A}^i(t, x), \\ \leq 0 & \text{if } Y^{*i}(t, x) = \bar{B}^i(t, x). \end{cases}$$

If i is not in both $\bar{S}_1$ and $\bar{S}_2$, then the appropriate modifications hold, as in the case of F_{z^i}.

If i belongs to S_1, let $j_1(i)$ denote the corresponding component of K; if i belongs to S_2, let $j_2(i)$ denote the corresponding component of K. Suppose that i belongs to S_1 and S_2. Then from (4.4) and the definition of separable constraints, we get

$$F_{z^i} = \sum_{j=1}^{p} \mu^j K^j_{z^i} = -\mu^{j_1(i)} + \mu^{j_2(i)}$$

The conclusion (4.13) now follows from (4.5), (4.6), and the observation that, since $A^i > B^i$, one and only one of the three conditions $Z^{*i} = A^i$, $Z^{*i} = B^i$, $B^i < Z^{*i} < A^i$ can hold at a point. The statement following (4.13) is a consequence of (4.5), (4.6), and the observation that if i belongs only to $S_1(S_2)$, then $\mu^{j_2(i)} = 0(\mu^{j_1(i)} = 0)$. Similar arguments yield (4.14).

Another special case in which a simplification of Theorem 1 can be effected is that in which the constraints are independent of the state variable x.

COROLLARY 2. If K is independent of x, i.e., $K = K(t, z)$, then at every (t, x) interior to some $\mathcal{R}_{jk}$,

$$F_z Z^*_x = 0. \tag{4.15}$$

If $\overline{K}$ is independent of x, i.e., $\overline{K} = \overline{K}(t, y)$, then at every (t, x) interior to some $\mathcal{R}_{jk}$,

$$F_y Y^*_x = 0. \tag{4.16}$$

If both K and $\overline{K}$ are independent of x, then along E^*,

$$\lambda'(t) = -F_x. \tag{4.17}$$

Equation (4.15) follows from (4.10) and the observation that since K is independent of x, we have $K_x = 0$. Equation (4.16) is established similarly. Equation (4.17) follows from (4.3), (4.15), and (4.17).

If the constraints are separable and the functions A^i, B^i, $\overline{A}^i$, $\overline{B}^i$ are all constants, or depend at most on t, then the conclusions of both corollaries apply.

4.4 Further Necessary Conditions

The next necessary condition is most easily stated in terms of a game that we now describe.

Let E^* be an optimal path and let $\lambda(t)$ be the function described in Theorem 2. At every point $Q = (t, x^*(t))$ of E^* with $t \neq t_{ij}$, we construct a game $\mathfrak{F}(t, x^*(t))$ in the following manner. The pure strategies for player I consist of (a) all vectors y such that $y = Y(t, x^*(t))$ for some Y in $\mathcal{Y}_1$ that is continuous at Q, and (b) all vectors y such that $y = Y^+(t, x^*(t))$ for some Y in $\mathcal{Y}_1$ that is discontinuous at Q. The pure strategies for player II consist of (a) all vectors z such that $z = Z(t, x^*(t))$ for some Z in $\mathcal{Z}_1$ that is continuous at Q, and (b) all vectors z such that $z = Z^+(t, x^*(t))$ for some Z in $\mathcal{Z}_1$ that is discontinuous at Q. The payoff is $F(t, x^*(t), y, z, \lambda(t))$, where F is defined in (4.1). If $t = t_{ij}$ at Q, we can construct $\mathfrak{F}$ as before, except that we take the payoff to be $F(t, x^*(t), y, z, \lambda^+(t))$.

THEOREM 3. At every point $(t, x^*(t))$, with $t \neq t_{ij}$, of an optimal path E^*, the game $\mathfrak{F}(t, x^*(t))$ has value $F(t, x^*(t), y^*(t), z^*(t), \lambda(t))$, where $y^*(t)$ and $z^*(t)$ are given by (2.15). An optimal pure strategy for player I is $y^*(t)$ and an optimal pure strategy for player II is $z^*(t)$. At $t = t_{ij}$, $\mathfrak{F}(t, x^*(t))$ has value $F(t, x^*(t), y^*(t)^+, z^*(t)^+, \lambda^+(t))$; an optimal pure strategy for player I is $y^*(t)^+$, and an optimal pure strategy for player II is $z^*(t)^+$.

This theorem follows from Condition II of Theorem 1 when that theorem is applied to the problems of maximizing against Z^* in the class $\mathcal{Y}_1$ and minimizing against Y^* in the class $\mathcal{Z}_1$.

The following theorem is a consequence of Condition III, applied to the preceding maximization and minimization problems.

THEOREM 4. Let E^* be an optimal path, and let $(t, x^*(t))$ be a point of E^* with $t \neq t_{ij}$. Let $\hat{K}$ denote the vector formed from K by taking those components of K that vanish at that point. Let $e = (e^1, \ldots, e^m)$ be a nonzero solution vector of the linear system $\hat{K}_z e = 0$. Then $e((F + \mu K)_{zz})e \geq 0$ at this point. Similarly, let $\hat{\bar{K}}$ denote the vector formed from $\bar{K}$ by taking these components of $\bar{K}$ that vanish at this point. Let $\bar{e}$ be an $\bar{s}$-dimensional nonzero solution vector of the system $\hat{\bar{K}}\,\bar{e} = 0$. Then $\bar{e}(F + \bar{\mu}\,\bar{K})_{yy})\bar{e} \leq 0$ at this point.

§ 5. THE VALUE $W(t, x)$

5.1 The Function $\Lambda(t, x)$

The principal purpose of this section is to study the differentiability properties of the value W and to deduce a partial differential equation that W satisfies in its regions of differentiability. For this purpose, it is convenient to introduce a function $\Lambda(t, x)$, related to the multiplier λ, and to derive some of its properties.

Consider the regular decomposition of $\mathcal{R}$ associated with (Y^*, Z^*) and consider a subregion $\mathcal{R}_i$, where $\mathcal{R}_i$ is as in (i) of the definition (see Sec. 2.4) of regular decomposition. In $\mathcal{R}_i$, we now consider $\mathcal{R}_{ij_i}$. To simplify notation, we set $j_i = m$.

Through each point of $\tilde{\mathcal{R}}^-_{im}$ there passes a unique optimal path terminating at $\mathfrak{T}_i$. Conversely, each point $(t_i(\sigma), x_i(\sigma))$ of $\mathfrak{T}_i$ is the terminal point of precisely one optimal path. Thus an optimal path in $\tilde{\mathcal{R}}_{im}$ terminating at $(T, x_T) = (t_i(\sigma), x_i(\sigma))$ can be written as

$$x^*(t, \sigma) = x^*(t, t_i(\sigma), x_i(\sigma)). \tag{5.1}$$

Since $x^*(t, \sigma)$ is the solution of (2.14) with $x(t_i(\sigma)) = x_i(\sigma)$, it follows that $x^*(t, \sigma)$ is of class $C^{(2)}$ on the region $\mathfrak{U}_{im}$ of (t, σ)-space defined by the requirements $\sigma \in \mathcal{K}_i$, and $t_{i,m-1}(\sigma) < t < t_i(\sigma)$. Moreover, $x^*(t, \sigma)$ is $C^{(2)}$ in $\bar{\mathfrak{U}}_{im}$. The argument to establish these statements uses (i) the continuity properties of G, Y^*, Z^*, (ii) extensions of Y^*_{im} and Z^*_{im} to functions that are of class $C^{(2)}$ in a region containing $\mathcal{R}_{im}$, $\mathfrak{T}_i$, and $\mathfrak{M}_{i,m-1}$ in its interior, (iii) the assumption that $t_i(\sigma)$ and $x_i(\sigma)$ are of class $C^{(2)}$, and (iv) standard theorems concerning the dependence of solutions of differential equations on initial conditions.

We now show that for (t, σ) in $\tilde{\mathfrak{U}}_{im}$, where $\tilde{\mathfrak{U}}_{im}$ consists of all (t, σ) such that σ in $\mathcal{K}_i$ and $t_{i,m-1}(\sigma) \le t \le t_i(\sigma)$,

$$\det\, x^*_\sigma(t, \sigma) \neq 0. \tag{5.2}$$

Let the vector $x^*_{t_i}$ obtained by differentiation of the function on the right-hand side of (5.1) be considered as an $(n \times 1)$ matrix. Let $x^*_{t_i}$ and the matrix $x^*_{x_i}$ obtained from (5.1) be evaluated at $(t, t_i(\sigma), x_i(\sigma))$. Let $t_{i\sigma}$ and $x_{i\sigma}$ be obtained from (2.4), and let $t_{i\sigma}$ be considered as a $(1 \times n)$ matrix. Then, from (5.1) we have

$$x^*_\sigma = x^*_{t_i} t_{i\sigma} + x^*_{x_i} x_{i\sigma}. \tag{5.3}$$

Since $x^*(t, \sigma)$ is a solution of (2.14), it follows from well known theorems that $x^*_{t_i} = -x^*_{x_i} G$, where G is evaluated at $(t_i(\sigma), x_i(\sigma))$. Substituting this into (5.3), gives $x^*_\sigma = -x^*_{x_i} M$, where M is the matrix (4.11). In the discussion following (4.11) we showed that $\det M \neq 0$. From standard theorems on differential equations, it follows that $\det x^*_{x_i} \neq 0$. Hence (5.2) is established.

From the manner in which it was defined, it is clear that

$$t = t \qquad x = x^*(t, \sigma) \tag{5.4}$$

yields a one to one mapping of $\widetilde{\mathfrak{U}}_{im}$ onto $\widetilde{\mathcal{R}}_{im}$. Let the inverse mapping be given as follows:

$$t = t \qquad \sigma = \sigma_0(t, x). \tag{5.5}$$

It follows from (5.2) and the implicit function theorem that σ_0 is of class $C^{(2)}$ in $\mathcal{R}_{im}$. By continuity we further get that σ_0 is of class $C^{(2)}$ in $\bar{\mathcal{R}}_{im}$.

Through each point of $\mathcal{R}_{i,m-1}$ there also passes a unique optimal path terminating at $\mathfrak{J}_i$. It is clear that these paths can also be written in the form (5.1) with σ in $\mathcal{K}_i$. In view of (5.2) and the nontangency assumption (2.16)-(i), it follows that we can reparametrize the manifold $\mathfrak{M}_{i,m-1}$ by σ in $\mathcal{K}_i$ instead of σ in $\mathcal{K}_{i,m-1}$. By arguments similar to those used in the preceding paragraph, with (2.16)-(ii) playing the role of (2.16)-(i), with (4.8) playing the role of (4.7), and using continuity to determine the initial data on $\mathfrak{M}_{i,m-1}$ for solutions of (2.14), we obtain that (5.2), (5.4) and (5.5) hold on a region $\widetilde{\mathfrak{U}}_{i,m-1}$. Here, $\widetilde{\mathfrak{U}}_{i,m-1}$ consists of all (t, σ) such that σ in $\mathcal{K}_i$ and $t_{i,m-2}(\sigma) \leq t \leq t_{i,m-1}(\sigma)$. We may proceed backwards and establish (5.1), (5.2), (5.4), and (5.5) on each of a succession of regions $\tilde{\mathcal{U}}_{i,j}$ corresponding to each $\tilde{\mathcal{R}}_{ij}$, $j = 1, \ldots, m$.

There is associated with each optimal path in $\tilde{\mathcal{R}}_{im}$ a unique multiplier $\lambda(t)$ that is obtained as a solution of (4.3). Let $\lambda(t, \sigma)$ denote the multiplier associated with the path E_σ defined by $x^*(t, \sigma)$. Let $h = h(\sigma)$ denote the terminal value $\lambda(t_i(\sigma), \sigma)$ of $\lambda(t, \sigma)$. The terminal value must be such that (4.7) holds. This, we saw, is equivalent to saying that h is the unique solution of (4.12). From the form of (4.12) and from the fact that $\det M \neq 0$, it is clear that $h(\sigma)$ is a $C^{(1)}$ function on $\mathcal{K}_i$.

Since $\lambda(t, \sigma)$ is a solution of (4.3) in which we have substituted $x = x^*(t, \sigma)$ into the right-hand side and that satisfies $\lambda(t_i, \sigma) = h(\sigma)$, we may write $\lambda(t, \sigma) = \lambda(t, t_i(\sigma), h(\sigma))$. From standard theorems on the dependence of solutions of differential equations on initial data, it follows that $\lambda(t, t_i, h)$ is $C^{(1)}$ on an appropriate region of (t, t_i, h)-space. From the properties of $t_i(\sigma)$ and $h(\sigma)$ it then follows that $\lambda(t, \sigma)$ is $C^{(1)}$ on $\bar{\mathcal{U}}_{im}$.

Let (t, x) and (t, σ) be related by (5.4) and (5.5). Define

$$\Lambda_{im}(t, x) = \lambda(t, \sigma) = \lambda(t, \sigma_0(t, x)).$$

The function Λ_{im} assigns to each point in $\tilde{\mathcal{R}}_{im}$ the value of the multiplier associated with the unique optimal path through the point. Since $\lambda(t, \sigma)$ is $C^{(1)}$ in $\bar{\mathcal{U}}_{im}$ and $\sigma_0(t, x)$ is $C^{(2)}$ on $\bar{\mathcal{R}}_{im}$, it follows that Λ_{im} is $C^{(1)}$ in $\mathcal{R}_{im}$.

A similar argument can now be carried out on $\mathcal{R}_{i,m-1}$, with $\mathfrak{M}_{i,m-1}$ playing the role of $\mathfrak{J}_i$ and $\mathfrak{M}_{i,m-2}$ playing the role of $\mathfrak{M}_{i,m-1}$. The initial conditions, however, are now determined by (4.8). Proceeding in this fashion, we obtain functions Λ_{ij} of class $C^{(1)}$ in $\mathcal{R}_{ij}$ for $j = 1, \ldots, m$. The truth of the following Lemma is now apparent.

LEMMA 2. There exists a function $\Lambda(t, x)$ on $\mathcal{R}$ with the following properties: (i) The function Λ is piecewise $C^{(1)}$ on $\mathcal{R}$, where the associated decomposition of $\mathcal{R}$ is the regular decomposition associated with (Y^*, Z^*). (ii) At each point (t, x) of $\mathcal{R}_{ij}$,

$i = 1, \ldots, \alpha$ and $j = 1, \ldots, j_i$, $\Lambda(t, x) = \lambda(t)$, where λ is the unique multiplier associated with the unique optimal path through (t, x). At points of the manifolds $\mathfrak{J}_i$, $\mathfrak{M}_{ij}$, $\mathfrak{N}_{i_1, \ldots, i_k}$, the statement holds for the appropriate one-sided limits. (iii) The function Λ is continuous across manifolds $\mathfrak{M}_{ij}$ corresponding to a discontinuity of only one of the functions Y^*, Z^*.

5.2 The Value $W(t, x)$

We can now state and prove our theorem about the differentiability of the value $W(t, x)$ and the differential equation it satisfies.

THEOREM 5. The value $W(t, x)$ of the game is continuous on $\mathcal{R}$ and is piecewise $C^{(2)}$ on $\mathcal{R}$; the associated decomposition is the regular decomposition associated with (Y^*, Z^*). On each $\mathcal{R}_{ij}$, $i = 1, \ldots, \alpha$ and $j = 1, \ldots, j_i$, we have

$$(5.6) \qquad \begin{aligned} W_t(t, x) &= - f(t, x, Y^*(t, x), Z^*(t, x)) \\ &\quad - \Lambda(t, x)G(t, x, Y^*(t, x), Z^*(t, x)), \\ W_x(t, x) &= \Lambda(t, x). \end{aligned}$$

The functions W_t and W_x are continuous across a manifold $\mathfrak{M}_{ij}$ if it is a manifold of discontinuity of only one of the functions Y^*, Z^*.

If we substitute W_x for Λ in the first equation in (5.6), we obtain the analogue of the Hamilton-Jacobi equation for our problem.

The continuity of $W(t, x)$ is an immediate consequence of the definition of $W(t, x)$. To establish the rest of the theorem, we shall study the behavior of $W(t, x)$ along curves in $\mathcal{R}_{ij}$.

Let

$$(5.7) \qquad t = T_0(v) \qquad x = X_0(v), \qquad 0 \leq v \leq 1$$

be functions of class $C^{(2)}$ that define a curve C that does not intersect itself and that, with the possible exception of end points, lies entirely within some $\mathcal{R}_{ij}$. For definiteness we take $j = 1$. It follows, with the help of (5.2) and the implicit function theorem, that the system of equations

$$(5.8) \qquad t = T_0(v) \qquad X_0(v) = x^*(t, \sigma)$$

defines a function $\sigma(v)$ that is $C^{(2)}$ on $[0, 1]$. Hence, as we traverse C as v goes from 0 to 1, we obtain a family of curves $E^*(v)$ by means of the function

$$x = x^*(t, \sigma(v)), \qquad T_0(v) \leq t \leq t_i(v).$$

Each $E(v)$ is an optimal path starting at $(T_0(v), X_0(v))$.

We have seen that each of the manifolds $\mathfrak{M}_{ij}$, $j = 1, \ldots, m-1$ can be parametrized by σ in $\mathcal{K}_i$; that is by the same parameter used to define $\mathfrak{J}_i$. For convenience of notation we will often designate $\mathfrak{J}_i$ by $\mathfrak{M}_{im}$. Since $x^*(t_{ij}(\sigma), \sigma) = x_{ij}(\sigma)$, $j = 1, \ldots, m$, it follows that the intersections of the curves $E(v)$ with the manifolds $\mathfrak{M}_{ij}$, $j = 1, \ldots, m$ are given by

$$(5.9) \qquad \begin{aligned} t &= T_j(v) \equiv t_{ij}(\sigma(v)) \\ x &= X_j(v) = x^*(t_{ij}(\sigma(v)), \sigma(v)). \end{aligned}$$

The functions T_j and X_i are clearly $C^{(2)}$ on $[0, 1]$ and trace out curves C_j in the manifolds $\mathfrak{M}_{ij}$. It follows from (5.9) that for $j = 1, \ldots, m-1$, we can compute dX_j/dv in two ways,

$$(5.10) \qquad \frac{dX_j}{dv} = x_t^{*+} \frac{dT_j}{dv} + x_\sigma^{*+} \frac{d\sigma}{dv} = x_t^{*-} \frac{dT_j}{dv} + x_\sigma^{*-} \frac{d\sigma}{dv},$$

where the superscript $+$ indicates that we are taking limits from the interior of $\mathcal{R}_{i,j+1}$, and the superscript $-$ indicates that we are taking limits from the interior of $\mathcal{R}_{ij}$. Equation (5.10) holds for $j = 0$, and

$j = m$ without the superscripts $+$ and $-$.

The associated functions (2.10) of the optimal paths (5.8) are given by the equations

$$y^*(t, v) = Y^*(t, x^*(t, \sigma(v))) \qquad Z^*(t, v) = Z^*(t, x^*(t, \sigma(v))).$$

The multipliers λ are related to Λ by the equation

$$\lambda(t, v) = \Lambda(t, x^*(t, \sigma(v))). \tag{5.11}$$

The functions $y^*(t, v)$ and $z^*(t, v)$ are $C^{(2)}$ on each region of (t, v) space defined by $0 \leq v \leq 1$, $T_j(v) \leq t \leq T_{j+1}(v)$, $j = 0, \ldots, m-1$; the function $\lambda(t, v)$ is $C^{(1)}$ on each of these regions.

We now consider the value $W(t, x)$ of the game as the initial point (t, x) of play moves along the curve C. We may write

$$W(T_o(v), X_0(v)) = g(T_m(v), X_m(v)) + \int_{T_0(v)}^{T_m(v)} f(t, x^*(t, \sigma(v)), y^*(t, v), z^*(t, v))dt.$$

Let

$$\delta_{ij} = f(t, x, Y^*(t, x)^-, Z^*(t, x)^-) - f(t, x, Y^*(t, x)^+, Z^*(t, x)^+),$$

where $t = T_j(v)$, $x = X_j(v)$, $j = 1, \ldots, m-1$. Then,

$$\frac{dW}{dv} = \left[g_t \frac{dT_m}{dv} + g_x \frac{dX_m}{dv} + f \frac{dT_m}{dv}\right]_{(T_m(v),\, X_m(v))} - \left[f \frac{dT_0}{dv}\right]_{(T_0(v),\, X_0(v))} + \sum_{j=1}^{m-1} \delta_{ij} \frac{dT_j}{dv} + \int_{T_0(v)}^{T_m(v)} \left(\frac{\partial f}{\partial x} x_\sigma^* \frac{d\sigma}{dv}\right)dt \tag{5.12}$$

where $\partial f/\partial x = f_x + f_y Y_x^* + f_z Z_x^*$, and where the integral is taken along $E(v)$. Since $E(v)$ is an optimal path, it follows from (4.3) and (4.1) that along $E(v)$

$$\frac{\partial f}{\partial x} = - \lambda_t - \lambda \frac{\partial G}{\partial x} ,$$

where $\partial G/\partial x = G_x + G_y Y_x^* + G_z Z_x^*$. From the equation

$$x_t^* = G(t, x^*, Y^*(t, x), Z^*(t, x^*)) \tag{5.13}$$

it follows that

$$x_{t\sigma}^* = \frac{\partial G}{\partial x} x_\sigma^* .$$

Moreover, on each $\mathcal{R}_{ij}$, $x_{t\sigma}^* = x_{\sigma t}^*$. Hence the integral in (5.12) can be written as follows:

$$- \frac{d\sigma}{dv} \int_{T_0(v)}^{T(v)} (\lambda x_\sigma^*)_t \, dt = - \frac{d\sigma}{dv} [\lambda x_\sigma^*]_{(T_0(v),\, X_0(v))}^{(T_m(v),\, X_m(v))}$$

$$- \frac{d\sigma}{dv} \sum_{j=1}^{m-1} \left[(\lambda x_\sigma^*)^- - (\lambda x_\sigma^*)^+ \right] .$$

If we substitute this expression for the integral into (5.12) and use (5.1), (5.10) and (5.13) we get

$$\frac{dW}{dv} = \left[(F + g_t) \frac{dT_m}{dv} - (\lambda - g_x) \frac{dX_m}{dv} \right]_{(T_m(v),\, X_m(v))}$$

$$- \left[F \frac{dT_0}{dv} - \lambda \frac{dX_0}{dv} \right]_{(T_0(v),\, X_0(v))}$$

$$- \sum_{j=1}^{m-1} \left[(F^+ - F^-) \frac{dT_j}{dv} - (\lambda^+ - \lambda^-) \frac{dX_j}{dv} \right]_{(T_j(v),\, X_j(v))} .$$

From (4.7) of Theorem 2, we get that the first term in square brackets, and from (4.8) we get that each square bracket, in $\Sigma_{j=1}^{m-1}$ vanishes. Since the curve C is arbitrary, we get, using (5.11), that

$$dW = - F dT + \Lambda dX \tag{5.14}$$

for arbitrary differentials dT, dX. The theorem is now an immediate consequence of the properties of f, G, Λ, Y^*, and Z^*.

5.3 Another Necessary Condition

Consider the manifold $\mathfrak{N}_{i_1, \ldots, i_k}$. For each i in the set $i_1, \ldots, i_k$, we may define a function $Y^*_{(i)}(t, x)$, on $\mathfrak{N}_{i_1, \ldots, i_k}$ as follows:

$$Y^*_{(i)}(t, x) = \lim_{(t', x') \to (t, x)} Y^*(t', x')$$

where (t', x') is in $\mathfrak{R}_i$. We may define functions $Z^*_{(i)}(t, x)$ and $\Lambda_{(i)}(t, x)$ on $\mathfrak{N}_{i_1, \ldots, i_k}$ in a similar fashion. Note that in terms of our previous definitions, we have $Y^*_{(i)}(t, x) = Y^+_{i,0}(t, x)$, $Z^*_{(i)} = Z^+_{i,0}(t, x)$ and $\Lambda_{(i)}(t, x) = \Lambda^+_{i,0}(t, x)$. Since, by definition (see Sec. 2.6), the value $W(t, x)$ is defined, single-valued, and continuous in all of $\mathfrak{R} \cup \mathfrak{J}$, it is clear that the arguments used to establish (5.14) carry over to curves C given by equations of the form (5.7) and lying on $\mathfrak{N}_{i_1, \ldots, i_k}$. The arguments of F in this case are $Y^*_{(i)}$, $Z^*_{(i)}$, $\Lambda_{(i)}$; and Λ is replaced by $\Lambda_{(i)}$. Since W is single-valued, the left-hand side of (5.14) is independent of i, for all $i = i_1, \ldots, i_k$. Moreover, since C is an arbitrary curve on $\mathfrak{N}_{i_1, \ldots, i_k}$, we obtain the following additional necessary condition.

THEOREM 6. At points (t, x) of a manifold $\mathfrak{N}_{i_1, \ldots, i_k}$, the following relation holds for all differentials dt, dx on the manifold, and all i, j belonging to the set $i_1, \ldots, i_k$:

$$F(t, x, Y^*_{(i)}, Z^*_{(i)}, \Lambda_{(i)}) dt - \Lambda_{(i)} dx$$

$$= F(t, x, Y^*_{(j)}, Z^*_{(j)}, \Lambda_{(j)}) dt - \Lambda_{(j)} dx,$$

where $Y^*_{(i)}$, $Z^*_{(i)}$, $\Lambda_{(i)}$, $Y^*_{(j)}$, $Z^*_{(j)}$, $\Lambda_{(j)}$ are evaluated at (t, x).

5.4 An Invariant Integral

Let C be a continuous curve in $\mathfrak{R} \cup \mathfrak{J}$ that is made up of a finite number of subarcs, each of which is given parametrically by equations of the form (5.7). Moreover, let each subarc be one of the following types: an admissible curve in some $\mathfrak{R}_{ij}$; a curve lying on a manifold $\mathfrak{M}_{ij}$; a curve lying on a manifold $\mathfrak{N}_{i_1, \ldots, i_k}$; a curve lying on a terminal manifold $\mathfrak{J}_i$. Such a curve will be called an admissible curve. Note that an admissible curve may have corners.

LEMMA 3. For all admissible curves C, the integral

$$I^*(C) = \int_C F\, dT - \Lambda\, dX \tag{5.16}$$

is independent of C.

For an admissible curve lying entirely in some $\mathfrak{R}_{ij}$, the lemma follows from (4.14). We saw in the proof of Theorem 6 that (5.14) holds for an admissible curve C lying on a manifold $\mathfrak{N}_{i_1, \ldots i_k}$. Moreover, it is immaterial which $Y^*_{(i)}$, $Z^*_{(i)}$, $\Lambda_{(i)}$ $i = i_1, \ldots, i_k$ we use in the integrand. It is also clear from the proof of Theorem 5 that (5.14) holds for an admissible curve lying on a manifold $\mathfrak{M}_{ij}$. Finally, it follows from (4.7), that for curves C lying entirely on a manifold $\mathfrak{J}_i$ we have

$$- F\, dT + \Lambda\, dX = g_t\, dT + g_x\, dX.$$

But on $\mathfrak{J}_i$, $dW = g_t\, dT + g_x\, dX$. Hence (5.14) holds for admissible curves on $\mathfrak{J}_i$, and the truth of the Lemma is evident.

§ 6. SUFFICIENT CONDITIONS

6.1 Fields

In this section, we shall define the notion of a field. This is a modification of the notion used in the calculus of variations.

Let $\mathcal{S}$, $\mathcal{G}$, $\mathfrak{I}$, and $\mathcal{R}$ be as in Section 2. Let Y^* and Z^* be two functions that are piecewise $C^{(2)}$ in $\bar{\mathcal{R}}$ and such that the resulting decomposition of $\bar{\mathcal{R}}$ is a regular decomposition. Let Λ be a function that is piecewise $C^{(1)}$ in $\mathcal{R}$ and such that a manifold of discontinuity of Λ is also a manifold of discontinuity of Y^* or Z^*. Then the region $\mathcal{R}$ and the functions Y^*, Z^*, Λ are said to constitute a <u>field</u> whenever the following hold:

(6.1) (i) For (t, x) in $\bar{\mathcal{R}}$, $(t, x, Y^*(t, x), Z^*(t, x))$ is in $\mathcal{S}$, $\bar{K}(t, x, Y^*(t, x)) \geq 0$, and $K(t, x, Z^*(t, x)) > 0$.

(ii) At a point (t_{ij}, x_{ij}) of a manifold $\mathfrak{M}_{ij}$, the directions $(1, G(t_{ij}, x_{ij}, Y^{*+}, Z^{*+}))$ and $(1, G(t_{ij}, x_{ij}, Y^{*-}, Z^{*-}))$ are not tangent to $\mathfrak{M}_{ij}$. At a point (T, x_T) of $\mathfrak{I}_i$, the direction $(1, G(T, x_T, Y^*, Z^*))$ is not tangent to $\mathfrak{I}_i$. At a point (t, x) of $\mathfrak{N}_{i_1, \ldots, i_k}$, the directions $(1, G(t, x, Y^*_{(i)}, Z^*_{(i)})$ are not tangent to $\mathfrak{N}_{i_1, \ldots, i_k}$, for each index i in the set $i_1, \ldots, i_k$.

(iii) If (t_0, x_0) is a point of $\tilde{\mathcal{R}}^-_{ij}$, then the solution of (2.10), with $Y_{ij} = Y^*_{ij}$, $Z_{ij} = Z^*_{ij}$ satisfying the initial condition $x(t_0) = x_0$, is interior to $\mathcal{R}_{ij}$ for $t_0 < t < t_{ij}$ and extends to $\mathfrak{M}_{ij}$, $j = 1, \ldots, j_i$. (Recall that $\mathfrak{M}_{ij_i} = \mathfrak{I}_i$.)*

(iv) The function

$$W(t, x) = g(T, x_T) + \int_C f(t, x, Y^*, Z^*)dt,$$

where C is a path from (t, x) to (T, x_T)

---

* We note here that (ii) and (iii) imply that (Y^*, Z^*) is a playable pair. Hence from any point (t_0, x_0) in $\bar{\mathcal{R}}_i \cap \mathcal{R}$ there is precisely one path $x^*(t)$ interior to $\mathcal{R}_i$ for $t_0 < t < T$ that terminates at $\mathfrak{I}_i$.

resulting from (Y^*, Z^*), is single-valued and continuous on $\mathcal{R} \cup \mathfrak{J}$.

(v) Along manifolds $\mathfrak{M}_{ij}$, $i = 1, \ldots, \alpha$ and $j = 1, \ldots, j_i - 1$,

$$\begin{aligned} &F(t, x, Y^{*+}, Z^{*+}, \Lambda^+)dt_{ij} - \Lambda^+ dx_{ij} \\ &= F(t, x, Y^{*-}, Z^{*-})dt_{ij} - \Lambda^- dx_{ij} \end{aligned}$$

for all differentials dt_{ij}, dx_{ij} on $\mathfrak{M}_{ij}$. If $\mathfrak{M}_{ij}$ is a manifold of discontinuity of only one of the functions Z^* or Y^*, then Λ and F are continuous across $\mathfrak{M}_{ij}$.

(vi) Along a manifold $\mathfrak{J}_i$,

$$F(t, x, Y^*, Z^*, \Lambda)dt - \Lambda dx = - dg.$$

(vii) In each $\widetilde{\mathcal{R}}_{ij}$, the integral $I^*(C)$, formed with the functions Y^*, Z^*, Λ, is independent of C for all admissible curves C in $\widetilde{\mathcal{R}}_{ij}$.

The functions Y^*, Z^*, Λ, are called the <u>field functions</u> and $\mathcal{R}$ is called the <u>field region</u>.

LEMMA 4. If the integral $I^*(C)$ of (5.16) is formed using the field functions Y^*, Z^*, Λ, then for all admissible curves C in $\mathcal{R}$, $I^*(C)$ is independent of the curve C, and (5.14) holds for W defined in (iv) above.

Since by (6.1)-(vii), $I^*(C)$ is independent of C in $\mathcal{R}_{ij}$, and $\mathcal{R}_{ij}$ is open, it follows that at each interior point of $\mathcal{R}_{ij}$,

$$\frac{\partial F}{\partial x}(t, x, Y^*(t, x), Z^*(t, x), \Lambda(t, x)) = - \Lambda_t.$$

Hence, using the relation $F_\lambda = G$, we have

$$F_x + F_y Y_x^* + F_z Z_x^* = -(\Lambda_t + \Lambda_x G).$$

Along a path $x^*(t)$ resulting from (Y^*, Z^*), we have $x^{*\prime} = G$. Therefore, if we define $\lambda(t) = \Lambda(t, x^*(t))$, we obtain the result that along a path $x^*(t)$, the relation (4.3) holds for all $t \neq t_{ij}$. At values of $t = t_{ij}$, the relation holds for the one-sided limits. From this, and from the definition of a field, it follows that each $\mathcal{R}_i$ is simply covered by paths that, except for conditions (4.4), (4.5), and (4.6), satisfy the conclusions of Theorem 2. Since W, defined in (6.1)-(iv), is continuous and single-valued, it follows by the arguments used to prove Theorem 6 that the conclusion of Theorem 6 holds along the manifolds $\mathfrak{N}_{i_1, \ldots, i_k}$. The arguments of Sec. 5.4 can now be used to show that $I^*(C)$ is independent of C.

Along a path $x^*(t)$ resulting from (Y^*, Z^*), we clearly have

$$W = g + \int f dt = g + \int F dt - \Lambda \mathbf{dx} .$$

From this relation and the independence of path of $I^*(C)$, we get that (5.14) holds for W defined in (6.1)-(iv).

6.2 A Sufficiency Theorem

Let the region $\mathcal{R}$ and the functions Y^*, Z^*, Λ constitute a field. Let $\mathcal{Y}_1$ and $\mathcal{Z}_1$ be two classes of strategies for the differential game, such that Y^* belongs to $\mathcal{Y}_1$ and Z^* belongs to $\mathcal{Z}_1$.

At each point (t, x) of $\mathcal{R}$ we may set up a game $\mathfrak{F}(t, x)$ similar to that of Theorem 2. The pure strategies for player I are all vectors y such that $y = Y(t, x)$ or $y = Y^+(t, x)$ for some Y in $\mathcal{Y}_1$. The pure strategies for player II are all vectors z such that $z = Z(t, x)$ or $z = Z^+(t, x)$ for some Z in $\mathcal{Z}_1$. The payoff is $F(t, x, y, z, \Lambda(t, x))$ if Λ is continuous at (t, x), and is $F(t, x, y, z, \Lambda(t, x)^+)$ if Λ is discontinuous at (t, x). In order to simplify the notation and exposition we shall always write $F(t, x, y, z, \Lambda)$, it being understood that Λ is interpreted as Λ^+ if necessary.

THEOREM 7. Let the region $\mathcal{R}$ and the functions Y^*, Z^*, Λ constitute a field. Let $\mathcal{Y}_1$ and $\mathcal{Z}_1$ be two classes of strategies having the properties listed in the first paragraph of this section. Let $\mathfrak{F}(t, x)$ be the game defined in the preceding paragraph. At each point (t, x) of $\mathcal{R}$, let y^* and z^*, where y^* is obtained from Y^* and z^* is obtained from Z^*, be pure strategy solutions of $\mathfrak{F}(t, x)$. Then the pair (Y^*, Z^*) is a saddle point relative to the classes $\mathcal{Y}_1$ and $\mathcal{Z}_1$, and the function $W(t, x)$ defined in (6.1)-(iv) is the value of the game.

In order to prove the theorem, we must show that for all (t, x) in $\mathcal{R}$, Y in $\mathcal{Y}_1$, and Z in $\mathcal{Z}_1$, all indices σ that may arise from (Y, Z^*) at (t, x) and all indices ν that may arise from the pairs (Y^*, Z) at (t, x), the inequalities (2.13) hold. We shall show that the right-hand inequality of (2.13) holds; the proof for the left-hand inequality is similar.

Let (t_0, x_0) be an arbitrary point of $\mathcal{R}$. Let C_ν be a path resulting from (Y^*, Z), starting at (t_0, x_0) and terminating at $(\overline{T}, \overline{x}_T)$. Since (Y^*, Z) is playable, it is readily seen from the various definitions that C_ν is an admissible curve in $\mathcal{R}$ relative to the decomposition associated with (Y^*, Z^*). Let E^* be a path resulting from (Y^*, Z^*), starting at (t_0, x_0) and terminating at (T, X_T). Let

$$\Delta_\nu = P_{(\nu)}(t_0, x_0; Y^*, Z) - W(t_0, x_0).$$

We must show that $\Delta_\nu \geq 0$.

Using (2.12) and (6.1)-(iv), we write

$$\Delta_\nu = \int_{C_\nu} f(t, x, Y^*, Z)dt - \int_{E^*} f(t, x, Y^*, Z^*)dt + g(\overline{T}, \overline{x}_T) - g(T, x_T). \tag{6.2}$$

Along C_ν, the relationship $x' = G(t, x, Y^*, Z)$ holds; and along E^*, (2.14) holds. We may therefore rewrite the first integral in (6.2) as

$$\int_{C_\nu} F(t, x, Y^*, Z, \Lambda)dt - \Lambda dx,$$

and we may rewrite the second integral as

$$I^*(E^*) = \int_{E^*} F(t, x, Y^*, Z^*, \Lambda)dt - \Lambda dx.$$

Since $\mathcal{R}$ is a region, there exists a curve that is admissible with respect to the regular decomposition of $\mathcal{R}$ associated with (Y^*, Z^*), that has its initial point at (T, x_T), that has its terminal point at $(\overline{T}, \overline{x}_T)$, and that otherwise lies in $\mathcal{R}$. Denote this curve by C. From Lemma 4 and (6.1)-(iv) it follows that

$$I^*(C) = g(T, x_T) - g(\overline{T}, \overline{x}_T).$$

We also get from Lemma 4 the result that

$$I^*(E^*) = I^*(C_\nu) - I^*(C).$$

Hence we may rewrite (6.3) as

$$(6.3) \qquad \Delta_\nu = \int_{C_\nu} [F(t, x, Y^*, Z, \Lambda) - F(t, x, Y^*, Z^*, \Lambda)]dt,$$

where on a subarc of C_ν lying on a manifold $\mathfrak{M}_{ij}$ or $\mathfrak{N}_{i_1, \ldots, i_k}$ we take $Z^{*\prime}$ and Λ' instead of Z and Λ. Since $\mathfrak{F}$ has y^* and z^* as optimal pure strategies, the integrand of (6.3) is nonnegative, and so $\Delta_\nu \geq 0$.

6.3 A Special Case

For each (t, x) in $\mathcal{R}$, let $f(t, x, y, z)$ be a concave function of y for each z and a convex function of z for each y. Further, for each (t, x) in $\mathcal{R}$, let each component of G be a bilinear function of y and z. These constraints are readily seen to be necessary and sufficient conditions that, for each (t, x) in $\mathcal{R}$ and all real vectors λ, the function $F(t, x, y, z, \lambda)$ be concave in y for each z and convex in z for each y. If, in addition, the constraints are separable, we have the following corollary to Theorem 7.

COROLLARY 1. Let f and G be as in the preceding paragraph, let the constraints be separable, and let $\mathcal{R}$, Y^*, Z^*, Λ, $\mathcal{Y}_1$, and $\mathcal{Z}_1$ be as in Theorem 7. Further, let $\mathcal{Y}_1$ and $\mathcal{Z}_1$ be such that at each (t, x) in $\mathcal{R}$, the set of pure strategies $\{y\}$ and the set of pure strategies $\{z\}$ for the game $\mathfrak{F}(t, x)$ are both convex. At every point of $\mathcal{R}$, let the conclusions of Corollary 1 of Theorem 2 hold, where the F_{z^i} and the F_{y^i} are evaluated at the argument $(t, x, Y^*(t, x), Z^*(t, x), \Lambda(t, x))$ if (t, x) is interior to some $\mathcal{R}_{ij}$, and at $(t, x, Y^{*+}, Z^{*+}, \Lambda^+)$ otherwise. Then the pair (Y^*, Z^*) is a saddle point relative to the classes $\mathcal{Y}_1$ and $\mathcal{Z}_1$.

To establish the corollary, we shall show that y^* and z^* are optimal pure strategies for the game $\mathfrak{F}(t, x)$. Since the set of pure strategies z is convex, we may write

$$(6.4) \qquad F(t, x, y^*, z, \Lambda) - F(t, x, y^*, z^*, \Lambda)$$
$$= \sum_{j=1}^{s} F_{z^j}(t, x, y^*, z^*, \Lambda)\Delta z^j + \frac{1}{2} \Delta z F_{zz}(t, x, y^*, \bar{z}, \Lambda)\Delta z,$$

where $\Delta z = z - z^*$ and $\bar{z}$ is a point on the line joining z^* and z. We have already noted that the conditions on f and G imply that F is concave in y and convex in z for all (t, x) in $\mathcal{R}$ and all λ. Hence, since the set of pure strategies z is convex, the quadratic form in (6.4) is nonnegative. It follows from (4.13) that each term in $\sum_{j=1}^{s}$ is nonnegative. Hence the left-hand member of (6.4) is nonnegative. In similar fashion, we can show that $F(t, x, y^*, z^*, \Lambda) - F(t, x, y, z^*, \Lambda) \geq 0$, and thus show that y^* and z^* are pure optimal strategies.

BIBLIOGRAPHY

[1] BERKOVITZ, L. D., and FLEMING, W. H., "On differential games with integral payoff," in M. Dresher, A. W. Tucker, and P. Wolfe (eds.) Contributions to the Theory of Games III, Annals of Mathematics Study 39, Princeton University Press, Princeton, N. J., 1957.

[2] BERKOVITZ, L. D., "Variational methods in problems of control and programming," Journal of Mathematical Analysis and Applications, Vol. 3, 1961, pp. 145–169.

[3] BLISS, G. A., Lectures on the Calculus of Variations, University of Chicago Press, Chicago, Ill., 1946.

[4] FLEMING, W. H., "A note on differential games of prescribed duration," in M. Dresher, A. W. Tucker, and P. Wolfe (eds.) Contributions to the Theory of Games III, Annals of Mathematics Study 39, Princeton University Press, Princeton, N. J., 1957.

[5] FLEMING, W. H., "The convergence problem for differential games," Journal of Mathematical Analysis and Applications, Vol. 3, 1961, pp. 102-116.

[6] ISAACS, R., "Differential Games I, II, III, IV, The RAND Corporation, Research Memoranda RM-1391, RM-1399, RM-1411, RM-1486, 1954.

[7] SCARF, H., "On differential games with survival payoff," in M. Dresher, A. W. Tucker, and P. Wolfe (eds.) Contributions to the Theory of Games III, Annals of Mathematics Study 39, Princeton University Press, Princeton, N. J., 1957.

[8] VALENTINE, F. A., "The problem of Lagrange with differential inequalities as added side conditions," Contributions to the Calculus of Variations, 1933-1937, University of Chicago Press, Chicago, Ill., 1937.

Leonard D. Berkovitz

The RAND Corporation

A DIFFERENTIAL GAME WITH NO PURE STRATEGY SOLUTION

Leonard D. Berkovitz

§ 1. INTRODUCTION

It is sometimes conjectured that a differential game with integral payoff or a continuous terminal payoff must have a solution in pure strategies. In this paper we give examples showing that this conjecture is false, if we accept the notion of pure strategy introduced in [1]. First, we shall use the methods of [1] to show that the continuous version of the multimove infinite game studied in [2] and [3] fails to have a pure strategy solution on an open set. This discussion will also serve to illustrate how the methods of [1] can be used to construct pure strategy solutions when they do exist. We assume, of course, that the reader is familiar with the formulation and results of [1].

From a purely mathematical point of view, a simpler example than the continuous version of [2] can, and will, be given to point out the general inadequacy of pure strategies in the study of differential games. Nevertheless, we present the continuous version of [2] because the game had its origin in an applied problem, and as such the example cannot be considered as unusual in any sense. Moreover, the analysis of the present example seems to indicate that the solution of our game may have certain features that are not present in the time-discrete version solved in [2].

§ 2. DESCRIPTION OF GAME

Let Blue be the maximizing player and let Red be the minimizing player. Let p and q be the state variables, where p denotes Blue's resources and q denotes Red's resources. The region of play $\mathcal{R}$ is defined by the inequalities

$$p > 0, \qquad q > 0, \qquad 0 < t < T,$$

where T is a fixed positive number. The strategy vectors for Blue are of the form $Y = (y, u)$, where y and u are piecewise $C^{(2)}$ in $\bar{\mathcal{R}}$ and satisfy the constraints

$$y \geq 0, \qquad u \geq 0, \qquad y + u \leq 1. \tag{2.1}$$

Similarly, the strategy vectors for Red are of the form $Z = (z, w)$, where z and w are piecewise $C^{(2)}$ on $\bar{\mathcal{R}}$ and satisfy the constraints

$$z \geq 0, \qquad w \geq 0, \qquad z + w \leq 1. \tag{2.2}$$

The state of the game is governed by the following system of differential equations

$$\begin{aligned} \frac{dp}{dt} &= -\varphi(zq - up), \qquad p(t_0) = p_0, \\ \frac{dq}{dt} &= -\psi(yp - wq), \qquad q(t_0) = q_0, \end{aligned} \tag{2.3}$$

where

$$\varphi(s) = \psi(s) = \max(0, s).$$

We use different designations for the same function to distinguish between the equation governing the state of Blue and the equation governing the state of Red. Play begins at any point (t_0, p_0, q_0) in $\mathcal{R}$ and terminates whenever one of the following occurs:

$$p = 0, \qquad q = 0, \qquad \text{or} \quad t = T.$$

Let (t_1, p_1, q_1) denote the boundary point of $\mathcal{R}$ at which play terminates. We define the payoff $P(t_0, p_0, q_0; Y, Z)$ to Blue for a play of the game starting at (t_0, p_0, q_0), in which Blue uses a strategy Y and Red uses a strategy Z, to be

$$(T - t_1)(p_1 - q_1) + \int_{t_0}^{t_1} [(1 - y - u)\, p - (1 - z - w)q]dt. \tag{2.4}$$

We tacitly assume here that (Y, Z) is a playable pair.

A comparison of the game just described with the game studied in [2] shows that we have formulated a version of [2] in which time is continuous. Since the functions φ and ψ of (2.3) are not $C^{(2)}$, the game just described does not conform exactly to the formulation in [1]. We shall be able to employ Theorem 1 of [1], however, in fixed point arguments to show that pure strategies are inadequate for the solution of the game.

Since the game is symmetric with respect to the roles of Red and Blue, we shall only carry out the analysis for the subregion of $\mathfrak{R}$ in which

$$p \geq q. \tag{2.5}$$

§ 3. TERMINATION AT $q_1 = 0$

In this section we shall establish the following theorem:

THEOREM 1. *There exists no optimal path E^* resulting from a pair of pure optimal strategies (Y^*, Z^*) and terminating at a point $\pi^* = (t_1, p_1, 0)$, where $0 < t_1 < T - 1$ and $p_1 > 0$.*

Let us suppose that there exist pure strategies (Y^*, Z^*), and an optimal path E^* resulting from (Y^*, Z^*) and terminating at π^*. If we take $Z = Z^*$ in (2.3) and (2.4), then Y^* maximizes (2.4) subject to (2.3) in the class of strategies $\mathcal{Y}_1$ of [1]. Clearly, the maximizing Y against Z^* will satisfy $z^*q - up \geq 0$. Furthermore, since E^* terminates at $q_1 = 0$, we cannot have $yp - w^*q < 0$ along E^* on a half open interval $I^o_\epsilon = [t_1 - \epsilon, t_1)$. Thus, for $Z = Z^*$, the problem of maximizing (2.4) subject to (2.3) in the class $\mathcal{Y}_1$ can be replaced by the following equivalent problem.

<u>Problem 1</u>. Maximize (2.4), subject to (2.3) with

(3.1) $$\varphi = z^*q - up, \qquad \psi = yp - w^*q,$$

over those Y in $\mathcal{Y}$ satisfying

(3.2) $$\bar{K}_1 = u \geq 0, \qquad \bar{K}_2 = 1 - y - u \geq 0,$$
$$\bar{K}_3 = yp - w^*q \geq 0, \qquad \bar{K}_4 = z^*q - up \geq 0.$$

To this problem we can apply the maximum problem version of Theorem 1 of [1]. We shall henceforth refer to this theorem in either its maximum or minimum version as Theorem A. Since we may take $\lambda_0 = 1$, our function H becomes

(3.3) $$H = (1 - y - u)p - (1 - z^* - w^*)q - \lambda_1\varphi(z^*q - up) - \lambda_2\psi(yp - w^*q).$$

Using (3.1), we may rewrite this as follows:

(3.4) $$H = A(p, q, \lambda_1, \lambda_2, Z^*) - yp(1 + \lambda_2) - up(1 - \lambda_1)$$

for appropriate A. Since $g = (T - t_1)(p_1 - q_1)$ and $q_1 = 0$ in the present problem, it follows from the $\bar{K}_4$ constraint in (3.2), and from (3.3), that (3.6) of Theorem A gives

(3.5) $$\lambda_1(t_1) = T - t_1, \qquad y(t_1)(1 + \lambda_2(t_1)) = 0.$$

LEMMA 1. An optimal Z^* cannot be such that $z^* = 0$ along E^* on an interval $I_\epsilon = [t_1 - \epsilon, t_1]$.

From the y component of (3.4) of Theorem A, we get

(3.6) $$-p(1 + \lambda_2) - \bar{\mu}_2 + \bar{\mu}_3\, p = 0.$$

From the q-component of (3.3) of Theorem A, we get

(3.7) $$\lambda_2' = 1 - z^*(1 - \lambda_1 + \bar{\mu}_4) - w^*(1 + \lambda_2 - \bar{\mu}_3).$$

If $z^* = 0$ on an interval I_ϵ, then since $\bar{K}_4 \geq 0$, the maximizing Y, which we shall denote by $\hat{Y}$, would have $\hat{u} = 0$. If $\hat{y} < 1$, then $\bar{K}_2 > 0$ and the corresponding multiplier $\bar{\mu}_2 = 0$, by (3.5) of Theorem A. Since E^* terminates at $q_1 = 0$, $\bar{K}_3 \neq 0$ in I_ϵ^o, and $\bar{\mu}_3 = 0$ by (3.5) of Theorem A. Hence from (3.6), we get that $(1 + \lambda_2) = 0$ on I_ϵ, and thus from (3.7), that $\lambda_2' = 1$. This is a contradiction, since $(1 + \lambda_2) = 0$ implies that $\lambda_2' = 0$ in I_ϵ. Thus, we must have $\hat{Y} = (1, 0)$.

We now let $Y = (1, 0)$ in (2.4) and (2.3) and seek a Z that minimizes (2.4) subject to (2.3). Since $p \geq q$, the Equations (2.3) become

$$\frac{dp}{dt} = - zq, \qquad \frac{dq}{dt} = - (p - wq),$$

and the constraints are given by (2.2). The function H for this minimization problem becomes

(3.8) $$H = q - \lambda_2 p + wq(1 + \lambda_2) + zq(1 - \lambda_1).$$

The terminal condition (3.6) of Theorem A now gives

$$\lambda_2 = - 1, \qquad \lambda_1 = T - t_1$$

at $t = t_1$. Hence if we assume that

(3.9) $$(T - t_1) > 1,$$

we get by continuity that on an interval I_ϵ along E^*, H is minimized by $z = 1$, $w = 0$. But then $Z^* = (0, w)$ cannot be optimal as originally supposed, and the lemma is proved.

LEMMA 2. If there exist optimal strategies (Y^*, Z^*) resulting in an optimal path E^* terminating at $(t_1, p_1, 0)$ with $0 < t_1 < T - 1$, then they are related as follows:

$$u^* = z^*q/p, \qquad y^* = (p - z^*q)/p,$$

where $z^* \neq 0$.

By Lemma 1, we have $z^* \neq 0$ on an interval I_ϵ. Thus, in Problem 1 we may take $z^* \neq 0$. Since $yp - w^*q > 0$ on I_ϵ we have $\bar{\mu}_3 = 0$. We next assert that on I_ϵ, $\hat{Y}$ must satisfy $\hat{y} + \hat{u} = 1$. For if $\hat{y} + \hat{u} < 1$, then $\bar{\mu}_2 = 0$ on I_ϵ, and from (3.6) we again get $\lambda_2' = 0$ on I_ϵ. Hence from (3.7) we obtain

$$0 = 1 - z^*(1 - \lambda_1 + \bar{\mu}_4).$$

On the other hand, the u component of (3.4) of Theorem A gives

$$- p(1 - \lambda_1 + \bar{\mu}_4) + \bar{\mu}_1 - \bar{\mu}_2 = 0. \tag{3.10}$$

Combining the last two equations and using $\bar{\mu}_2 = 0$ we have

$$0 = 1 - (z^*\bar{\mu}_1)/p.$$

Since $z^* \neq 0$, we have $\bar{\mu}_1 = p/z^*$. Hence by (3.5) of Theorem A, $\hat{u} = 0$ on I_ϵ. Accordingly, from (3.2), and (3.5) of Theorem A, $\bar{\mu}_4 = 0$. It now follows from this discussion and (3.10) that along E^*, $- z^*(1 - \lambda_1) + 1 = 0$. From this and (3.5), it follows that at $t = t_1$,

$$z^* = 1/(t_1 - T + 1).$$

Since $z^* \geq 0$, we have a contradiction if $t_1 < T - 1$. Thus we have shown that $\hat{y} + \hat{u} = 1$.

Since from $\bar{K}_4$ of (3.2) we have $\hat{u}(t_1) = 0$, it follows that $\hat{y}(t_1) = 1$. From (3.5) and (3.9), at $t = t_1$ we now have $(\lambda_1 - 1) > -(\lambda_2 + 1)$. Hence, by continuity, this relation holds on an interval I_ϵ. Since $\hat{Y}$ maximizes (3.4), the conclusion now follows from (3.4) and the constraints (3.2).

LEMMA 3. Let $Y^* = (1 - \theta q/p, \theta q/p)$, where $0 \leq \theta \leq 1$. Then the minimum of (2.4), subject to (2.3), over all $Z = (z, w)$ in $\mathcal{Z}_1$ such that $z \neq 0$, is given by $\hat{Z} = (1, 0)$.

With Y^* as in the hypothesis, (2.4) becomes

$$(T - t_1)p_1 - \int_t^{t_1} (1 - z - w)q\,dt,$$

and the Equations (2.3) become

$$\frac{dp}{dt} = -\varphi((z - \theta)q), \qquad \frac{dq}{dt} = -\psi(p - (\theta + w)q). \tag{3.11}$$

Clearly, if $z \neq 0$, we have that the minimizing z must be such that $z - \theta \geq 0$. Since $z + w \leq 1$ and $p \geq q$, we immediately get that $p - (\theta + w)q \geq 0$ always holds. Thus, we can remove the φ and ψ from Equations (3.11) and impose the following constraints

$$z - \theta \geq 0, \qquad w \geq 0, \qquad 1 - z - w \geq 0.$$

The function H becomes

$$H = A(p, q, \theta, \lambda_1, \lambda_2) + wq(1 + \lambda_2) + zq(1 - \lambda_1). \tag{3.12}$$

Equation (3.6) of Theorem A now yields

$$\lambda_2(t_1) = -1, \qquad \lambda_1(t_1) = (T - t_1).$$

Thus, at $t = t_1$, the coefficient of z in (3.12) is negative and the coefficient of w is zero. Hence, by continuity, $\hat{z} = 1$, $\hat{w} = 0$ minimize H on an interval I_ϵ, and the Lemma is proved.

If we combine the results of Lemmas 1 to 3, we get the following lemma.

LEMMA 4. If there exists an optimal path E^* as described in Theorem 1 and resulting from a pair of pure strategies (Y^*, Z^*), then we must have:

$$Y^* = (1 - q/p, q/p), \qquad Z^* = (1, 0). \tag{3.13}$$

Let $\pi_0 = (t, p, q)$ be a point of $\mathcal{R}$ such that the optimal path E^*, resulting from (Y^*, Z^*) as in (3.13) and starting at π_0, terminates at $\pi_1 = (t_1, p_1, 0)$, where $T - t_1 > 1$. From (2.4) and (3.13) we get

$$P(t, p, q; Y^*, Z^*) = (T - t_1)p_1. \tag{3.14}$$

The Equations (2.3) become

$$\frac{dp}{dt} = 0, \qquad \frac{dq}{dt} = -(p - q).$$

From the first of these we conclude that p is constant along E^*, and so $p = p_1$. Integration of the second of these equations from q to 0 and t to t_1 gives

$$T - t_1 = (t - t) + \log(1 - \frac{q}{p}). \tag{3.15}$$

The requirement that $T - t_1 > 1$ imposes the following condition on π_0:

$$q < p(1 - e^{-\tau}), \tag{3.16}$$

where $\tau = T - 1 - t$. Substituting (3.14) into (3.15), and using $p_1 = p$, gives

$$P(t, p, q; Y^*, Z^*) = [T - t + \log(1 - \frac{q}{p})]p \tag{3.17}$$

for all points π_0 whose coordinates satisfy (3.16).

Let Y^* be as in (3.13) and let $\hat{Z} = (0, 1)$. Let π_0 now be a point of $\mathcal{R}$ such that the path $\hat{E}$, resulting from $(Y^*, \hat{Z})$ and starting at

π_0, terminates at a point $\pi_1' = (t_1', p_1', 0)$, where $T - t_1' > 1$. From (2.4) we get

$$P(t, p, q; Y^*, \hat{Z}) = (T - t_1')p_1'. \tag{3.18}$$

The Equations (2.3) become

$$\frac{dp}{dt} = 0, \qquad \frac{dq}{dt} = -(p - 2q).$$

From the first of these equations we get that $p = p_1'$ along $\hat{E}$. Let us now suppose that π_0 is such that $q/p \leq 1/2$. Then we may integrate the second equation and get

$$T - t_1' = [T - t + \tfrac{1}{2}\log(1 - \tfrac{2q}{2})]. \tag{3.19}$$

The requirement $T - t_1' > 1$ now imposes the following condition on π_0:

$$q < \tfrac{p}{2}(1 - e^{-2\tau}), \tag{3.20}$$

where $\tau = T - 1 - t$. Substituting (3.19) into (3.18), and using $p_1' = p$, gives

$$P(t, p, q; Y^*, \hat{Z}) = [T - t + \tfrac{1}{2}\log(1 - \tfrac{2q}{p})]p \tag{3.21}$$

for all points π_0 whose coordinates satisfy (3.20).

Since the right-hand side of (3.20) does not exceed the right-hand side of (3.16), it follows that for all points π_0 whose coordinates satisfy (3.20), both (3.21) and (3.17) are valid. For such points we get

$$P(Y^*, \hat{Z}) - P(Y^*, Z^*) = \tfrac{p}{2}\log\left[\frac{p^2 - 2pq}{(p - q)^2}\right], \tag{3.22}$$

where we have omitted the arguments (t, p, q) in P. Since the term in square brackets is less than 1, the right-hand side of (3.22) is negative. Hence (Y^*, Z^*) cannot be a saddle point, and Theorem 1 is proved.

§ 4. THE SUBREGION $T - 1 \leq t \leq T$

THEOREM 2. In the region defined by $p > 0$, $q > 0$, $T - 1 \leq t < T$, there exists a unique pure strategy solution given by

$$Y^* = Y_0 \equiv (0, 0), \qquad Z^* = Z_0 \equiv (0, 0). \tag{4.1}$$

In (2.4), set $t_1 = T$, $Y = Y_0$, and consider the problem of minimizing (2.4) subject to (2.3). The functional (2.4) becomes

$$\int_t^T [p - q(1 - y - w)]dt,$$

and the Equations (2.3) become

$$\frac{dp}{dt} = - zq, \qquad \frac{dq}{dt} = 0.$$

Clearly, the minimizing Z will be such that $w = 0$. Hence the constraints on $Z = (z, 0)$ become $z \geq 0$, $1 - z \geq 0$. To this minimum problem, Theorem A is applicable. The function H becomes

$$H = p - q + zq(1 - \lambda_1), \tag{4.2}$$

and from (3.6) of Theorem A, we get $\lambda_1(t_1) = 0$. From (3.3) of Theorem A we get $\lambda_1' = - 1$, whence $\lambda_1(t) = (T - t)$. Substituting this expression for λ into (4.2) and applying Condition II of Theorem A shows that the minimizing Z must be such that $z = 0$ for $T - 1 \leq t \leq T$, and $z = 1$ for $t \leq T - 1$. Furthermore, by the arguments of Theorem 7 of [1], it can be readily shown that $Z = (0, 0)$ for $T - 1 \leq t \leq T$, and that $Z = (1, 0)$ for $t < T - 1$, does minimize against Y_0.

In a similar fashion we get that against $Z = Z_0$, the strategy $Y = (0, 0)$ for $T - 1 \leq t \leq T$ and $Y = (1, 0)$ for $t < T - 1$ maximize (2.4).

The arguments of Sec. 3 can be used to show that if there exist pure strategies leading to termination of play at points $\pi_1 = (t_1, p_1, 0)$, where $T - 1 < t_1 < T$, then in a neighborhood of $q_1 = 0$ with $t_1 > T - 1$, these strategies must be

$$Y = Y_1 = (1, 0), \qquad Z = Z_2 = (0, 1). \tag{4.3}$$

We now show that these cannot be optimal. It follows from (2.4) that

$$P(t, p, q; Y_1, Z_2) = p_1(T - t_1).$$

If we substitute (4.3) into (2.3) and integrate the resulting equations from (t, p, q) to $(t_1, p_1, 0)$, we get

$$p = p_1, \qquad \log \frac{p}{p - q} = t_1 - t.$$

Hence

$$P(t, p, q; Y_1, Z_2) = p\left[T - t - \log \frac{p}{p - q}\right]. \tag{4.4}$$

On the other hand, it is readily calculated that

$$P(t, p, q; Y_0, Z_2) = p(T - t). \tag{4.5}$$

A comparison of (4.4) and (4.5) immediately shows that (Y_1, Z_2) cannot be optimal.

Since a path must either terminate at $q_1 = 0$ or $t_1 = T$, the theorem will be proved if we can show that (Y^*, Z^*) is a unique pair leading to termination at $t = T$.

To show the uniqueness of (4.1) we proceed as follows. Suppose that there exists an optimal strategy $Y^* = (y^*, u^*)$ different from $(0, 0)$. We substitute Y^* into (2.3) and (2.4) and consider the problem of minimizing (2.3) subject to (2.4) over Z in $\mathcal{Z}_1$. It is immediately seen that a minimizing Z will satisfy the constraint $y^*p - wq \geq 0$, and that either $z = 0$ or

$$zq - u^*p \geq 0. \tag{4.6}$$

If we add the constraint (4.6) to the minimum problem, then the right-hand sides of (2.3) become $(zq - u^*p)$ and $(y^*p - wq)$. We now have a problem to which Theorem A is applicable.

The function H becomes

$$H = A(p, q, \lambda_1, \lambda_2, y^*, u^*) + zq(1 - \lambda_1) + wq(1 + \lambda_2),$$

where A is an appropriate expression. From (3.6) we get that $\lambda_1(T) = \lambda_2(T) = 0$. Hence, in a neighborhood of $t = T$, H is minimized by $z = u^*p/q$, $w = 0$. For these values of z and w, the differential Equations (2.3) become

$$\frac{dp}{dt} = 0, \qquad \frac{dq}{dt} = - y^*q.$$

For $Z = Z_0 = (0, 0)$, the Equations (2.3) also take on this form. If we now compare (2.4) in which we have set $Y = (y^*, u^*)$, $Z = Z_0$, with (2.4) in which we have set $Y = (y^*, u^*)$, $Z = (u^*p/q, 0)$, we see that $Z = Z_0$ provides (2.4) with a lower value for initial points in a neighborhood of $t = T$.

We have already shown that the maximum of (2.4) with $Z = Z_0$ is given by $Y = Y_0$. Hence no Y different from Y_0 can be a pure optimal strategy.

§ 5. THE SUBREGION $\mathcal{R}_{12}$

Let

$$\tau = T - 1 - t, \tag{5.1}$$

and let the regions $\mathcal{R}_{11}$ and $\mathcal{R}_{12}$ be defined as follows:

$$\mathcal{R}_{11}\colon \quad T - 1 < t < T, \quad p > q, \quad q > 0,$$

$$\mathcal{R}_{12}\colon \quad 0 < t < T - 1, \quad p > q, \quad q > p \tanh \tau.$$

Let us consider the strategies

$$Y_1 = (1, 0), \qquad Z_1 = (1, 0), \tag{5.2}$$

in the region $\hat{\mathcal{R}}$ defined as follows:

$$\hat{\mathcal{R}}: \quad 0 < t < T - 1, \quad p > q, \quad q > 0.$$

If we substitute (5.2) into (2.3) and integrate from an initial point (t_0, p_0, q_0) in $\hat{\mathcal{R}}$, we get

$$\begin{aligned} p &= \alpha e^{(t-t_0)} + \beta e^{-(t-t_0)}, \\ q &= \alpha e^{(t-t_0)} + \beta e^{-(t-t_0)}, \end{aligned} \tag{5.3}$$

where $\alpha = (p_0 - q_0)/2$ and $\beta = (p_0 + q_0)/2$. The path defined by (5.3) will reach the boundary of $\hat{\mathcal{R}}$ at $t = T - 1$ if and only if $q(T - 1) > 0$. From the second equation in (5.3), we see that $q(T - 1) \geq 0$ if and only if (t_0, p_0, q_0) belongs to $\mathcal{R}_{12}$, or is a boundary point of $\mathcal{R}_{12}$ such that $t = 0$ or $q = p \tanh \tau$.

Let us now suppose that there exist optimal pure strategies (Y^*, Z^*) defined on a subregion $\mathcal{R}^*$ of $\hat{\mathcal{R}}$ and on $\mathcal{R}_{11}$ such that a path starting at a point of $\mathcal{R}^* \cup \mathcal{R}_{11}$ and resulting from (Y^*, Z^*) terminates at $t = T$. By Theorem 2, we know that on $\mathcal{R}_{11}$, $Y^* = Z^* = (0, 0)$. Let $\mathcal{Z}'$ denote the subclass of $\mathcal{Z}_1$ consisting of those functions Z in $\mathcal{Z}_1$ such that $Z = (0, 0)$ on $\mathcal{R}_{11}$ and such that a path resulting from (Y^*, Z) and starting at a point of $\mathcal{R}^*$ reaches the boundary of $\mathcal{R}^*$ at $t = T - 1$. Then, for (t, p, q) in $\mathcal{R}^*$,

$$\begin{aligned} P(t, p, q; Y^*, Z^*) &= \min_{\mathcal{Z}'} P(t, p, q; Y^*, Z) \\ &= \min_{\mathcal{Z}'} \Big[(p_2 - q_2) + \int_t^{T-1} \{ (1 - y^* - u^*)p \\ &\qquad - (1 - z - w)q \} dt \Big], \end{aligned} \tag{5.4}$$

where the subscript 2 denotes points of the manifold defined by $t = T - 1$. Let $\mathcal{Y}'$ denote the subclass of $\mathcal{Y}_1$ consisting of those Y in $\mathcal{Y}_1$ such that $Y = (0, 0)$ in $\mathcal{R}_{11}$ and such that a path resulting from (Y, Z^*) and starting at a point of $\mathcal{R}^*$ reaches the boundary of $\mathcal{R}^*$ at $t = T - 1$. Then, for (t, p, q) in $\mathcal{R}^*$,

$$(5.5) \qquad P(t, p, q; Y^*, Z^*) = \max_{\mathcal{Y}'} P(t, p, q; Y, Z^*) = \max_{\mathcal{Y}'} \left[(p_2 - q_2) + \int_t^{T-1} \{(1 - y - u)p - (1 - z^* - w^*)q\} dt \right].$$

In the square brackets of (5.4) let us take $(y^*, u^*) = (1, 0) = Y_1$. Then the class $\mathcal{Z}'$ corresponding to Y_1 is not void if we take $\mathcal{R}^* = \mathcal{R}_{12}$, for we saw above that Z_1 belongs to $\mathcal{Z}'$. The differential Equations (2.3) governing the state now become

$$\frac{dp}{dt} = -zq, \qquad \frac{dq}{dt} = -(p - wq),$$

and the constraints are given by (2.2). Theorem A is applicable to this problem, and it can be shown that a minimizing Z must satisfy the condition $z + w = 1$. Moreover, it can also be shown that any such Z is minimizing. We omit the details, since they are straightforward and several samples of such determinations have already been given.

Similarly, in the maximizing problem (5.5), if we take $(z^*, w^*) = (1, 0) = Z_1$, then it can be shown that any Y in $\mathcal{Y}'$ such that $y + u = 1$ will maximize. Hence, if we take

$$Y^* = Y_1, \qquad Z^* = Z_1,$$

both (5.4) and (5.5) hold simultaneously. Thus, if there exist optimal pure strategies (Y^*, Z^*) on $\mathcal{R}_{12}$ such that a path starting at a point of $(\bar{\mathcal{R}}_{12} \cup \bar{\mathcal{R}}_{11}) \cap \mathcal{R}$ and resulting from (Y^*, Z^*) terminates at $t = T$, then the strategies (5.2) are one such pair. Moreover, it can be shown that this is the only pair possible.

The proof of uniqueness involves a series of successive maximization and minimization problems. It is rather straightforward, but tedious. We merely outline the major steps. We first show that a strategy $Z^* = (0, w^*)$ cannot be optimal. If $(0, w^*)$ is substituted into (2.4), the resulting maximum problem on Y is readily seen to involve two possibilities. Either $\hat{Y}$ is such that $\hat{y} = 0$, or $\hat{y}p - w^*q \geq 0$. In the first instance, it is seen that $\hat{w} = 0$ must also hold. If we now minimize Z against $\hat{Y} = Y_0 = (0, 0)$, we get $\hat{Z} = (1, 0)$. On the other hand, if $\hat{y}p - w^*q \geq 0$ holds, then $\hat{Y} = (1, 0)$. The minimum Z against $Y = (1, 0)$ such that $z = 0$ is $Z = (0, 1)$. The maximum Y against $Z = (0, 1)$, however, is $Y = (0, 0)$. Hence $Z^* = (0, w^*)$ cannot be optimal.

On the other hand if we assume that the optimal Z is $Z^* = (z^*, w^*)$, where $z^* > 0$, then the following holds: The maximum Y is either $Y = (0, \theta q/p)$ or $Y = (1 - \theta q/p, \theta q/p)$, where $0 \leq \theta \leq 1$. If the maximum Y is $(0, \theta q/p)$, then the minimum Z against this Y is either $Z = (0, w)$ or $Z = (1, 0)$. The strategy $Z = (0, w)$ was already shown to be impossible for a saddle point. The maximum Y against $Z = (1, 0)$, however, is $Y = (y, u)$ with $y + u = 1$. Hence, $Y = (0, \theta q/p)$ is impossible for a saddle point.

Let us now consider $Y = (1 - \theta q/p, \theta q/p)$, where $0 \leq \theta \leq 1$. The minimizing Z is seen to be either of the form $Z = (0, w)$, or $Z = (z, 1 - z)$, where $z \geq 0$. By a direct comparison, however, it is easily shown that unless $\theta = 0$, $Z = (0, 1 - \theta)$ yields a smaller value of (2.4) than does $Z = (z, 1 - z)$, where $z \geq \theta$. Hence, we must have $Y = (1, 0)$ in a saddle point. The minimum against $Y = (1, 0)$ is furnished by any Z of one form $Z = (z, 1 - z)$, $z \geq 0$. It is then seen that if $z < 1$, then for appropriate w, $Y = (0, w)$ will furnish the maximum against this Z. Hence, we must have $Z = (1, 0)$ in a saddle point.

We summarize the results of this section in the following theorem.

THEOREM 3. If the region over which there exist optimal pure strategies (Y^*, Z^*) leading to termination at $t = T$ includes points such that $t < T - 1$, then it must include $\mathcal{R}_{12}$. In the region $\mathcal{R}_{12}$ the strategies (Y^*, Z^*) are unique and are as follows:

$$Y^* = Y_1 = (1, 0), \qquad Z^* = Z_1 = (1, 0).$$

It is clear from Theorems 2 and 3 and Equations (2.4) and (5.3) that if the hypotheses of Theorem 3 are fulfilled then the value of the game in $\mathcal{R}_{12}$ is

$$W(t, p, q) = (p - q)e^{\tau}, \tag{5.6}$$

where τ is defined in (5.1).

§ 6. THE PRINCIPAL THEOREM

Let $\mathcal{R}_2$ be defined as follows:

$$\tau > 0, \quad p > 0, \qquad 0 < q < p \tanh \tau.$$

Let us suppose that there exist optimal pure strategies $(\overline{Y}^*, \overline{Z}^*)$ defined on a subregion $\mathcal{R}_2^*$ of $\mathcal{R}_2$ and on $(\overline{\mathcal{R}}_{11} \cup \overline{\mathcal{R}}_{12}) \cap \mathcal{R}$ such that a path starting at a point of $\mathcal{R}_2^*$ and resulting from $(\overline{Y}^*, Z^*)$ terminates at $t = T$. From the preceding discussion it is clear that such a path must leave $\mathcal{R}_2^*$ at a point (t_0, p_0, q_0) of the manifold $\mathfrak{M}$ defined by the equation

$$q_0 = p_0 \tanh \tau_0, \qquad (\tau_0, 0, p_0 > 0).$$

Let $\mathcal{Z}'$ now denote the subclass of $\mathcal{Z}_1$ consisting of those functions Z in $\mathcal{Z}_1$ such that $Z = (1, 0)$ on $\mathcal{R}_{12}$, $Z = (0, 0)$ on $\mathcal{R}_{11}$, and such that a path starting at a point of $\mathcal{R}_2^*$ and resulting from $(\overline{Y}^*, Z)$ leaves $\mathcal{R}_2^*$ at a point $\mathfrak{M}$. Let $\mathcal{Y}'$ denote the subclass of $\mathcal{Y}_1$ consisting of those functions Y in $\mathcal{Y}_1$ such that $Y = (1, 0)$ in $\mathcal{R}_{12}$, $Y = (0, 0)$ on $\mathcal{R}_{11}$ and such that a path starting at a point of $\mathcal{R}_2^*$ and resulting from $(Y, \overline{Z}^*)$ leaves $\mathcal{R}_2^*$ at a point of $\mathfrak{M}$. Then, it follows from (5.6) that

$$
\begin{aligned}
(6.1) \qquad P(t, p, q; \overline{Y}^*, \overline{Z}^*) &= \min_{\mathcal{Z}'} \Big[(p_0 - q_0)e^{\tau_0} + \int_t^{t_0} \{(1 - \overline{y}^* - \overline{u}^*)p \\
&\qquad - (1 - z - w)q\} dt \Big] \\
&= \max_{\mathcal{Y}'} \Big[(p_0 - q_0)e^{\tau_0} + \int_t^{t_0} \{(1 - y - u)p \\
&\qquad - (1 - \overline{z}^* - \overline{w}^*)q\} \Big] .
\end{aligned}
$$

It can be shown, by arguments that are similar in principle to the ones used to establish Theorem 1, that there exists in $\mathcal{R}_2$ a neighborhood $\mathcal{U}$ of $\mathcal{M}$ such that for no point (t, p, q) in $\mathcal{U}$ does (6.1) hold. The calculations involved to establish this are quite tedious, but are fairly straightforward. We shall not present them here.

The preceding conclusion can be restated as follows. There exists in $\mathcal{R}_2$ a neighborhood $\mathcal{U}$ of $\mathcal{M}$ such that if there exist pure optimal strategies (Y^*, Z^*) in $\mathcal{U}$, then the paths resulting from this pair and starting in $\mathcal{U}$ cannot leave $\mathcal{R}_2$ at a point of $\mathcal{M}$. On the other hand, Theorem 1 shows that in $\mathcal{R}_2$ there is a neighborhood $\mathcal{U}'$ of the manifold $q_1 = 0$ such that if there exist optimal pure strategies (Y^*, Z^*) in $\mathcal{U}'$, then the paths resulting from these strategies and starting at points of $\mathcal{U}'$ do not terminate at $q_1 = 0$. If, however, there exist optimal pure strategies (Y^*, Z^*) on all of $\mathcal{R}_2$, then the paths resulting from these strategies must either terminate at $q_1 = 0$ or leave $\mathcal{R}_2$ at $\mathcal{M}$. Thus, in at least one of the subregions $\mathcal{U}$, $\mathcal{U}'$, there cannot exist optimal pure strategies (Y^*, Z^*). This gives our principal theorem.

THEOREM 4. There exists an open subset of $\mathcal{R}_2$ on which there can exist no pair of optimal pure strategies.

REMARKS. We showed in Theorem 2 that in the region $\mathcal{R}_{11}$ there does exist a pure strategy solution. This is in agreement with the results of the discrete-time version [2]. Theorem 3 shows that in the subclass of strategies leading to termination at $t = T$, there does exist a pure strategy solution in $\mathcal{R}_{12}$. It is an open question whether the strategies (5.2) are indeed a saddle point in the full strategy class. If the answer is positive, then the solution of the continuous game would differ from the discrete game.

§ 7. ANOTHER EXAMPLE

In this section we give a simple example of a game in the plane for which there exists no pure strategy solution. Blue chooses $y = Y(t, x)$, where Y has the requisite continuity and differentiability properties and $0 \leq Y(t, x) \leq 1$. Red chooses $z = Z(t, x)$ where Z has the requisite continuity and differentiability properties and $0 \leq Z(t, x) \leq 1$. The state x is determined by the differential equation

$$\frac{dx}{dt} = (y - z)^2, \qquad x(0) = x_0.$$

Play terminates at a fixed time $t = T$. Blue wishes to maximize

$$\int_0^T x\,dt \tag{7.1}$$

and Red wishes to minimize (7.1).

If there exist optimal pure strategies (Y^*, Z^*) then, along a path E^* resulting from these strategies the necessary conditions of Chapter 4 of [1] must hold. The function F becomes

$$F = x + \lambda(y - z)^2. \tag{7.2}$$

The Equations (4.3) become

$$\lambda' = -1,$$

and the transversality condition (4.7) becomes

$$-\lambda dx = 0$$

for all dx. Hence $\lambda(T) = 0$, and so $\lambda(t) = (T - t)$ along E^*.

Substituting this into (7.2) gives

$$F = x + (T - t)(y - z)^2. \tag{7.3}$$

Clearly, the game $\mathfrak{F}$ with payoff given by (7.3) does not have pure strategy solutions. Hence the differential game cannot have pure strategy solutions.

It should be noted that one can show without reference to [1] that the game considered here has no pure strategy solutions. For given a choice of Y by Blue, Red can choose the same strategy and can thus guarantee $dx/dt = 0$ on the interval $0 \leq t \leq T$. Hence

$$\max_y \min_z \int_0^T x\,dt \leq x_o T.$$

But since $\int_0^T x\,dt \geq x_o T$ for any pair of strategies (Y, Z), it follows that

$$\max_y \min_z \int_0^T x\,dt = x_o T \tag{7.4}$$

On the other hand, if Red chooses a strategy Z, then Blue can choose a strategy Y defined as follows

$$\begin{aligned} Y(t, x) &= 1 \qquad \text{if} \quad Z(t, x) \leq 1/2 \\ &= 0 \qquad \text{if} \quad Z(t, x) > 1/2, \end{aligned}$$

and can guarantee $dx/dt \geq 1/4$ for $0 \leq t \leq T$. Hence

$$\min_z \max_y \int_0^T x\,dt \geq x_o T + \frac{1}{8} T^2. \tag{7.5}$$

Let Red now choose the strategy $Z(t, x) \equiv 1/2$. Then, for any Y, $dx/dt \leq 1/4$ on $0 \leq t \leq T$. Upon integrating this inequality and comparing with (7.5) we get that

$$\min_z \max_y \int_0^T x\,dt = x_o T + \frac{1}{8} T^2. \tag{7.6}$$

Comparing (7.6) and (7.4) shows that there can be no pure strategy solution.

Moreover, the game has no value in pure strateties.

BIBLIOGRAPHY

[1] BERKOVITZ, Leonard D., "A variational approach to differential games," This study.

[2] BERKOVITZ, Leonard D.,and DRESHER, Melvin, "A multimove infinite game with linear payoff," Pacific Journal of Mathematics, Vol. 10, 1960, pp. 743-765.

[3] BERKOVITZ, Leonard D. and DRESHER, Melvin, "A game theory analysis of tactical air war," Operations Research, Vol. 7, 1959, pp. 599-620.

Leonard D. Berkovitz

The RAND Corporation
Santa Monica, California

THE CONVERGENCE PROBLEM FOR DIFFERENTIAL GAMES, II

Wendell H. Fleming[1]

§ 1. INTRODUCTION

In a recent paper [3] with the same title as this one we considered differential games in which (speaking imprecisely) the position $\xi = (\xi^1, \ldots, \xi^r)$ is a vector-valued function of time governed by a system of r ordinary differential equations

$$(1.1) \qquad \dot{\xi} = g(\xi, y, z), \qquad 0 \leq t \leq T,$$

together with initial data

$$(1.2) \qquad \xi(0) = x.$$

The variable y is continuously controlled by the first player I and z by the second player II. The control variables y and z are selected respectively from given compact sets Y and Z in some euclidean space (or, equally well, in any metric space). Each player has complete instantaneous information about the position ξ. Play stops at time T, and player I receives

$$(1.3) \qquad U_0[\xi\,(T)] + \int_0^T f(\xi, y, z)\, dt$$

It is easy to formulate this problem precisely if one discretizes time. As in [3] let us fix $T_0 > 0$ and let

$$\delta_n = 2^{-n}\, T_0, \qquad n = 1, 2, \ldots .$$

1 This work was supported by a grant from the National Science Foundation.

Let us consider only those T such that $0 \leq T \leq T_0$ and T is a dyadic rational number times T_0,

$$T = j\,\delta_n, \qquad j = 0, 1, \ldots, 2^n$$

for some sufficiently large n. Let $n(T)$ be the smallest n for which T can be so expressed. Replacing (1.1) by a difference equation and the integral in (1.3) by a finite sum, we obtain the game described in [3, Section 2]. According to [3, Theorem 2] the value $V_n(x, T)$ of this game tends to a limit as $n \to \infty$ provided f and g have a certain special form. The purpose of the present paper is to prove the convergence of $V_n(x, T)$ without this restriction. To avoid certain technical complications we shall consider only games of prescribed duration T. This amounts to considering an initial value problem rather than a mixed initial-boundary value problem for the basic partial differential equation (1.4), below, of the differential game.

Let us assume as in [2] that f and $g = (g^1, \ldots, g^r)$ are continuous and satisfy a Lipschitz condition in x, with constant K independent of y and z. We also assume that U_0 satisfies a Lipschitz condition on every bounded set.

THEOREM. $\lim_{n \to \infty} V_n(x, T)$ exists, uniformly on bounded sets in (x, T)-space.

This result was proved in [2] under the highly restrictive assumption that the first order non-linear partial differential equation

(1.4) $$\frac{\partial V}{\partial T} = H(x, \text{grad } V), \text{ where}$$

$$H(x, p) = \operatorname*{val}_{Y,Z} [f(x, y, z) + p \cdot g(x, y, z)]$$

has a continuously differentiable solution for $0 \leq T \leq T_0$ satisfying the initial data

(1.5) $$V(x, 0) = U_0(x).$$

The symbol val denotes the value of the game over $Y \times Z$ with the expression in brackets as payoff, and grad V the gradient in the "space variables" $x = (x^1, \ldots, x^r)$. The proof in [2], as well as that in [3] for special f and g, is elementary. However the present one is not. The idea is to replace (1.4) by the second order nonlinear parabolic equation

$$\frac{\partial V}{\partial T} = \frac{a^2}{2} \Delta V + H(x, \text{grad } V), \tag{1.6}$$

where a is a small positive parameter and Δ is the Laplacian. This amounts to adding a small Brownian motion process on the right side of the integrated form of (1.1) and taking as payoff the expected value of (1.3). Games of this type are themselves of interest in connection with stochastic variational problems [4]. By some recent results of A. Friedman [5] and S. Kaplan [6] for parabolic equations, (1.6) has a solution $V^{(a)}$ in bounded domains satisfying the initial data (1.5) provided U_0 is sufficiently smooth. The two main lemmas needed in the proof are then: (1) Define the value $V_n^{(a)}(x, T)$ for the corresponding time-discrete game and show that, for each fixed $a > 0$, $V_n^{(a)}(x, T)$ tends to $V^{(a)}(x, T)$ as $n \to \infty$; (2) show that $|V_n^{(a)}(x, T) - V_n(x, T)|$ is uniformly small if a is small.

§ 2. THE FIRST LEMMA

Let B be a bounded set in r-space. The difference equation for the successive positions $x_1, x_2, \ldots$ of the nth game with discrete time is

$$x_{i+1} = x_i + \delta_n g(x, y, z), \qquad x_1 = x. \tag{2.1}$$

Since $g(x, y, z)$ satisfies a uniform Lipschitz condition in x, a standard estimate shows that there is an R such that $|x_i| \leq R$ for $i = 1, 2, \ldots, 2^n$ whenever $x \in B$. Let $R_1 = 2R$, and $R_2 > R_1$. Since the values of f and U_0 for $|x_i| \geq R$ do not affect $V_n(x, T)$ when $x \in B$, we may arrange that $f(\xi, y, z) = U_0(\xi) = 0$ when $|\xi| \geq R_1$. Let us assume that U_0 has continuous third derivatives. If not we take a smooth approximation $U_0^{(a)}$, which adds a small extra term on the right side in Lemma 2.

We consider the parabolic equation (1.6) together with the mixed initial-boundary data $V(x, T) = U_0(x)$ for $T = 0$, $|x| \le R_2$ and for $0 \le T \le T_0$, $|x| = R_2$. By a theorem of S. Kaplan [6] this initial-boundary value problem has a unique solution $U^{(a)}(x, T)$, where $U^{(a)}$ together with all of its first partial derivatives and its second partial derivatives in the "space variables" x are Hölder continuous in (x, T) for $|x| \le R_2$ and $0 \le T \le T_0$. In fact, if u denotes $U^{(a)}$ or any of its partial derivatives in question, there is an estimate of the form

$$|u(x', T') - u(x, T)| \le c\, [|x' - x|^{\beta} + |T' - T|^{\frac{\beta}{2}}],$$

where c and β are constants (depending on a) with $0 < \beta \le 1$.

Kaplan's theorem makes use of results of A. Friedman [5]. Since [6] is not yet published in journal form, we have added at the referee's suggestion an Appendix in which the needed existence theorem is derived from [5]. As Kaplan pointed out to the author, Oleĭnik and Kruzhkov [7, Section 4] independently obtained similar results which can be used to show the existence of $U^{(a)}$.

Let $h(x, T)$ be any infinitely differentiable function such that $0 \le h \le 1$, $h = 1$ for $|x| \le R_1$, and $h = 0$ for $|x| \ge R_2$. Let

$$V^{(a)}(x, T) = \begin{cases} h(x, T)U^{(a)}(x, T), & \text{if } |x| \le R_2, \\ 0, & \text{if } |x| \ge R_2. \end{cases}$$

Then $V^{(a)}$ satisfies 1.5 and 1.6 if $|x| \le R_1$.

Let us refer to the games previously described as deterministic and consider now corresponding nondeterministic games, depending on $n = 1, 2, \ldots$ and the parameter $a > 0$. To avoid certain analytical questions of integrability and the existence of a value, we shall arrange that each of these games is finite. Let y_i and z_i be chosen for $i = 1, \ldots, j$, $j\delta_n = T$, respectively from finite sets Y_n, Z_n, where

$$Y_1 \subset Y_2 \subset \ldots \subset Y, \quad Z_1 \subset Z_2 \subset \ldots \subset Z$$

and the union of the sets Y_n (sets Z_n) is dense in Y (in Z). In place of the sequence $x_1, x_2, \ldots, x_{j+1}$ defined by (2.1) we now have a sequence $X_1, X_2, \ldots, X_{j+1}$ of random variables, where

$$X_{i+1} = X_i + \delta_n g(X_i, y_i, z_i) + a\delta^{\frac{1}{2}} \eta_i, \; X_1 = x. \tag{2.2}$$

Here $\eta_i = (\eta_i^1, \ldots, \eta_i^r)$ and the random variables η_i^u, $i = 1, 2, \ldots, j$, $u = 1, \ldots, r$ are mutually independent normalized random variables (mean 0, variance 1), which assume only the values 1 and −1. (Equally well, η_i^u could assume finitely many values. The central limit theorem suggests that the form of the distribution of η_i^u is unimportant for large n. Were it not for the integrability problems mentioned before, we would let η_i^u be Gaussian.)

Just as in the deterministic case $(a = 0)$ both players move simultaneously. At the ith move each player knows all previous moves y_k, z_k and all values of η_k for $k = 1, \ldots, i-1$, as well as the initial position x. The game stops at the first time $t_F = \delta_n F$ when either $t_F = T$ (i.e., $F = j$) or $|X_{F+1}| \geq R_1$. The payoff is the expected value

$$\exp\left\{\delta_n \sum_{i=1}^{F} f(X_i, y_i, z_i) + V^{(a)}(X_{F+1}, t_F)\right\}.$$

If $t_F < T$ (i.e., $F < j$) we agree that play continues until time T, even though the moves after the Fth one do not affect the outcome.

For each $a > 0$ and $n \geq n(T)$ this game is defined and has a value $V_n^{(a)}(x, T)$ satisfying the functional equation

$$V_n^{(a)}(x, T) = \operatorname*{val}_{Y_n, Z_n} \operatorname*{exp}_{\eta} [\delta_n f(x, y, z) + V_n^{(a)}(X, T - \delta_n)], \tag{2.3}$$

if $|x| < R_1$ and $0 < T \leq T_0$, where the components η^u of η are mutually independent random variables with the same distribution as η_i^u,

$$X = x + \delta_n g(x, y, z) + a\delta_n^{\frac{1}{2}} \eta; \tag{2.4}$$

and

$$V_n^{(a)}(x, T) = V^{(a)}(x, T) \tag{2.5}$$

if $|x| \geq R_1$ or $T = 0$. The game has a solution in mixed strategies $\mu(x, T; y)$, $\nu(x, T; z)$, where for each (x, T), $\mu(x, T;\ )$ is a probability measure on Y_n and $\nu(x, T;\ )$ a probability measure on Z_n. A larger class of mixed strategies which depend on complete past history will be used in the proof of Lemma 2.

LEMMA 1. There exist constants $A > 0$ and $0 < \alpha \leq 1$ (both depending on a) such that the following is true: given $\epsilon > 0$ there exists n_0 such that

$$|V_n^{(a)}(x, T) - V^{(a)}(x, T)| \leq (\epsilon + A\ \delta_n^{\alpha})T$$

for every (x, T) and $n \geq \max\ (n_0, n(T))$.

PROOF. For short let us write $\delta_n = \delta$, and define X as in (2.4). By Taylor's formula

$$V^{(a)}(X, T-\delta) = V^{(a)}(X, T) - \delta\ \frac{\partial V^{(a)}}{\partial T}\ (X, T) + \theta_1$$

$$V^{(a)}(X, T) = V^{(a)}(x, T) + \text{grad}\ V^{(a)} \cdot (X - x)$$

$$+ \frac{1}{2} \sum_{u,v=1}^{r} \frac{\partial^2 V^{(a)}}{\partial x^u \partial x^v}\ (X^u - x^u)(X^v - x^v) + \theta_2\ .$$

In the last expression the partials are evaluated at (x, T). For suitable constants A_1, A_2,

$$|\theta_1| \leq A_1\ \delta^{1+\frac{\beta}{2}},\ |\theta_2| \leq A_2\ |X - x|^{2+\beta}\ .$$

Since

$$|\frac{\partial V^{(a)}}{\partial T}\ (X, T) - \frac{\partial V^{(a)}}{\partial T}\ (x, T)| \leq A_3\ |X - x|^{\beta}$$

we have

$$V^{(a)}(X,\ T-\delta) = V^{(a)}(x,\ T) - \delta \frac{\partial V^{(a)}}{\partial T} + \text{grad } V^{(a)} \cdot (X - x) + \frac{1}{2} \sum_{u,v=1}^{r} \frac{\partial^2 V^{(a)}}{\partial x^u \partial x^v} (X^u - x^u)(X^v - x^v) + \theta, \tag{2.6}$$

where

$$|\theta| \leq A_4 \, [\delta^{1+\frac{\beta}{2}} + \delta |X - x|^{\beta} + |X - x|^{2+\beta}].$$

Taking expected values, we have

$$\underset{\eta}{\exp}\ V^{(a)}(X,\ T-\delta) = V^{(a)}(x,\ T) - \delta \frac{\partial V^{(a)}}{\partial T} + \delta \text{ grad } V^{(a)} \cdot g + \frac{\delta a^2}{2} \Delta V^{(a)} + \exp\ \theta\ . \tag{2.7}$$

Given $\varepsilon > 0$ there exist n_0 and a function $\mu^*(x,\ T;\ \ )$ defined for $|x| \leq R_1$ and $0 \leq T \leq T_0$ with values probability measures on Y_{n_0}, such that $\mu^*(x,\ T;\ \ )$ is an ε-effective mixed strategy for the game over $Y \times Z$ with payoff $f + \text{grad } V^{(a)} \cdot g$. Thus, for every probability measure ν on z,

$$\int_Y \int_Z [f(x,y,z) + \text{grad } V^{(a)}(x,T) \cdot g(x,y,z)] d\nu(z) d\mu^*(x,T;y) \geq H(x,\ \text{grad } V^{(a)}(x,T)) - \varepsilon .$$

If $\gamma \leq 3$ and $|x| \leq R_1$,

$$|X - x|^{\gamma} = |\delta g + a\delta^{\frac{1}{2}}\eta|^{\gamma} \leq A_5 [\delta^{\gamma} + (a\delta^{\frac{1}{2}}\eta)^{\gamma}],$$

and

$$\underset{\eta}{\exp}\ |\eta|^{\gamma} = 1\ .$$

We have

$$|\underset{\eta}{\exp}\ \theta| \leq \underset{\eta}{\exp}\ |\theta| \leq A_6 [\delta^{1+\frac{\beta}{2}} + \delta^{1+\beta} + \delta^{2+\beta}].$$

If $\delta < 1$ (as is the case for n_0 large enough) we have

$$|\underset{\eta}{\exp}\ \theta| \leq A\delta^{1+\alpha},$$

where $\alpha = \beta/2$ and $A = 3A_6$.

Since $V^{(a)}$ satisfies (1.6) for $|x| \leq R_1$ and $0 \leq T \leq T_0$ we have

$$\int_Y \int_Z \exp_\eta \left\{\delta f + V^{(a)}(X, T-\delta)\right\} d\nu(z)\, d\mu^*(x, T; y) \tag{2.8}$$

$$\geq V^{(a)}(x, T) - \delta\, \epsilon - A\delta^{1+\alpha} .$$

Let n be as in the lemma, and let I use mixed strategy $\mu^*(x, T; \quad)$ and II any mixed strategy $\nu(x, T; \quad)$. The payoff using these strategies for a game of duration T starting at position x satisfies the functional equation

$$P(x, T) = \int_Y \int_Z \exp \left\{\delta f + P(X, T-\delta)\right\} d\nu(x, T; z)\, d\mu^*(x, T; y),$$

if $|x| < R_1$, $T > 0$, and

$$P(x, T) = V^{(a)}(x, T), \text{ if } |x| \geq R_1 \text{ or } T = 0.$$

Let us show by induction that, for every x and T of the form $j\delta$,

$$P(x, T) \geq V^{(a)}(x, T) - (\epsilon + A\delta^{\alpha})T. \tag{2.9}$$

This is clearly true for $T = 0$. Suppose that it is true for time $T - \delta$ and all x. Then

$$P(X, T-\delta) \geq V^{(a)}(X, T-\delta) - (\epsilon + A\delta^{\alpha})(T - \delta),$$

and (2.9) follows from this and (2.8).

By (2.9)

$$V_n^{(a)}(x, T) \geq V^{(a)}(x, T) - (\epsilon + A\delta^{\alpha})T.$$

By considering an ϵ-effective $\nu^*(x, T; \quad)$ we obtain the corresponding estimate

$$V_n^{(a)}(x, T) \leq V^{(a)}(x, T) + (\varepsilon + A\delta^{\alpha})T.$$

This proves Lemma 1.

§ 3. THE SECOND LEMMA

In order to estimate $V_n^{(a)} - V_n$ we need first to estimate $|X_i - x_i|$, which with high probability is uniformly small if a is small. Let us fix (x, T) with $x \in B$, $0 \leq T \leq T_0$, and $n \geq n(T)$. We again write δ for δ_n and $j = \delta^{-1}T$. Let

$$\zeta = (\eta_1, \ldots, \eta_j),$$

$$w_i = \eta_1 + \ldots + \eta_i, \qquad i = 1, \ldots, j,$$

$$\|\zeta\| = \max\,[|w_1|, |w_2|, \ldots, |w_j|]\,.$$

For the components w_i^u, $u = 1, \ldots, r$, of w_i one has by [1, p. 106] the estimate

$$\exp\left\{\max_{i\leq j} |w_i^u|\right\} \leq 2 \exp |w_j^u|.$$

Since $j^{-\frac{1}{2}}w_j^u$ has normalized binomial distribution

$$\exp |w_j^u| \leq \frac{pj^{\frac{1}{2}}}{2}, \qquad u = 1, \ldots, r,$$

where p is a constant independent of n. Hence

$$\exp a\delta^{\frac{1}{2}} \|\zeta\| \leq p\, r\, a\, T^{\frac{1}{2}} \tag{3.1}$$

Given any y_i, z_i, η_i for $i = 1, \ldots, j$ we have

$$X_i - x_i = \sum_{k=1}^{i-1} \delta[g(X_k, y_k, z_k) - g(x_k, y_k, z_k) + a\delta^{\frac{1}{2}}w_{i-1},$$

$$|X_i - x_i| \leq K \sum_{k=1}^{i-1} \delta|X_k - x_k| + a\delta^{\frac{1}{2}} \|\zeta\|, \quad X_1 = x_1\,.$$

Then $|X_i - x_i| \leq b_i$, where $b_1 = 0$ and for $i > 1$

$$b_i = K \sum_{k=1}^{i-1} \delta b_k + m, \quad m = a\delta^{\frac{1}{2}} \|\zeta\| .$$

By an elementary estimate in differential equations,

$$\sum_{k=1}^{i-1} \delta b_k \leq \Psi(i\delta),$$

where

$$\dot{\Psi} = K\Psi + m, \quad \Psi(0) = 0,$$

and hence $\Psi(t) = mK^{-1}(e^{Kt} - 1)$. From this, for each $i = 1, \ldots, j$,

$$|X_i - x_i| \leq e^{Ki\delta} a\delta^{\frac{1}{2}}\|\zeta\| \leq e^{KT_0} a\delta^{\frac{1}{2}}\|\zeta\|. \tag{3.2}$$

Since $x \in B$, $|x_i| \leq R$ for each $i = 1, \ldots, j$. Therefore $F = j$ provided $a\delta^{\frac{1}{2}}\|\zeta\| < e^{-KT_0}R$. For the probability that this event does not occur we have the estimate

$$\Pr\left\{ a\delta^{\frac{1}{2}}\|\zeta\| \geq e^{-KT_0} R\right\} \leq \frac{e^{KT_0}}{R} \exp a\delta^{\frac{1}{2}}\|\zeta\| . \tag{3.3}$$

If $F = j$ then $V^{(a)}(X_{F+1}, t_F) = U_0(X_{F+1})$, and if L is a Lipschitz constant for U_0,

$$|\delta \sum_{i=1}^{F} f(X_i, y_i, z_i) - \delta \sum_{i=1}^{F} f(x_i, y_i, z_i)| \leq KT_0 e^{KT_0} a\delta^{\frac{1}{2}}\|\zeta\| \tag{3.4a}$$

$$|V^{(a)}(X_{F+1}, t_F) - U_0(x_{F+1})| \leq Le^{KT_0} a\delta^{\frac{1}{2}}\|\zeta\| . \tag{3.4b}$$

From the maximum principle $V^{(a)}$ is bounded independent of a. See for instance [7, p. 133]. Then there is an M such that

$$|\delta \sum_{i=1}^{j} f(x_i, y_i, z_i) + U_0(x_{j+1})| \leq M, \tag{3.5a}$$

$$|\delta \sum_{i=1}^{F} f(X_i, y_i, z_i) + V^{(a)}(X_{F+1}, t_F)| \leq M. \tag{3.5b}$$

Let

$$C = pre^{KT_0}(KT_0 + L + \frac{2M}{R}) . \tag{3.6}$$

LEMMA 2. For every $x \in B$, $0 \leq T \leq T_0$, $a > 0$, and $n \geq n(T)$,

$$|V_n^{(a)}(x, T) - V_n(x, T)| \leq C\, a\, T^{\frac{1}{2}} .$$

PROOF. Let x, T, a, and n be given as in the lemma. It is convenient to use mixed strategies which depend on complete past history in the following way. Let $\Theta_1 = Y_n$, $\Omega_1 = Z_n$. For $i = 2, \ldots, j$ let Θ_i denote the set of all functions $y(y_k, z_k, k = 1, \ldots, i-1)$ with values in the finite set Y_n, and Ω_i the set of all functions $z(y_k, z_k, k = 1, \ldots, i-1)$ with values in Z_n. The elements of Θ_i, Ω_i are pure strategies for the ith move of the deterministic game. A mixed strategy for player I in the deterministic game is a probability measure ϕ on the cartesian product $\Theta = \Theta_1 \times \cdots \times \Theta_j$. Mixed strategies Ψ for player II are probability measures on $\Omega = \Omega_1 \times \cdots \times \Omega_j$.

For the non-deterministic game we replace Θ_i by $\tilde{\Theta}_i$, the set of all functions $y(y_k, z_k, \eta_k, k = 1, \ldots, i-1)$ with values in Y_n, and similarly replace Ω_i by $\tilde{\Omega}_i$, ϕ by $\tilde{\phi}$, Ψ by $\tilde{\Psi}$.

A strategy ϕ for the deterministic game defines a strategy (again denoted by ϕ) for the non-deterministic game, namely, ϕ assigns probability 1 to those j-tuples of functions which do not depend on $\zeta = (\eta_1, \ldots, \eta_j)$. On the other hand, given ζ a strategy $\tilde{\phi}$ for the non-deterministic game induces a strategy (denoted by $\tilde{\phi}|\zeta$) for the deterministic game, since prescribing ζ maps $\tilde{\Theta}$ into Θ in the obvious way, and this mapping induces a mapping of probability measures. Similar remarks apply to the second player.

Now let ϕ^* be an optimal strategy for player I in the deterministic game, and $\tilde{\Psi}$ any strategy for player II in the non-deterministic game. Let $\tilde{p}$ be the payoff if I uses ϕ^* in the non-deterministic game and II uses $\tilde{\Psi}$, and $\tilde{p}|\zeta$ the payoff given ζ. Then

$$\tilde{p} = \exp \tilde{p}|\zeta .$$

Let $p|\zeta$ be the payoff in the deterministic game for the strategies ϕ^* and $\tilde{\Psi}|\zeta$. Since ϕ^* is optimal

$$p|\zeta \geq V_n(x, T) \tag{3.7}$$

for every ζ. If $a\delta^{\frac{1}{2}} \|\zeta\| \leq Re^{-KT_0}$, then from (3.4a) and (3.4b), which hold for pure strategies and hence also for mixed strategies,

$$\tilde{p}|\zeta \geq V_n(x, T) - e^{KT_0}(KT_0 + L)\, a\delta^{\frac{1}{2}} \|\zeta\| . \tag{3.8}$$

In any case from (3.5a) and (3.5b)

$$\tilde{p}|\zeta \geq -M, \quad V_n(x, T) \geq -M . \tag{3.9}$$

Let E be the event $a\delta^{\frac{1}{2}} \|\zeta\| \leq e^{-KT_0} R$, E' the complementary event, and γ the probability distribution of ζ. Then

$$\tilde{p} = \int_E \tilde{p}|\zeta \, d\gamma + \int_{E'} \tilde{p}|\zeta \, d\gamma \geq \int_E [V_n(x, T) - e^{KT_0}(KT_0 + L)a\delta^{\frac{1}{2}}\|\zeta\|]d\gamma - M\,Pr(E'),$$

$$p \geq V_n(x, T)(1 - Pr(E')) - e^{KT_0}(KT_0 + L) \exp a\delta^{\frac{1}{2}}\|\zeta\| - M\,Pr(E') .$$

Using (3.1) and (3.3)

$$\tilde{p} \geq V_n(x, T) - C\, a\, T^{\frac{1}{2}} .$$

Since $\tilde{\Psi}$ is arbitrary

$$V_n^{(a)}(x, T) \geq V_n(x, T) - C\, a\, T^{\frac{1}{2}} .$$

Similarly, choosing an optimal mixed strategy Ψ^* for II, we get

$$V_n^{(a)}(x, T) \leq V_n(x, T) + C\, a\, T^{\frac{1}{2}} .$$

These two inequalities give Lemma 2.

§ 4. PROOF OF THE THEOREM

By Lemmas 1 and 2, for any $a > 0$, $\epsilon > 0$, $x \in B$, and $0 \leq T \leq T_0$ such that T is a dyadic rational number times T_0,

$$\limsup_{n \to \infty} V_n(x, T) \leq V^{(a)}(x, T) + \epsilon T + C a T^{\frac{1}{2}},$$

$$\liminf_{n \to \infty} V_n(x, T) \geq V^{(a)}(x, T) - \epsilon T - C a T^{\frac{1}{2}}.$$

Since ϵ and a are arbitrary, $V_n(x, T)$ tends to a limit $V(x, T)$ as $n \to \infty$. By the same proof as for [3, Lemma 1], the functions V_n satisfy a uniform Lipschitz condition on B. Hence the sequence V_n converges uniformly on B. Since B is an arbitrary bounded set the theorem is proved.

§ 5. REMARKS

The limit $V(x, T)$ of the sequence $V_n(x, T)$ has a unique extension to values of T in the interval $[0, T_0]$ which are not of the form a dyadic rational times T_0. It is conjectured that this extended function $V(x, T)$ is the value of the differential game with continuous time, in the sense described in [2, part 3] and [3, part 6]. Since V satisfies a Lipschitz condition on bounded sets, V is totally differentiable almost everywhere by a theorem of Rademacher. It is conjectured that V satisfies (1.4) almost everywhere. The author hopes to return later to these questions, as well as a discussion of games whose duration T is not prescribed in advance.

APPENDIX

The following argument is a variant of one due to Kaplan [6], and shows that the equation

$$(1.6) \qquad \frac{\partial V}{\partial T} = \frac{a^2}{2} \Delta V + H(x, \text{grad } V), \quad \text{where}$$

$$H(x,p) = \operatorname*{val}_{Y,Z} [f(x, y, z) + p \cdot g(x, y, z)],$$

has a unique solution $U^{(a)}$ satisfying the given initial-boundary data. First of all, H satisfies a local Lipschitz condition in x of the form

$$|H(x', p) - H(x, p)| \leq K(1 + |p|)\ |x' - x|$$

for $|x|, |x'| \leq R_2$ and all p, and a Lipschitz condition in p,

$$|H(x, p') - H(x, p)| \leq c_1\ |p' - p|$$

where c_1 is an upper bound for $|g(x, y, z)|$ on the compact set $(|x| \leq R_2) \times Y \times Z$. Let us write

$$H(x, p) = G_0(x) + \sum_{i=1}^{r} G_i(x, p)p^i ,$$

where

$$G_0(x) = H(x, 0)$$

$$G_i(x, p) = \begin{cases} \dfrac{H(x, P_i) - H(x, P_{i-1})}{p^i}, & \text{if } p^i \neq 0 \\ c_1 & \text{if } p^i = 0 \end{cases}$$

for $i = 1, \ldots, r$, where $P_0 = 0$, $P_1 = (p^1, 0, 0, \ldots, 0)$, $P_2 = (p^1, p^2, 0, \ldots, 0), \ldots, P_r = p$. The functions $G_1, \ldots, G_r$ need not be continuous, but at any rate are Baire functions. Since $P_i - P_{i-1} = p^i$, we have the estimate

$$|G_i(x, p)| \leq c_1 \quad \text{for } i = 1, \ldots, r .$$

According to Friedman [5, part II, p. 541, 552], if $V(x, T)$ is a solution of

$$\frac{\partial V}{\partial T} = \frac{a^2}{2} \Delta V + \sum_{i=1}^{r} b_i(x, T) \frac{\partial V}{\partial x^i} + \varphi(x, T) ,$$

there is a constant c depending only on the initial-boundary data, on a, and on

$$s = \sup_{(x,T)} [|b_1(x, T)| + \ldots + |b_r(x, T)| + |\varphi(x, T)|] ,$$

such that $|\text{grad } V(x, T)| \leq c$ for every (x, T). We suppose of course that $|x| \leq R_2$ and $0 \leq T \leq T_0$. Friedman assumes that $b_1, \ldots, b_r$ and φ are continuous. However, an approximation argument shows that the estimate remains true if $b_1, \ldots, b_r$ and φ are merely bounded and measurable [4, Section 3].

Now let

$$s_0 = rc_1 + \max |G_0(x)| ,$$

and c_0 Friedman's constant with $s = s_0$. Let $\Psi(p)$ be an infinitely differentiable function such that $\Psi(p) = 1$ for $|p| \leq c_0$, $\Psi(p) = 0$ for $|p| \geq 2c_0$, and $0 \leq \Psi \leq 1$. The function ΨH is bounded, and the existence theorem [5, part II, p. 555] applies to show that the equation

$$\frac{\partial V}{\partial T} = \frac{a^2}{2} \Delta V + \Psi H$$

has a unique solution $U^{(a)}$ satisfying the given initial-boundary data. Let

$$b_i(x, T) = \Psi[\nabla U^{(a)}(x, T)]G_i[x, \nabla U^{(a)}(x, T)]$$

$$\varphi(x, T) = \Psi[\nabla U^{(a)}(x, T)]G_0(x) ,$$

where for short we set $\text{grad } V = \nabla V$. Then

$$\frac{\partial U^{(a)}}{\partial T} = \frac{a^2}{2} \Delta U^{(a)} + \sum_{i=1}^{r} b_i(x, T) \frac{\partial U^{(a)}}{\partial x^i} + \varphi(x, T) ,$$

and $s \leq s_0$. Hence $|\text{grad } U^{(a)}(x, T)| \leq c_0$. But $\Psi(p) = 1$ for $|p| \leq c_0$. Consequently,

$$\frac{\partial U^{(a)}}{\partial T} = \frac{a^2}{2} \Delta U^{(a)} + H(x, \nabla U^{(a)})$$

as required.

To show uniqueness let U be any solution of (1.6) and the given initial-boundary data. Then choose $c_2 > c_0$ such that $c_2 \geq \max |\text{grad } U(x, T)|$, and Ψ as before with c_0 replaced by c_2. By Friedman's uniqueness theorem $U = U^{(a)}$.

ADDED IN PROOF

From Theorem 14 of [7] the initial value problem (1.5)-(1.6) has a solution. Using this solution in place of the function $V^{(a)}$ one need not introduce the artificial bound $|X_i| \leq R_1$. The proof of Lemma 2 can be simplified accordingly.

BIBLIOGRAPHY

[1] DOOB, J. L., Stochastic Processes, New York, 1953.

[2] FLEMING, W. H., "A note on differential games of prescribed duration," Contributions to the Theory of Games, Vol. III (1957), pp. 407-416.

[3] FLEMING, W. H., "The convergence problem for differential games," Journal Math. Anal. And Appl., Vol. 3 (1961), pp. 102-116.

[4] FLEMING, W. H., "Some Markovian optimization problems," Journal Math. and Mech. Vol. 12 (1963).

[5] FRIEDMAN, A., "On quasi-linear parabolic equations of the second order," Journal Math. and Mech. Part I, Vol. 7 (1958), pp. 793-809.

[6] KAPLAN, S., On the Global Existence of Solutions of Nonlinear Parabolic Equations. Harvard thesis, 1961; see also Abstract No. 583-29 in Notices AMS, Vol. 8 (1961).

[7] OLEINIK, O. A. and Kruzhkov, S. N., "Quasi-linear parabolic equations of the second order with many independent variables," (in Russian), Uspekhi Mat. Nauk, Vol. 16 (1961), pp. 115-155.

Wendell H. Fleming

MARKOV GAMES

Lars Erik Zachrisson

§ 1. INTRODUCTION

The concept of a Markov game can be described roughly as follows:

1. A two-indexed set of Markov processes $M_{x,y}$ is given.
2. To every realization of a Markov process (each state history) is attached a payoff from player B to player A.
3. Player A chooses x to maximize the payoff and player B chooses y to minimize it.

The following introduction will describe the situation in more detail.

A Markov process is a set of stochastic variables $z(\tau)$ generated by transition probabilities. In such simple kinds of Markov processes as we are going to treat here, $z(\tau)$ can take one of the values $0, 1, \ldots, N$, for each possible <u>time</u> τ, and the domain of τ is either the set of integers $1, \ldots, T$ (the <u>discrete</u> case) or the closed interval $[0, T]$ (the <u>continuous</u> case). The value of $z(\tau)$ is the <u>state</u> of the system considered, and at each time there is a probability distribution $q_i(\tau)$ on the states $i (i = 1, 2, \ldots, N; \Sigma\, q_i(\tau) = 1,\ q_i(\tau) \geq 0)$.

In the discrete case there is a conditional probability $p_{ij}(\tau)$ that, the state being <u>i</u> at time τ, the state will be <u>j</u> at time $\tau + 1$.

$$\sum_j p_{ij}(\tau) = 1; \qquad p_{ij}(\tau) \geq 0 \,. \tag{1.1}$$

The following relationship (the "forward" Kolmogorov equation) between $q_i(\tau + 1)$ and $q_j(\tau)$ will ensue

$$q_i(\tau + 1) = \sum_j q_j(\tau)\, p_{ji}(\tau) \,. \tag{1.2}$$

If we define $q(\tau)$ as the row-vector

$$q(\tau) = \left\{q_1(\tau),\ q_2(\tau),\ \ldots,\ q_n(\tau)\right\} \tag{1.3}$$

and $P(\tau)$ as the matrix

$$P(\tau) = \left\{p_{ij}(\tau)\right\} \tag{1.4}$$

the relation (1.2) can be written

$$q(\tau + 1) = q(\tau)\ P(\tau)\ . \tag{1.5}$$

Given the probability vector $q(0)$, all the following $q(\tau)$ are determined by (1.5).

The joint probability of the states $i_{\tau_1}(\tau_1),\ i_{\tau_2}(\tau_2),\ \ldots,\ i_{\tau_n}(\tau_n)$ $(\tau_1 < \tau_2 \cdots < \tau_n)$ is

$$\Sigma\ q_{i_0}(0)\ p_{i_0 i_1}(1)\ p_{i_1 i_2}(2) \cdots p_{i_{\tau_n - 1} i_{\tau_n}}(\tau_n)$$

where the summation is extended over all $i_0,\ i_1 \ldots$ except $i_{\tau_1}, i_{\tau_2}, \ldots, i_{\tau_n}$. These latter indices are fixed. Hence all probability distributions necessary for the qualification of the Markov process as a stochastic process in the sense of probabilistic theory are defined.

In the continuous case it is assumed, that the conditional probability, the state being $\underline{i}$ at time τ will be $\underline{j}$ at time $\tau + \Delta\tau$, satisfies

$$p_{ij} \cdot \Delta\tau + o(\Delta\tau) \qquad (j \neq i)$$

$$1 - \sum_{j \neq i} p_{ij} \cdot \Delta\tau + o(\Delta\tau) \qquad (j = i)$$

and we get

$$\frac{dq_i}{d\tau} = \sum_j q_j(\tau)\ p_{ji}(\tau) \tag{1.6}$$

where

$$p_{ii} = - \sum_{j \neq i} p_{ij} \; .$$

In matrix formulation (1.6) is

$$\frac{dq}{d\tau} = q(\tau)\; P(\tau). \tag{1.7}$$

In the matrix P each row has the sum zero, and all its elements except the diagonal ones are nonnegative.

In the <u>game case</u> we suppose, that, at each time τ, there is available a set of transition probability matrices

$$P(\tau;\; x,\; y)$$

where x takes values from some compact, generally convex set $X(\tau)$ in a finite dimensional Euclidean space and y takes values from a compact (convex) set $Y(\tau)$ in some other Euclidean space. We have two players, A and B, and player A chooses some function $x(\tau)$ with values from $X(\tau)$ and player B chooses a function $y(\tau)$ from $Y(\tau)$. This choice will pick a definite Markov process out of the number of possible ones and will thus determine a probability distribution on the possible state histories. Suppose that to each state history there is attached a certain payoff K from player B to player A. Player A wants to maximize K and player B wants to minimize it. Evidently we have here a zero-sum two-person game, and it may be called a <u>Markov game</u> because of its intimate relation with the concept of a Markov process.

To complete the picture of the above game situation it is necessary to describe the information available to each player in the course of the play. We have a game of <u>perfect state information</u> if each player knows the state of the system at the current time. In this case it is possible to give a rather complete theory with the main assertion, that in a certain sense local optimization implies total optimization, at least if we restrict the payoff to "additive" payoff-functions. In the discrete case this means that the payoff of a time history passing through the states $i_0(0),\; i_1(1), \ldots, i_T(T)$

can be expressed in the form

$$w_{i_0}(0) + w_{i_1}(1) + \dots + w_{i_T}(T) .$$

However, in the battle or duel situations the assumption of a perfect state information may be rather unrealistic. In such cases the state is a composite one, described by a pair of numbers $(\underline{i}, \underline{l})$, where $\underline{i}$ is the state of player A and $\underline{l}$ is the state of player B. In many cases A may know his own state $\underline{i}$ but knows nothing of his opponent's state. He may be able to give a probability distribution on B's states. With reference to Kuhn's work on Extensive Games [1] we call a game, where each player knows his own state history and can make the choice of his function $x(\tau)$ or $y(\tau)$ dependent on this time history a Markov game of perfect state recall. The theory of such games, has, however, hardly left its starting point, viz. its definition.

The research reported in this contribution was mainly done in 1954 and 1955, when the theory of Markov games of perfect state information was developed and applied to several operations research problems in the Research Institute of National Defence in Sweden [2, 3, 4]. Due to other tasks it has not been possible for me to go on with my work in this field since that date. Therefore the theory as it now stands is limited almost exclusively to the case of perfect state information. However, I am sincerely grateful to Dr. Melvin Dresher for the invitation extended to me to contribute to the present volume.

§ 2. DISCRETE TIME. FUNDAMENTAL CONCEPTS

In the rest of the paper we shall not use natural time τ to indicate position along the time axis but reverse time t.

$$t = T - \tau .$$

Thus t denotes the interval of time that is left from the present time until the game is ended. The beginning of the play, as the players see it, is at $t = T$ and the end at $t = 0$. However, as will be seen below, the mathematical solution of the game essentially depends on an integration of a difference equation, starting with boundary values at $t = 0$ and working backwards in natural time but forward in reverse time. This is the reason for the choice of time indication. From now on "time" will generally mean "reverse time". In ambiguous cases we shall use "reverse time" or "natural time".

In the discrete case, t takes the values $0, 1, 2, \ldots, T$. The system can be in N different states, $1, 2, \ldots, N$. At each reverse time t a matrix of transition probabilities is given:

$$P(t; x, y) = p_{ij}(t; x, y)$$

where $p_{ij}(t; x, y)$ is the probability of a transition from state $\underline{i}$ at reverse time $t + 1$ to state $\underline{j}$ at reverse time t. x and y are two sets of parameters; x will be determined by player A and y by player B.

Now we suppose, that if the state is $\underline{i}$ at time t, then B will have to pay A $w_i(t)$. In many applications, $w_i(t)$ will be zero for all t except $t = 0$. In that case, the states at the end of the play $(t = 0)$ determine the outcome of the play. $w(t)$ is defined as the column vector with the components $w_i(t)$. We will call $w(t)$ the current payoff vector.

If the state is $\underline{i}$ at time t and x and y are given functions of t, then there will exist an expected value $v_i(t)$ of what B will have to pay A from reverse time t (inclusive) to the end of the game (inclusive). The $v_i(t)$ $(i = 1, 2, \ldots, N)$ will form a new column vector $v(t)$. We will call $v(t)$ the value vector at (reverse) time t.

It is evident that

$$v(0) = w(0) \ . \tag{2.1}$$

$v(t + 1)$ can be determined from $v(t)$ by means of the almost self-evident recursive relation

$$\boxed{v(t + 1) = w(t + 1) + P(t; x, y)\, v(t)} \tag{2.2}$$

or

(2.3) $$v_i(t+1) = w_i(t+1) + \sum_j p_{ij}(t)\, v_j(t) .$$

This important recursive relation, essentially equivalent to the Kolmogorov "backward" equation, is the starting point of the theory of Markov games with perfect state information. Given the strategy functions x and y it determines the value vector at every time t from the initial values (2.1) (end values in natural time).

The rest of this paragraph (§ 2) has no importance in the case of perfect state information. However, in other cases it can not be neglected.

Suppose, that the states at the beginning of the play $(t = T)$ are distributed according to a distribution function $q_i(T)$, where $q_i(T)$ is the probability of the state $\underline{i}$. Let $Q(T)$ be the row vector formed by the $q_i(T)$ $(i = 1, 2, \ldots, N)$. Then the payoff of the game will be

(2.4) $$q(T)\, v(T) = \sum_{i=1}^{N} q_i(T)\, v_i(T) .$$

Player A chooses x in order to maximize $q(T)\, v(T)$ and player B chooses y in order to minimize it. The game is solvable if there is a saddle-point.

There is a recursive relation (the "forward" equation) between $q(t)$ and $q(t + 1)$:

(2.5) $$q(t) = q(t+1)\, P(t;\, x,\, y)$$

or

(2.6) $$q_j(t) = \sum_i q_i(t+1)\, p_{ij}(t;\, x,\, y).$$

This equation is adjoint to equation (2.2). It will be called the "<u>distribution relation</u>". We get, using (2.2) and (2.5),

(2.7) $$q(t+1)\, v(t+1) = q(t+1)\, w(t+1) + q(t+1)\, P(t;\, x,\, y)\, v(t) =$$
$$= q(t+1)\, w(t+1) + q(t)\, v(t) .$$

Hence, if the current payoff vector is zero, $q(t)\, v(t)$ is independent of time, as it should be.

As $q(T)$ is given, all $q(t)$ with $0 \leq t \leq T$ can be computed from (2.5), given x and y. This "integration" is performed in the natural direction of increasing time, which is reverse to the integration direction of (2.2).

§ 3. POSITIVITY

All the matrices $P(t;\, x,\, y)$ occurring in our considerations, have nonnegative elements. This fact has the immediate consequence, that if two column vectors $z(t)$ and $z(t + 1)$ have the relationship

$$z(t + 1) = P(t;\, x,\, y)\, z(t) \tag{2.8}$$

and if $z(t)$ is positive, that means if all components of $z(t)$ are nonnegative, so is $z(t + 1)$. In other words, P can be looked upon as a <u>positivity-preserving transformation</u>. Equation (2.8) is the value relation specialized to the case with the current payoff vector equal to zero.

The concept of positivity in the above sense makes it possible to introduce an order relation in the space of all vectors z. The order relation

$$z_1 \geq z_2$$

(z_1 dominates z_2) means that the vector $z_1 - z_2$ is positive.

This relation has the properties

$$z_1 \geq z_2,\; z_2 \geq z_3 \quad \text{implies} \quad z_1 \geq z_3$$

$$z_1 \geq z_2,\; z_3 \geq z_4 \quad \text{implies} \quad z_1 + z_3 \geq z_2 + z_4 \;.$$

Supremum (Sup) of a set of vectors $z(\chi)$ $(\chi \in K)$ is the vector, whose i^{th} component $(i = 1, 2, \ldots, N)$ is the Sup of the set of ith components of the vectors in the set.

$$\text{Sup } z(\chi) = \left\{\text{Sup } z_1(\chi), \ \text{Sup } z_2(\chi), \ \ldots\right\}.$$

Hence Sup $z(\chi)$ dominates every $z(\chi)$ and is dominated by all other vectors dominating every $z(\chi)$.

Inf $z(\chi)$ is defined in a similar way.

When Sup $z(\chi)$ belongs to the set of vectors, we can use "Max" for "Sup". In this case the N maximum problems:

> To find a $\chi_{i,0} \in K$ $(i = 1, 2, \ldots, N)$ which maximizes the ith component $z_i(\chi)$,

have a common solution χ_0, for which

$$z(\chi_0) = \text{Max } z(\chi).$$

A point χ_0, λ_0 is a (vector) saddle-point in the set of vectors $z(\chi, \lambda)$ if

$$z(\chi, \lambda_0) \leq z(\chi_0, \lambda_0) \leq z(\chi_0, \lambda).$$

This notion is illustrated for a two-dimensional vector space in Figure 1.

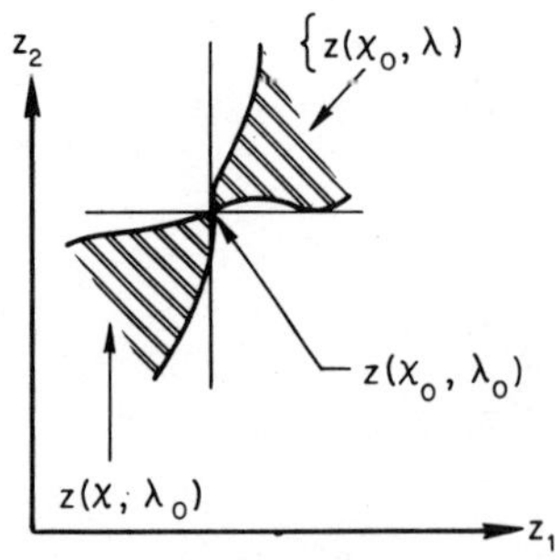

FIGURE 1

To emphasize the fact that a vector saddle-point is simultaneously a saddle-point for each of the vector components, we can call it a <u>unanimous saddle-point</u>.

The norm in the ordered vector space is the Max-norm

$$\|z\| = \operatorname*{Max}_i |z_i|$$

and the norm of the transformation is the maximal row sum:

$$\|P\| = \max_i \sum_j p_{ij} .$$

If the matrix is a complete transition matrix, as we have assumed, $\|P\| = 1$. (If there are a number of trapped states (see § 7) it may be advantageous to consider "incomplete" transition matrices with $\|P\| < 1$.)

The norm of the distribution vectors Q will be

$$\|q\| = \sum_i |q_i| .$$

All complete distribution vectors will have the norm equal to one.

We shall need some lemmas on recurrence equations (2.2) when P is positivity-preserving. The transformation $z(t+1) = P(t)\,z(t)$ is (locally) order-preserving, i.e., $z_1(t) \geq z_2(t)$ implies that $z_1(t+1) \geq z_2(t+1)$. A generalization to nonlinear transformations (functions from the space of vectors z to the same space) is given by the

DEFINITION. The transformation $z(t+1) = R[t, z(t)]$ is order-preserving if and only if

$$z_1(t) \geq z_2(t) \text{ implies that } z_1(t+1) \geq z_2(t+1).$$

The following lemma is evident by recursion.

LEMMA 1. Consider the recursive relation

$$z_i(t) = f_i(t) + R[t-1, z_i(t-1)]$$

$$z_i(0) = f_i(0)$$

with an order-preserving R for $i = 1, 2$. If $f_1(t) \geq f_2(t)$ for all t, then

$$z_1(t) \geq z_2(t)$$

for all t.

We can express the content of Lemma 1 by saying that a locally order-preserving transformation is globally order-preserving with respect to both initial values and to "driving functions" $f_i(t)$.

A special case is the following Lemma 2.

LEMMA 2. If $P(t)$ is order-preserving and all vectors $f(t)$ $(t = 0, 1, \ldots)$ are positive (negative), then all vectors $z(t)$ determined by the relations

$$z(t) = f(t) + P(t-1)\, z(t-1): \quad (t = 1, 2, \ldots) \tag{2.9}$$
$$z(0) = f(0)$$

are positive (negative).

We want to stress that according to our definition a vector is positive (negative) if all its components are nonnegative (non-positive). Thus the null-vector is both positive and negative.

It will prove expedient to <u>introduce an order relation for transformations</u>. We will say that R_1 dominates R_2 on some sub-set S of the z-space if

$$R_1(z) \geq R_2(z)$$

for every $z \in S$.

This notion will give us the following generalization of Lemma 1.

LEMMA 3. If

$$z_i(t) = f_i(t) + R_i[t-1, z_i(t-1)];$$
$$(t = 1, 2, 3, \ldots, i = 1, 2)$$
$$z_i(0) = f_i(0) \qquad (i = 1, 2)$$

and if $f_1(t)$ dominates $f_2(t)$ and R_1 and R_2 are order-preserving transformations and $R_1(t, \ldots)$ dominates $R_2(t, \ldots)$ $(t = 0, 1, \ldots)$ on some $z_0(t)$

with

$$\text{Inf}\ \{z_1, z_2\} \leq z_0 \leq \text{Sup}\ (z_1, z_2)$$

then $z_1(t)$ will dominate $z_2(t)$.

PROOF. We shall prove the lemma by induction. $z_1(0) \geq z_2(0)$ according to assumption.

Suppose $z_1(n-1) \geq z_2(n-1)$. That means

$$\text{Inf}\ [z_1(n-1), z_2(n-1)] = z_2(n-1)$$

$$\text{Sup}\ [z_1(n-1), z_2(n-1)] = z_1(n-1)$$

and

$$z_2(n-1) \leq z_0(n-1) \leq z_1(n-1)\ .$$

Now

$$\begin{aligned} z_1(n) - z_2(n) &= f_1(n) - f_2(n) + R_1[n-1, z_1(n-1)] \\ &\quad - R_2[n-1, z_2(n-1)] \\ &= \{f_1(n) - f_2(n)\} \\ &\quad + \{R_1[n-1, z_1(n-1)] - R_1[n-1, z_0(n-1)]\} \\ &\quad + \{R_1[n-1, z_0(n-1)] - R_2[n-1, z_0(n-1)]\} \\ &\quad + \{R_2[n-1, z_0(n-1)] - R_2[n-1, z_2(n-1)]\}\ . \end{aligned}$$

Here, according to the assumptions, all four differences after the last sign of equality are positive vectors and the lemma is proved.

As is evident from the assumptions, z_0 may coincide with z_1 or z_2.

§ 4. PERFECT STATE INFORMATION AND LOCAL UNANIMOUS EXTREMA

In the game situation, the transformations P depend at time t upon parameters $x(t)$ and $y(t)$. Player A chooses the value of x and player B that of y.

Now, if the players know the state of the game, they can make their choice dependent on that knowledge. A convenient way to describe this situation is to let each player have a separate set $X_i(t)$ resp. $Y_i(t)$ of parameters for every state $\underline{i}$.

This means, that equation (2.3) takes the form

$$
\begin{aligned}
&v_i(t+1) = w_i(t+1) + \sum_j p_{ij}[t, x_i(t), y_i(t)]\, v_j(t) \\
(2.10) \qquad &\qquad\qquad\qquad\qquad (t = 0, 1, 2, \dots) \\
&v_i(0) = w_i(0) \\
&x_i(t \in X_i(t); \quad y_i(t) \in Y_i(t) \, .
\end{aligned}
$$

The p_{ij} are assumed to be continuous in x_i and y_i simultaneously. The set $X(t)$ above is the Cartesian product of the sets $X_i(t)$ and $Y(t)$ is the Cartesian product of $Y_i(t)$. X_i and Y_i are supposed to be compact.

In Section 3 we have defined Max and Min of a set of vectors in the ordered Z-space. We say, that the set is unanimous if Max and Min exist. Let $x \in X(t)$ and $y \in Y(t)$ be the vectors, which have $x_i(t) \in X_i(t)$ and $y_i(t) \in Y_i(t)$ as i^{th} components. Then (2.11) defines a set of vectors $v(t+1)$ with x and y as parameters. This set is unanimous as a consequence of the assumption of perfect state information. For every component of v, only the corresponding components of x and y have to be considered and for every value of i the right member of (2.10) has a maximum with respect to x_i and a minimum with respect to y_i in the ordinary sense because of the assumptions of compactness and continuity.

Hence the transformations defined below exist in the ordered Z-space and are continuous with respect to their arguments.

DEFINITION. $R_1(t;\, y,\, v) = \underset{x\in X(t)}{\text{Max}} \left\{w(t+1) + P(t;\, x,\, y)v\right\}$

$$R_2(t;\, x,\, v) = \underset{y\in Y(t)}{\text{Min}} \left\{w(t+1) + P(t;\, x,\, y)v\right\}$$

$$R_3(t;\, v) = \underset{y\in Y(t)}{\text{Min}}\ \underset{x\in X(t)}{\text{Max}} \left\{w(t+1) + P(t;\, x,\, y)v\right\}$$

$$= \underset{y\in Y(t)}{\text{Min}}\ R_1(t;\, y,\, v)$$

$$R_4(t;\, v) = \underset{x\in X(t)}{\text{Max}}\ \underset{y\in Y(t)}{\text{Min}} \left\{w(t+1) + P(t;\, x,\, y)v\right\}$$

$$= \underset{x\in X(t)}{\text{Max}}\ R_2(t;\, x,\, v)\ .$$

These transformations are generally <u>nonlinear</u> in v.

They are, as the generating linear transformations, <u>order-preserving contractions</u>. R_1 evidently form an unanimous set, as do R_2. (A contraction is a transformation $R(v)$ with $\|R(v_1) - R(v_2)\| \leq \|v_1 - v_2\|$.)

This fact is a consequence of the following two lemmas (applied once in the case of R_1 and R_2, twice in the case of R_3 and R_4).

LEMMA 4. If the transformations $R(x, v)$ are order-preserving and form a unanimous set, then $S(v) = \text{Max}\ R(x, v)$ is order-preserving. The lemma holds when Max is replaced by Min.

PROOF. According to assumptions $v_1 \geq v_2$ implies $R(x_1, v_1) \geq R(x_1, v_2)$.

Hence

$$\underset{\xi}{\text{Max}}\ R(\xi,\, v_1) \geq R(x,\, v_2)$$

is true for every x, which implies

$$\underset{x}{\text{Max}}\ R(x,\, v_1) \geq \underset{x}{\text{Max}}\ R(x,\, v_2)\ .$$

Using $R(x, v_1) \geq \text{Min}\ R(\xi, v_2)$ the second part of the lemma is proved in a similar way.

LEMMA 5. With the assumptions of Lemma 4,

$$\|R(x_1, v_1) - R(x_1, v_2)\| \leq A\|v_1 - v_2\|$$

implies that

$$\|S(v_1) - S(v_2)\| \leq A\|v_1 - v_2\| .$$

PROOF. The proof depends on the fact, that in an ordered space the inequality

$$\|Z\| \leq B$$

is equivalent to

$$- B\ e \leq Z \leq B\ e$$

where e is the vector with all its components equal to one.

Hence, the assumption of the lemma is

$$- A\|v_1 - v_2\| \cdot e \leq R(x_1, v_1) - R(x_1, v_2) \leq A\|v_1 - v_2\| \cdot e.$$

The method of proof of Lemma 4 will then immediately give the result

$$- A\|v_1 - v_2\| \cdot e \leq S(v_1) - S(v_2) \leq A\|v_1 - v_2\| \cdot e$$

or

$$\|S(v_1)\| - S(v_2) \leq A\|v_1 - v_2\|$$

and the proof is complete.

§ 5. LOCAL AND GLOBAL OPTIMIZATION

We shall consider the sequence of vectors $v^*(0) = w(0)$, $v^*(1)$, $v^*(2)$, ..., $v^*(t)$, ... which is obtained by local optimization, that is by solving the equations

$$(2.11)\quad \begin{aligned} v^*(t+1) &= R_3[t, v^*(t)] \qquad (t = 0, 1, 2, \ldots) \\ v^*(0) &= w(0) \end{aligned}$$

with R_3 equal to the optimized transformation

$$(2.12)\qquad R_3(t, v) = \operatorname*{Min}_{y=Y(t)} \operatorname*{Max}_{x=X(t)} \left\{w(t+1) + P(t; x, y)v\right\} .$$

Evidently, the sequence is well defined and unique.

The sequence determines, not necessarily uniquely, two strategy-functions $x^*(t)$ and $y^*(t)$ in the following way.

An arbitrary $y(t) \in Y(t)$ determines an $x(t) = x_1[y(t)]$ such that

$$\begin{aligned} R_1[t; y(t), v^*(t)] &= \operatorname*{Max}_{x \in X(t)} \left\{w(t+1) + P[t; x(t), y(t)]v^*\right\} \\ &= w(t+1) + P\left\{t; x_1[y(t)], y(t)\right\}v^*(t) . \end{aligned}$$

Then $y^*(t)$ is determined such that

$$R_3(t, v^*) = \operatorname*{Min}_{y \in Y(t)} R_1[t; y(t), v^*(t)] = R_1[t; y^*(t), v^*(t)] .$$

At last $x^*(t)$ is defined as

$$x^*(t) = x_1[y^*(t)] .$$

Now, evidently,

$$R_3[t, v^*(t)] = w(t+1) + P[t, x^*(t), y^*(t)]v^*(t) .$$

Thus we have defined the matrix $P(t; x^*, y^*)$ of a linear transformation, such that the transformation $w(t + 1) + P(t; x^*, y^*)v$ in a sense coincides with the nonlinear transformation $R_3(t, v)$ at v^*.

We will call it the <u>coincident linear transformation</u> and denote it by $P^*(t)$:

$$P^*(t) = P[t, x^*(t), y^*(t)] \ . \tag{2.13}$$

Hence, $v^*(t)$ is also a solution of the equations

$$\begin{aligned} v^*(t + 1) &= w(t + 1) + P^*(t)v^*(t) \qquad (t = 0, 1, 2, \ldots) \\ v^*(0) &= w(0) \end{aligned} \tag{2.14}$$

that should be compared to (2.11).

Summing up, by solving the local optimization problem we have defined <u>one optimal value function</u> $v^*(t)$, <u>two optimal strategy functions</u> $x^*(t)$ and $y^*(t)$ and <u>one optimal transition matrix or transformation</u> $P^*(t)$.

What will be proved now is that <u>the solution of the local optimization problem is also a solution of the global one</u> in the following sense.

Compare v^* with the solution v' belonging to an arbitrary x:

$$\begin{aligned} v'(t + 1) &= w(t + 1) + P[t; x(t), y^*(t)]v'(t) \\ v'(0) &= w(0) . \end{aligned} \tag{2.15}$$

The local inequality

$$P[t; x(t), y^*(t)]v^*(t) \leq P[t; x^*(t), y^*(t)]v^*(t) \tag{2.16}$$

is valid, as

$$\operatorname*{Max}_{x} P[t, x, y^*(t)]v^*(t) = P[t, x^*(t), y^*(t)]v^*(t) \ .$$

Expressed in the terminology introduced in Sec. 3, $P(t, x^*, y^*)$ dominates $P_3(t, x, y^*)$ on v^*.

This gives the global inequality

$$v'(t) \leq v^*(t) \qquad (t = 0, 1, \ldots) ,$$

as a simple consequence of Lemma 3. We have only to identify R_1 of the lemma with $w + P^* v$, R_2 with $w + P[t, x, y^*]v$, F_1 and F_2 with zero and Z_0 with v^* to obtain the result.

We formulate this result as a theorem.

THEOREM 1. The value vector $v^*(t)$ obtained from the local Min Max-problem dominates for each t any value vector $v'(t)$ obtained by keeping the B-strategy fixed but changing the A-strategy.

It should be observed that in deriving the optimal strategies $x^*(t)$ and $y^*(t)$ we have used the value vector $v^*(t)$, but after their derivation they can be looked upon as functions of t only and the players have no need to know them in any other way. It should also be observed that the length of the play, T, does not influence the <u>mathematical</u> solution of the game in any way. (This statement, of course, refers to reverse time. The optimal behaviour of the players will in general be different in identical situations at different "times to go.")

Similarly we obtain

THEOREM 2. The value vector $v^{**}(t)$ obtained from the local Max Min-problem is, for each t, dominated by the value vector $v''(t)$ obtained by keeping the A-strategy fixed but varying the B-strategy.

Finally Lemma 3 and the fact that, locally, Min Max $\geq$ Max Min show the truth of

THEOREM 3. $v^*(t)$ dominates $v^{**}(t)$.

Now it is time to consider saddle-points.

If, at least along v^* or v^{**}, $R_3(t, v)$ coincides with $R_4(t, v)$, that is, if the set of transformations $P(t; x, y)v$ has an unanimous saddle-point, then v^* and v^{**} are identical. Moreover, if x_1^* and y_1^* are a couple of saddle-point strategies as well as x_2^* and y_2^*, then both x_1^* and y_2^* and x_2^* and y_1^* form two other couples, all four couples forming the same transformation P^*.

This constitutes the main theorem on Markov games in discrete time with perfect state information:

> THEOREM 4. Local unanimous saddle-points of $P(t, x, y)v$ along v^* or v^{**} guarantee the existence of a global unanimous saddle-point with the value vector $v^* = v^{**}$. This means that, for any pair of saddle-point strategies x^*, y^*, the inequalities
>
> $$v(t, x, y^*) \leq v(t, x^*, y^*) \leq v(t, x^*, y)$$
>
> are true for every t.

The strategy function $x^*(t)$ (not necessarily unique) is an indication to player A, that, if the game is in state i at reverse time $t + 1$, he should choose the values of the parameters x_i available to him in that situation as $x_i^*(t)$. Correspondingly $y^*(t)$ is an indication to player B.

§ 6. FUNCTION CLASSES GIVING UNANIMOUS SADDLE-POINTS

We are interested in classes of functions of two vectors x and y such that arbitrary linear combinations of them have a saddle-point. If the transition probabilities are such functions, then the right-hand members in (2.10) have saddle-points and the transformations of interest to us have unanimous saddle-points.

Such a class K consists of functions $f(x, y)$ that are linear in x and y separately, where x belongs to a compact, convex set X in some Euclidean space and y in the same way belongs to a compact, convex set Y

in some other Euclidean space, [5].

This fact guarantees, that any Markov game with perfect state information can be made solvable by local mixing of the strategies. The connection with the concept of behaviour strategies is close.

Another such set consists of the functions $f(x, y)/f_0(x, y)$, $f \in K$, if f_0 cannot take the value zero and belongs to K.

As this is probably not well known, a short proof will be given. It is an immediate consequence of

LEMMA 5. If $C(x, y, \lambda) = A(x, y) - \lambda B(x, y)$ has a saddle-point in x and y for every λ, if A and B are continuous and if $B > 0$, then $A(x, y)/B(x, y)$ has a saddle-point.

PROOF. Define the function $f(\lambda)$ as

$$f(\lambda) = \min_{y \in Y} \max_{x \in X} C(x, y, \lambda) .$$

It is evident that $f(\lambda)$ is continuous, and that it has the same sign as λ if $|\lambda| > \text{Max } A/\text{Min } B$. Hence there exists a λ_0 such that $f(\lambda_0) = 0$.

Suppose that (x^*, y^*) is the saddle-point of $C(x, y, \lambda_0)$. That is

$$C(x^*, y^*, \lambda_0) = 0$$

and

$$A(x, y^*) - \lambda_0 B(x, y^*) \leq 0 \leq A(x^*, y) - \lambda_0 B(x^*, y) .$$

Hence

$$\frac{A(x, y^*)}{B(x, y^*)} \leq \lambda_0 \leq \frac{A(x^*, y)}{B(x^*, y)} .$$

Thus, (x^*, y^*) is a saddle-point of $A(x, y)/B(x, y)$ and the value is λ_0.

At last we want to remark, that if we do not have the restriction $\sum_j p_{ij}(x, y) = \text{const.}$ although $p_{ij} \geq 0$, and if we are interested of such cases only where v is positive, a wider class can be given, viz. that consisting of functions, convex in x and concave in y and divided by a positive function that is concave in x and convex in y.

§ 7. THE CASE $T = \infty$. TRAPPED STATES

It is of interest to study cases, where $v^*(t)$ tends to some fixed value as t tends to infinity.

Suppose that the current payoff $w(t)$ is zero if $t \neq 0$ and that the transition probabilities and the sets $X(t)$ and $Y(t)$ do not depend on t. Suppose further that there are a number of trapped states [6], that is states $M + 1, M + 2, \ldots, N$ from which no transition to other states is possible. That means that the result of the play is determined as soon as some of these states is attained. This means that $v_{M+1}, v_{M+2}, \ldots, v_N$ are independent of time and given by $w_{M+1}(0), w_{M+2}(0), \ldots, w_N(0)$. Because of that we can restrict our considerations to an M-dimensional space.

Define $q_i(x, y)$ by

$$\sum_{j=M+1}^{N} p_{ij}(x, y) w_j(0) = q_i(x, y) \qquad (i = 1, 2, \ldots, M).$$

Then (2.10) is transformed into

$$(2.16)\quad \begin{aligned} & v_i(t + 1) = q_i(x, y) + \sum_{j=1}^{M} p_{ij}[x_i(t), y_i(t)] v_j(t) \qquad \begin{matrix}(i = 1, 2, \ldots, M)\\(t = 0, 1, 2, \ldots)\end{matrix} \\ & v_i(0) = w_i(0);\ x_i(t) \in X_i,\ y_i(t) \in Y_i . \end{aligned}$$

Equation (2.16) differs from (2.10) in two respects: q_i in (2.16) depends on x, y but the corresponding w_i in (2.10) does not, and the sum $\sum_j p_{ij}$ can be smaller than one. Neither of these changes are, however, essential for the proofs in the theory for Equation (2.10). Thus the theory can be applied here too.

The theory can be completed with results on the behaviour of $v^*(t)$, when $t \to \infty$, if the transformation

$$v_i(t+1) = q_i(x, y) + \sum_{j=1}^{M} p_{ij}[x_i(t), y_i(t)]v_j(t)$$

is a contraction in the strict sense, which means that, uniformly in t,

$$\|v_i(t+1) - u_i(t+1)\| < \|v_i(t) - u_i(t)\|$$

for any v and $u \neq v$, that is $\|P\| < 1$ uniformly.

That is the case, if

$$\min_{x,y} \sum_{j=M+1}^{M} p_{ij}(x, y) = r > 0.$$

Then

$$\|P\| \leq 1 - r.$$

As is shown by the proof of Lemma 5, this implies

$$\|R_3(v_1) - R_3(v_2)\| \leq (1 - r)\ \|v_1 - v_2\|. \tag{2.17}$$

It is well known that the solution of

$$v^*(t+1) = R_3[v^*(t)] \tag{2.18}$$

has a unique limit as $t \to \infty$. We reproduce the easy proofs.

From (2.17) follows that

$$\|v^*(t+1) - v^*(t)\| = \|R_3[v^*(t)] - R_3[v^*(t-1)]\| \leq (1-r)\|v^*(t) - v^*(t-1)\|$$

and the series

$$w(0) + [v^*(1) - v^*(0)] + [v^*(2) - v^*(1)] + \ldots + [v^*(t+1) - v^*(t)] + \ldots$$

is absolutely convergent, that is $v^*(t)$ tends to a fixed value $v^*(\infty)$.

This value is independent of the initial conditions. For suppose, that both v_1^* and v_2^* satisfy (2.18). Then

$$\|v_1^*(t + 1) - v_2^*(t + 1)\| = \|R_3[v_1^*(t)] - R_3[v_2^*(t)]\| \leq (1 - r)\|v_1^*(t) - v_2^*(t)\|$$

and $\|v_1^*(t) - v_2^*(t)\|$ tends to zero as t tends to infinity. Especially, if $w_i(0) = v_i^*(\infty)$, $v_i^*(t)$ is independent of t and

$$v^*(\infty) = R_3[v^*(\infty)]$$

or

$$v_i^*(\infty) = \operatorname*{Min}_y \operatorname*{Max}_x \left\{q_i(x, y) + \sum_{j=1}^{M} p_{ij}(x_i, y_i) \cdot v_j^*(\infty)\right\} .$$

$v^*(\infty)$ is the only solution of this equation.

These results are essentially equivalent to those of [7].

§ 8. CONTINUOUS TIME. FUNDAMENTAL ASSUMPTIONS AND EQUATIONS

The detailed preceding exposition should have made the logical structure of the theory of Markov games in the case of perfect state information clear. If we pass from the simple spaces and transformations considered there to more advanced ones, the essential additions that have to be made will be in the area of existence theorems. Probably much advanced mathematical work could be spent in (or — according to tast — lost in) this field.

A first step in this generalization procedure, viz. that of passing from discrete to continuous time, will be made below.

Suppose we have an $N \times N$-matrix $P(t; x, y) = \left\{p_{ij}(t; x_i, y_i)\right\}$ which is continuous in t, x_i, y_i simultaneously in domains that will be specified later on. All its elements are nonnegative except the diagonal ones. For the theory to be exposed here they, too, could be positive, but in problems originating in Markov processes they are not. In fact, for "complete" processes $\sum_j p_{ij} = 0$ and for processes with trapped states $\sum_j p_{ij} \leq 0$.

There is also given a current payoff vector with the components $w_i(t; x_i, y_i)$. For the sake of completeness we have made the current payoff dependent on the strategy parameters x and y.

If these are given as functions of time, the expected payoff vector will be given by the set of differential equations and boundary conditions:

$$\begin{aligned} \frac{dv_i}{dt} &= w_i(t;\, x_i,\, y_i) + \sum_{j=1}^{N} p_{ij}[t,\, x_i(t),\, y_i(t)]v_j \\ v_i(0) &= a_i\,. \end{aligned} \tag{3.1}$$

Here, as in Section 2, t is <u>reverse time</u>.

These equations constitute the <u>value relation</u> in this case.

The formal adjoint of the homogenous part of (3.1)

$$\frac{dq_i}{dt} = -\sum_{j=1}^{N} p_{ji}[t;\, x_i(t),\, y_i(t)]q_j \tag{3.2}$$

gives a relation between the distributions of the states at different times. If q_i is given at the beginning of the play (as the players see it), that is at $t = T$, then the integration of (3.2) with decreasing t (increasing τ) gives the state distribution at all times of interest.

The Lagrangean equation gives

$$\sum_i v_i(t_2)\, q_i(t_2) - \sum_i v_i(t_1)\, q_i(t_1) = \int_{t_1}^{t_2} \sum_i w_i(t)\, q_i(t)\, dt. \tag{3.3}$$

This equation should prove most important in any attempt to tackle Markov games with non-perfect state information. In the case of perfect state information it is not equally important.

The values $x_i(t)$ could be any point from a compact set $X_i(t)$ in an Euclidean space and $y_i(t)$ any point from a compact set $Y_i(t)$. The Cartesian product of all $x_i(t)$ is a compact set $X(t)$ in an Euclidean space. Similarly $Y(t)$ is defined. A function from t to sets in an Euclidean space will be called an <u>S-function</u> (shorthand for set-function; the use of this word is, however, precluded due to its use in more familiar contexts). If the sets are compact we shall call the S-function a <u>compact S-function</u>. Hence $X_i(t)$, $X(t)$ and $Y_i(t)$, $Y(t)$ are compact S-functions. Furthermore, they are supposed to be <u>continuous</u> S-functions, if we define continuity by means of the Hausdorff metric of sets in a metric space [8]. As is easily seen, the graph of a continuous, compact set function is closed. [The graph of the set function $X(t)$ is defined as the set of all pairs $[t, x(t)]$ with $x(t) \in X(t)$.] If the graph is closed, we shall say that the S-function is closed.

Some regularity restrictions on $x_i(t)$ and $y_i(t)$ are necessary. The restriction to be used is that $x_i(t)$ and $y_i(t)$ should be Lebesgue measurable functions in the sense that every component of the vector function $x_i(t)$ should be measurable in the usual sense. We shall call such functions admissible strategy functions.

Now a strict formulation of the game problem to be solved can be given.

Derive, if possible, two admissible strategy functions $x^*(t)$ and $y^*(t)$, such that the associated solution v^* of (3.1) is an unanimous saddle-point in v-space, that is such that for any admissible strategy functions $x(t)$ and $y(t)$

$$v_i[t;\ x_i(t),\ y_i^*(t)] \leq v_i[t;\ x_i^*(t),\ y_i^*(t)] \leq v_i[t;\ x_i^*(t),\ y_i(t)]$$

holds true for all i and all t.

In that case

$$v[t, x^*(t), y^*(t)] = \underset{y(t)\in Y(t)}{\text{Min}}\ \underset{x(t)\in X(t)}{\text{Max}}\ v[t, x(t), y(t)]$$

$$= \underset{x(t)\in X(t)}{\text{Max}}\ \underset{y(t)\in Y(t)}{\text{Min}}\ v[t, x(t), y(t)]$$

in the vector sense.

The natural assumption to secure this behaviour is, as in discrete time, to prescribe the existence of local, unanimous saddle-points, that is to condition the right number of (3.1) to have saddle-points. As before, however, we shall treat the Min Max- and Max Min-problems first, without any assumptions on saddle-points, and afterwards make the final step on their existence. For that purpose the following definition is made.

DEFINITION. $U_0(t;\ x,\ y,\ v) = w(t;\ x,\ y) + P(t;\ x,\ y)v$

$$U_1(t;\ y,\ v) = \operatorname*{Max}_{x\in X} U_0(t;\ x,\ y,\ v)$$

$$U_2(t;\ x,\ v) = \operatorname*{Min}_{y\in Y} U_0(t;\ x,\ y,\ v)$$

$$U_3(t;\ v) = \operatorname*{Min}_{y\in Y}\ \operatorname*{Max}_{x\in X} U_0(t;\ x,\ y,\ v) = \operatorname*{Min}_{y\in Y} U_1(t;\ y,\ v)$$

$$U_4(t;\ v) = \operatorname*{Max}_{x\in X}\ \operatorname*{Min}_{y\in Y} U_0(t;\ x,\ y,\ v) = \operatorname*{Max}_{x\in X} U_2(t;\ x,\ v).$$

These operators exist, due to the assumption on perfect information. We shall be interested in the solutions, if existing, of the differential equations

$$\frac{dv}{dt} = U_k \qquad (k = 0,\ 3,\ 4).$$

The full equivalent of the operators R_k in Section 2 are not the operators U_k but the operators $(d/dt - U_k)^{-1}$. It is the latter operators that are order-preserving, not the operators U_k themselves.

§ 9. EXISTENCE AND UNIQUENESS OF SOLUTIONS OF THE RELEVANT DIFFERENTIAL EQUATIONS

As $p_{ij}[t,\ x_i(t),\ y_i(t)]v_j$ is bounded and measurable for any admissible strategy-functions and any bounded and continuous (even measurable) v_j, the usual iteration scheme of Picard proves existence and uniqueness of the solutions of

$$\text{(3.4)} \qquad \frac{dv}{dt} = U_0[t;\ x(t),\ y(t),\ v]\ .$$

By the methods of proof of Lemma 5 is easily shown that the following theorem is true.

THEOREM 5. 1. $U_3(t,\ v)$ and $U_4(t,\ v)$ are continuous in t and v simultaneously.

2. $\|U_i(t,\ v)\| \leq a + A\|v\| \qquad (i = 3,\ 4).$

3. $\|U_i(t,\ v_1) - U_i(t,\ v_2)\| \leq A\|v_1 - v_2\| \ (i = 3,\ 4).$

Here a and A are positive numbers such that

$$a = \max_{t,x,y} \|w(x, y)\| \quad \text{and} \quad A = \max_{t,x,y} \|P(t; x, y\|.$$

The starting point of the proof of the statements is the vector inequalities (e is the vector with all components equal to one).

1. $-\varepsilon e \leq U_0(t_1, x, y, v_1) - U_0(t_2, x, y, v_2) \leq \varepsilon \cdot e$
$|t_1 - t_2| < \delta, \|v_1 - v_2\| < \delta,$
(ε arbitrarily small
δ sufficiently small)

2. $-(a + A\|v\|)\, e \leq U_0(t, x, y, v) \leq (a + A\|v\|)\, e$

3. $-A\|v_1 - v_2\|\, e \leq U_0(t, x, y, v_1) - U_0(t, x, y, v_2)$
$\leq A\|v_1 - v_2\|\, e.$

Afterwards the Max- and Min-operations are carried through in the expressions between the signs of inequality.

The theorem shows that the simplest sufficient conditions for existence and uniqueness of the solutions of the differential equations

$$\frac{dv^*}{dt} = U_3(t, v^*) \tag{3.5}$$

$$\frac{dv^{**}}{dt} = U_4(t, v^{**}) \tag{3.6}$$

are fulfilled. The usual Lipschitz-condition is statement 3 of Theorem 5. It is important to note, that v, v^* and v^{**} are absolutely continuous in t and depend continously on the initial conditions and on w.

§ 10. ORDER-PRESERVATION

A necessary condition, that a system of linear differential operations

$$\frac{dv_i}{dt} = w_i(t) + \sum_j p_{ij}(t) v_j(t) \tag{3.7}$$

be order-preserving is that no coefficient p_{ij} with $i \neq j$ is negative. For suppose that $p_{i1}(t)$ is negative almost everywhere (a.e.) in a neighbourhood of $t = t_0$. Suppose that $v_j''(t_0) = v_j'(t_0)$ $(j \neq 1)$ but

$v_1''(t_0) = v_1'(t_0) + a$ (a positive).

This means that $v''(t_0) \geq v'(t_0)$. Equation (3.7) shows that

$$\frac{d}{dt}[v_i''(t) - v_i'(t)] = p_{i1}\, a + \varepsilon \qquad (\varepsilon \text{ arbitrarily small})$$

is negative a.e., in a neighbourhood of $t = t_0$. This means that although $v''(t)$ does dominate $v'(t)$ at $t = t_0$, it does not at any t in the interval (t_0, δ), δ sufficiently small. Hence the differential equation is not order-preserving.

In this paragraph will be shown, that the condition $p_{ij}(t) \geq 0$ $(i \neq j)$, that we will call the <u>local order-preservation condition</u>, guarantees global order-preservation according to the following theorem, that is an analogue of Lemma 3.

THEOREM 6. If

1. P' and P'' are locally order-preserving

2. $\frac{dz'}{dt} = w' + P'\, z'$; $z'(0) = a'$

   $\frac{dz''}{dt} = w'' + P''\, z''$; $z''(0) = a''$

3. $a' \geq a''$

4. $w' + P'\, z'$ dominates $w'' + P''\, z'$

   or

   $w' + P'\, z''$ dominates $w'' + P''\, z''$

   then

   $z' > z''$.

The formulation of condition 4 should be observed. Domination is required on z' (or z'') only, not that $w' + P'\, z'$ dominates $w'' + P''\, z''$.

PROOF. Assume first the strengthened version of condition 3:

$$a_i' > a_i'' \qquad (i = 1, 2, \ldots, N).$$

The proof is performed by establishing a contradiction. Suppose for that purpose that order is not preserved. This means that there exists a t and an i such that

$$z_i''(t) > z_i'(t).$$

As z' and z'' are continuous, there exists a minimal time t_0 and an 1 such that

$$z_1''(t_0) = z_1'(t_0).$$

and

$$z_i''(t) < z_i'(t) \qquad (i = 1, 2, \ldots, N)$$

for every t in the half open interval $I: [0, t_0)$.

The following two equalities are true

$$\frac{d}{dt}(z' - z'') = [(w' + P'z') - (w'' + P''z')] + P''(z' - z'')$$

$$\frac{d}{dt}(z' - z'') = [(w' + P'z'') - (w'' + P''z'')] + P'(z' - z'').$$

The first condition 4 of the theorem shows, that <u>in I</u>

$$\frac{d}{dt}(z' - z'') \geq P''(z' - z'')$$

and the second that

$$\frac{d}{dt}(z' - z'') \geq P'(z' - z'').$$

Write z instead of $z' - z''$, a instead of $a' - a''$, and P instead of P' or P''. Thus we have the following situation in I:

$$\begin{aligned} &\frac{d}{dt} z \geq P z \\ &z_i(t) > 0 \qquad (i = 1, 2, \ldots, N) \\ &z_i(0) = a_i \\ &z_1(t_0) = 0 \end{aligned} \tag{3.8}$$

P is locally order-preserving.

This situation is impossible. For consider the 1th component of (3.8)

$$\frac{d}{dt} z_1 - p_{11}(t)\, z_1 = \sum_{j \neq 1} p_{1j}\, z_j \tag{3.9}$$

and define

$$r(t) = \int_0^t p_{11}(\tau)\, d\tau$$

and

$$S(t) = \sum_{j \neq 1} p_{1j}(t)\, z_j(t).$$

As all p_{1j} $(j \neq 1)$ and all $z_1(t)$ are nonnegative, $S(t) \geq 0$.

Equation (3.9) can be rewritten as

$$\frac{d}{dt} \left\{ z_1\, e^{-r(t)} \right\} = S(t)\, e^{-r(t)}.$$

Here the right member is nonnegative. Therefore

$$\int_0^{t_0} \frac{d}{dt} \left\{ z_1(t)\, e^{-r(t)} \right\} dt \geq 0$$

or

$$- a_1 \geq 0.$$

This contradicts the assumption

$$a_i > 0 \qquad (i = 1, 2, \ldots, N)$$

and the theorem is proved in the case of strict inequalities between the initial values. Due to the continuous dependence of $z(t)$ on a, it remains true even for the weaker condition 3 of the theorem. This concludes the proof.

A corollary of the theorem evidently is that order-relations in the initial conditions and in "driving functions" (w) are preserved in the solutions of

$$\frac{dz}{dt} = w + P\, z$$

if P is locally order-preserving.

§ 11. EXISTENCE OF COINCIDENT LINEAR TRANSFORMATIONS

After the proof of the existence of v^* and the definition of the operations U_1 and U_3 we know, that at each separate time t there exist parameter values $x^*(t)$ and $y^*(t)$ such that

$$U_1[t;\, y^*(t),\, v^*(t)] = \operatorname*{Min}_{y \in Y(t)} U_1[t;\, y,\, v^*(t)] = U_3[t;\, v^*(t)]$$

$$U_0[t;\, x^*(t),\, y^*(t),\, v^*(t)] = \operatorname*{Max}_{x \in X(t)} U_0[t;\, x,\, y^*(t),\, v^*(t)]$$

$$= U_1[t;\, y^*(t),\, v^*(t)].$$

This existence is due to the fact, which has not been explicitly proved but which is an evident analogue of the statements of Theorem 5, that U_1 is continuous in y and U_0 is continuous in x. The possible couples $[x^*(t),\, y^*(t)] = z^*(t)$ form a set $Z^*(t)$ in the Cartesian product space of x and y. It is not necessary that $Z^*(t)$ consists of a single point.

It remains to prove, that $z^*(t)$ can be "chosen" in such a way, for each t, that the function $z^*(\cdot)$ is admissible. If this is the case, all the coefficients $p_{ij}(t)$, $x^*(t)$, $y^*(t)$ are measurable, and the solution of the equation

$$\frac{dv}{dt} = U_0[t; x^*(t), y^*(t), v(t)]$$

$$v(0) = a$$

exists and is unique and identical to $v^*(t)$. $U_0[t; x^*(t), y^*(t), v(t)]$ is the coincident linear transformation. Thus, the problem now is to pick an admissible function $z^*(t)$ out of the S-function $Z^*(t)$, which can be looked upon as an ensemble of functions $z^*(t)$.

It is this possibility of realizing the coincident linear transformation in an admissible way that is to be proved in the present paragraph. The first step is the proof of

LEMMA 7. The graph of the S-function $Z^*(t)$ is a closed set in the space which is the Cartesian product of the spaces t, x, and y.

We have to prove, that any convergent sequence $\{t_n, x^*(t_n), y^*(t_n)\}$ converges towards a point $[\tau, \xi^*(\tau), \eta^*(\tau)]$ that belongs to the graph.

First, due to the continuity of the S-functions $X(t)$ and $Y(t)$, $\xi^*(\tau)$ belongs to $X(\tau)$ and $\eta^*(\tau)$ belongs to $Y(\tau)$.

Second, $\eta^*(\tau)$ belongs to $Y^*(\tau)$, i.e., $\eta^*(\tau)$ is a minimum point of $U_1[\tau; y, v^*(\tau)]$. Else, there would exist an y such that at least one component of the vector $U_1[\tau; \eta^*(\tau), v^*(\tau)]$ exceeds the same component of $U_1[\tau; y, v^*(\tau)]$ by a positive amount. This is impossible, as the minimum of $U_1[t_n; \cdot, v^*(t_n)]$ converges towards the minimum of $U_1[\tau; \cdot, v^*(\tau)]$ and at the same time, due to the continuity of U_1 with respect to its three arguments, towards $U_1[\tau, \eta^*(\tau), v^*(\tau)]$.

Third, $\xi^*(\tau)$ is a maximal point of $U_0[t; \cdot, \eta^*(\tau), v^*(\tau)]$. Again, the sequence $\{U_0[t_n; x_n^*(t_n), y_n^*(t_n), v^*(t_n)]\} = \{\operatorname{Max}_x U_0[t_n; x, y_n^*(t_n), v_n^*(t_n)]\}$ converges both to $U_0[\tau, \xi^*(\tau), \eta^*(\tau), v^*(\tau)]$ and to $\operatorname{Max} U_0[\tau, x, y^*(\tau), v^*(\tau)]$. This is impossible if these values are not equal. Hence $\xi^*(\tau)$ is a maximal point. The lemma has been proved.

A simple corollary of the lemma is, that if $x^*(t)$, $y^*(t)$ are unique, that is if the sets $X^*(t)$ and $Y^*(t)$ consist of one single point each, then $x^*(\cdot)$ and $y^*(\cdot)$ are continuous and therefore admissible.

We proceed to the construction of an admissible function $z^*(t)$ in the closed interval $[0, T]$. For the construction we shall use <u>box-functions</u>, $B(t)$, which are S-functions of a special kind. A <u>box-function $B(t)$ with the width h</u> has the following properties:

1. For every fixed t, $0 \leq t \leq T$, the set $B(t)$ is a half open box B with the side h in the space of $z = \{z_1, \dots z_K\}$. This means that there are K numbers $a_k (k = 1, \dots, K)$ such that B consists of all points z with

$$a_k < z_k \leq a_k + h, \qquad k = 1, \dots, K.$$

2. The function $B(t)$ can take only a finite number of "values", that is, for any fixed t, $B(t)$ is either B_1 or B_2 or $\dots B_M$, where M is some finite number.
3. If $i \neq j$, the boxes B_i and B_j have no point in common.
4. The subset T_i of $[0, T]$ where $B(t)$ takes the value B_i, is measurable.

A box-function $B^2(t)$ with the width h_2 will be said to be a <u>refinement</u> of the box-function $B^1(t)$ with the width h_1 if the following two conditions are satisfied:

1. $h_2 < h_1$
2. $B^2(t) \subset B^1(t) \qquad (0 \leq t \leq T)$.

Consider a sequence of box-functions $\{B^n(t)\}_0^\infty$, each one of them a refinement of the preceding one, having widths h_n that tend to zero with n^{-1}. Such a sequence will be called a <u>regular sequence</u>. For a fixed value of t, the sequence of sets $\{B^n(t)\}$ determines a single point $z(t)$ as the intersection of all the sets $B^n(t)$. In the sense of vector ordering we define $z^n_{Inf}(t) = \operatorname{Inf}_{m \geq n} B^m(t)$ and $z^n_{Sup}(t) = \operatorname{Sup}_{m \geq n} B^m(t)$. The components of $z^n_{Inf}(t)$ are increasing as functions of n, those of $z^n_{Sup}(t)$ are decreasing, in both cases towards the components of $z(t)$. Due to the properties 2 and 4 above, the components of $z^n_{Inf}(t)$ are measurable step-functions, which are bounded since $z^n_{Inf}(t) \leq z^0_{Sup}(t)$. Hence their limits are measurable bounded functions. Thus, a regular sequence of box-functions determines a measurable function as its point-wise limit.

To carry through the construction of the admissible function $z^*(t)$ we have to associate with the closed S-function $Z^*(t)$ a regular sequence of box-functions such that the limit function $z(t)$ of the sequence has the property $z(t) \in Z^*(t)$ for any t in $[0, T]$. In that case, we choose $z^*(t)$ as $z(t)$.

This link is easily supplied. Suppose that we have already constructed the box-function $B^n(t)$ with the width $h_n = \frac{a}{2^n}$, having the extra property that the intersection $B^n(t) \cap Z^*(t)$ is non-void for any t in $[0, T]$. ($B^0(t)$ can be chosen as a function with one value B^0_1 only, the box B^0_1 being big enough to make $Z^*(t) \subset B^0_1$ for any t in $[0, T]$.) In order to proceed to $B^{n+1}(t)$ subdivide each box in $B^n(t)$ in 2^K smaller ones (where K is the dimension of the z-space) by dividing each side in two halves. The boxes for $B^{n+1}(t)$ are to be chosen among these smaller boxes with the side $h_{n+1} = \frac{1}{2} h_n$. These boxes are finite in number. We enumerate them in an arbitrary order. Consider the first box B^{n+1}_1. Let the function $B^{n+1}(t)$ take the value B^{n+1}_1 for every t such that $B^{n+1}_1 \cap B^n(t) \cap Z^*(t)$ is non-void. Due to the fact that the graph of $Z^*(t)$ is closed and that the set where $B^n(t)$ takes the value consisting of the box surrounding B^{n+1}_1, is measurable, the set of [illegible] thus selected form a measurable set T^{n+1}_1. $B^{n+1}(t)$ has not yet been defined on the set $[0, T] - T^{n+1}_1$. Now consider the next box in the enumeration B^{n+1}_2. Let $B^{n+1}(t)$ take the value B^{n+1}_2 for every $t \in [0, T] - T^{n+1}_1$ such that $B^{n+1}_2 \cap B^n(t) \cap Z^*(t)$ is non-void. These t again form a measurable set T^{n+1}_2. At this stage $B^{n+1}(t)$ is undefined on $[0, T] - T^{n+1}_1 - T^{n+1}_2$. Consider the next box B^3_{n+1}, choose T^3_{n+1} from $[0, T] - T^{n+1}_1 - T^{n+1}_2$ and carry on in the same way until all the boxes have been treated. Then the new box-function $B^{n+1}(t)$ has been defined as a function taking the value B^{n+1}_1 on T^{n+1}_1, B^{n+1}_2 on T^{n+1}_2 and so on. As $B^n(t) \cap Z^*(t)$ is non-void for any t in $[0, T]$ and the box $B^n(t)$ is a subset of the union $\cup B^{n+1}_i$, $B^{n+1}(t)$ has been defined for every t in $[0, T]$, in other words $\cup T^{n+1}_i = [0, T]$. According to construction, all the

T_i^{n+1} are disjoint. Hence, every t in $[0, T]$ has been coupled with one and only one set B_1^{n+1}, and a new S-function $B^{n+1}(t)$ has been defined.

It is evident that $B^{n+1}(t)$ is a box-function which is a refinement of $B^n(t)$ and that for every t in $[0, T]$ the intersection $B^{n+1}(t) \cap Z^*(t)$ is non-void. Thus we have constructed a regular sequence $\{B^n(t)\}$ which defines in a unique way the limit function $z(t)$. Since $Z^*(t)$ is compact it holds that $z(t) \in Z^*(t)$ and $z^*(t) = z(t)$ is an admissible function. Thus the following theorem is true.

THEOREM 7a. There exist admissible strategy-functions $x^*(t)$, $y^*(t)$ that realize the linear transformation U_0 that coincides with U_3 on v^*. For any admissible $x(t)$

$$U_0[t;\, x(t),\, y^*(t),\, v^*(t)] \le U_0[t;\, x^*(t),\, y^*(t),\, v^*(t)]$$

$$= U_3[t;\, v^*(t)].$$

In the same way the corresponding theorem on the Max Min-problem is shown:

THEOREM 7b. There exist admissible strategy-functions $x^{**}(t)$, $y^{**}(t)$ that realize the linear transformation U_0 that coincides with U_4 on v^{**}. For any admissible $y(t)$

$$U_0[t;\, x^{**}(t),\, y(t),\, v^{**}(t)] \ge U_0[t;\, x^{**}(t),\, y^{**}(t),\, v^{**}(t)]$$

$$= U_4[t;\, v^{**}(t)].$$

Now we take saddle-points in consideration. In the same way as in the discrete case (cf. Theorem 4) it is clear that

THEOREM 8. Local unanimous saddle-points of $P(t;\, x,\, y)v$ along v^* or v^{**} guarantee that for any admissible x^*, y^*; x^{**}, y^{**} the following four pairs form admissible saddle-point strategies realizing coincidence of U_0 with

with $U_3 = U_4$ along $v^* = v^{**}$:

$$(x^*, y^*),\ (x^*, y^{**}),\ (x^{**}, y^*),\ (x^{**}, y^{**}).$$

§ 12. MEASURABLE S-FUNCTIONS

This paragraph is a side issue but gives a mathematical rounding-off of the last paragraph. A measurable vector function $z(t)$ can be defined as a function where all the sets

$$M_k(a) = \{t \mid z_k(t) < a\}$$

are measurable. Denote the set $z_k < a$ by $A_k(a)$. Then the definition of the set $M_k(a)$ can be rewritten as

$$M_k(a) = \{t \mid z_k(t) \cap A_k(a) \neq 0\}.$$

In the same way a measurable S-function $Z(t)$ is defined as an S-function where all the sets

$$M_k(a) = \{t \mid Z_k(t) \cap A_k(a) \neq 0\}$$

are measurable.

With very small additions and modifications of the description of the construction procedure of the last paragraph it is shown that the following theorem is true.

THEOREM 9. A measurable function can be selected from a measurable compact S-function.

§ 13. GLOBAL OPTIMIZATION

By use of the coincident linear transformations of Theorem 7 and the order-preservation Theorem 6 for linear transformations the following theorem is clear.

THEOREM 10. $v^* = v[x^*(t), y^*(t), t]$ dominates any $v' = v[x(t), y^*(t), t]$ for all t. $v^{**} = v[x^{**}(t), y^{**}(t), t]$ is dominated by any $v'' = v[x^{**}(t), y(t), t]$ for all t.

We have not stated in the theorem that v^* dominates v^{**}, because the domination Theorem 6 as formulated is not strong enough to prove that statement. Nevertheless it is true.

By applying Theorem 8 we get our _main theorem on Markov games in continuous time with perfect state information_.

THEOREM 11. Local unanimous saddle-points of $w + P v$ along v^* or v^{**} guarantee the existence of a saddle-point vector $v^* = v^{**}$ and of admissible saddle-point strategy-functions $x^*(t)$ and $y^*(t)$. Here $v^* = v[t; x^*(t), y^*(t)]$ and

$$v[t; x(t), y^*(t)] \leq v^*(t) \leq v[t; x^*(t), y(t)]$$

in the vector sense.

§ 14. COMMENTS ON COMPLETE STATE INFORMATION

In games with perfect state information the players know the state of the game at the times of decision and have separate sets of control parameters in the different states. Due to the Markov property they cannot improve their decisions by taking account of the past history of the game. They perform equally well if they forget all about the past without losing their information on the current state.

In games with incomplete information the states appear to each player as members of a number of classes, different for the two players. Such a class includes such states that are indistinguishable to the player. We shall call it an _ignorance class_. In the states of the same ignorance class the player must use the same set of control parameters, but when the game changes from a state of one class to that of another class, the control set also is changed.

About such a game it is natural to suppose that a play is started (at reverse time $t = T$) with a state, that is determined through a chance move according to a probability distribution prescribed by the rules of the game. Thus the probability of state i at reverse time t is $q_i(T)$. Evidently it is in general necessary for the player to take this distribution in account for effective performance. By noting the ignorance class of states at time T, he can form the conditional probability distribution of the states of the class, and this should form an essential decision element.

For behaving rationally in the course of a play the player should try to make as good estimates as possible of the probability distribution of the indistinguishable states in the current ignorance class, and for that purpose he must in general take account of the past in the form it is known to him, which means the past sequence of ignorance classes and his past control parameter settings. It may happen, for instance, that it is impossible to end up in a certain state at the present time if the ignorance class at the immediately preceding time (we consider discrete time) was C_1 but not if it was C_2.

A game of the kind described above is what we have called in the introduction a *Markov game with perfect state recall*. (State here stands for the players own state, which is equivalent to his ignorance class.) The fact that the players have to consider the past for optimal performance, imposes the conclusion that a *correctly played solvable Markov game of this kind is no Markov process* because it has lost the Markov property. ("Solvable" means "having saddle points.") Expressed in another way, an extraneous observer of a great number of plays, keeping record of the change of states with time, may find that the estimated values of a certain transition probability $p_{ij}(t)$ will be significantly different if based on two different populations, for instance one where the state before i is m, another where it is n. On the contrary, a correctly played Markov game with perfect state information is still a Markov process. A Markov game that is a Markov process on the original set of states, might be called a *proper Markov game*.

For a Markov game to be proper it is evidently necessary, that the optimal strategies are functions of time and state and nothing more. By forcing a clause of forgetfulness on the players (if they are not naturally oblivious) the play can be transformed into a proper one. Let the rules of the game prescribe that the players shall not know the time histories, but that they shall know the identity of the current ignorance class. Such a game may be called a Markov game with oblivion, and it evidently is a proper Markov game. If it is prescribed in the discrete time case, that the players remember the last n-time points, including the present one, then the play can be made a proper one by making a new definition of state: a new state is a sequence of n old states.

In this way a state with perfect state recall played from reverse time T to zero can be transformed to a one step Markov game by making $n = T$. This transformation is an essential step in transforming the game from a game in extensive form to one in normal form. This transformation also shows, that the class of solvable Markov games with perfect state recall is not void, as the Kuhn [1] behaviour strategies can be looked upon as forming the set of control parameters and there exist behaviour strategies giving a saddle-point.

§ 15. GAMES WITH OBLIVION

Consider discrete time. For the fundamental equations we refer to § 2. The row vector $q(T)$ with its ith component $(i = 1, 2, \ldots, N)$ defined as the probability $q_i(T)$ of the ith state at reverse time T is prescribed according to the last paragraph. The expected payoff of the game for fixed strategy functions $x(t)$, $y(t)$ is the scalar quantity

$$q(T) \quad v(T; x, y) \tag{4.1}$$

where $v(T; x, y)$ is determined recursively by the equations (we assume for simplicity $w(t) = 0$ for $t \neq 0$).

$$v(t + 1; x, y) = P[t; x(t), y(t)]\, v(t, x, y) \qquad (t = 0, 1, \ldots) \tag{4.2}$$

$$v(0; x, y) = w(0)$$

$x(t)$ is a vector from a compact set $X(t)$ and $y(t)$ is a vector from a compact set $Y(t)$.

The problem to be solved, if possible, is to determine saddle-point strategies x^* and y^* that are functions of time only and a saddle-point value C such that

$$C = \underset{y}{\mathrm{Min}}\ \underset{x}{\mathrm{Max}}\ q(T)\ v(T;\ x,\ y) = \underset{x}{\mathrm{Max}}\ \underset{y}{\mathrm{Min}}\ q(T)\ v(T,\ x,\ y) = q(T)\ v(T,\ x^*,\ y^*).$$

That is, (x^*, y^*) is a maximum in x and a minimum in y of $q(T)\ v(t, x, y)$. This can be given the following geometric interpretation, which is illustrated in Figure 2.

The set of $v(T, x, y^*)$ for all x, $\{v(T, x, y^*)\}$, is a set in N-dimensional vector space that is situated on the negative side of the hyperplane

$$qv = C$$

and has at least the point $v(T, x^*, y^*) = v^*(T)$ in common with it. In the same way $\{v(T, x^*, y)\}$ is situated on the positive side, and it has the same point in common with the hyperplane.

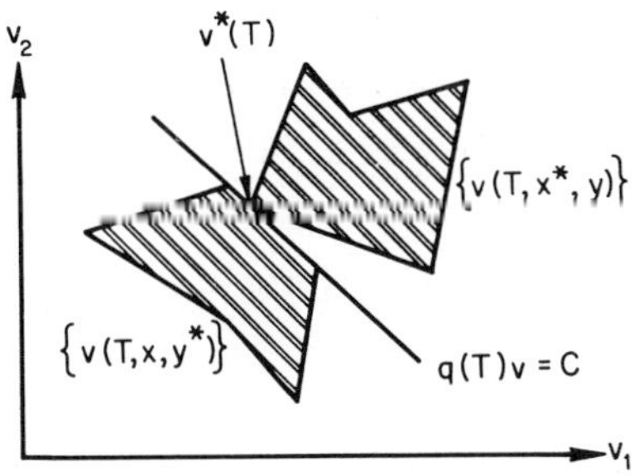

FIGURE 2

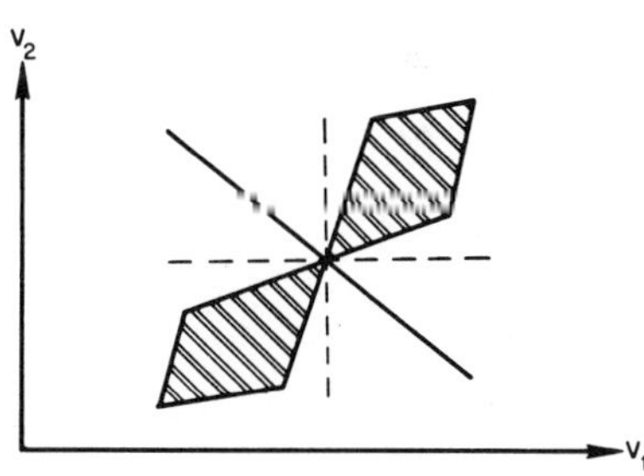

FIGURE 3

The situation according to Figure 2 should be compared to that of Figure 3, illustrating the saddle-point case of perfect state information. Here we have saddle-points of $q(T)\ v$ <u>irrespective of the choice of</u> $q(T)$

within the limits allowed by the conditions $q_i(T) \geq 0$.

It is interesting to note, that if the sets $[v(t, x, y^*)]$ and $v[t, x^*, y]$ are separated from each other by a hyperplane $q(T)$ when $t = T$, then they are separated for all positive $t \leq T$ by hyperplanes $q^*(t)$ that are defined recursively from the given $q(T)$ by

$$q^*(t-1) = q^*(t)\ P[t-1, x^*(t-1), y^*(t-1)]. \tag{4.3}$$

This means that $q^*(t)$ is the distribution vector of the states at time t if $x = x^*$ and $y = y^*$. A simple proof of the stated separation property shall be given.

The distribution in case the strategies are x and y, is denoted by $q(t; x, y)$; we have the fundamental relation (cf. (2.7)).

$$q(T)\ v(T; x, y) = q(\tau; x, y)\ v(\tau; x, y) \tag{4.4}$$

By making $x = x^*$ and $y = y^*$ for $t = \tau, \tau + 1, \ldots, T - 1$ we obtain the effect that $q(\tau; x, y) = q(\tau; x^*, y^*) = q^*(\tau)$. Hence

$$q(T)\ v(T; x, y) = q^*(\tau)\ v(\tau, x, y). \tag{4.5}$$

In the left member the point (x^*, y^*) is a maximal point in x and a minimal point in y. Then the same is true for the right member too. As x and y are restricted only for $t = \tau, \tau + 1, \ldots, T - 1$ and their values at these points do not influence the value of $v(\tau, x, y)$, the plane

$$q^*(\tau)\ v = C$$

separates the two sets $v(\tau, x, y^*)$ and $v(\tau, x^*, y)$. The proof is concluded.

A similar consideration gives <u>necessary local conditions</u> for the separation or saddle-point property. They are closely connected with the <u>Pontryagin maximum principle</u>. [9].

Define x and y as equal to x^* and y^* respectively everywhere except at the point τ. Thereby we obtain

$$q(\tau + 1, x, y) = q^*(\tau - 1); \quad v(\tau; x, y) = v^*(\tau)$$

and

$$q(\tau + 1, x, y)\, v(\tau + 1, x, y) = q(\tau + 1, x, y)\, P[\tau; x(\tau), y(\tau)]\, v(\tau, x, y)$$

or, considering (4.4),

$$q(T)\, v(T; x, y) = q^*(\tau + 1)\, P[\tau; x(\tau)\, y(\tau)]\, v^*(\tau). \tag{4.6}$$

Still, the point $x^*(\tau), y^*(\tau)$ is a maximum in x and a minimum in y.

Geometrically speaking, this means that local sets

$$\{P[\tau; x(\tau), y^*(\tau)]\, v^*(\tau)\} \quad \text{and} \quad \{P[\tau; x^*(\tau), y(\tau)]\, v^*(\tau)\}$$

are separated by the plane

$$q^*(\tau + 1)\, v = C.$$

The word "local" is explained by the fact that these sets are the result of a one step transformation of the point $v^*(\tau)$ by the matrix P.

We get therefore the <u>necessary local conditions</u>

$$\begin{aligned} C &= \underset{y(\tau)}{\text{Min}}\ \underset{x(\tau)}{\text{Max}}\ q^*(\tau + 1)\, P[\tau, x(\tau), y(\tau)]\, v^*(\tau) \\ &= \underset{x(\tau)}{\text{Max}}\ \underset{y(\tau)}{\text{Min}}\ q^*(\tau + 1)\, P[\tau, x(\tau), y(\tau)]\, v^*(\tau) \\ &\quad (\tau = 0, 1, \ldots, T - 1). \end{aligned} \tag{4.7}$$

These conditions are very similar to those used in the case of perfect state information. In fact, those conditions are equivalent to (4.7) with the addition that the <u>same</u> vector $(x^*(\tau), y^*(\tau))$ shall form the saddle-point of q^*Pv^* <u>irrespective</u> of which positive vector q^* we choose.

In the case of perfect state information the local conditions are sufficient. This is not generally the case here. A problem of some interest is to state such extra conditions that guarantee the existence of a global saddle-point. We do not want to probe further into this field now. We observe, that the right hand member of (4.6) is a linear combination of the $p_{ij}[\tau, x(\tau), y(\tau)]$. Therefore the necessary local conditions are satisfied by the function classes discussed in §6.

At last we remark, that local conditions of a similar kind seem to be necessary for the existence of a global saddle-point in Markov games with perfect state recall, too. An open question is how they should be complemented to form a sufficient set of conditions, and still another how and if they could be used for efficient computing.

§ 16. CONCLUSION

The work reported in the present contribution in no way represents an isolated trend in contemporary game theory development. Cf. [10] – [14], which treat similar problems. In particular Everett's definition of recursive games in [1] seems to have much in common with our definition of a Markov game with perfect state information.

No applications on concrete examples have been given in the paper. Such examples (duels, fire distribution) are treated in [2]. Finally I want to acknowledge my debt to Magnus Tideman, Saab Aircraft Co., for his constructive criticism in the course of the preparation of the manuscript.

BIBLIOGRAPHY

[1] KUHN, H. W., "Extensive games and the problem of information," Annals of Mathematics Studies, No. 28, Princeton University Press, Princeton, N. J., 1953, pp. 193-216.

[2] ZACHRISSON, L. E., "En stridsvagnsduell med spelteoretiska konsekvenser (A tank duel with game theoretic implications)," (Swedish) Research Inst. Nat. Defence (Div 2) Report No. 2355-2990, 1955, 81 pages.

[3] ZACHRISSON, L. E., "En stridsvagnsduell med spelteoretiska konsekvenser," Artilleritidskrift (Stockholm) 84, No. 3 (1955), pp. 112-121.

[4] ZACHRISSON, L. E., "A tank duel with game theoretic Implications," (Translation of [3] by James J. Robbins), Naval Res. Logist. Quart. 4 (1957) pp. 131-138.

[5] DRESHER, M., and KARLIN, S., "Solutions of convex games as fixed-points," Annals of Mathematics Studies, No. 28, Princeton University Press, Princeton, N. J., 1953, pp. 75-86.

[6] SMITH, N. M., WALTERS, S. S., BROOKS, F. C., BLACKWELL, D. H., "The theory of value and the science of decision," a Summary, J. Op. Res. Soc. Am., 1 (1953), pp. 103-113.

[7] SHAPLEY, L. S., "Stochastic games," Proc. Nat. Acad. Sci. 39 (1953) p. 1095.

[8] HAUSDORFF, F., Mengenlehre, 3 rd Ed., Dover, p. 145.

[9] BOLTYANSKI, V. G., GAMKRELIDZE, R. V., MISCHCHENKO, E. P., PONTRYAGIN, L. S.; "The Maximum Principle in the Theory of Optimal Processes of Control," Automation and Remote Control (Proc. 1st Int. Congr. Int. Fed. Aut. Contr. Moscow 1960), Vol. I, London 1961, pp. 454-459.

[10] MILNOR, J. and SHAPLEY, L. S., "On games of survival," Annals of Mathematics Studies, No. 39, Princeton University Press, Princeton, N. J., 1957, pp. 15-45.

[11] EVERETT, H., "Recursive games," Annals of Mathematics Studies, No. 39, Princeton University Press, Princeton, N. J., 1957, pp. 47-78.

[12] SCARF, H. E., "On differential games with survival payoff," Annals of Mathematics Studies, No. 39, Princeton University Press, Princeton, N. J., 1957, pp. 393-405.

[13] FLEMING, W. H., "A note on differential games of prescribed duration," Annals of Mathematics Studies, No. 39, Princeton University Press, Princeton, N. J., 1957, pp. 407-412.

[14] BERKOVITZ, L. S. and FLEMING, W. H., "On differential games with integral payoff," Annals of Mathematics Studies, No. 39, Princeton University Press, Princeton, N. J., 1957, pp. 413-435.

Lars Erik Zachrisson

Saab Aircraft Co., Linköping, Sweden
Royal Institute of Technology, Stockholm, Sweden
Research Institute of National Defence, Stockholm, Sweden

HOMOGENEOUS GAMES. III

J. R. Isbell

§ 1. INTRODUCTION

A strong simple game on a set N of players is called <u>homogeneous</u> if it admits a transitive group of permutations of N. This series of papers concerns the problem: for what n do there exist homogeneous (strong simple) n-player games? However, the first two papers (which are completely superseded by this one) presented little more than the facts, which seemed surprising at the time, that these games exist (a) for many even numbers n and (b) even for some numbers 4p, p a prime. The results now given here remain very far from complete, but they may be described as the first substantial progress on the problem.

The simplest conjecture consistent with the known results would be the following. <u>Let</u> $n = 2^k m$, <u>with</u> m <u>odd</u>. <u>There exist homogeneous</u> n-<u>player games if and only if</u> $m > 2^k$. What is actually proved here is (i) for each odd $m > 1$, for the existence of a homogeneous game of $2^k m$ players, it is necessary that $k < 2m - 4$; (ii) for infinitely many m (e.g., for $m = 2^a \pm 1$) it is sufficient that $2^k < m$; and some marginal improvements of (i) and (ii). The numbers $n \leq 100$ for which the question remains open* are 40, 56, 80, 88.

This work was done at the Combinatorial Symposium sponsored by The RAND Corporation in the summer of 1961, and was advanced by comments and suggestions from several other participants and visitors, especially Marshall Hall, Jr., E. T. Parker, and Walter Feit.

* The numbers 44, 52, 76 are not settled in this paper, but are settled by the following recent result of N. Ito (unpublished): <u>For every prime</u> $p > 4$, <u>there is a homogeneous game of</u> 4p <u>players</u>. I have left in this paper the proof of a special case of Ito's theorem, since the method is different from his and involves some structure theory which, if further developed, seems likely to be useful for higher powers of 2.

A bibliographical note: the first paper "Homogeneous games" [1] actually raised and partially answered three different questions. The other two are completely solved in [3] and [4]. Hence [1] as well as [2] is completely superseded.

§ 2. THE GROUP AND THE FIXED POINTS

A strong simple game on a set N of players may be defined briefly as a maximal family W of subsets of N no two of which are disjoint. The _group_ of the game is the group G of all permutations of N which take W to itself. The game is called _homogeneous_ if its group is transitive.

The group of a homogeneous strong simple game clearly cannot transform any subset of N to its complement; for exactly one of these two sets must be in W. This means that every permutation in G, when written as a product of disjoint cycles, has at least one cycle of odd length. Let us call such a group an _odd transitive group_. Conversely, suppose given an odd transitive group G on a set N. Let W be a maximal family of subsets of N which (1) contains no two disjoint sets, (2) is invariant under G, and (3) contains no set consisting of less than half the elements of N. One readily verifies that any such W is a homogeneous game.

The existence of homogeneous games on n players is equivalent to the existence of odd transitive groups on n letters.

For the problem of odd transitive groups, a moment's thought should suggest that we must concentrate first on the Sylow 2-subgroups. (The suggestion actually came, after a moment's thought, from Marshall Hall, Jr.)

Obviously any subgroup of an odd permutation group is odd. This means _every element of a 2-subgroup must leave some points fixed_.

Miller has shown [5] that if G is a group acting transitively on n points, and p^k divides n (p a prime), then for any Sylow p-subgroup of G the sizes of the transitive constituents are divisible by p^k. Writing n as $p^k m$, this means there are at most m transitive constituents, and _elements of the Sylow group have on the average m (or fewer) fixed points_. Since the number of fixed points must be a multiple of p, we can certainly conclude that there is an element leaving no point fixed in case $m < p$. With

$p = 2$ this gives us nothing but the powers of 2. (It was already known [1] that there are no homogeneous games of 2^k players.) We shall have to look more closely at the Sylow groups.

Apparently one may conjecture that when $n = p^k m$ with $m < p^k$, then in any transitive group of degree n there is an element whose order is a power of p and which leaves no point fixed. This would be far stronger than what will be proved here, and would establish one half of the conjecture offered in the Introduction.

We shall work from the

LEMMA. In a transitive 2-group on 2^k points, the fraction of the elements which have fixed points does not exceed u_k, where u_k is given by a Sylow subgroup of the symmetric group, with $u_1 = 1/2$, $u_{k+1} = u_k - u_k^2/2$. In particular, $u_k \to 0$.

PROOF. The result is evident for $k = 1$; suppose it proved for k, and consider $k + 1$. Now the transitive 2-group H is contained in a Sylow subgroup of the symmetric group, and therefore H is imprimitive with classes S_1, S_2 of 2^k points each. H has a subgroup J of index 2 which is a subdirect product of transitive groups J_1 on S_1, J_2 on S_2. As for the elements of H which are not in J, first, none of them has a fixed point. Second, any one of them establishes a one-one correspondence between S_1 and S_2 which gives a permutation isomorphism of J_1 and J_2, associating the subgroups K_{12}, K_{21}, paired to the identity in the construction of the subdirect product; corresponding cosets $xK_{12} = C_i$, $yK_{21} = D_i$ are also associated. In particular, the number n_i of elements of C_i having no fixed points is the same as for D_i. The number of elements of $J = \cup (C_i \times D_i)$ which have no fixed points is Σn_i^2. Let s denote the number of elements of K_{12}, st the number of elements of J (so that the summation index runs from 1 to t). By Cauchy's inequality, $(\Sigma n_i)^2 \leq t\Sigma n_i^2$. Since Σn_i is the number of elements of J_1 leaving no point fixed, it is at least $(1 - u_k)st$ by the inductive hypothesis. These inequalities show that the fraction of the elements of J which have fixed points is at most $1 - (1 - u_k^2)$. In H it is only half as great, as was to be shown.

But now consider $n = 2^k m$, where k is so large that $u_k < 1/m$. For any transitive group on n points, a Sylow 2-subgroup has m or fewer transitive constituents, on each of which less than the mth part of its elements have fixed points. We conclude:

THEOREM 1. For each odd m there is an integer h such that for $k \geq h$ there is no odd transitive group, nor homogeneous game, on $2^k m$ points.

ADDENDUM. Here for $m = 1$, $h = 1$; for $m = 3$, $h = 2$; for $m = 5$, $h \leq 6$. Always $h \leq 2m - 4$.

The addendum for $m = 1$ was already noted; for $m = 5$, one computes it easily; and past 5 one can compare with the solution of the differential equation $y' = -y^2$. For $m = 3$, the lemma gives $h \leq 3$. The case of $12 = 2^2 3$ has previously been ruled out [1] by use of Miller's complete list of transitive groups of degree 12; but it seems worthwhile to sketch a proof along the lines of the present discussion.

We are concerned with the dihedral group D_4, 3/8 of whose elements have fixed points; its transitive subgroups $V(Z_2 + Z_2)$ and $Z(Z_4)$, with 1/4; and its only four-element subgroup more than one of whose elements have fixed points, call it $K = \{1, (ab), (cd), (ab)(cd)\}$. To build a 2-group H on 12 points with 3 transitive constituents S_1, S_2, S_3, we must have H acting as $D_4 = D^i$ on each S_i; H is a subdirect product of D^1, D^2, D^3. But $H \cap (V^1 \times D^2 \times D^3)$ also has no element without fixed points; hence the 1/4 attached to V^1 must be balanced by some fraction larger than 3/8, so V^1 must be associated with K^2 or K^3. Similar conclusions hold for V^2 and V^3, and also for Z^1, Z^2, Z^3; but there are not enough K's to go around. Finally, an easier argument disposes of the case of two transitive constituents of sizes 8 and 4.

§ 3. THE CONSTRUCTION

All values of n for which homogeneous n-player games are now known to exist can be gotten from a single type of construction which is due to Walter Feit. Feit's construction replaced several related but less efficient ones, and removed infinitely many n from the "unsolved" list, beginning with 20, 72, 136. It involves vectors, matrices, and polynomials, <u>all of which are over the Galois field of two elements</u>.

For any odd number m, let P_m be the polynomial of degree $\varphi(m)$ (Euler's φ-function) whose roots are the primitive m^{th} roots of unity. Suppose for some d, in the affine space E_d of dimension d, we have a linear automorphism A satisfying $P_m(A) = 0$ and a subspace E_h such that $\cup_i A^i E_h = E_d$. Then let G be the group of all transformations of E_d of the form $x \to A^i x + b$; $i = 0, 1, \ldots, m-1$; $b \in E_d$. Let H be the subgroup of G consisting of the translations $x \to x + b$, with b in E_h. G, or some quotient group of G, acts transitively on the $2^k m$ cosets of G/H, where k is the deficiency $d - h$. We shall show that the action is odd.

Every element of G except the translations has odd order. For consider the transformation $x \to Cx + b$, where C is a power of A, and the order of C is $r > 1$. The rth power of $x \to Cx + b$ is $x \to x + (\Sigma_{i=o}^{r-1} C^i)b$. Since r is a divisor of m, and $C = A^s$ with rs a multiple of m, the equation $\Sigma_{i=o}^{r-1} x^{is} = 0$ is satisfied by all primitive m^{th} roots of unity. Hence A satisfies it, and the transformation has odd order r.

For a translation $x \to x + b$, write b as $A^i \beta$ with β in E_h. Computation shows that if we take $g \in G$ to be $x \to A^i x$, the coset gH is fixed by $x \to x + b$. Thus every element of G has an odd cycle in its action on cosets mod H. We conclude:

> THEOREM 2 (Feit). When an affine space over the two-element field has a subspace of deficiency k which is swept over the whole space by powers of a non-singular matrix which satisfies every polynomial equation satisfied by all primitive m^{th} roots of unity, there exist odd transitive groups on $2^k m$ points.

The simplest application is to a subfield E_h of a field E_d. The multiplicative group of E_d is cyclic of order $2^d - 1$. Write $2^d - 1 = (2^h - 1)m$. If m and $2^h - 1$ are relatively prime, then the multiplicative group of E_h is a direct summand. In this case Theorem 2 applies; for A we take the linear operator on E_d which is multiplication by any primitive m^{th} root of unity.

In this case $k = d - h$ is the largest integer such that $2^k < m$. Note that in any case <u>the hypothesis of Theorem</u> 2 <u>implies</u> $m > 2^k$. (For a space with $2^k 2^h$ points is covered by at most m copies of a subspace of 2^h points.) Thus Theorem 2 is at its strongest for these numbers m. This includes all the numbers $2^d - 1$ (E_h being the prime field) and $2^h + 1$ (d being 2h).

If m and $2^h - 1$ are not relatively prime, this argument fails. In case $2^h - 1 = 3|m$, it can be repaired. Fix A: multiplication by a primitive m^{th} root of unity, α. The multiplicative group of E_d splits into 3 cosets modulo the subgroup of powers of α. We need a line of the projective space, whose 3 points lie in the 3 cosets. Every pair of points in different cosets lies on a line, and counting shows that some line must meet all 3. Hence for these m also (for $m = (2^{6e} - 1)/3$), there are $2^k m$-player homogeneous games for all $2^k < m$. The first instance for which this argument seems necessary is 336 = 16.21.

For $m = (2^{21e} - 1)/7$, I do not know what can be done. Of course we can find a 6—(even 5—) dimensional projective subspace meeting the crucial 7 cosets, giving $2^k m$-player games for $k \leq 21e - 6$. This can be generalized, but let us omit the details.

For odd numbers m not of this special form $((2^d - 1)/(2^h - 1))$ we have much less. We have at least $2m (m > 1)$. This was shown in [1] by a simple special construction. For completeness I wish to point out that Feit's construction works too; but let us omit the proof. If E_d is the smallest field containing a primitive m^{th} root of unity α, it turns out that any subspace of deficiency 1 meets all cosets of the multiplicative group modulo powers of α. Moreover, one can write down the sizes of the intersections of the various cosets with the subspace E_{d-1}; they are all the odd numbers from 1 through the size of a coset, with 1 occurring exactly once.

Another result from [1] must be recalled. <u>If there exist odd transitive groups on n_1 points and on n_2 points, then there exists an odd transitive group on $n_1 n_2$ points</u>. The natural action of the direct product group on the product set is transitive and odd. The first n for which this construction seems to be needed is 100.

Let us summarize the known instances of Theorem 2, including one whose proof will be given below. In every case the form is this: for certain odd m there is $k_o = k(m)$ such that $2^k m$-player homogeneous games exist for all $k \leq k_o$. In no case is k_o known to exceed $\log_2 m$.

(a) k_o takes the maximum value $[\log_2 m]$ if (i) $(2^h - 1)m = 2^d - 1$, where $(2^h - 1, m) = 1$; or (ii) $m = (2^{6e} - 1)/3$.

(b) For other numbers m of the form $(2^d - 1)/(2^h - 1)$, k_o is not less than $[\log_2 m] - 2^h + h + 2$.

(c) For every m, k_o is not less than the number of prime factors of m, counting multiplicity.

(d) For a prime p modulo which 2 is not a primitive root, $k_o = k(p) \geq 2$.

Case (d) covers at least "half" the primes, namely those of the form $8a \pm 1$. The rest of the paper concerns case (d) and the machinery for it.

§ 4. BOOLEAN DESIGNS

Given a prime p modulo which 2 is not a primitive root, the field (everything is still of characteristic 2) E_e generated by a pth root of unity has degree less than $p - 1$. We shall treat the space E_{p-1} as a direct sum $E_e + E_r$, and we shall find a suitable matrix A acting on the sum, satisfying $P_p(A) = \Sigma_{i=o}^{p-1} A^i = 0$.

It will turn out that we are really working with E_{p-1} considered as a direct sum of copies of the field E_e, and A is a direct sum of multiplications by various p^{th} roots of unity. Two of them would suffice. (Further arguments, which will not be given here, show that if e is even we may use any two non-conjugate primitive p^{th} roots; if e is odd, any two conjugate ones.)

Crucial properties of the matrix A will be (1) $A^p = 1$; (2) A fixes no non-zero vector. This implies that $A - 1$ is non-singular; so $P_p(A) = 0$. Further, the minimum polynomial of A has no repeated factor, and the Jordan canonical form of A is a diagonal array of e by e blocks. Each block A_j is a root of an irreducible factor of P_p; hence A_j

generates a field E_e and is (as we wished to show) multiplication by a primitive p^{th} root of unity.

Each such A decomposes the space on which it operates into a number of p-point orbits, and 0. The orbit M, with the vector space D of functions on M which are the linear functionals on the containing space E, forms what we shall call a Boolean design.

For the present construction on $E_e + E_r$, each of the orbits of A will have non-empty intersection with every subspace E_{p-3} of the form $S + T$, where S is a hyperplane in E_e and T is a hyperplane in E_r. Hence $\cup A^i E_{p-3} = E_{p-1}$, and Theorem 2 applies.

In general we may define a Boolean design on a set M as a vector space of 2-valued functions on M which is invariant under some transitive group of permutations of M. This permits trivialities such as the zero design consisting of one function vanishing identically; but it simplifies the following definitions. If D is a Boolean design of dimension (rank) r on a set M of m points, the polar of D is the design D^* of rank $m - r$ consisting of all functions g such that for every f in D, $\Sigma g(x)f(x) = 0$. The even section of D is the design $D_o \subset D$ consisting of all f in D such that $\Sigma f(x) = 0$. The rank of D_o is, of course, either r or $r - 1$.

The supports of the functions in a Boolean design D determine D; they may be called simply supports of D. The even section consists of the even supports. We may call D even if $D_o = D$.

The point of these notions for the present construction is this: if D is even, then for any f in D and g in $(D^*)_o$ the supports of f and g are even and have even intersection. Hence they have even union. If m is odd, this means there must be a point where both f and g vanish.

A Boolean design on M is separated if its functions separate the points of M. If D is a separated design of rank r, we can embed M in the conjugate space E_r of D so that the functions in D are represented by the linear functionals on E_r. Then the permutations of M which preserve D may be extended uniquely to non-singular linear transformations of E_r. (Of course M is embedded in $E_r - \{0\}$.)

Note that a Boolean design D on a prime number p of points necessarily admits a cyclic permutation group Z_p. If the rank r exceeds

1, D is separated. We have $2^r \equiv 1$ or 2 (mod p) according as D is even or not (assuming $p \neq 2$), since Z_p decomposes the non-constant functions in D into classes of size p.

Thus if 2 is a primitive root modulo p, there are no Boolean designs on p points except the four trivial ones. We may note that for any odd number $m(m > 1)$, the trivial design $(0^*)_o$ on m points yields a representation in E_{m-1} and a cyclic group Z_m acting on E_{m-1} suitable for use in Theorem 2; this gives the result we have already noted, that $k(m) \geq 1$. To go further in this direction we now resume the assumption that m is a prime p, and the exponent e of 2 modulo p is less than $p - 1$.

Any two Boolean designs of rank e are isomorphic. This is clear from the representation, where the cyclic group must consist of pth roots of unity A^i; any two of the orbits under this group define isomorphic Boolean designs, since multiplication by some element of the field E_e will take one orbit to the other.

Consider the set M of p points with a definite action of Z_p on it. There is more than one Boolean design of minimum (non-trivial) rank e admitting this group; in fact, one easily verifies that the number is $(p - 1)/e$. Here, or whenever we have several designs D_i simultaneously admitting a transitive group of permutations, we may form the <u>sum</u> of the D_i, or of any set of them; that is the smallest vector space of functions on M containing all D_i. In the present case the linear representation of the sum design embeds M in a sum $E_e^1 + \cdots + E_e^s$, and Z_p is represented by powers of a matrix A which is a diagonal array of multiplications by non-conjugate p^{th} roots of unity. As for the embedding, some point x of M must go to $x^* = (x_1, \ldots, x_s)$ in $E_e^1 + \cdots + E_e^s$, arbitrary except that no x_i is 0; then for each j in Z_p, j(x) goes to $A^j(x^*)$. To prove this it suffices to observe that for any x^* and y^* satisfying this condition, there is a non-singular matrix B commuting with A and satisfying $B(x^*) = y^*$.

Specifically, once we have such a matrix A acting on a space E_{es}, every orbit under A (except 0) is a representation either of the sum $D_1 + \cdots + D_s$ or of some subdesign (in fact, subsum).

Moreover, <u>every even Boolean design on a prime number</u> p <u>of points is a sum of several</u> (pairwise isomorphic) <u>Boolean designs whose rank is the exponent of</u> 2 <u>modulo</u> p. This follows easily from the linear representation and Jordan canonical form.

Now to construct the desired application of Theorem 2, let A_1 be the matrix of multiplication by a primitive p^{th} root of unity. Let M be a non-zero orbit under the group Z_p of powers of A_1, and D the Boolean design of linear functionals on M. Represent $(D^*)_o$ by embedding M in $E_r (r = p - e - 1)$; Z_p is represented in r by r matrices, with A_1 going to some A_2. Then let

$$A = \begin{pmatrix} A_1 & 0 \\ 0 & A_2 \end{pmatrix}; \qquad P_p(A) = 0.$$

Moreover, every orbit under A, projected into E_e and into E_r, yields simultaneous representations of D and $(D^*)_o$ or of some subdesigns of these; consequently for any hyperplanes S of E_e, T of E_r, each orbit meets S + T. As we have already noted, this suffices. Hence homogeneous games of 4p players exist.

Two concluding remarks are called for. First, the use of Boolean designs is not limited to subspaces of deficiency 2. We may define the depth of a Boolean design on a set M as the largest integer d such that M cannot be covered by d supports; then if there is a design of depth k on p points, there is a 2^kp-player homogeneous game. (The use of Boolean designs on composite numbers of points involves extra restrictions. Some special cases were treated in [2] — actually just m = 15 and m = 63, but one can easily go on to all $2^d - 1$.)

Second, it is of interest to know the depths of particular Boolean designs, though there should be more powerful methods of constructing homogeneous games. I have computed the depths for primes $p \leq 67$; none exceeds 2 except for p = 31. On 73 points (and probably on 71 and on 89) there is a Boolean design of depth 3. (Compare the result $k(73) \geq 6$, already established by means of subfields.) Determining the depth is quite tedious. (The design of rank 9 on 73 points has depth 3 because its 511 non-zero supports, split into 7 classes by Z_{73}, each consist of at most 40 points; if 3 sets cover m points, some member of the group of 8 sets they generate contains at least 4m/7 points.) At least determining the design need not always be tedious. <u>For every odd number</u> m <u>modulo which not all numbers are powers of</u> 2, <u>the set</u> S <u>of powers of</u> 2 <u>modulo</u> m <u>and its translates generate a non-trivial Boolean design</u>. One support of the polar design is $-S \cup \{0\}$. Unfortunately, when S is a small set, these designs have higher rank than is necessary, and one must find out how to decompose them.

BIBLIOGRAPHY

[1] ISBELL, J. R., "Homogeneous games," Math. Student, Vol. 25 (1957) pp. 123-128.

[2] ISBELL, J. R., "Homogeneous games. II," Proc. Amer. Math. Soc., Vol. 11 (1960), pp. 159-161.

[3] ISBELL, J. R., "A class of simple games," Duke Math. J., Vol. 25 (1958), pp. 423-439.

[4] ISBELL, J. R., "An inequality for incidence matrices," Proc. Amer. Math. Soc., Vol. 10 (1959), pp. 216-218.

[5] MILLER, G. A., "A fundamental theorem with respect to transitive substitution groups," Bull. Amer. Math. Soc., Vol. 9 (1903), pp. 543-544. Also in Collected Works, Vol. 2, Urbana, 1938.

J. R. Isbell

University of Washington

SOLUTIONS OF COMPOUND SIMPLE GAMES

L. S. Shapley

INTRODUCTION

§ 1. COMPOUNDING OF GAMES

A game is called "simple" if every coalition of players either can win outright or is completely helpless. A simple game can therefore be described by specifying its players and its winning coalitions. By a "compound simple game," we mean one that is built up out of two or more "component" simple games, linked together by another simple game called the "quotient." In effect, the compound is a disaggregated version of the quotient, played by committees instead of individuals. The committees are in turn "played" (controlled) by their members through the formation of component winning coalitions. We shall assume in this paper that the committee memberships do not overlap.

A good example of a compound simple game is the multimillion-person game of "Presidential Election." The quotient is a 51-person weighted majority game (the Electoral College), and the components are symmetric majority games of assorted sizes (the electorates of the fifty states and the District of Columbia).

Given two components, there are four possible ways to compound them, depending on which one-person sets win in the quotient. Two of the cases are uninteresting, with the compound essentially equivalent to one of the components alone, the other players appearing just as dummies. The two nontrivial cases are the <u>product</u>, in which a winning coalition in the compound must include winning contingents from both components, and the <u>sum</u>, in which winning strength in either one of the components is enough to win in the compound.

One example of a product is a bicameral legislature; another is any organization in which some member has veto power. Sums, which are "improper" (nonsuperadditive) games, are not found so readily in practice; but they do appear as components of larger "proper" games. An example would be a situation in which either of two groups is able to initiate action, but both are restrained by the veto of a third body. Such a compound structure would be indicated by $(G_1 \oplus G_2) \otimes G_3$ in our notation.

Compound simple games were apparently first introduced by the author (1953 lecture notes (unpublished); see also [6b, d, e] and [7]). Gurk and Isbell treat them briefly in [3] (p. 259), under the name of "committee compositions." Isbell uses them also in [4], to prove the existence of n-person simple games of a certain type; in his definition, but not his application, he permits the component player-sets to overlap.*

§ 2. COMPOUNDING OF SOLUTIONS

In Part I of this paper we investigate the behavior of the von Neumann-Morgenstern solutions of sums and products. In Part II we do the same for the case of arbitrary quotients. At the end of each Part (§ 8 and § 13 respectively) we analyze several examples that are small enough to permit the use of sketches; the reader may wish to glance at these in advance.

A verbal statement of the fundamental theorems of Part I (Theorems 1 and 2, in § 5) will help convey the nature of the results. We find that in a sum or product (of two or more components), there are solutions that distribute the proceeds within each component according to a solution of that component; the amounts must be scaled down, however, to allow for what the players in the other component(s) receive. In the case of a sum, the allocation between components is arbitrary, but fixed; each possible allocation gives rise to a different solution. In the case of a product, all possible allocations occur in the same solution. Solutions of products are therefore larger sets, generally speaking, than solutions of sums, reflecting the fact that products have fewer winning coalitions.

---

* More recently (and independently), Owen [5] has formulated compound games as "tensor products," in a way that avoids the requirement that the components and quotient be simple.

In similar vein, the fundamental theorem of Part II (Theorem 7, § 10) states that if one first divides the proceeds of a compound game among the components, in accordance with a solution of the quotient, and then subdivides them among the players of each component according to a scaled-down solution of that component, then the resulting set of imputations is a solution of the compound. This theorem includes as special cases the sum and product theorems just described, as well as a result of Gurk and Isbell [3] on the composition of "simple" solutions. It is essentially equivalent to a theorem of Owen [5].

It is natural to attempt to extend the theory to include nonsimple games — especially nonsimple quotients.* One soon finds that the rule given above for compounding solutions is not universally valid, although some special cases of interest can be handled in that way. The full extension — to nonsimple components as well — presents still further difficulties to solution theory.**

§ 3. THE PROBLEM OF AGGREGATION

An important question in the application of n-person game theory is the extent to which it is permissible to treat firms, committees, political parties, labor unions, nations, etc., as though they were individual players. Behind every game model played by such aggregates, there lies another, more detailed model: a compound game of which the original is the quotient. Given any solution concept, it is legitimate to ask how well it behaves under the twin processes of aggregation and disagregation. In what respects are the theoretical predictions dependent on the level of detail adopted in constructing the model?***

From our verbal accounts, it is already clear that the solutions in

* The extension to nonsimple quotients is not as straightforward as one might expect; for a discussion see [6e].

** See [5] and [6e].

*** Owen [5] presents some favorable results for the "core" and "reasonable outcome" concepts. On the other hand, it can be shown that the "Shapley value" is not stable under aggregation. The game of "Presidential Election," for example, distributes value among the states quite differently from its quotient, "Electoral College," giving substantially more to the populous states.

the von Neumann-Morgenstern theory do stand up well under <u>dis</u>aggregation. A solution of the grosser model (the quotient), with details added at the component level, becomes a solution of the refined model (the compound game). The situation with respect to aggregation is more complex, since we are not dealing with a theory in which solutions are unique. Two things could conceivably go wrong. The refined solution might reduce to a <u>non</u>solution of the gross model, or it might have such a structure that it does not reduce at all.*

We therefore ask, first, whether it is possible for a compound game to have a solution with the same internal compound structure that would arise from the fundamental composition theorems, but based on elements that are not necessarily solutions of the respective component and quotient games. In short, can nonsolutions take the place of solutions in the aggregation process? We suspect that the answer is no, but can prove it only for sume (§ 6) and products (§ 7). In the general case, we show that if nonsolutions occur, then they must occur simultaneously in the quotient and in at least one component; moremover, they are circumscribed by some rather stringent conditions which may turn out on further investigation to be impossible to satisfy (see § 11).

To complete the picture, we inquire into the existence of solutions that are not "factorable" along the lines of the compound structure of the game. In Part I we discover that sums never have such noncompound solutions (§ 6), while products usually do (§ 7, § 8). In the more general context of Part II, we find that it is the sums that are exceptional; most compound simple games, it appears, have both compound and noncompound solutions (§ 12). Heuristically, this means that possibilities for stable social configurations often exist that are undetectable in an aggregated model.

The moral seems to be, then, that in applying von Neumann-Morgenstern solution theory to games between aggregates, we should seek a level of detail at which the decision-making and payoff-receiving units are also social units, with, ideally, no "fraternization" permitted (by custom) between subordinate members of different units. Failing that, we should disaggregate down to a level where fraternization cannot seriously affect the solutions in the large.

---

* Methodologically the aggregation problem is similar at many points to the "embedding" problem, as treated in [9] (see esp. pp. 341, 366, 398-401).

PART I: SUMS AND PRODUCTS

§ 4. DEFINITIONS

SIMPLE GAMES. We shall denote a simple game by the symbol $\Gamma(P, \mathcal{W})$, where P is a finite set (the players) and $\mathcal{W}$ is a collection of subsets of P (the winning coalitions). We shall require of $\mathcal{W}$ that it include P and exclude 0 (the empty set), and that it be monotonic in the following sense:

$$S \in \mathcal{W},\ T \supseteq S \Longrightarrow T \in \mathcal{W}. \tag{1}$$

In view of (1), the collection of <u>minimal winning coalitions</u>, denoted by $\mathcal{W}^m$, is sufficient to specify the game. If $\mathcal{W}$ has the further property that

$$S \in \mathcal{W} \Longrightarrow P-S \notin \mathcal{W}, \tag{2}$$

which implies (with (1)) that every two winning coalitions intersect, then the game is said to be <u>proper</u>. If (2) fails for some S, then the game is termed <u>improper</u>. The characteristic function of an improper game in the classical theory is not superadditive.

If $\mathcal{W}$ enjoys the converse property:

$$S \notin \mathcal{W} \Longrightarrow P-S \in \mathcal{W}, \tag{3}$$

which implies that there are no "blocking" coalitions, then the game is said to be <u>strong</u>. The proper strong games are "constant-sum" in the classical theory; they comprise the class of simple games that was originally defined and investigated by von Neumann and Morgenstern.*

If we count the winning coalitions, we find that in a strong game $|\mathcal{W}| \geq 2^{|P|-1}$, while in a proper game $|\mathcal{W}| \leq 2^{|P|-1}$. The fewer the winning coalitions in a game, the "weaker" it is said to be.

SUMS AND PRODUCTS. A pair of simple games with different players

---

* See [9], Chapter X.

can be combined into a larger game in two natural ways, related to the Boolean operations of addition and multiplication.

Let $P_1 \cup P_2 = P$ and $P_1 \cap P_2 = 0$. Then the sum

$$\Gamma(P_1, \mathcal{W}_1) \oplus \Gamma(P_2, \mathcal{W}_2)$$

is defined as the game $\Gamma(P, \mathcal{W})$, where $\mathcal{W}$ consists of all $S \subseteq P$ such that either $S \cap P_1 \in \mathcal{W}_1$ or $S \cap P_2 \in \mathcal{W}_2$. In terms of minimal winning coalitions we have $\mathcal{W}^m = \mathcal{W}_1^m \cup \mathcal{W}_2^m$, and hence $|\mathcal{W}^m| = |\mathcal{W}_1^m| + |\mathcal{W}_2^m|$.

Similarly, the product

$$\Gamma(P_1, \mathcal{W}_1) \otimes \Gamma(P_2, \mathcal{W}_2)$$

is defined as the game $\Gamma(P, \mathcal{W})$, where $\mathcal{W}$ consists of all $S \subseteq P$ such that both $S \cap P_1 \in \mathcal{W}_1$ and $S \cap P_2 \in \mathcal{W}_2$. In terms of minimal winning coalitions we have $\mathcal{W}^m = \text{set}\{S \mid S \cap P_1 \in \mathcal{W}_1^m, S \cap P_2 \in \mathcal{W}_2^m\}$, and hence $|\mathcal{W}^m| = |\mathcal{W}_1^m| \cdot |\mathcal{W}_2^m|$.

Products are always weaker than the corresponding sums, despite the fact that they have more minimal winning coalitions. In general, sums are improper and products are nonstrong. This may explain why these operations do not appear in the simple-game theory of von Neumann and Morgenstern [9], which is limited to the constant-sum (i.e., strong and proper) case. Addition and multiplication are commutative and associative, and yield a unique decomposition (except for order) of any simple game into components that are not sums or products. A further discussion of these operations will be found in [8].

IMPUTATIONS. Let A_P denote the simplex of real nonnegative vectors x such that $\Sigma_P x_i = 1$. These vectors are traditionally called "imputations" when P is the set of players in a simple game. By convention, we shall say that $x_j = 0$ for any and all $j \notin P$. This is tantamount to having A_P embedded in a simplex A_Q, where Q is some large, unspecified set including all the players that might ever come under consideration.

Let us write $x(S)$ for $\Sigma_S x_i$. Let $R_S x$ be the "restriction" of x to S, thus:

$$(R_S x)_i = \begin{cases} x_i & \text{if } i \in S, \\ 0 & \text{if } i \notin S. \end{cases}$$

The barycentric projection of x on A_S is then given by

$$B_S x = \frac{1}{x(S)} R_S x.$$

This is not well defined if $x(S) = 0$; otherwise it is an element of A_S.

Let S and T be fixed, disjoint sets, and let X and Y be subsets of A_S and A_T, respectively. Given $0 \leq \alpha \leq 1$, we define an operation $\underset{\alpha}{\times}$ by

$$X \underset{\alpha}{\times} Y = \text{set}\{z \mid z = \alpha x + (1 - \alpha)y \text{ for some } x \in X,\ y \in Y\}.$$

Thus we have $X \underset{1}{\times} Y = X$ and $X \underset{0}{\times} Y = Y$. For $0 < \alpha < 1$, $X \underset{\alpha}{\times} Y$ is a scaled-down cartesian product of X and Y (see Fig. 1).

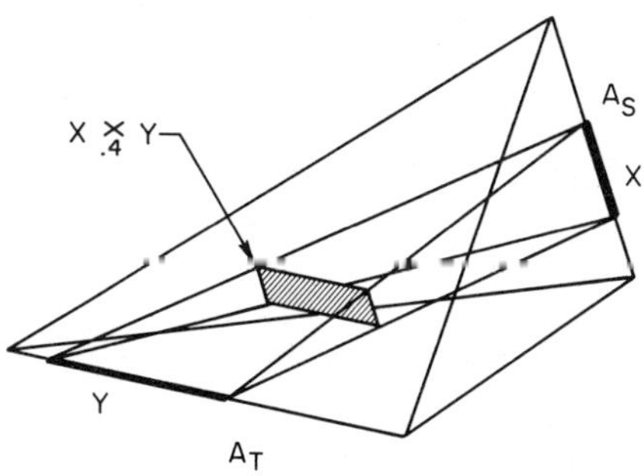

Fig. 1 — Illustrating the operations $\underset{\alpha}{\times}$ and $\times$.

Finally, we define an operation $\times$ by

$$X \times Y = \bigcup_{0 \leq \alpha \leq 1} X \underset{\alpha}{\times} Y.$$

This set consists of all imputations z such that $B_S z$ and $B_T z$ are in X and Y respectively, together with the sets X and Y themselves. It is sometimes called the "join" of X and Y, being the union of all straight lines joining X and Y. If X and Y are convex, as in Fig. 1, then $X \times Y$ is the convex hull of $X \cup Y$.

§ 5. TWO BASIC THEOREMS

In this section we shall state and prove two theorems on the composition of solutions — one dealing with sums, the other with products.

We recall that a <u>solution</u> of the game $\Gamma(P, \mathcal{W})$ is a set X of imputations such that

$$X = A_P - \text{dom } X.$$

Here "dom X" denotes the set of all y such that, for some $x \in X$, the set $\{i \mid x_i > y_i\}$ is an element of $\mathcal{W}$. The notations "dom_1" and "dom_2" will be used for domination with respect to special classes $\mathcal{W}_1$, $\mathcal{W}_2$.

Solutions are known to exist for all simple games. For example, A_S is a solution whenever $S \in \mathcal{W}^m$.

THEOREM 1. If X_i is a solution of $\Gamma(P_i, \mathcal{W}_i)$, $i = 1, 2$, then $X_1 \times X_2$ is a solution of $\Gamma(P_1, \mathcal{W}_1) \otimes \Gamma(P_2, \mathcal{W}_2)$.

THEOREM 2. If X_i is a solution of $\Gamma(P_i, \mathcal{W}_i)$, $i = 1, 2$, and if $0 \le \alpha \le 1$, then $X_1 \underset{\alpha}{\times} X_2$ is a solution of $\Gamma(P_1, \mathcal{W}_1) \oplus \Gamma(P_2, \mathcal{W}_2)$.

The reader may now wish to read again the paragraph in § 2 that gives a verbal description of the content of these theorems. As indicated there, both theorems generalize directly to sums or products of more than two components.

If we set $\alpha = 0$ or 1 in Theorem 2, we see that any solution of a component also solves the sum. Note also that the solutions obtained by choosing different values of α in Theorem 2 are all proper subsets of the solution given in Theorem 1. This is in accord with the intuitive idea that a weaker game should have a less determinate outcome.

PROOF OF THEOREM 1. Let $\Gamma(P, \mathcal{W}) = \Gamma(P_1, \mathcal{W}_1) \otimes \Gamma(P_2, \mathcal{W}_2)$, and let $X = X_1 \times X_2$. (A) External stability. The first part of the proof is devoted to showing that $X \cup \text{dom}\, X = A_P$. Let y be an arbitrary point in A_P, and let $\beta_1 = y(P_1)$ and $\beta_2 = y(P_2) = 1 - \beta_1$. Let y_i denote* the barycentric projection of y on A_{P_i}, $i = 1, 2$, if this is well defined, i.e., if $\beta_i > 0$. There are a number of cases; our object each time will be to show that either $y \in X$ or $y \in \text{dom}\, X$.

Case 1: $0 < \beta_1 < 1$.

Case 1a: $y_1 \in X_1$, $y_2 \in X_2$. Here $y \in X$.

Case 1b: $y_1 \notin X_1$, $y_2 \notin X_2$. Find a $z_1 \in X_1$ and a $z_2 \in X_2$ such that $y_1 \in \text{dom}_1 z_1$ and $y_2 \in \text{dom}_2 z_2$. Define the imputation z by

$$z = \beta_1 z_1 + \beta_2 z_2.$$

Then it is clear that $y \in \text{dom}\, z$. Moreover, $z \in X$. Hence $y \in \text{dom}\, X$.

Case 1c: $y_1 \in X_1$, $y_2 \notin X_2$. Find a $z_2 \in X_2$ such that $y_2 \in \text{dom}_2 z_2$. Denote by S_2 the set of players on which $z_2 - y_2$ is positive; we have $S_2 \in \mathcal{W}_2$. Choose $\varepsilon > 0$ so small that $(\beta_2 - \varepsilon)z_2 - \beta_2 y_2$ is also positive on S_2. Let u_1 be any interior point of A_{P_1}, that is, a vector that is positive on P_1 and zero on P_2, and has $u(P) = 1$. Then we assert that the vector x, defined by

$$x = \beta_1 y_1 + \varepsilon u_1 + (\beta_2 - \varepsilon) z_2,$$

is an imputation, since it is a weighted average of three imputations, and that it dominates y, since $x - y$ is positive on the winning set $P_1 \cup S_2$. There are now two possibilities. If it happens that $x \in X$ then we are finished with this case, since we have $y \in \text{dom}\, X$. Suppose then that $x \notin X$.

---

* Henceforth we shall avoid direct reference to the components of vectors, and use subscripted symbols x_i, y_i, etc., to denote vectors also. Usually they will be related in some stated way to their unsubscripted counterparts.

Then it must be the projection on A_{P_1} that fails to be in X_1, since the other projection is in X_2 by construction. Hence we can find $z_1 \in X_1$ such that $B_{P_1}x \in \mathrm{dom}_1 z_1$. Let $S_1 \in \mathcal{W}_1$ be the set of players on which $z_1 - B_{P_1}x$ is positive. Then the imputation z, defined by

$$z = (\beta_1 + \epsilon)z_1 + (\beta_2 - \epsilon)z_2,$$

dominates y, since $z - y$ is positive on the winning set $S_1 \cup S_2$. Moreover, $z \in X$. Hence again $y \in \mathrm{dom}\, X$.

<u>Case 1d:</u> $y_1 \notin X_1$, $y_2 \in X_2$. Like Case 1c.

<u>Case 2:</u> $\beta_1 = 0$.

<u>Case 2a</u>: $y_2 \in X_2$. Then $y \in X$.

<u>Case 2b</u>: $y_2 \notin X_2$. The argument of Case 1c may be repeated here, with the understanding (since y_1 is no longer well defined) that $\beta_1 y_1$ is to be taken equal to zero wherever it appears.

<u>Case 3</u>: $\beta_1 = 1$. Like Case 2. This completes the proof of external stability.

(B) Internal stability. It remains to show that $X \cap \mathrm{dom}\, X = 0$. Suppose the contrary. Then imputations $x, y \in X$ exist with $x - y$ positive on some winning coalition S. Without loss of generality we may assume that $y(P_1) \geq x(P_1)$. But $x(P_1) \geq x(S \cap P_1) > y(S \cap P_1) \geq 0$. Hence the barycentric projections of both x and y on A_{P_1} are well defined; denote them by x_1 and y_1, respectively. As a result of our assumed inequality, the difference $x_1 - y_1$ is positive on at least the set $S \cap P_1$, which is winning in $\Gamma(P_1, \mathcal{W}_1)$. Hence $y_1 \in \mathrm{dom}_1 x_1$. But x_1 and y_1 are both elements of the "subsolution" X_1; we have therefore reached a contradiction. This completes the proof of Theorem 1.

PROOF OF THEOREM 2. Let $\Gamma(P, \mathcal{W}) = \Gamma(P_1, \mathcal{W}_1) \oplus \Gamma(P_2, \mathcal{W}_2)$, and let $X^\alpha = X_1 \underset{\alpha}{\times} X_2$. (A) External stability. Let y be an arbitrary point in A_P, with β_i and y_i, $i = 1, 2$, defined as in the previous proof. Without loss of generality we may assume that $\beta_1 \leq \alpha$. Let u_1 be an arbitrary interior point of A_{P_1}, and construct an imputation z:

$$z = \beta_1 y_1 + (\alpha - \beta_1)u_1 + (1 - \alpha)x_2,$$

where x_2 is an arbitrary element of the subsolution X_2, and the term $\beta_1 y_1$ is understood to be zero if y_1 is not well defined. There are three cases to be distinguished at this juncture.

Case 1: $z \notin X^\alpha$. Then $\alpha > 0$, and the barycentric projection of z on A_P is well-defined; call it z_1. Then we have

$$z_1 = \frac{\beta_1}{\alpha} y_1 + \frac{\alpha - \beta_1}{\alpha} u_1.$$

If $z_1 \in X_1$, then we would have $z \in X^\alpha$, contrary to assumption; hence $z_1 \notin X_1$. Hence $z_1 \in \text{dom}_1 X_1$, and there exists $x_1 \in X_1$ such that $x_1 - z_1$ is positive over some $S_1 \in \mathcal{W}_1$. Therefore, multiplying by α, we see that $\alpha x_1 - \beta_1 y_1$ is also positive over S_1. But S_1 is also winning in the full game; hence the imputation x, defined by

$$x = \alpha x_1 + (1 - \alpha)x_2,$$

dominates y. Since $x \in X^\alpha$, we have $y \in \text{dom}\, X^\alpha$.

Case 2: $z \in X^\alpha$, $\beta_1 < \alpha$. Then $z - y$ is positive on the winning coalition $P_1 \in \mathcal{W}$, and we have $y \in \text{dom}\, X^\alpha$.

Case 3: $z \in X^\alpha$, $\beta_1 = \alpha$. Then $y_1 \in X_1$ (or is undefined). If $y_2 \in X_2$ (or is undefined) as well, then y is an element of X^α, and we are through. On the other hand, if $y_2 \notin X_2$ we can reverse the roles of P_1 and P_2 from the beginning, and the new z that will be constructed will not belong to X^α — i.e., Case 1 will apply. This completes the proof of (A).

(B) Internal stability. Suppose $x, y \in X^\alpha$, with $x - y$ positive on some winning coalition $S \in \mathcal{W}$. Then either $S \cap P_1 \in \mathcal{W}_1$ or $S \cap P_2 \in \mathcal{W}_2$; without loss of generality, assume the former. Then $\alpha = x(P_1) > 0$, so that the barycentric projections x_1 and y_1 of x and y on A_{P_1} are well defined. Obviously $x_1 - y_1$ is positive on $S \cap P_1$; hence $y_1 \in \text{dom}_1 x_1$. This contradicts the internal stability of the subsolution X_1, and completes the proof of Theorem 2.

§ 6. THE SOLUTIONS OF SUMS: A COMPLETE CHARACTERIZATION

The first two theorems gave sufficient conditions for a set of imputations to be a solution of a compound simple game. It is natural to inquire whether the conditions are necessary as well — whether every solution is of one of the types described. The answer, curiously enough, is "yes" for sums, "no" for products. We shall consider first the case of sums.

THEOREM 3. Let X be any solution of the sum

$$\Gamma(P, \mathcal{W}) = \Gamma(P_1, \mathcal{W}_1) \oplus \Gamma(P_2, \mathcal{W}_2).$$

Then there exist solutions X_1 and X_2 of $\Gamma(P_1, \mathcal{W}_1)$ and $\Gamma(P_2, \mathcal{W}_2)$, respectively, and a real number α, $0 \leq \alpha \leq 1$, such that $X = X_1 \underset{\alpha}{\times} X_2$.

PROOF. We shall first show that X is a cartesian product — i.e., a set of the form $X_1 \underset{\alpha}{\times} X_2$ for some $X_1 \subseteq A_{P_1}$, $X_2 \subseteq A_{P_2}$, and $0 \leq \alpha \leq 1$. Then we shall show that X_1 and X_2 are solutions of their respective subgames. Take any $x, y \in X$, and let x_i, y_i denote their barycentric projections on the face A_{P_i}, $i = 1, 2$, if well defined. Let u_1, u_2 be arbitrary interior points of A_{P_1}, A_{P_2}, respectively. Define $\alpha_i = x(P_i)$, $\beta_i = y(P_i)$, $i = 1, 2$, and assume without loss of generality that $\beta_1 \geq \alpha_1$. Then the imputation z, defined by*

$$z = \alpha_1 x_1 + \beta_2 y_2 + (\beta_1 - \alpha_1)u_2,$$

is necessarily in X, since any imputation dominating z also dominates either x or y. Similarly, the imputation z', defined by

$$z' = \alpha_1 x_1 + (\beta_1 - \alpha_1)u_1 + \beta_2 y_2,$$

* We follow our usual convention in regard to ill-defined barycentric projections.

is in X. But if $\beta_1 - \alpha_1$ were strictly positive, z and z' would dominate each other via the winning coalitions P_1 and P_2, in violation of the internal stability of X. Hence

$$\alpha_1 = \beta_1$$

(making z and z' the same), and we have

$$\alpha_1 x_1 + \alpha_2 y_2 \in X.$$

By symmetry,

$$\alpha_1 y_1 + \alpha_2 x_2 \in X.$$

Since x and y were arbitrary points in X, these facts show that X is of the required "cartesian" form $X_1 \underset{\alpha}{\times} X_2$, with of course $\alpha = \alpha_1$.

It remains to show that each X_i is, or can be chosen to be, a solution of its corresponding subgame $\Gamma(P_i, \mathcal{W}_i)$. If $\alpha_i = 0$ then X_i is irrevelant to X, and we can use any solution we please to satisfy the literal requirements of the theorem. On the other hand, if $\alpha_i > 0$ then X determines a unique X_i, and we must prove that the latter is a solution of $\Gamma(P_i, \mathcal{W}_i)$. As before, we do this in two parts. It will suffice to consider the case $i = 1$.

(A) External stability. Let x_1, x_2 be arbitrary elements of X_1, X_2, respectively, and take any $y_1 \in A_{P_1} - X_1$. Then the imputation y, defined by

$$y = \alpha_1 y_1 + \alpha_2 x_2,$$

is not in X, since we are assuming $\alpha_1 > 0$. Hence, some $z \in X$ dominates y. Let $S \in \mathcal{W}$ denote the set of players on which $z - y$ is positive. If $S \cap P_2$ were in $\mathcal{W}_2$, then z would dominate $\alpha_1 x_1 + \alpha_2 x_2 \in X$, violating internal stability. Therefore we must have $S \cap P_1 \in \mathcal{W}_1$. Let $z_1 = B_{P_1} z$. Then $z_1 - y_1$ is also positive on $S \cap P_1$. Hence $y_1 \in \mathrm{dom}_1 z_1$. But z_1 is clearly an element of X_1. Hence $y_1 \in \mathrm{dom}_1 X_1$.

(B) Internal stability. Suppose that $y_1 \in \mathrm{dom}_1 x_1$ for some x_1, $y_1 \in X_1$. Then we would also have $\alpha_1 y_1 + \alpha_2 x_2 \in \mathrm{dom}(\alpha_1 x_1 + \alpha_2 x_2)$, for arbitrary $x_2 \in X_2$, since $\mathcal{W}_1 \subseteq \mathcal{W}$. This contradicts the internal stability of X. This completes the proof of the theorem.

There is a strong formal resemblance between our results for sums (Theorems 2 and 3) and the results of von Neumann and Morgenstern [9] and Gillies [2], concerning "decomposable" games. It is worth emphasizing that the two theories apply to different situations. The "composition" of two (essential) simple games is never a simple game, since some coalitions will win in only one component. On the other hand, the "sum" of two simple games is not even a game, in the classical sense, since it has disjoint winning coalitions and hence a nonsuperadditive characteristic function.*

§ 7. THE SOLUTIONS OF PRODUCTS: FURTHER RESULTS

The striking formal similarity between Theorems 1 and 2 (see § 5) unfortunately does not extend to their converses. There is no product counterpart to Theorem 3, i.e., a full-fledged converse to Theorem 1. Instead, we must be content with the following, weaker result, which states that <u>if</u> a solution of a product is of the form $X_1 \times X_2$, <u>then</u> the sets X_1 and X_2 are solutions of their respective components. Though not uninteresting in itself, this contributes little to the complete characterization of the solutions of products.

THEOREM 4. Let X be a solution of $\Gamma(P, \mathcal{W}) = \Gamma(P_1, \mathcal{W}_1) \otimes \Gamma(P_2, \mathcal{W}_2)$, and let $X_1 \subseteq A_{P_1}$, $X_2 \subseteq A_{P_2}$ be such that $X = X_1 \times X_2$. Then X_1 is a solution of $\Gamma(P_1, \mathcal{W}_1)$ and X_2 is a solution of $\Gamma(P_2, \mathcal{W}_2)$.

PROOF. It will suffice to show that X_1 is a solution of $\Gamma(P_1, \mathcal{W}_1)$. (A) External stability. Take $y_1 \in A_{P_1} - X_1$. Then $y_1 \notin X$. Hence $y_1 \in \mathrm{dom}\, z$ for some $z \in X$. Let $S \in \mathcal{W}$ be the set of players on which $z - y_1$ is positive. Then $S \cap P_1 \in \mathcal{W}_1$. This incidentally gives us

* A roundabout connection between the two theories can be established, however, by means of the equivalence relation of [2], p. 66.

$z(P_1) > 0$, so that the barycentric projection of z on A_{P_1} is well defined, denote it by z_1. Clearly $z_1 \in X_1$. Moreover, the set on which $z_1 - y_1$ is positive must include $S \cap P_1$, since $z_1 > z$ on P_1. Therefore we have $y_1 \in \text{dom}_1 z_1 \subseteq \text{dom}_1 X_1$, as required.

(B) Internal stability. Take $y_1 \in X_1$, and suppose that $y_1 \in \text{dom}_1 x_1$ for some $x_1 \in X_1$. Let $S_1 \in \mathcal{W}_1$ denote the set of players on which $x_1 - y_1$ is positive. We assert (proof below) that we can find $x_2 \in X_2$ and $S_2 \in \mathcal{W}_2$ such that x_2 is strictly positive on S_2. The imputation x, defined by

$$x = (1 - \epsilon)x_1 + \epsilon x_2,$$

is then in X, and for small $\epsilon > 0$ the difference $x - y_1$ is positive on both S_1 and S_2. Hence $y_1 \in \text{dom } X$. Since y is an element of X, this violates internal stability of X, and refutes the supposition $y_1 \in \text{dom}_1 X_1$.

The assertion in the third sentence of (B) must still be proved. Part (A) permits us to assume that X_2 is externally stable. In particular, $A_{P_2} \cap \text{dom}_2 X_2$ is not empty unless $X_2 = A_{P_2}$. If the latter, we may take $S_2 = P_2$ and x_2 any interior point of A_{P_2}. Otherwise, let z_2 be any point in $A_{P_2} \cap \text{dom}_2 X_2$, let x_2 be a point of X_2 that dominates z_2, and let S_2 be the set on which $x_2 - z_2$ is positive. In both instances we obtain $x_2 \in X_2$ and $S_2 \in \mathcal{W}_2$ such that x_2 is strictly positive on S_2, as required. This completes the proof of the theorem.

SEMIMONOTONIC FAMILIES OF SETS. The three-person game "G" at the beginning of § 8 (see Fig. 2) shows that the technique of combining subsolutions by means of the "join" operation $\times$ has been unnecessarily rigid. A possible relaxation, suggested by the game "G", is to allow the subsolutions to vary — but not too rapidly — as a function of the parameter α. How best to constrain the variation, i.e., to generalize the monotonicity condition satisfied by the solutions of "G", is a rather delicate problem. We shall handle it in two slightly different ways in the next two theorems, which are both substantial generalizations of Theorem 1, and which both account for all the solutions of the game "G".

A parametrized family of sets of imputations:

$$\text{set}\{X(\alpha) \mid 0 \leq \alpha \leq 1\},$$

will be called semimonotonic if, for every α, β, x such that $0 \leq \beta \leq \alpha \leq 1$ and $x \in X(\alpha)$, there exists $y \in X(\beta)$ such that $\beta y \leq \alpha x$.

One example of a semimonotonic family* is a constant family: $X(\alpha) \equiv X$. Another is a family that is monotonically decreasing in the sense of set-inclusion: $\alpha > \beta$ implies $X(\alpha) \subseteq X(\beta)$. A third example can be derived from an arbitrary "monotonic" curve of the kind illustrated in Fig. 2, by taking $X(\alpha)$ to be the barycentric projection of the single point comprising the "α" cross section of the curve.

THEOREM 5. Let $\{X_1(\alpha)\}$, $\{X_2(\alpha)\}$ be semimonotonic families of solutions of $\Gamma(P_1, \mathcal{W}_1)$, $\Gamma(P_2, \mathcal{W}_2)$, respectively. Then

$$X = \bigcup_{0 \leq \alpha \leq 1} X_1(\alpha) \underset{\alpha}{\times} X_2(1 - \alpha)$$

is a solution of $\Gamma(P, \mathcal{W}) = \Gamma(P_1, \mathcal{W}_1) \otimes \Gamma(P_2, \mathcal{W}_2)$.

PROOF. The proof that follows parallels that of Theorem 1. (A) External stability. Take any $y \in A_P$ and define $\beta_i = y(P_i)$, $i = 1, 2$.

Case 1: $0 < \beta_1 < 1$.

Case 1a: $y_1 \in X_1(\beta_1)$, $y_2 \in X_2(\beta_2)$. Then $y \in X$.

Case 1b: $y_1 \notin X_1(\beta_1)$, $y_2 \notin X_2(\beta_2)$. Find $z_1 \in X_1(\beta_1)$ and $z_2 \in X_2(\beta_2)$ such that $y_1 \in \mathrm{dom}_1 z_1$ and $y_2 \in \mathrm{dom}_2 z_2$. Then the imputation z, defined by

$$z = \beta_1 z_1 + \beta_2 z_2,$$

which is in X, clearly dominates y. Hence $y \in \mathrm{dom}\, X$.

Case 1c: $y_1 \in X_1(\beta_1)$, $y_2 \notin X_2(\beta_2)$. Find $z_2 \in X_2(\beta_2)$ such that $y_2 \in \mathrm{dom}_2 z_2$. Let $S_2 \in \mathcal{W}_2$ be the set of players on which $z_2 - y_2$ is positive. Choose $\epsilon > 0$ so that $\beta_2 z_2 - \beta_2 y_2$ is greater than ϵ on S_2. Using

---

* Our terminology is not ideal, since the vectors in $X(\alpha)$ must be multiplied by α before anything "monotonic" in the ordinary sense emerges. The "semi" refers to the one-sideness of the hypothesis $\beta \leq \alpha$. (Compare the definition of "monotonic" later in this section.)

the semimonotonic property of $\{X_2(\alpha)\}$, find z_2' in $X_2(\beta_2 - \epsilon)$ such that $\beta_2 z_2 - (\beta_2 - \epsilon)z_2'$ is nonnegative. This vector must be $\leq \epsilon$ in all components, since no component of a nonnegative vector can exceed the sum of all components. It follows that we have

$$(\beta_2 - \epsilon)z_2' > \beta_2 y_2 \quad \text{on} \quad S_2.$$

Let u_1 be any interior point of A_{P_1}. Then the imputation x, defined by

$$x = \beta_1 y_1 + \epsilon u_1 + (\beta_2 - \epsilon)z_2',$$

dominates y, because $x - y$ is positive on the winning set $P_1 \cup S_2$. If $x \in X$ then $y \in \text{dom } X$, and we are finished with this case. Suppose $x \notin X$. Then it must be the barycentric projection of x on A_{P_1}, namely

$$x_1 = \frac{\beta_1}{\beta_1 + \epsilon} y_1 + \frac{\epsilon}{\beta_1 + \epsilon} u_1,$$

that fails to lie in the corresponding subsolution $X_1(\beta_1 + \epsilon)$. Hence we can find $z_1 \in X_1(\beta_1 + \epsilon)$ such that $x_1 \in \text{dom}_1 z_1$. Then it is clear that the imputation z, defined by

$$z = (\beta_1 + \epsilon)z_1 + (\beta_2 - \epsilon)z_2',$$

dominates y. Moreover $z \in X$. Hence $y \in \text{dom } X$.

<u>Case 1d</u>: $y_1 \notin X_1(\beta_1)$, $y_2 \in X_2(\beta_2)$. Like Case 1c.

<u>Case 2</u>: $\beta_1 = 0$.

<u>Case 2a</u>: $y_2 \in X_2(1)$. Then $y \in X$.

<u>Case 2b</u>: $y_2 \notin X_2(1)$. Then the argument of Case 1c may be repeated, with the usual understanding that $\beta_1 y_1$ is zero.

<u>Case 3</u>: $\beta_1 = 1$. Like Case 2. This completes part (A) of the proof.

(B) Internal stability. Suppose that for some $x, y \in X$, x dominates y — i.e., $x - y$ is positive on a winning set $S \in \mathcal{W}$. With no loss of generality we may assume that $y(P_1) \geq x(P_1)$. Define $\alpha_1 = x(P_1)$, $\beta_1 = y(P_1)$, $S_1 = S \cap P_1 \in \mathcal{W}_1$. We have

$$\beta_1 \geq \alpha_1 \geq x(S_1) > y(S_1) \geq 0;$$

hence, the projections $x_1 = B_{P_1}x$, $y_1 = B_{P_1}y$ are well defined. Moreover, $x_1 \in X_1(\alpha_1)$ and $y_1 \in X_1(\beta_1)$. Using the semimonotonicity of $\{X_1(\alpha)\}$, we can find $z_1 \in X_1(\alpha_1)$ such that $\alpha_1 z_1 \leq \beta_1 y_1$. But by our original assumption, $\alpha_1 x_1 - \beta_1 y_1$ is positive on S_1. Hence $z_1 \in \mathrm{dom}_1 x_1$, contrary to the internal stability of $X_1(\alpha_1)$. This contradiction establishes the internal stability of X, and completes the proof of the theorem.

MONOTONIC FAMILIES OF SETS. The alert reader may have observed that the internal stability of the sets $X_i(1)$, $i = 1, 2$, was not used in the preceding proof. (Note that the α_1 in (B) is always less than 1, since α_2 is necessarily positive.) This observation by itself does not enlarge the scope of the theorem, since, as it happens, the internal stability of nearby sets $X_i(1 - \epsilon)$ implies the internal stability of $X_i(1)$ as well. To obtain a stronger result in this direction, we must tighten some of the other hypotheses. Because of this adjustment, the following theorem is logically independent of Theorem 5: each one yields solutions that the other misses.

A semimonotonic family:

$$\mathrm{set}\{Y(\alpha) \mid 0 \leq \alpha \leq 1\},$$

will be called <u>monotonic</u> if for every α, β, x such that $0 \leq \alpha \leq \beta \leq 1$ and $x \in Y(\alpha)$, there exists $y \in Y(\beta)$ such that $\beta y \geq \alpha x$. (Compare the previous definition of "semimonotonic.")

THEOREM 6. For $i = 1, 2$, let $\{Y_i(\alpha)\}$ be a monotonic family of solutions of $\Gamma(P_i, \mathcal{W}_i)$, except that $Y_i(1)$ need not be externally stable, and let $X_i(\alpha) = A_{P_i} - \mathrm{dom}_i Y_i(\alpha)$. Then

$$X = \bigcup_{0 \leq \alpha \leq 1} X_1(\alpha) \underset{\alpha}{\times} X_2(1 - \alpha)$$

is a solution of $\Gamma(P, \mathcal{W}) = \Gamma(P_1, \mathcal{W}_1) \otimes \Gamma(P_2, \mathcal{W}_2)$.

Note that X_i and Y_i are identical except at $\alpha = 1$. In the latter case, X_i is obtained from Y_i by adding the external points that Y_i fails to dominate, if any. Thus, $X_i(1)$ will be externally stable, but perhaps not internally stable; moreover $X_i(1)$ may contain points that are not monotonically related to nearby sets $X_i(1 - \epsilon)$. (For an illustration, see Fig. 4c.)

Theorem 6 is certainly not a major extension of the theory, after Theorem 5; it does give us, however, a first indication of the occurrence of nonsolutions in the cross sections of a solution of a product game. The proof, which we omit, entails some minor modifications of the preceding proof, but should give no trouble to the aforementioned alert reader.

We might remark that the word "monotonic" in Theorem 6 cannot be replaced by "semimonotonic," since this would permit the deletion of an arbitrary subset of $Y_i(1)$. Whether the requirement of full monotonicity can be relaxed outside a neighborhood of $\alpha = 1$ is not known.

In the next section we shall provide examples to demonstrate the logical independence of Theorems 5 and 6. In one case we shall find a semimonotonic, nonmonotonic family of solutions of a certain game "J". Games of the form $J \otimes K$, for any K, will therefore have solutions that are given by Theorem 5 but not Theorem 6. In another case we shall find an application of Theorem 6 in which $X_i(1)$ and $Y_i(1)$ are different, and which therefore yields solutions not given by Theorem 5.

§ 8. EXAMPLES OF PRODUCT GAMES

FIRST EXAMPLE. Consider the three-person simple game "G":*

$$G = \Gamma(\overline{123}, \{\overline{13}, \overline{23}, \overline{123}\}).$$

This decomposes into a product, one term of which is a sum:

$$G = (B_1^1 \oplus B_1^2) \otimes B_1^3.$$

* For brevity, we write $\overline{123}$ for $\{1, 2, 3\}$, etc.

Here B_1^i denotes the one-person simple game $\Gamma(\{i\}, \{\{i\}\})$. The solutions of G are well known. They include all the straight lines in the triangle of imputations joining the vertex $A_{\overline{3}}$ to the opposite edge $A_{\overline{12}}$ — the solutions predicted by Theorem 1. They also include all other arcs running monotonically from $A_{\overline{3}}$ to $A_{\overline{12}}$, that is, curves along which the share of

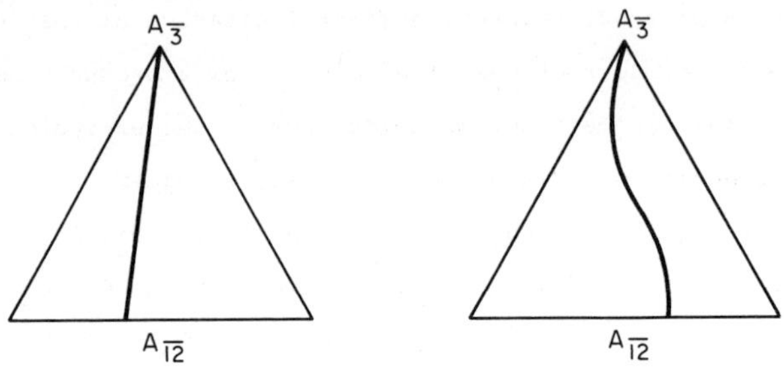

Fig. 2 — Typical solutions of $(B_1^1 \oplus B_1^2) \otimes B_1^3$

player 3 decreases steadily from 1 to 0, while the shares of players 1 and 2 increase or remain constant (see Fig. 2). These solutions are given by Theorem 5 (or 6). There are no other solutions.

SECOND EXAMPLE. Our next example is a four-person game "H":

$$H = \Gamma(\overline{1234}, \{\overline{134}, \overline{234}, \overline{1234}\}).$$

This game is a product of three components, namely $(B_1^1 \oplus B_1^2)$, B_1^3, and B_1^4. Denote the first by B_2^*. The factors can be associated in two essentially different ways, as follows:

(a) $$H = B_2^* \otimes (B_1^3 \otimes B_1^4),$$

(b) $$H = (B_2^* \otimes B_1^3) \otimes B_1^4.$$

We shall discover that our methods give different sets of solutions to "H", depending on which grouping is used.

In the case (a) there is no difficulty in solving completely the two two-person factors: a solution of B_2^* is an arbitrary single point of $A_{\overline{12}}$, while the unique solution of $B_1^3 \otimes B_1^4$ is the whole set $A_{\overline{34}}$. Application of Theorem 5 (or 6) yields the class of product solutions exemplified in Fig. 3a. They are three-cornered surfaces spanning the tetrahedron $A_{\overline{1234}}$, made up of linear elements parallel to the edge $A_{\overline{34}}$. (The dashed lines outline a typical cross-section of constant α.) The curvature of the surface

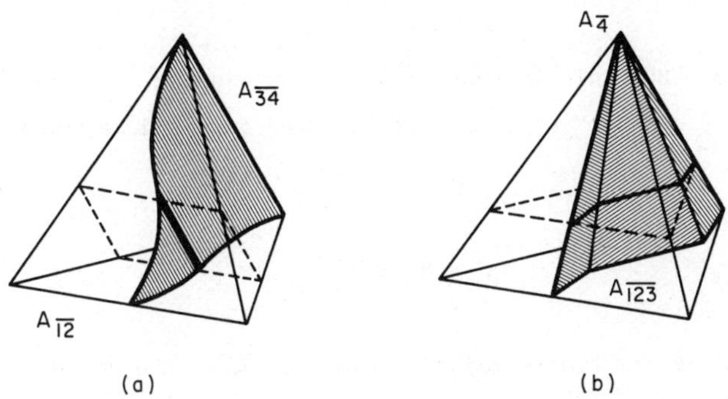

Fig. 3 — Typical solutions of $(B_1^1 \oplus B_1^2) \otimes B_1^3 \otimes B_1^4$

is thus restricted to one dimension, and is further constrained by the monotonicity condition, which requires that the separate payoffs to players 1 and 2 must decrease (or remain constant) as the combined payoff to players 3 and 4 goes up.

Now consider the grouping (b). It consists of the one-person game B_1^4 multiplied by the game "G" already considered (see Fig. 2). Theorem 5 (or 6) again yields a class of three-cornered surfaces spanning the simplex $A_{\overline{1234}}$, but this time they are generated by arbitrary, monotonically-varying, monotonic curves in the cross sections parallel to the base $A_{\overline{123}}$. The curvature of the surface will in general be two-dimensional. In Fig. 3b we have been constant to depict a conical surface (vertex at $A_{\overline{4}}$), such as would result from an application of Theorem 1 at the last stage, and leave further wrinkles to the imagination.

A comparison of the two classes of solutions reveals that "(b)" includes "(a)" but not vice versa. Class "(b)" in fact contains all the solutions of the game. (We omit the proof.) It is the failure of "(a)" to do the same that is the provocative feature of this example. The trouble seems to be that most of the solutions (including the one shown in Fig. 3b) simply do not decompose in accordance with the partition $\{\overline{12}, \overline{34}\}$; instead, the distribution of proceeds inside the two subgames are interdependent.

THIRD EXAMPLE. The four-person game that we shall call "J" is defined by

$$J = \Gamma(\overline{1234}, \{\overline{124}, \overline{134}, \overline{234}, \overline{1234}\}).$$

It differs from "H" only in that $\overline{124}$ is now winning. Player 4, the only remaining player with a "veto," can be factored out, and we obtain the decomposition

$$J = M_3 \otimes B_1^4,$$

where M_3 denotes the simple majority game on $\overline{123}$. The solutions of M_3 are well known:*

(a) The finite set

$$\{(\tfrac{1}{2}, \tfrac{1}{2}, 0), (\tfrac{1}{2}, 0, \tfrac{1}{2}), (0, \tfrac{1}{2}, \tfrac{1}{2})\}.$$

(b_δ) The line-segment

$$\text{set}\{(\delta, t, 1 - \delta - t) \mid 0 < t \leq 1 - \delta\},$$

for each $0 \leq \delta < \frac{1}{2}$.

(c_δ) The line-segment obtained by permuting players 1 and 2 in (b_δ), for each $0 \leq \delta < \frac{1}{2}$.

(d_δ) The line-segment obtained by permuting players 1 and 3 in (b_δ), for each $0 \leq \delta < \frac{1}{2}$.

No semimonotonic family drawn from this list can include representatives from more than one of the four categories (a)-(d); hence the only possible variation within such a family is in the value of δ. Theorem 5 therefore

* See [9], pp. 282-288.

generates only solutions of the types exemplified in Figs. 4a and 4b.

Theorem 6, for once, gives something new. Setting $\delta = 1/2$ in (b) (or (c) or (d)) produces an internally stable set (not a solution) that is monotonically related to the solutions nearby, and that is therefore available

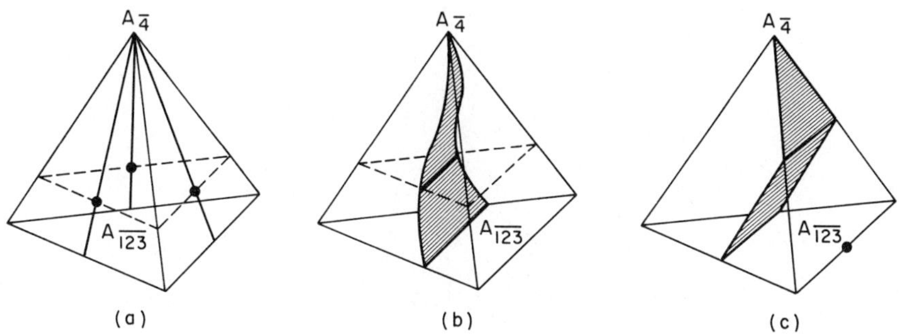

Fig. 4 — Typical solutions of $M_3 \otimes B_1$

as a "$Y_i(1)$" in Theorem 6. The associated "$X_i(1)$" will contain an additional point. Figure 4c illustrates one of the resulting solutions of "J", Its intersection with the base of the tetrahedron is internally unstable in the component game M_3.* This is one of the examples promised at the end of § 7.

Continuing, we observe that the solution of Fig. 4c, without the isolated point, can be represented as a limit of the kind of solutions typified by Fig. 4b. This enables us to construct a parametrized family of solutions of "J" that is semimonotonic, but not monotonic. For example, we can make $X(\alpha)$ like Fig. 4c for $\alpha \leq 1/2$, and like Fig. 4b for $\alpha > 1/2$, keeping everything perfectly continuous and monotonic except for the abrupt appearance of the isolated point as $\alpha \downarrow 1/2$. This family can then be used to produce solutions to arbitrary games of the form $J \otimes K$ — solutions which Theorem 5 would predict, but not Theorem 6. This is the other example promised at the end of § 7.

The game "J", innocent as it appears, actually has an enormous variety of solutions. It has been shown [7] that one may choose an entirely

* The isolated point has coordinates (0, 1/2, 1/2, 0).

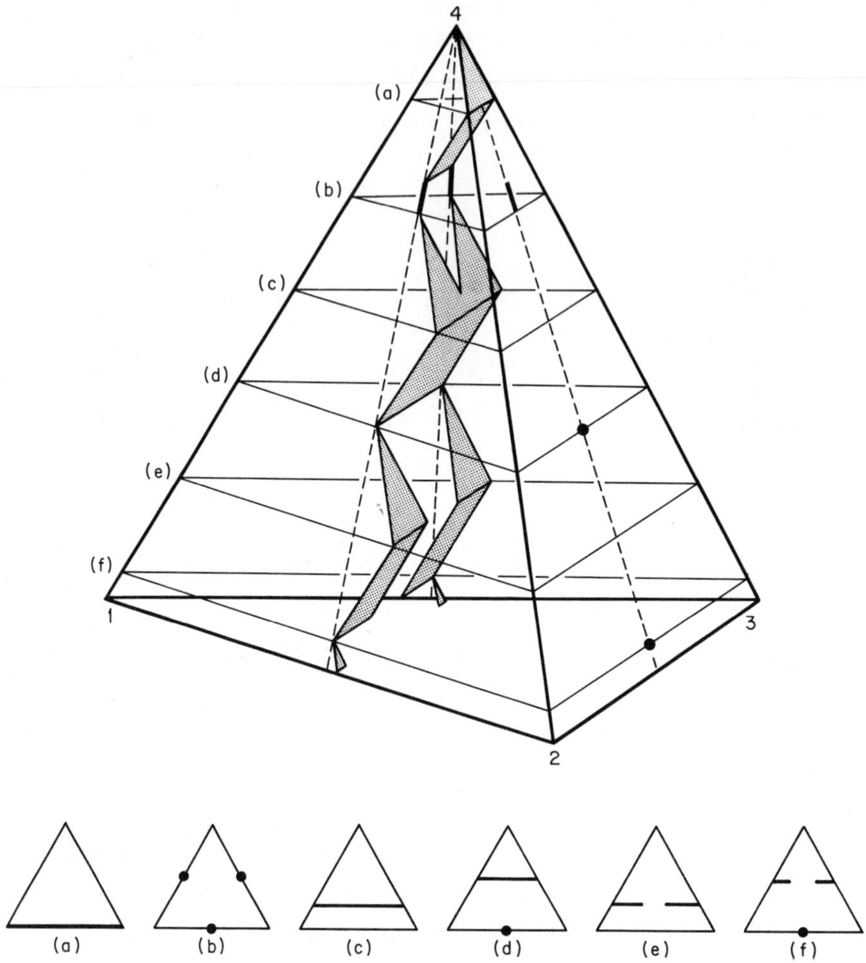

Fig. 5 — An irregular solution of $M_3 \otimes B_1$

arbitrary closed subset of the line connecting $A_{\overline{4}}$ to the midpoint of $A_{\overline{23}}$, and then construct a solution containing this set and no other nearby points, except in the neighborhood of $A_{\overline{4}}$ if the latter was an element of the chosen set. Figure 5 illustrates the construction.*

* The figure is reproduced from [7]. The arbitrary set here consists of the "dash" and two "dots" lying on the indicated median of $A_{\overline{234}}$.

Still other solutions exist to this game. For example, it is possible to "twist" a solution so as to rotate the odd man's role among players 1, 2, 3 at different levels. In particular, it is possible to include all of the categories (a)-(d) among the cross sections of the same solution. A limit to the complexity of this game may be in sight, however, since the possible cross sections of its solutions appear to be restricted to the six types shown in Fig. 5 and their rotations. Three of these types represent nonsolutions of M_3: specifically, "e" is externally unstable and "d" and "f" are internally unstable.

PART II: GENERAL COMPOUNDS

§ 9. COMPOUND SIMPLE GAMES

We recall from Part I that the notation $\Gamma(P, \mathcal{W})$ denotes the simple game with players P and winning coalitions $\mathcal{W}$. The elements of $\mathcal{W}$ are nonvoid subsets of P, and $\mathcal{W}$ itself is nonvoid and contains all the supersets of each of its elements.

Let there be given m <u>component</u> games:

$$G_i = \Gamma(P_i, \mathcal{W}_i), \qquad i = 1, 2, \ldots, m,$$

together with an m-person <u>quotient</u> game:

$$Q = \Gamma(\overline{12\ldots m}, \mathcal{U}).$$

The sets P_i are assumed disjoint, and their union will denoted by P. The players of Q are identified with the integers $1, 2, \ldots, m$; they are, of course, of a different species from the players of the G_i. For $S \subseteq P$ we define $K(S)$ to be the set of players of Q that are "controlled" by S, thus:

$$K(S) = \text{set}\{i \mid S \cap P_i \in \mathcal{W}_i\}.$$

Then the <u>compound</u> game denoted by

$$G = Q[G_1, \dots, G_m] = \Gamma(P, \mathcal{W}),$$

is defined by the condition

$$S \in \mathcal{W} \Longleftrightarrow K(S) \in \mathcal{U}.$$

Thus, a coalition wins in the compound if and only if <u>it includes winning coalitions from enough of the components to make up a winning coalition in the quotient</u>.

Another way to describe $\mathcal{W}$ is by the formula

$$\mathcal{W} = \bigcup_{R \in \mathcal{U}} \bigcap_{i \in R} \mathcal{W}_i^+,$$

where $\mathcal{W}_i^+$ consists of all subsets of P that are supersets of elements of $\mathcal{W}_i$. In other words, $\Gamma(P, \mathcal{W}_i^+)$ is the game $\Gamma(P_i, \mathcal{W}_i)$ with the members of $P - P_i$ thrown in as dummies.

ELEMENTARY PROPERTIES. Let B_k denote the k-person "pure bargaining" game, in which only the all-player set wins, and let B_k^* denote its "dual", in which only the empty set loses. To begin with, we have the trivial identities:

$$G = B_1[G] = G[B_1, \dots, B_1].$$

We shall not call G "compound" unless it can be decomposed in some less trivial way.

Sums and products may be expressed as follows:

$$G_1 \oplus G_2 = B_2^*[G_1, G_2],$$

$$G_1 \otimes G_2 = B_2[G_1, G_2].$$

A sum of three games corresponds to the quotient B_3^*, etc. Polynomials in $\oplus$ and $\otimes$ can be expressed in our present notation by substituting B_1 into the polynomial to determine the quotient. For example:

$$G_1 \otimes (G_2 \oplus G_3) = (B_1 \otimes B_2^*)[G_1, G_2, G_3].$$

In fact, iterated compound structures can always be collapsed to a single level, i.e., a quotient — possibly compound — operating on irreducible components.*

The general formula for the number of minimal winning coalitions in a compound game is

$$|\mathscr{W}^m| = \sum_{R \in \mathfrak{U}^m} \prod_{i \in R} |\mathscr{W}_i^m| .$$

This bears an obvious resemblance to our second definition of $\mathscr{W}$, above.

§ 10. THE FUNDAMENTAL COMPOSITION THEOREM

We shall use A to denote the imputation simplex of the compound game $G = Q[G_1, \ldots, G_m]$ with player-set $P = \cup_1^m P_i$, and we shall use A_i to denote what we previously called A_{P_i} — the face of A on which all players in $P - P_i$ receive zero. As before, the A_i will be identified in the natural way with the imputation spaces of the components G_i. The imputation space of Q will be denoted by A.

Elements of A will be denoted by latin letters x, y, etc., sometimes with subscripts to indicate the face of A to which they belong. (The coordinates of these "latin" vectors will not be referred to individually.) Elements of A will be denoted by greek letters α, β, etc., and their coordinates will be indicated by subscripts in the usual way.

Let X, $X_1, \ldots, X_m$ be subsets of A, $A_1, \ldots, A_m$, respectively. Then the expression

$$\mathsf{X}[X_1, \ldots, X_m]$$

will denote the set of all $x \in A$ of the form

---

* For a further discussion, see [8].

$$x = \sum_{i=1}^{m} \alpha_i x_i, \quad \alpha \in \mathbf{X}, \quad x_i \in X_i, \quad i = 1, \ldots, m.$$

Such a set will be called <u>compound</u>.

THEOREM 7.* Let $X_1, \ldots, X_m$ be solutions of $G_1, \ldots, G_m$, respectively, and let $\mathbf{X}$ be a solution of Q. Then $X = \mathbf{X}[X_1, \ldots, X_m]$ is a solution of $G = Q[G_1, \ldots, G_m]$.

PROOF. (A) External stability. Consider an imputation y in $A - X$. It can be represented in the form

$$y = \sum_{i=1}^{m} \beta_i y_i,$$

where $\beta \in \mathbf{A}$ and $y_i \in A_i$, $i = 1, \ldots, m$. The object is to find a vector $x \in X$ that dominates y. We shall first construct a vector $z = \sum \gamma_i z_i$, not necessarily in X, which dominates y, and then we shall modify z to obtain the desired vector x. There are two cases to be distinguished.

<u>Case 1</u>: $\beta \in \mathbf{X}$. Since $y \notin X$, there must be an i^* such that $\beta_{i^*} > 0$ and $y_{i^*} \notin X_{i^*}$. To simplify notation, suppose $i^* = 1$. Find a $z_1 \in X_1$ such that $y_1 \in \mathrm{dom}_1 z_1$, and let $S_1 \in \mathcal{W}_1$ denote the effective set for the domination. For a sufficiently small $\epsilon > 0$, we then have

$$(\beta_1 - \epsilon) z_1 - \beta_1 y_1 > 0 \quad \text{on} \quad S_1.$$

Let $\gamma_1 = \beta_1 - \epsilon$, and for $i = 2, \ldots, m$ define

$$\begin{cases} \gamma_i = \beta_i + \frac{\epsilon}{m-1} \\ z_i = \frac{1}{\gamma_i} (\beta_i y_i + (\gamma_i - \beta_i) u_i), \end{cases}$$

---

* A special case of this theorem, in which all solutions are "simple" (i.e., allot to each player at most one non-zero payoff), was noted by Gurk and Isbell [3]. The special cases $Q = B_2$ and $Q = B_2^*$ correspond to Theorems 1 and 2 of Part I.

where the u_i are arbitrary interior vectors in the corresponding A_i. It is clear that $\gamma \in \mathbf{A}$ and $z_i \in A_i$, $i = 1, \ldots, m$. This completes the construction of z.

It can easily be shown that z dominates y, but we have no assurance that $z \in X$. We therefore make certain adjustments. To start, take $x_1 = z_1$. For each $i > 1$, if $z_i \notin X_i$, find a dominating $x_i \in X_i$ with $x_i - z_i$ positive on some $S_i \in \mathcal{W}_i$; but if $z_i \in X_i$, simply take $x_i = z_i$ and $S_i = P_i$. In similar fashion if $\gamma \notin \mathbf{X}$, find a dominating $\alpha \in \mathbf{X}$ with $\alpha - \gamma$ positive on some $T \in \mathcal{U}$; but if $\gamma \in \mathbf{X}$, simply take $\alpha = \gamma$ and $T = \overline{12\ldots m}$. Now we can assemble the pieces, by defining

$$x = \sum_1^m \alpha_i x_i \quad \text{and} \quad S = \bigcup_{i \in T} S_i .$$

By construction, $x \in X$ and $S \in \mathcal{W}$. To show that x dominates y, we shall verify that $x - y$ is positive on S. Applying the various definitions, we see that the chain of inequalities

$$\alpha_i x_i \geq \alpha_i z_i \geq \gamma_i z_i > \beta_i y_i$$

is valid on S_i, for every $i \in T$. Hence $y \in \text{dom}\, X$, as required.

<u>Case 2</u>. $\beta \notin \mathbf{X}$. Find a dominating $\alpha \in \mathbf{X}$ with $\alpha - \beta$ positive on some $T \in \mathcal{U}$. Define

$$z_i = \begin{cases} \frac{1}{\alpha_i}(\beta_i y_i + (\alpha_i - \beta_i)u_i) & \text{for } i \in T, \\ \\ u_i & \text{for } i \notin T, \end{cases}$$

where the u_i are arbitrary interior vectors of the corresponding A_i. Then $z_i \in A_i$ for every i. If $z_i \notin X_i$, find a dominating $x_i \in X_i$ with $x_i - z_i$ positive on some $S_i \in \mathcal{W}_i$; but if $z_i \in X_i$, simply take $x_i = z_i$ and $S_i = P_i$. Assembling x and S as in Case 1, we see that $x - y$ is again positive on S, by virtue of the inequalities

$$\alpha_i x_i \geq \alpha_i z_i > \beta_i y_i ,$$

which are valid on S_i for all $i \in T$. Since $x \in X$ and $S \in \mathcal{W}$, this means that $y \in \text{dom } X$ as before. This completes part (A) of the proof.

(B) Internal stability. We must show that $y \in \text{dom } x$ never occurs for $x, y \in X$. Suppose it did occur. Then there would exist $\alpha, \beta \in \mathsf{X}$, $T \in \mathfrak{U}$, $x_i, y_i \in X_i$, and $S_i \in \mathcal{W}_i$, such that $\alpha_i x_i - \beta_i y_i$ is positive on S_i for all $i \in T$. There are two cases:

Case 1: $\alpha_i > \beta_i$ for all $i \in T$. Then $\beta \in \text{dom } \alpha$, contradicting the internal stability of X.

Case 2: $\alpha_i \leq \beta_i$ for at least one $i \in T$. Then we have

$$\beta_i x_i \geq \alpha_i x_i > \beta_i y_i \quad \text{on} \quad S_i.$$

Dividing through by the positive number β_i, we see that this entails $y_i \in \text{dom}_i x_i$, contradicting the internal stability of X_i. This completes the proof.

The application of Theorem 7 to some simple examples will be found in § 13 below. A verbal description of its content was given in § 2.

§ 11. A PARTIAL CONVERSE

In this section, we consider the following question: If a solution to $Q[G_1, \ldots, G_m]$ is the compound set $\mathsf{X}[X_1, \ldots, X_m]$, then are $X_1, \ldots, X_m$, and X necessarily solutions of the games $G_1, \ldots, G_m$, and Q, respectively? This question bears on the problem of aggregation in the application of game theory (see § 3).

Given a compound set $X = \mathsf{X}[X_1, \ldots, X_m] \neq 0$, we shall say that X_i (or i) is relevant if α_i does not vanish identically on X. It is clear that X and all relevant X_i are uniquely determined when we know X. On the other hand, the nonrelevant X_i, if any, can be quite arbitrary, since their only function is to fill out the notational form.

THEOREM 8. Let $X = \mathsf{X}[X_1, \ldots, X_m]$ be a solution of $G = Q[G_1, \ldots, G_m]$. Then either it is the case that

(a) X is a solution of Q and each relevant X_i is a solution of the corresponding G_i, *or* the following four statements are all true:

(b) $\mathbf{X}$ is internally stable for Q;

(c) $\mathbf{X}$ is externally unstable for Q;

(d) each relevant X_i is externally stable for its G_i;

(e) some relevant X_i is internally *totally* unstable for its G_i, in the sense that $X_i \subseteq \mathrm{dom}_i X_i$.

This theorem generalizes Theorem 4 of Part I, which in effect asserts (a) for the case $Q = B_2$, $\mathbf{X} = \mathbf{A}$. We should remark at once that we have found no instances where (a) does not hold, despite much searching. Note that (b)-(e) imply that both $\mathbf{X}$ and at least one of the X_i are nonsolutions. We therefore have the following corollary, with the aid of Theorem 7.

COROLLARY. Let $G = Q[G_1, \ldots, G_m]$ and let $X = \mathbf{X}[X_1, \ldots, X_m]$. Then any two of the following statements imply the third:

(i) X solves G;

(ii) $\mathbf{X}$ solves Q;

(iii) each relevant X_i solves its G_i.

It remains an open question whether (i) by itself implies both (ii) and (iii).

PROOF OF THEOREM 8. The theorem can be broken down into the following four propositions:*

(A) $\mathbf{X}$ is internally stable.

(B) Each relevant X_i is externally stable.

(C) $\mathbf{X}$ is externally stable, provided no relevant X_i is internally totally unstable.

(D) Each relevant X_i is internally stable, provided $\mathbf{X}$ is externally stable.

---

* To see this, observe that if the provisos in (C) and (D) are not violated, then we have (a). But violation of either proviso implies violation of the other, and we have (b)-(e).

We shall take these up in order. First we establish a useful lemma.

LEMMA. If i is relevant, then there exist $x_i \in X_i$ and $S_i \in \mathcal{W}_i$ such that $x_i > 0$ on S_i.

Suppose no such x_i, S_i exist. Then no member of P_i is essential to any domination by any $x \in \mathbf{X}$. Hence X solves not only G, but also the weaker game $G' = \Gamma(P, \mathcal{W}')$, defined by

$$S \in \mathcal{W}' \Longleftrightarrow S - P_i \in \mathcal{W}.$$

In this game, all members of P_i are dummies. But dummies always get zero in a solution;* hence i is not relevant, and the lemma is proved.

(A) Internal stability of $\mathbf{X}$. Take $\alpha, \beta \in \mathbf{X}$, and suppose $\alpha > \beta$ on some $R \in \mathcal{U}$. Then $\alpha > 0$ on R. Using the lemma, for each $i \in R$ choose $x_i \in X_i$ such that $x_i > 0$ on some $S_i \in \mathcal{W}_i$. For $i \notin R$, choose $x_i \in X_i$ arbitrarily. Then

$$\alpha_i x_i > \beta_i x_i \quad \text{on} \quad S_i, \quad \text{for all} \quad i \in R.$$

Hence

$$\sum \alpha_i x_i > \sum \beta_i x_i \quad \text{on} \quad \bigcup_R S_i \in \mathcal{W}.$$

This violates the internal stability of X. Hence $\mathbf{X}$ is internally stable.

(B) External stability of relevant X_i. Without loss of generality, suppose X_1 relevant. Take $\alpha \in \mathbf{X}$ such that $\alpha_1 > 0$. Let y_1 be an arbitrary element of $A_1 - X_1$. Using the lemma, choose $x_1 \in X_1$ such that $x_1 > 0$ on some $S_1 \in \mathcal{W}_1$. For $i \neq 1$, choose $x_i \in X_i$ arbitrarily. Define

$$x = \sum \alpha_i x_i$$

---

* See [9], p. 397-398.

and

$$y = \alpha_1 y_1 + \sum_{i \neq 1} \alpha_i x_i .$$

Then $x \in X$ and $y \notin X$. By the external stability of X, there is a $z \in X$ that dominates y via some winning set $T = \cup_R T_i$, where $R \in \mathfrak{U}$ and $T_i \in \mathcal{W}_i$. for all $i \in R$. Let us represent z as follows:

$$z = \sum \gamma_i z_i, \quad \text{with} \quad \gamma \in \mathbf{X} \quad \text{and all} \quad z_i \in X_i .$$

Also define

$$w = \gamma_1 x_1 + \sum_{i \neq 1} \gamma_i z_i ;$$

this is also in X. Now z dominates y, but must not dominate x. Hence $1 \in R$ and

$$\gamma_1 z_1 > \alpha_1 y_1 \quad \text{on} \quad T_1 .$$

There are two possibilities. If $\gamma_1 > \alpha_1$, then w would dominate x via the winning set $(T - T_1) \cup S_1$, in violation of the internal stability of X. Thus $\gamma_1 \leq \alpha_1$, and we have $z_1 > y_1$ on T_1, so that $y_1 \in \text{dom}_1 z_1$. But z_1 is in X_1, and y_1 was arbitrary in $A_1 - X_1$; hence X_1 is externally stable.

(C) External stability of $\mathbf{X}$, provided no relevant X_i is internally totally unstable. Let β be arbitrary in $\mathbf{A} - \mathbf{X}$. Using the proviso, choose $y_i \in A_i - \text{dom}_i X_i$ for each relevant i; for other values of i, if any, let y_i be any element of A_i. Define

$$y = \sum \beta_i y_i .$$

Then $y \notin X$, and there is an $x \in X$ that dominates y via some winning set $S = \cup_R S_i$, where $R \in \mathfrak{U}$ and $S_i \in \mathcal{W}_i$ for all $i \in R$. Let us represent x as follows:

$$x = \sum \alpha_i x_i, \quad \text{with} \quad \alpha \in \mathbf{X} \quad \text{and all} \quad x_i \in X_i.$$

There are two possibilities to consider. If $\alpha > \beta$ on R, then we have $\beta \in \text{dom}\,\mathbf{X}$, as required. But if $\alpha_i \leq \beta_i$ for some $i \in R$, then

$$\alpha_i x_i > \beta_i y_i \geq \alpha_i y_i \quad \text{on} \quad S_i.$$

This inequality implies that $x_i > y_i$ on S_i, giving us $y \in \text{dom}_i X_i$. But it also implies that $\alpha_i > 0$, making i relevant and giving us $y \notin \text{dom}_i X_i$, by the definition of y. This contradiction establishes the result.

(D) Internal stability of each relevant X_i, provided $\mathbf{X}$ is externally stable. Suppose X_1 (say) is relevant but internally unstable. Choose $\alpha \in \mathbf{X}$ such that $\alpha_1 > 0$, and choose $x_1, y_1 \in X_1$ such that $x_1 > y_1$ on some $S_1 \in \mathcal{W}_1$. Take $\epsilon > 0$ so small that

$$(\alpha_1 - \epsilon)x_1 > \alpha_1 y_1 \quad \text{on} \quad S_1,$$

and define $\beta \in \mathbf{A}$ by

$$\beta_1 = \alpha_1 - \epsilon$$

$$\beta_i = \alpha_i + \epsilon', \quad \text{for} \quad i \neq 1,$$

where $\epsilon' = \epsilon/(m-1)$. Note that $\beta > 0$. There are two possibilities. Suppose first that $\beta \notin \mathbf{X}$. Then, using the proviso, we may choose $\gamma \in \mathbf{X}$ such that $\gamma > \beta$ on some $R \in \mathfrak{U}$. Suppose instead that $\beta \in \mathbf{X}$. Then define $\gamma = \beta$ and $R = 12\ldots m$. In either case, we have $\gamma \geq \beta$ on R. Observe that $\gamma > 0$ on R; hence, by the lemma, for each $i \in R$ there is an $x_i \in X_i$ that is positive on some $S_i \in \mathcal{W}_i$. (Note that R may or may not contain 1; if it does, we shall use the x_1, S_1 previously specified, which have the stated property.) For any remaining values of i, choose $x_i \in X_i$ arbitrarily. Now define

$$x = \sum \gamma_i x_i$$

and

$$y = \alpha_1 y_1 + \sum_{i \neq 1} \alpha_i x_i .$$

Both x and y are in X. For every $i \in R$ we have

$$\gamma_i x_i \geq \beta_i x_i = \begin{cases} (\alpha_1 - \varepsilon) x_1 > \alpha_1 y_1 & \text{on } S_1 \quad (\text{if } i = 1), \\ (\alpha_i + \varepsilon') x_i > \alpha_i x_i & \text{on } S_i \quad (\text{if } i \neq 1). \end{cases}$$

It follows that $x > y$ on the winning set $\cup_R S_i$, violating the internal stability of X. Hence X_1 is internally stable. This completes the proof.*

§ 12. THE EXISTENCE OF NONCOMPOUND SOLUTIONS

In Part I we found ways to construct noncompound solutions to product games, using the idea of monotonic (or semi-monotonic) variation of the component solutions. By a simple trick, we can extend these techniques to general compound games, and produce, in most cases, a wealth of noncompound solutions.

THEOREM 9. Let $G = Q[G_1, \ldots, G_m]$, and let R be a minimal winning coalition of Q. Then any solution of the product game

$$H = \bigotimes_{i \in R} G_i$$

is also a solution of G, it being understood that the players of G that are not players of H receive zero.

PROOF. In any simple game, if the players outside a given winning coalition T are replaced by dummies, i.e., if $\mathcal{W}$ is replaced by

---

* A further tightening of the conditions (b)-(e) in Theorem 8, leading to an interesting extension of the notion of solution, is described in [6e].

$$\mathcal{W}_T = \text{set}\{S \mid S \cap T \in \mathcal{W}\},$$

for some $T \in \mathcal{W}$, then the solutions of the resulting, weaker game are precisely those solutions of the original game that give nothing to the players outside T.* In the present instance, the game H (plus dummies) is related to the game G in exactly this way, with $T = \cup_R P_i$. Q.E.D.

It was shown in Part I (Theorem 3) that sums do not have noncompound solutions. If we attempt to apply Theorem 9 to a sum, setting $Q = B_m^*$, we always have $|R| = 1$, and hence a trivial product H. Another case in which the solutions are necessarily of compound form is typified by the first four-person example in § 13 (see Fig. 6). Here H is trivial in a different sense, since every factor — and hence H itself — is of the form B_k, with a unique solution.

Generalizing from these cases, we are led to conjecture that every solution of a given compound game is a compound set if and only if (in our usual notation) for each $R \in \mathfrak{U}^m$, either $|R| = 1$, or $|\mathcal{W}_i^m| = 1$ for all $i \in R$. However, we have not yet succeeded in establishing either the necessity or sufficiency of these conditions.

Applied to the second four-person game in § 13, Theorem 9 produces some, but not all, of the noncompound solutions described there. The others can be obtained in this case by the technique that Gillies called "inflation" (see [1], p. 335), but this method is no longer sufficient in the case of larger compound games.

§ 13. EXAMPLES

The smallest compound simple games that are not sums or products are the two four-person games:

$$M_3[B_1, B_1, B_2],$$

(with minimal winning coalitions $\overline{12}$, $\overline{134}$, $\overline{234}$), and

---

* This "projection" principle also works for general (nonsimple) games, provided that there is a winning coalition T to project on. (See [6c]; also 56:I:a in [9].)

$$M_3[B_1, B_1, B_2^*]$$

(with minimal winning coalitions $\overline{12}$, $\overline{13}$, $\overline{14}$, $\overline{23}$, $\overline{24}$). Applying Theorem 7, we obtain the solutions illustrated in Figures 6 and 7. The labels (a), ..., (d) match the listing of solutions to M_3 in § 8 of Part I. In cases (b), (c), and (d) there is a variable parameter, indicated by the outside arrows. In the second game (Fig. 7) there is another variable parameter, indicated by the inside arrows, which arises from the multiple solutions of B_2^*.

The first of these games has no other solutions. The second has only those additional solutions that are obtained by distorting the straight lines of (b) and (c) into monotonic curves, in the style of Figure 2 in Part I.

The smallest <u>constant-sum</u> compound simple game has five players:

$$M_3[B_1, B_1, M_3].$$

This proves to be a homogeneous weighted majority game,* and its main simple solution is the finite, compound set generated by the unique finite solutions

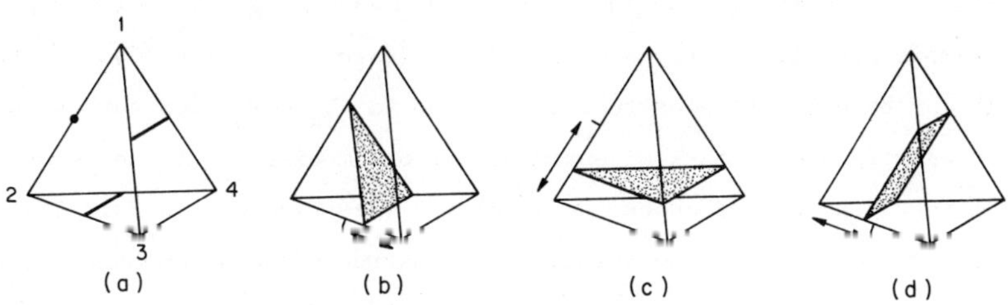

Fig. 6 — Solutions of $M_3 [B_1, B_1, B_2]$

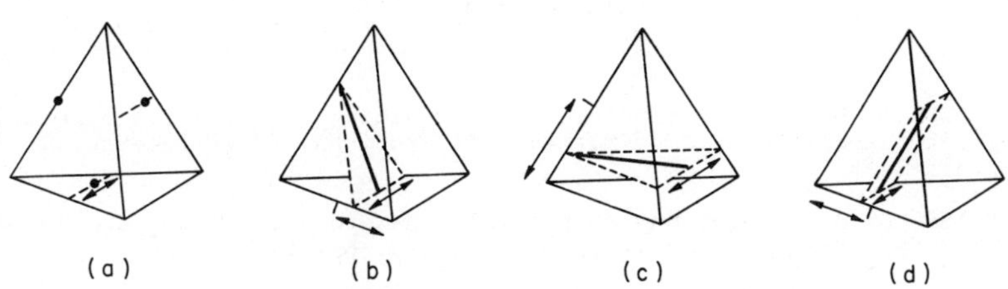

Fig. 7 — Solutions of $M_3 [B_1, B_1, B_2^*]$

---

* Its symbol is $[2, 2, 1, 1, 1]_h$ (see [9], pp. 444, 462-3).

of M_3 and B_1, consisting of the seven imputations

$$\begin{array}{lll} (1/2;\ 1/2;\ 0,\ 0,\ 0) & (1/2;\ 0;\ 1/4,\ 1/4,\ 0) & (0;\ 1/2;\ 1/4,\ 1/4,\ 0) \\ & (1/2;\ 0;\ 1/4,\ 0,\ 1/4) & (0;\ 1/2;\ 1/4,\ 0,\ 1/4) \\ & (1/2;\ 0;\ 0,\ 1/4,\ 1/4) & (0;\ 1/2;\ 0,\ 1/4,\ 1/4), \end{array}$$

corresponding to the seven minimal winning coalitions.

Similarly, the highly symmetric, nine-person game

$$M_3[M_3,\ M_3,\ M_3]$$

has a main simple solution consisting of twenty-seven imputations of the form (1/4, 1/4, 0; 1/4, 1/4, 0; 0, 0, 0). These finite solutions illustrate the composition principle described by Gurk and Isbell [3]. There are of course many infinite compound solutions.

By the "projection" principle used in the proof of Theorem 9, it can be shown that the last two games are generously supplied with noncompound solutions. Indeed, among the solutions of the five-person game are all solutions of $B_1 \otimes M_3$ (take $T = \overline{1345}$ or $\overline{2345}$), and among those of the nine-person example are all solutions of $B_2 \otimes M_3$ (take, e.g., $T = \overline{12456}$). But in [7] it is shown how to construct a solution to $B_k \otimes M_3$, for any k, whose intersection with a certain k-dimensional set of imputations is any closed subset of that set that we choose to specify. (See Figure 5 in Part I.) Only in extreme cases will the solution so constructed be a compound set with respect to any larger compound games of which $B_k \otimes M_3$ is the "projection".

BIBLIOGRAPHY

[1] GILLIES, D. B., "Discriminatory and bargaining solutions to symmetric majority games," Annals of Mathematics Study No. 28 (Princeton, 1953), pp. 325-342.

[2] GILLIES, D. B., "Solutions to general non-zero-sum games," Annals of Mathematics Study No. 40. (Princeton, 1959), pp. 47-85.

[3] GURK, H. M., and ISBELL, J. R., "Simple solutions," Annals of Mathematics Study No. 40 (Princeton, 1959), pp. 247-265.

[4] ISBELL, J. R., "Homogeneous games," Mathematics Student, Vol. 25 (1957), pp. 159-161.

[5] OWEN, G., "Tensor composition of non-negative games," this Study.

[6a] SHAPLEY, L. S., "Some variants of the von Neumann-Morgenstern definition of solution," The RAND Corporation, Research Memorandum RM-817 (1952).

[6b] SHAPLEY, L. S., "Simple games: An outline of the descriptive theory," The RAND Corporation, Research Memorandum RM-1384 (1954).

[6c] SHAPLEY, L. S., "On solutions that exclude one or more players," The RAND Corporation, Research Memorandum RM-2533 (1960).

[6d] SHAPLEY, L. S., "Compound simple games I: Solutions of sums and products," The RAND Corporation, Research Memorandum RM-3192 (1962).

[6e] SHAPLEY, L. S., "Compound simple games II: Some general composition theorems," The RAND Corporation, Research Memorandum RM-3643 (1963).

[7] SHAPLEY, L. S., "A solution containing an arbitrary closed component," Annals of Mathematics Study No. 40 (Princeton, 1959), pp. 87-93.

[8] SHAPLEY, L. S., "Simple games: An outline of the descriptive theory," Behavioral Science, Vol. 7 (1962), pp. 59-66.

[9] VON NEUMANN, J., and MORGENSTERN, O., Theory of Games and Economic Behavior, Princeton, 1944, 1947, 1953.

L. S. Shapley

The RAND Corporation

TENSOR COMPOSITION OF NONNEGATIVE GAMES*

Guillermo Owen

§ 1. PRELIMINARIES

As treated in this paper, an n-person game is simply a super-additive set function, v, defined on the subsets of the set $N = \{1, \ldots, n\}$. For a given game v, imputations, the domination relation, and solutions are defined as in von Neumann and Morgenstern [3]. In this paper, we shall write $E(v)$ for the set of all imputations of a game v, and shall, moreover, deal only with nonnegative games, i.e., games v which satisfy

$$v(S) \geq 0 \tag{1}$$

for all coalitions S.

§ 2. THE TENSOR COMPOSITION OF GAMES

By analogy with vectors, we would like to define a tensor product operation on nonnegative games. We shall later show that the analysis of games aids in the analysis of their tensor products.

Given an m-vector, $x = (x_1, \ldots, x_m)$, and an n-vector, $y = (y_1, \ldots, y_n)$, the tensor product of x and y can be defined as the $m \times n$ matrix (a_{ij}) defined by:

$$a_{ij} = x_i y_j \qquad i = 1, 2, \ldots, m; \quad j = 1, 2, \ldots, n \tag{2}$$

This is actually, however, a special case of a more general idea, that of a composition of vectors.

* This is a revised version of part of the author's doctorate thesis, which was directed by Professor Harold W. Kuhn and accepted by the Faculty of Princeton University in May, 1962.

DEFINITION 1. Given an n-vector y and n m-vectors, $x^1, x^2, \ldots, x^n$, then the y-composition of $(x^1, \ldots, x^n)$ is the matrix (a_{ij}) defined by:

$$(3) \qquad a_{ij} = x_i^j y_j \qquad i = 1, 2, \ldots, m; \quad j = 1, 2, \ldots, n.$$

We shall use the notation $(x^j) \otimes y$ for this composition. If $x^1 = x^2 = \ldots = x^n$, then the composition reduces to the tensor product $x \otimes y$.

In defining the analogue of this composition for games, we shall consider (heuristically) what this is supposed to be.

Essentially, an n-person game consists of n players, each trying to form a coalition with some of the other players and thus receive more than his "minimum level", $v(\{i\})$. Suppose, however, that each "player" of this game were actually a group of people, playing a "subgame" among themselves to determine their action in the larger game. Then, if their opinion is nearly unanimous, we may expect them to act decidedly, but, if it is divided, they shall probably act only half-heartedly. This should, however, depend on the subgame: if, for example, a particular subgame is one of majority rule, then the majority should decide entirely the action taken.

These considerations lead us to the definition of a composition for a special class of games, viz., simple games in (0, 1) normalization.

DEFINITION 2. Let v be a nonnegative n-person game, and let $w_1, \ldots, w_n$, be simple m-person games in (0, 1) normalization. Then by the v-composition of $(w_1, \ldots, w_n)$, we mean the mn-person game u, given by:

$$(4) \qquad u : 2^{M \times N} \to R$$

such that, for $S \subset M \times N$,

$$(5) \qquad u(S) = v(T),$$

where T is the maximal subset of N such that, for $j \in T$,

(6) $$w_j(S_j) = 1,$$

where

(7) $$S_j = \{i \mid (i, j) \in S\}.$$

To get a graphic idea of such a game, we may think of a legislature. The senators, representatives, and such can be thought of as playing a game — a 537-person game, in the case of the present Congress. But they themselves are elected by means of a simple game among their constituents. So, assuming that Congressmen are the true voice of the people, and vote only as their constituents would have them vote, this becomes a game among the voters. Note, however, that not every subgame has the same number of players, as some states and districts have more voters than others. We overcome this difficulty, and shall in future do so in general, by expanding the smaller subgames through the addition of dummies. A greater difficulty is presented by the fact that the subgames overlap: in fact, each voter helps to elect two senators and one (or, possibly, two) representatives and thus participates in three or four of the subgames.

A better example, perhaps, is the electoral college, where the electors are usually pledged to support a certain candidate. In this case we have a 51-person game among the states, while there are simple majority subgames in the states to choose the electors.

When the games w_j are not simple, the composition is not so clearly defined. Let us, however, assume that the games w_j are nonnegative m-person games such that $w_j(M) = 1$, and suppose that a game u is the composition

(8) $$u = (w_j) \otimes v.$$

Now if this is really a generalization of the case where the w_j are simple, and if it is really to act according to our previous considerations, this game U must satisfy certain relations. Clearly we must have

(9) $$u(M \times T) = v(T) \quad \text{for all} \quad T \subset N,$$

as this simply means that unanimity of the players in a subgame will cause decided action by that group of players. It is clear that where the w_j are simple, the composition defined above (by 4-7) satisfies Equation (9).

Let us see what other relations our composition whould satisfy. Supposing that S is a set such that $S_j = \phi$, then, in general,

$$u(S \cup M \times \{j\}) \geq u(S)$$

due to the fact that $u(M \times \{j\}) = v(\{j\}) \geq 0$. On the other hand, if we add only a subset of $M \times \{j\}$, say $R \times \{j\}$, to S, this subset will not raise $u(S)$ as much as $M \times \{j\}$ will. We should have

$$u(S) \leq u(S \cup R \times \{j\}) \leq u(S \cup M \times \{j\}).$$

Now, what is the middle term in this (double) inequality? Reason tells us that it should be near the left term if R is "weak" in the game w_j, and near the right hand term if R is "strong". How near to each side must depend on the exact strength of R. Now, since by hypothesis, $w_j(M) = 1$, we shall write our new condition in the form

$$u(S \cup R \times \{j\}) - u(S) = w_j(R)\,[u(S \cup M \times \{j\}) - u(S)] \quad \text{if} \quad S_j = \phi.$$

Then

$$u(S \cup R \times \{j\} \cup \{(i,j)\}) - u(S) = w_j(R \cup \{i\})\,[u(S \cup M \times \{j\}) - u(S)].$$

Taking the difference between these two, we obtain

$$u(S \cup R \times \{j\} \cup \{(i,j)\}) - u(S \cup R \times \{j\})$$
$$= \left[w_j(R \cup \{i\}) - w_j(R)\right]\left[u(S \cup M \times \{j\}) - u(S)\right] \quad \text{if} \quad S_j = \phi.$$

Now we change S by letting $S_j = R$, and thus get our new relation

$$(10) \quad u(S \cup \{(i,j)\}) - u(S) = \left[w_j(S_j \cup \{i\}) - w_j(S_j)\right]\left[u(S \cup M \times \{j\}) - u(S - M \times \{j\})\right].$$

This means, approximately, that the marginal value of player (i, j) (that is, the ith player in the jth subgame) to the coalition S is proportional to the marginal value of player i to the coalition S_j in w_j, and to the marginal value of all the players in the jth subgame to the coalition S.

The question, now, is as to the existence and uniqueness of a function satisfying the functional equation (10) as well as the boundary condition (9). If no such u exists, then our two conditions are too strict; if it is not unique, then they are not strict enough. It is, fortunately, possible to state

THEOREM 1. Given v and w_j as above, there exists a unique function u satisfying (9) and (10).

PROOF. Uniqueness: Given a set $S \subset M \times N$, we shall say that (i, j) is "loose" in S if $(i, j) \in S$ but $M \times \{j\} \not\subset S$.

Clearly, there is at most one function satisfying (9) for sets S with no loose elements. By induction, suppose that there is at most one function satisfying (9) and (10) for sets with not more than k loose elements. Let $S \cup (i,j)$ be a set with $k + 1$ loose elements, one of which is (i, j). Then (10) defines $u(S \{(i,j)\})$ in terms of $u(S)$, $u(S \cup M \times \{j\})$, $u(S - M \times \{j\})$, and w_j. Now S contains exactly k loose elements, while $S \cup M \times \{j\}$ and $S - M \times \{j\}$ both contain not more than k loose elements. Thus there is at most one u defined on these sets. Moreover, w_j is well-defined by hypothesis. Thus there is at most one u defined as desired on $S \cup \{(i,j)\}$.

Existence: Consider the function u defined by

$$u(S) = \sum_{T \subset N} \left\{ \prod_{j \in T} w_j(S_j) \prod_{j \notin T} \left[1 - w_j(S_j)\right] v(T) \right\}. \tag{11}$$

If $S = M \times R$, then

$$w_j(S_j) = \begin{cases} 0 & \text{if } j \notin R \\ 1 & \text{if } j \in R \end{cases}.$$

Thus, in (11), all terms in the sum would vanish except for $T = R$, and (11) reduces to

$$u(M\times R) = v(R),$$

which is (9).

On the other hand,

$$u(S\cup M\times\{j_o\}) = \sum_{\substack{T\subset N\\ j_o\notin T}} \left\{ \prod_{\substack{j\in T\\ j\neq j_o}} w_j(S_j) \prod_{j\notin T} \left[1 - w_j(S_j)\right] v(T) \right\}.$$

(since $(S\cup M\times\{j_o\})_{j_o} = M$), or, equivalently,

$$u(S\cup M\times\{j_o\}) = \sum_{\substack{T\subset N\\ j_o\notin T}} \left\{ \prod_{j\in T} w_j(S_j) \prod_{\substack{j\notin T\\ j\neq j_o}} \left[1 - w_j(S_j)\right] v(T\cup\{j_o\}) \right\},$$

whereas

$$u(S - M\times\{j_o\}) = \sum_{\substack{T\subset N\\ j_o\notin T}} \left\{ \prod_{j\in T} w_j(S_j) \prod_{\substack{j\notin T\\ j\neq j_o}} \left[1 - w_j(S_j)\right] v(T) \right\}$$

(since $(S - M\times\{j_o\})_{j_o} = \phi$), so that

(12) $$u(S\cup M\times\{j_o\}) - u(S - M\times\{j_o\})$$

$$= \sum_{\substack{T\subset N\\ j_o\notin T}} \left\{ \prod_{j\in T} w_j(S_j) \prod_{\substack{j\notin T\\ j\neq j_o}} \left[1 - w_j(S_j)\right]\left[v(T\cup\{j_o\}) - v(T)\right] \right\}.$$

On the other hand, letting $S' = S\cup\{(i,j_o)\}$, we have

$$u(S') = \sum_{T\subset N} \left\{ \prod_{j\in T} w_j(S_j') \prod_{j\notin T} \left[1 - w_j(S_j')\right] v(T) \right\}$$

$$= \sum_{\substack{T\subset N\\ j_o\in T}} \left\{ \prod_{j\in T} w_j(S_j') \prod_{j\notin T} \left[1 - w_j(S_j')\right] v(T) \right\}$$

$$+ \sum_{\substack{T\subset N\\ j_o\notin T}} \left\{ \prod_{j\in T} w_j(S_j') \prod_{j\notin T} \left[1 - w_j(S_j')\right] v(T) \right\}$$

$$= w_{j_o}(S_{j_o} \cup \{i\}) \sum_{\substack{T \subset N \\ j_o \in T}} \left\{ \prod_{\substack{j \in T \\ j \neq j_o}} w_j(S_j) \prod_{j \notin T} \left[1 - w_j(S_j)\right] v(T) \right\}$$

$$+ \left[1 - w_{j_o}(S_{j_o} \cup \{i\})\right] \sum_{\substack{T \subset N \\ j_o \notin T}} \left\{ \prod_{j \in T} w_j(S_j) \prod_{\substack{j \notin T \\ j \neq j_o}} \left[1 - w_j(S_j)\right] v(T) \right\}$$

$$= w_{j_o}(S_{j_o} \cup \{i\}) \sum_{\substack{T \subset N \\ j_o \notin T}} \left\{ \prod_{j \in T} w_j(S_j \prod_{\substack{j \notin T \\ j \neq j_o}} \left[1 - w_j(S_j)\right] v(T \cup \{j_o\}) \right\}$$

$$+ \left[1 - w_{j_o}(S_{j_o} \cup \{i\})\right] \sum_{\substack{T \subset N \\ j_o \notin T}} \left\{ \prod_{j \in T} w_j(S_j) \prod_{\substack{j \notin T \\ j \neq j_o}} \left[1 - w_j(S_j)\right] v(T) \right\}$$

$$= w_{j_o}(S_{j_o} \cup \{i\}) \sum_{\substack{T \subset N \\ j_o \notin T}} \left\{ \prod_{j \in T} w_j(S_j) \prod_{\substack{j \notin T \\ j \neq j_o}} \left[1 - w_j(S_j)\right] v(T \cup \{j_o\}) - v(T) \right\}$$

$$+ \sum_{\substack{T \subset N \\ j_o \notin T}} \left\{ \prod_{j \in T} w_j(S_j) \prod_{\substack{j \notin T \\ j \neq j_o}} \left[1 - w_j(S_j)\right] v(T) \right\}.$$

Similarly,

$$u(S) = w_j(S_j) \sum_{\substack{T \subset N \\ j_o \notin T}} \left\{ \prod_{j \in T} \left[w_j(S_j)\right] \prod_{\substack{j \notin T \\ j \neq j_o}} \left[1 - w_j(S_j)\right] \left[v(T \cup \{j_o\}) - v(T)\right] \right\}$$

$$+ \sum_{\substack{T \subset N \\ j_o \notin T}} \left\{ \prod_{j \in T} w_j(S_j) \prod_{\substack{j \notin T \\ j \neq j_o}} \left[1 - w_j(S_j)\right] v(T) \right\}.$$

Hence

$$u(S') - u(S) = \left[w_{j_o}(S_{j_o} \cup \{i\}) - w_{j_o}(S_{j_o})\right]$$

$$\sum_{\substack{T \subset N \\ j_o \neq T}} \left\{ \prod_{j \in T} w_j(S_j) \prod_{\substack{j \notin T \\ j \neq j_o}} \left[1 - w_j(S_j)\right] \left[v(T \cup \{j_o\}) - v(T)\right] \right\}.$$

This, together with (12), gives (10). Thus, the u defined by (11) is the unique solution to the difference equation (10) with boundary conditions (9), and we may properly write $u = (w_j) \otimes v$. Q.E.D.

§ 3. ANALYSIS OF THE COMPOSED GAME

It is our desire, now, to show that an analysis of the subgames w_j and v will help us to analyze the larger game u; the advantage of this is that, in general, the difficulties inherent in analyzing an n-person game increase very rapidly as n increases (cf. [3], p. 339). For example, all solutions to three-person games have been characterized, but it is not even known whether all four-person games have solutions.

As a preliminary step, we shall prove

LEMMA 1. Let $x_1, \ldots, x_n$ be numbers (real or complex). Then, if $N = \{1, \ldots, n\}$,

$$\sum_{T \subset N} \left\{ \prod_{j \in T} x_j \prod_{j \notin T} (1 - x_j) \right\} = 1. \tag{14}$$

PROOF. We prove by induction on n. For $n = 1$, the only subsets of N are N itself and ϕ. Then

$$\sum_{T \subset N} \left\{ \prod_{j \in T} x_j \prod_{j \notin T} (1 - x_j) \right\} = x_1 + (1 - x_1) = 1.$$

Suppose, now, by induction, that this lemma is true for $n - 1$. Then

$$\begin{aligned}
\sum_{T \subset N} \left\{ \prod_{j \in T} x_j \prod_{j \notin T} (1 - x_j) \right\} &= \sum_{\substack{T \subset N \\ n \in T}} \left\{ \prod_{j \in T} x_j \prod_{j \notin T} (1 - x_j) \right\} + \sum_{\substack{T \subset N \\ n \notin T}} \left\{ \prod_{j \in T} x_j \prod_{j \notin T} (1 - x_j) \right\} \\
&= x_n \sum_{\substack{T \subset N \\ n \notin T}} \left\{ \prod_{j \in T} x_j \prod_{\substack{j \notin T \\ j \neq n}} (1 - x_j) \right\} \\
&\qquad + (1 - x_n) \sum_{\substack{T \subset N \\ n \notin T}} \left\{ \prod_{j \in T} x_j \prod_{\substack{j \notin T \\ j \neq n}} (1 - x_j) \right\} \\
&= \sum_{\substack{T \subset N \\ n \notin T}} \left\{ \prod_{j \in T} x_j \prod_{\substack{j \notin T \\ j \neq n}} (1 - x_j) \right\} .
\end{aligned}$$

But this, by our induction hypothesis, is equal to 1. Q.E.D.

Remembering that an n-person game v is constant-sum if and only if

$$v(S) + v(N-S) = v(N) \quad \text{for all} \quad S \subset N \tag{15}$$

we can now use Lemma 1 to prove the following theorem.

THEOREM 2. If v and w_j, $j = 1, \ldots, n$, are constant-sum games, then so is their composition $u = (w_j$

PROOF. Let $S \subset M \times N$, and $U = M \times N - S$. Note that $U_j = M - S_j$, and

$$w_j(U_j) = 1 - w_j(S_j),$$

since w_j is constant-sum with $w_j(M) = 1$. Then

$$u(S) = \sum_{T \subset N} \left\{ \prod_{j \in T} w_j(S_j) \prod_{j \notin T} \left[1 - w_j(S_j)\right] v(T) \right\}$$

$$= \sum_{T \subset N} \left\{ \prod_{j \notin T} w_j(S_j) \prod_{j \in T} \left[1 - w_j(S_j)\right] v(N-T) \right\}$$

$$= \sum_{T \subset N} \left\{ \prod_{j \in T} w_j(U_j) \prod_{j \notin T} \left[1 - w_j(U_j)\right] v(N-T) \right\} .$$

Also

$$u(U) = \sum_{T \subset N} \left\{ \prod_{j \in T} w_j(U_j) \prod_{j \notin T} \left[1 - w_j(U_j)\right] v(T) \right\} .$$

so that

$$u(S) + u(U) = \sum_{T \subset N} \left\{ \prod_{j \in T} w_j(U_j) \prod_{j \notin T} \left[1 - w_j(U_j)\right] \left[v(N-T) + v(T)\right] \right\}$$

$$= v(N)$$

$$= u(M \times N).$$

Q.E.D.

In case v and the w_j are all simple games, we can prove the following result:

THEOREM 3. Let v and w_j be simple games in (0, 1)-normalization, and let $u = (w_j) \otimes v$. Let V and W_j, $j = 1, \ldots, n$, be solutions to v and w_j respectively. Then

$$U = (W_j) \otimes v = \{(x^j) \otimes y \mid x^j \in W_j, \; y \in V\}$$

is a solution to u.

PROOF. We must show, essentially, that

$$\text{(i)} \quad U \cap \text{dom } U = \phi$$
$$\text{(ii)} \quad U \cup \text{dom } U = E(u).$$

(i) Suppose that $U \cap \text{dom } U \neq \phi$. This means that there exist $x, y \in U$, and $S \subset M \times N$, $S \neq \phi$, such that

$$x_{ij} > y_{ij} \qquad \text{for all} \quad (i, j) \in S$$

and

$$u(S) \geq \sum_{(i,j) \in S} x_{ij}.$$

Since $y_{ij} \geq 0$, this means that $u(S) = 1$ (as u can take on only the values 0 and 1), i.e., that there is $T \subset N$ such that $v(T) = 1$ and $w_j(S_j) = 1$ for all $j \in T$. Now,

$$x = (\bar{x}^j) \otimes x^* \tag{16}$$

$$y = (\bar{y}^j) \otimes y^*, \tag{17}$$

where

$$\bar{x}^j,\ \bar{y}^j \in W_j \text{ and } x^*,\ y^* \in V,$$

and

(18) $$x^* = x\cdot 1,\ y^* = y\cdot 1;$$

whereas

(19) $$\bar{x}^j = \frac{x_{\cdot j}}{(x\cdot 1)_j}, \qquad \bar{y}^j = \frac{y_{\cdot j}}{(y\cdot 1)_j}$$

(where 1 denotes the m-vector all of whose components are 1, and $x_{\cdot j}$ the jth row of the matrix x).

Now, as V is a solution, it follows that we cannot have $x^* \vdash y^*$. Therefore, for some $j \in T$, we must have

$$x_j^* \leq y_j^* \text{ and } (x\cdot 1)_j \leq (y\cdot 1)_j.$$

But

$$x_{ij} > y_{ij} \text{ for all } i \in S_j;$$

hence

$$\frac{x_{ij}}{(x\cdot 1)_j} > \frac{y_{ij}}{(y\cdot 1)_j} \text{ for all } i \in S_j,$$

$$\bar{x}_i^j > \bar{y}_i^j \qquad \text{for all } i \in S_j,$$

and certainly,

$$\sum_{i\in S_j} \bar{x}_i^j \leq 1 = w_j(S_j).$$

Hence

$$\bar{x}^j \vdash \bar{y}^j$$

But this contradicts the fact that $\bar{x}^j$ and $\bar{y}^j$ are both elements of W_j, which is a solution. Hence $U \cap \text{dom } U = \phi$.

(ii) Suppose $y \in E(u) - U$. We shall show $y \in \text{dom } U$. Since $y \in E(u)$, we know that $y = (y_{ij})$, with $y_{ij} \geq 0$ for $(i, j) \in M \times N$, and

$$\sum_{(i,j) \in M \times N} y_{ij} = 1.$$

Thus we may write

$$y = (\bar{y}^j) \otimes y^*,$$

where y^*, $\bar{y}^j$ are as in (18) and (19). Note that this specifies $\bar{y}^j$ except for such $j \in N$ that $y_j^* = 0$; we may, for such j, let $\bar{y}^j$ be any element we desire in $E(w_j)$.

Since $y \notin U$, we know that either

$$y^* \notin V$$

or

$$y_j \notin W_j \quad \text{for some } j \text{ such that } y_j^* \neq 0.$$

Let us consider the first case: $y^* \notin V$. Since V is a solution, this means that there exists $x^* \in V$ such that $x^* \succ y^*$, i.e., for some $T \subset N$ with $v(T) = 1$, we have $x_j^* > y_j^*$ for all $j \in T$. Now, for every $j \in T$, consider the vector

$$\bar{\bar{y}}^j = \frac{y_{\cdot j}}{x_j^*}.$$

We have

$$\sum_{i \in M} \bar{\bar{y}}_i^j = \frac{y_j^*}{x_j^*} < 1.$$

Therefore, by Theorem 56.11.2 in [3], there exists $\bar{x}^j \in W_j$ such that, for some $S_j \subset N$ with $w_j(S_j) = 1$,

$$\bar{x}_i^j > \bar{\bar{y}}_i^j \quad \text{for all} \quad i \in S_j.$$

Take now, for $j \in N-T$, any $\bar{x}^j \in W_j$. Consider the vector

$$x = (\bar{x}^j) \otimes x^*.$$

For any $j \in T$, and any $i \in S_j$, we have

$$x_{ij} = \bar{x}_i^j x_j^* > \bar{\bar{y}}_i^j x_j^* = \bar{y}_i^j y_j^* = y_{ij}.$$

Setting

$$S = \bigcup_{j \in T} (S_j \times \{j\})$$

it is now clear that

$$x_{ij} > y_{ij} \quad \text{for all} \quad (i,j) \in S,$$

and that

$$u(S) = 1,$$

so that $x \vdash y$. Hence $y \in \text{dom } U$.

Let us, then, consider the second case. We assume that $y = (\bar{y}^j) \otimes y^*$, where, for some j_o such that $y_{j_o}^* \neq 0$,

$$\bar{y}^{j_o} \notin W_{j_o}.$$

This means that there exists $\bar{z}^{j_o} \in W_{j_o}$ such that, for some S_{j_o} with $w_{j_o}(S_{j_o}) = 1$,

$$\bar{z}_i^{j_o} > \bar{y}_i^{j_o} \quad \text{for all} \quad i \in S_{j_o}.$$

As S_{j_o} is finite, we may actually write

$$(1-r)\bar{z}_i^{j_o} > \bar{y}_i^{j_o} \quad \text{for all} \quad i \in S_{j_o}$$

for some $r > 0$. Note this means $r < 1$.

We now define z^* by:

(20) $$z^*_{j_o} = y^*_{j_o} - ry^*_{j_o} = y^*_{j_o}(1-r),$$

(21) $$z^*_j = y^*_j + \frac{r}{n-1} y^*_{j_o} \quad \text{for } j \neq j_o.$$

Clearly, $z^*_j \geq 0$ for all $j \in N$. Moreover,

$$\sum_{j \in N} z^*_j = \sum_{j \in N} y^*_j = 1,$$

so that $z^* \in E(v)$.

Now, if $z^* \in V$, we let $x^* = z^*$. If, on the other hand, $z^* \notin V$, we let x^* be some element of V which dominates z^*. For $j \neq j_o$, we have

$$y^*_j < z^*_j.$$

Hence

$$\sum_{i \in N} \frac{y_{ij}}{z^*_j} < 1.$$

Therefore, there exists $\bar{x}^j \in W_j$ such that, for some S_j with $w_j(S_j) = 1$,

$$\bar{x}^j_i > \frac{y_{ij}}{z^*_j} \quad \text{for all } i \in S_j.$$

Since $x^* \vdash z^*$, there exists $T \subset N$, with $v(T) = 1$, such that

$$x^*_j > z^*_j \quad \text{for all } j \in T.$$

Set now $x = (\bar{x}^j) \otimes x^*$, where $\bar{x}^{j_o} = \bar{z}^{j_o}$. Let

$$S = \bigcup_{j \in T} (S_j \times \{j\}).$$

Since $v(T) = 1$, and $w_j(S_j) = 1$ for all j, we have $u(S) = 1$. Also, for $(i, j) \in S$, and $j \neq j_o$,

$$x_{ij} = \bar{x}_i^j x_j^* \geq \bar{x}_i^j z_j^* > y_{ij},$$

while, for $i \in S_{j_o}$ (if $j_o \in T$),

$$x_{ij_o} = \bar{x}_i^{j_o} x_{j_o}^* = \bar{z}_i^{j_o} x_{j_o}^* \geq \bar{z}_i^{j_o} z_{j_o}^* > \frac{\bar{y}_i^{j_o}(1-r)y_{j_o}^*}{(1-r)} = y_{ij_o}.$$

Hence

$$x_{ij} > y_{ij} \quad \text{for all} \quad (i,j) \in S;$$

therefore $x \succ y$. Also, as $x^* \in V$, $\bar{x}^j \in W_j$, and $x = (\bar{x}^j) \otimes x^*$, we have $x \in U$. Hence $y \in \text{dom}\, U$, and U is a solution for u. Q.E.D.

The theorem, as it stands, can be generalized to the case where v is a simple game satisfying

(22) $$v(\{i\}) = 0 \quad \text{for all} \quad i,$$

(23) $$v(N) = t.$$

Such a game is said to be in $(0,t)$-normalization, and the theorem generalizes directly.

In case v is a simple nonnegative game (not in $(0,t)$ normalization) the theorem will hold if the solution, V, touches all sides of the simplex of imputations $E(v)$.

In case v is not simple, or the w_j are not simple in $(0,1)$ normalization, the theorem will not, in general, hold.

EXAMPLE 1. Let v be a symmetric simple constant-sum 3-person game in $(0,1)$ normalization. This means

(24) $$v(S) = \begin{cases} 0 & \text{if } S \text{ has } 0 \text{ or } 1 \text{ element} \\ 1 & \text{if } S \text{ has } 2 \text{ or } 3 \text{ elements} \end{cases}$$

We shall let w_j be the same game for $j = 1, 2, 3$.

Now this game has a symmetric solution, consisting of the three imputations (1/2, 1/2, 0), (1/2, 0, 1/2), and (0, 1/2, 1/2). Besides this solution, there also exist a continuum of discriminatory solutions, in each of which one player receives a fixed amount less than 1/2 while the other two split the remainder in all possible ways.

The composite game, u, is a simple nine-person game which contains twenty-seven minimal winning coalitions, each a four-person coalition. We can get a solution for u by simply taking a solution to each of v, w_1, w_2, and w_3, and composing them.

If, in particular, we take the symmetric solution for each of our four games, we obtain a (non-symmetric) 27-point solution to the game u, each point corresponding to an equal division of the profits among the four members of a minimal winning coalition. (That is, each such point has four components equal to 1/4, and the other five components equal to 0.)

If, on the other hand, we take a discriminatory solution to each of the games, we obtain a discriminatory solution to u, in which the four members of a minimal winning coalition split most of the profits, while two other pairs of players split smaller amounts, and the last player receives a small fixed amount (possibly 0, certainly not more than 1/4). Besides this, we may combine symmetric and discriminatory solutions to the smaller games, giving us several other types of solution to the composite game u. There is, finally, the possibility of solutions to u which are not of the form $(W_j) \otimes V$.

EXAMPLE 2. Once again let v and w_j be three-person games. For we take the symmetric game defined by

$$v(S) = \begin{cases} 0 & \text{if } S \text{ has } 0 \text{ or } 1 \text{ member} \\ 2/3 & \text{if } S \text{ has } 2 \text{ members} \\ 1 & \text{if } S \text{ has } 3 \text{ members} \end{cases} \tag{25}$$

This game has a core consisting of the single point (1/3, 1/3, 1/3). A solution consists of three "bargaining curves," beginning at this point and heading towards each of the three sides of the simplex of imputations, subject only to the constraint that for points on the same curve, the shares of the two players who are receiving more than 1/3 must decrease or increase together. In particular, a solution V exists containing the points $(2/5, 2/5, 1/5)$ and $(3/8, 3/8, 1/4)$.

For our w_j, we take the same three-person game as in our previous example — a simple game. We may take all our solutions W_j to be the same, and to contain the imputation $(1/4, 1/4, 1/2)$. (Every imputation of this game belongs to some solution.)

Now, the coalition $S = \{1,2\} \times \{1,2\}$ has value

$$u(S) = v(\{1,2\}) = 2/3.$$

The set $W_j \otimes V$ contains the points

$$x = (1/4, 1/4, 1/2) \otimes (2/5, 2/5, 1/5) = \begin{pmatrix} 1/10 & 1/10 & 1/5 \\ 1/10 & 1/10 & 1/5 \\ 1/20 & 1/20 & 1/10 \end{pmatrix}$$

and

$$y = (1/4, 1/4, 1/2) \otimes (3/8, 3/8, 1/4) = \begin{pmatrix} 3/32 & 3/32 & 3/16 \\ 3/32 & 3/32 & 3/16 \\ 1/16 & 1/16 & 1/8 \end{pmatrix}.$$

Clearly, $x_{ij} > y_{ij}$ for $i, j = 1, 2$. Moreover,

$$\sum_{(i,j)\in S} x_{ij} = 2/5 < u(S).$$

Therefore, $x \vdash y$, and $(W_j) \otimes V$ is not a solution to u.

While the theorem does not generalize to non-simple games, the form of the counter-example suggests that this trouble is due to the fact that coalitions are permitted to form among proper subsets of the members of one subgame and members of the other subgames. It has been suggested that this could be avoided by using ψ-stability in the analysis of these games, but such an approach does not seem to give interesting results. However, we may say something about the core:

THEOREM 4. Let v be an n-person game, let w_j be m-person games such that $w_j(M) = 1$ for all j, and suppose that x^* is in the core of v and $\bar{x}^j$ is in the core of w_j for each j. Then the imputation

$$x = (\bar{x}^j) \otimes x^*$$

is in the core of the game $u = (w_j) \otimes v$.

PROOF. We must show that

$$\sum_{(i,j)\in S} x_{ij} \geq u(S).$$

for each $S \subset M{\times}N$. As $\bar{x}^j$ is in the core of w_j, we have for each j

$$\sum_{i\in S_j} \bar{x}_i^j \geq w_j(S_j);$$

similarly, for all $T \subset N$,

$$\sum_{j\in T} x_j^* \geq v(T).$$

Now,

$$\begin{aligned}
u(S) &= \sum_{T\subset N} \left\{ \prod_{j\in T} w_j(S_j) \prod_{j\notin T} \left[1 - w_j(S_j)\right] v(T) \right\} \\
&\leq \sum_{T\subset N} \left\{ \prod_{j\in T} w_j(S_j) \prod_{j\notin T} \left[1 - w_j(S_j)\right] \sum_{j\in T} x_j^* \right\} \\
&= \sum_{T\subset N} \left\{ \left(\sum_{j\in T} x_j^* \right) \prod_{j\in T} w_j(S_j) \prod_{j\notin T} \left[1 - w_j(S_j)\right] \right\} \\
&= \sum_{j\in N} \left\{ x_j^* \sum_{\substack{T\subset N \\ j\in T}} \left(\prod_{k\in T} w_k(S_k) \prod_{k\notin T} \left[1 - w_k(S_k)\right] \right) \right\} \\
&= \sum_{j\in N} \left\{ x_j^* w_j(S_j) \left(\sum_{\substack{T\subset N \\ j\in T}} \prod_{\substack{k\in T \\ k\neq j}} w_k(S_k) \prod_{k\notin T} \left[1 - w_k(S_k)\right] \right) \right\} \\
&= \sum_{j\in N} x_j^* w_j(S_j) \qquad \text{(by Lemma 1)}
\end{aligned}$$

$$\leq \sum_{j \in N} \left(x_j^* \sum_{i \in S_j} \bar{x}_i^j \right)$$

$$= \sum_{(i,j) \in S} x_{ij}. \qquad \text{Q.E.D.}$$

The converse of Theorem 4 is not true in general, though it is trivially so for simple games. But simple games with cores are usually not interesting enough to warrant study.

For a final theorem on this type of games, we shall define the function b_v, for an n-person game v, by

$$b_v(j) = \max_{T \subset N} \left\{ v(T \cup \{j\}) - v(T) \right\}.$$

The number $b_v(j)$ is a measure of the "best possible" outlook for player j in the game v; it was first formulated by Milnor [2], and it is true that no imputation x with $x_j > b_v(j)$ is in the core, nor can it belong to any solution.

THEOREM 5. Let v and w_j be games as above and let $u = (w_j) \otimes v$. Then

$$b_u(i,j) = b_{w_j}(i)\, b_v(j).$$

PROOF. There is some $T \subset N$ such that

$$v(T \cup \{j\}) - v(T) = b_v(j),$$

and some $S_j \subset M$ such that

$$w_j(S_j \cup \{i\}) - w_j(S_j) = b_{w_j}(i).$$

Now, define S^* by $S^* = M \times T \cup S_j \times \{j\}$. Then

$$u(S^*\cup\{(i,j)\}) - u(S^*) = \left[w_j(S_j^*\cup\{i\}) - w_j(S_j^*)\right]\left[u(S^*\cup M\times\{j\}) - u(S^* - M\times\{j\})\right]$$

$$= \left[w_j(S_j\cup\{i\}) - w_j(S_j)\right]\left[u(M\times T\cup\{j\}) - u(M\times T)\right]$$

$$= b_{w_j}(i)\, b_v(j)$$

Hence, $b_u(i,j) \geq b_{w_j}(i)\, b_v(j)$.

On the other hand, for any S, by (13),

$$u(S\cup\{(i,j)\}) - u(S)$$

$$= \left[w_j(S_j\cup\{i\}) - w_j(S_j)\right]\sum_{\substack{T\subset N \\ j\notin T}}\left\{\prod_{k\in T} w_k(S_k)\prod_{\substack{k\notin T \\ k\neq j}}\left[1 - w_k(S_k)\right]\left[v(T\cup\{j\}) - v(T)\right]\right\}$$

$$\leq b_{w_j}(i)\sum_{\substack{T\subset N \\ j\notin T}}\left\{\prod_{k\in T} w_k(S_k)\prod_{\substack{k\notin T \\ k\neq j}}\left[1 - w_k(S_k)\right] b_v(j)\right\}$$

$$= b_{w_j}(i)\, b_v(j)\ .$$

Hence equality holds. Q.E.D.

BIBLIOGRAPHY

[1] LUCE, R. Duncan, and RAIFFA, Howard, Games and Decisions, New York, John Wiley and Sons, 1957.

[2] MILNOR, John W., "Reasonable outcomes for n-person games," Research Memorandum RM-679, The RAND Corporation, Santa Monica, 1952.

[3] VON NEUMANN, John and MORGENSTERN, Oskar, Theory of Games and Economic Behavior, Princeton University Press, Princeton, 1953.

Guillermo Owen

Fordham University

ON THE CARDINALITY OF SOLUTIONS OF FOUR-PERSON CONSTANT-SUM GAMES*

Alberto Raul Galmarino

§ 1. INTRODUCTION

This paper aims at the study of two open problems mentioned by Shapley [2], regarding the cardinality of solutions of n-person constant-sum games in von Neumann's theory: (1) Do countably infinite solutions ever occur? (2) Is there an upper bound to the number of imputations in a finite solution? Here these questions are considered and solved for $n = 4$.

Section 2 presents the basic notations and a summary review of the main concepts in von Neumann's theory. Section 3 contains a theorem relevant to the investigation of points of accumulation in solutions of n-person constant-sum games. Section 4 proves the non-existence of denumerable infinite solutions for $n = 4$, and Section 5 the existence of an upper bound for the number of imputations in a finite solution for the same case. Some final remarks are included in Section 6.

§ 2. BASIC DEFINITIONS AND NOTATIONS

Let $I = (1, 2, \ldots, n)$ be the (finite) set of players of an n-person constant-sum game. According to von Neumann and Morgenstern [1] all strategic properties of the game are determined by its <u>characteristic function</u> v. This function assigns a real number to all subsets $S \subset I$ and satisfies the relations

$$v(\emptyset) = 0 \tag{2.1}$$

$$v(S) + v(I - S) = v(I) \tag{2.2}$$

* This research was sponsored by the Air Force Cambridge Research Laboratories, OAR (USAF), Detection Physics Laboratory, contract AF 19(604)-4573.

for all $S \subset I$, and

$$(2.3) \qquad v(S \cup T) \geq v(S) + v(T)$$

for all $S \subset I$, $T \subset I$, $S \cap T = \emptyset$. Conversely any real-valued function $v(S)$ satisfying (2.1), (2.2), (2.3) is the characteristic function of an n-person constant-sum game.

An imputation is any n-tuple of real numbers $\alpha = (\alpha_1, \alpha_2, \ldots, \alpha_n)$ such that

$$\sum_{i=1}^{n} \alpha_i = v(I), \text{ and } \alpha_i \geq v((i)) \text{ for all } i \in I.$$

The set of all imputations will be always designated by H.

A set $S \subset I$ is said to be effective for an imputation α when

$$(2.4) \qquad \sum_{i \in S} \alpha_i \leq v(S)$$

If (2.4) holds as a strict inequality then S is said to be strictly effective for α. If there is equality in (2.4) then S is exactly effective for α. If (2.4) does not hold then S is said to be ineffective for α.

An imputation α dominates another imputation β via a subset $S \subset I$ when $\alpha_i > \beta_i$ for $i \in S$, and S is effective for α. In symbols,

$$\alpha \underset{S}{\vdash\!\!\!\dashv} \beta \, .$$

If there is some S for which $\alpha \underset{S}{\vdash\!\!\!\dashv} \beta$ it will be often written as $\alpha \vdash\!\!\!\dashv \beta$. If there is no such S, $\alpha \not\vdash\!\!\!\dashv \beta$.

A solution is a set of imputations $V \subset H$ such that

(2.5) (a) for all $\alpha, \beta \in V$, $\alpha \not\vdash\!\!\!\dashv \beta$

(b) for all $\alpha \in H - V$, there exists $\beta \in V$ such that $\beta \vdash\!\!\!\dashv \alpha$.

An n-person constant-sum game is essential if at least one of the relations (2.3) holds with strict inequality. If a game is inessential it has only one solution, which is trivial. If it is essential, then H can be represented as an (n–1)-dimensional hypertetrahedron (simplex of imputations) of an n-dimensional Euclidean space endowed with the usual topology.

For a more detailed explanation of these concepts the reader is referred to [1] and [3].

§ 3. A PROPERTY OF ACCUMULATION POINTS OF A SOLUTION

THEOREM 3.1. Let V be a solution of an n-person constant-sum game. For each $\alpha \in V$ let $N(\alpha)$ be the set of all imputations $\beta \in H$ satisfying, for all $S \subset I$, the following condition:

If S is strictly effective for α then S is also strictly effective for β.

Then $N(\alpha)$ is an open neighborhood of α in the topology referred to at the end of Sec. 2, and for all $\beta \in V \cap N(\alpha)$,

$$(i : \alpha_i = \beta_i) \neq \emptyset.$$

PROOF. $N(\alpha)$ is an intersection of a finite number of open sets of the type $\Sigma_S \beta_i < v(S)$. Hence $N(\alpha)$ is open. Clearly $\alpha \in N(\alpha)$. This proves the first assertion.

Suppose now that there exists $\beta \in N(\alpha) \cap V$ such that

$$(i : \alpha_i = \beta_i) = \emptyset. \tag{3.1}$$

Let $T = (i : \alpha_i > \beta_i)$. If T is effective for α, then $\alpha \underset{T}{\vdash} \beta$, which contradicts (2.5a). But if T is not effective for α, then $I - T$ is strictly effective for α, and hence for $\beta \in N(\alpha)$. Therefore $\beta \vdash \alpha$ via $I - T$, again in contradiction to (2.5a). Therefore (3.1) is false and the proof is complete.

COROLLARY 3.1. A solution V of an n-person constant-sum game is contained in at most $(2^{2^{n-1}} - {}^{n-1})n$ hyperplanes of the form β_i = constant, parallel to the faces of the hypertetrahedron of imputations.

PROOF. By Theorem 3.1, for each of the open sets $N(\alpha)$ introduced by that theorem, the subset $N(\alpha) \cap V$ is contained in at most n hyperplanes. Clearly each $N(\alpha)$ is a convex open polyhedron whose boundary consists of subsets of the planes $\Sigma_S \beta_i = v(S)$ for $S \subset I$. It is clear that $\cup_{\alpha \in V} N(\alpha) \supset V$. The maximum number of polyhedra of such type which are maximal with respect to inclusion is, by a simple combinatorial argument, $2^{2^{n-1}} - n-1$, corresponding to the case in which no point of V is in any plane $\Sigma_S \beta_i = v(S)$. Corollary 3.1 thus follows.

§ 4. NON-EXISTENCE OF A DENUMERABLE INFINITE SOLUTION FOR $n = 4$

The discussion in this section will prove the following result:

THEOREM 4.1. A four-person constant-sum game has no infinite denumerable solution.

Suppose V is a denumerable infinite solution of a four-person constant-sum game. As V is bounded, it has at least one accumulation point, say, α^*. As V is closed (see proof in [1] or [3]), $\alpha^* \in V$.

The proof will consist in showing the impossibility of the existence of a sequence $(\alpha^n) \subset V$, satisfying $\alpha^n \to \alpha^*$ and any of the following three conditions:

$$\alpha_1^n \downarrow \alpha_1^*, \quad \alpha_2^n \uparrow \alpha_2^*, \quad \alpha_3^n \uparrow \alpha_3^*, \quad \alpha_4^n \equiv \alpha_4^*, \tag{4.1}$$

$$\alpha_1^n \downarrow \alpha_1^*, \quad \alpha_2^n \equiv \alpha_2^*, \quad \alpha_3^n \uparrow \alpha_3^*, \quad \alpha_4^n \equiv \alpha_4^*, \tag{4.2}$$

$$\alpha_1^n \downarrow \alpha_1^*, \quad \alpha_2^n \downarrow \alpha_2^*, \quad \alpha_3^n \uparrow \alpha_3^*, \quad \alpha_4^n \equiv \alpha_4^* . \tag{4.3}$$

By Theorem 3.1 and considerations of symmetry, the impossibility of (4.1), (4.2) and (4.3) implies the impossibility of such a solution V.

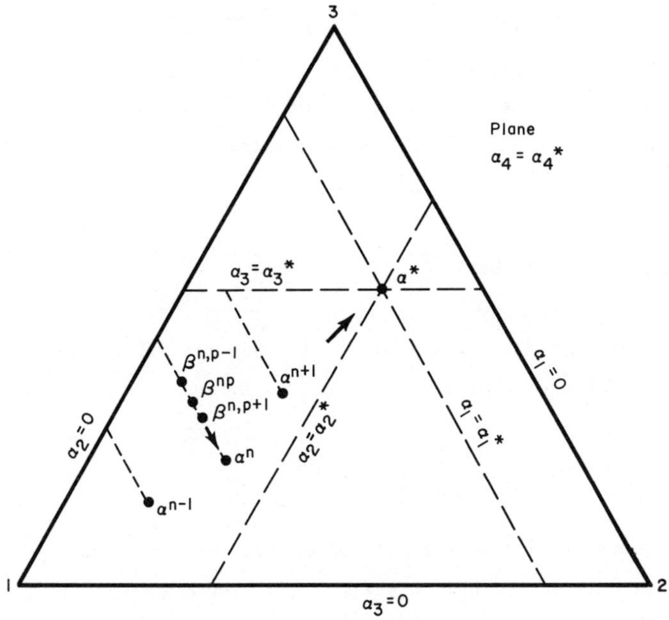

Fig. 1 — Sequences (α^n) and (β^{np}) in the proof for (4.1)

PROOF. FOR (4.1). Let (α^n) be a sequence in V satisfying (4.1). As V is denumerable, for each $\alpha^n \in (\alpha^n)$ there exists a sequence $(\beta^{np}) \subset H - V$ such that, as $p \to \infty$,

$$\beta_1^{np} = \alpha_1^n, \quad \beta_2^{np} \uparrow \alpha_2^n, \quad \beta_3^{np} \downarrow \alpha_3^n, \quad \beta_4^{np} \equiv \alpha_4^n.$$

As $(\beta^{np}) \subset H - V$, there exists $(\omega^{np}) \subset V$ such that $\omega^{np} \mapsto \beta^{np}$. As the number of subsets of I is finite, there exists, for n fixed, some $S \subset I$ such that, at least through a subsequence of p,

(4.4) $$\omega^{np} \underset{S}{\mapsto} \beta^{np} .$$

As V is compact, there will be a subsequence of (ω^{np}), for fixed n, such that, for some $\omega^{n*} \in V$:

(4.5) $$\text{either } \omega^{np} \to \omega^{n*} \text{ or } \omega^{np} \equiv \omega^{n*}$$

as $p \to \infty$. The different possibilities for S will now be discussed.

First notice that $S \ni 2$, for otherwise $\omega^{np} \underset{S}{\mapsto} \alpha^n$. It follows that for each n

(4.6) $$\text{either } \omega_2^{np} \uparrow \alpha_2^n \text{ or } \omega_2^{np} \equiv \alpha_2^n .$$

Next notice that $S \neq (2, 3), (2, 3, 4), (2, 4)$, for otherwise $\omega^{np} \underset{S}{\mapsto} \alpha^{n-1}$ (for large p). Hence it must hold that $S \ni 1$, and

$$\alpha_1^{n-1} \geq \omega_1^{np}.$$

Next notice that $(1, 4)$ is strictly effective for (α^n), for, if not, $\alpha^n \mapsto \alpha^{n-1}$ via $(2, 3)$. Hence $S \neq (1, 2, 3)$, for otherwise $\alpha^{n-2} \mapsto \omega^{np}$ via $(1, 4)$. The only possibilities left for S are therefore $(1, 2)$ and $(1, 2, 4)$.

If $S = (1, 2)$, it must be the case that $(1, 2)$ is strictly effective for (α^n), by $\omega_1^{np} > \beta_1^{np} = \alpha_1^n$ and (4.6).

If $S = (1, 2, 4)$, it must be the case that $(2, 3)$ is strictly effective for (ω^{np}), for otherwise $\omega^{np} \mapsto \alpha^n$ via $(1, 4)$. As $(1, 4)$ is strictly effective for α^*, all ω^{np} are outside some neighborhood of α^*, and, in fact,

(4.7) $$\omega_3^{np} < \alpha_3^* - \epsilon, \text{ for some } \epsilon > 0.$$

Also $\alpha_1^{n-1} \geq \omega_1^{np}$, for otherwise $\omega^{np} \mapsto \alpha^{n-1}$ via $(1, 2, 4)$ for large p. Hence (2.4) is strictly effective for (α^n), for otherwise there would be some n, n', such that $\alpha^n \mapsto \omega^{n'p}$ via $(1, 3)$. It follows that one of the following statements must be true:

(4.8) $(1, 2)$ is strictly effective for (α^n)

(4.9) (2.4) is strictly effective for (α^n).

As case (4.1) is symmetric with respect to the subindexes 2 and 3, one of the next two statements must also be true:

(4.10) (1, 3) is strictly effective for (α^n)

(4.11) (3, 4) is strictly effective for (α^n).

Clearly (4.8) is incompatible with (4.11), and (4.9) with (4.10). The foregoing discussion thus proves that one (and only one) of the next two statements is true for (α^n):

(4.12) (1, 2), (1, 3), (1, 4) are strictly effective for (α^n) and $S = (1, 2)$ for <u>all</u> α^n in some neighborhood of α^*.

(4.13) (1, 4), (2, 4), (3, 4) are strictly effective for (α^n) and $S = (1, 2, 4)$ for <u>all</u> α^n in some neighborhood of α^*.

<u>Discussion for</u> $S = (1, 2)$. This corresponds to case (4.12), which excludes $S = (1, 2, 4)$. Hence $\omega_4^{np} = \beta_4^{np} = \alpha_4^n = \alpha_4^*$. It is easily checked that

(4.14) If the second alternative in (4.6) holds, there exists $\alpha'^n \in V$ such that $\alpha_1'^n > \alpha_1^n$, $\alpha_2'^n = \alpha_2'^n$, $\alpha_3'^n < \alpha_3^n$, $\alpha_4'^n = \alpha_4^n$ and (1, 2), (1, 3) and (1, 4) are strictly effective for (α'^n).

The following reasoning is decisive in the proof.

Let $\bar{\alpha} \in V$ be an imputation satisfying

(4.15) $$\bar{\alpha}_1 > \alpha_1^*, \quad \bar{\alpha}_2 < \alpha_2^*, \quad \bar{\alpha}_3 < \alpha_3^*, \quad \bar{\alpha}_4 = \alpha_4^* .$$

It is clear that, if $\bar{\alpha}_1$ is close enough to α_1^*, so that (1, 2), (1, 3) and (1, 4) are strictly effective for $\bar{\alpha}$, then $\bar{\alpha}$ can be identified with some element, say α^n (not the first), of a sequence (α^n) satisfying (4.1) and (4.12). So to $\bar{\alpha}$ there corresponds a sequence (ω^{np}) as described before. If the second alternative in (4.6) holds for $\bar{\alpha}$, then (4.14) shows the existence of $\bar{\alpha}' \in V$ such that $\bar{\alpha}_1' > \bar{\alpha}_1$, $\bar{\alpha}_2' = \bar{\alpha}_2$, $\bar{\alpha}_3' < \bar{\alpha}_3$, $\bar{\alpha}_4' = \bar{\alpha}_4$. By induction, if the second alternative in (4.6) holds for $\bar{\alpha}, \bar{\alpha}', \bar{\alpha}'', \ldots, \bar{\alpha}^{(i-1)}$, there exists $\bar{\alpha}^{(i)}$ such that $\bar{\alpha}_1^{(i)} > \bar{\alpha}^{(i-1)}$, $\bar{\alpha}_2^{(i)} = \bar{\alpha}_2^{(i-1)}$, and (1, 2), (1, 3) and (1, 4) are strictly effective for $\bar{\alpha}^{(i)}$.

If this procedure could be carried on up to some $\bar{\alpha}^{(m)}$ with $\bar{\alpha}_1^{(m)} > \alpha_1^{n-1}$ (where α^{n-1} belongs to the sequence (α^n) in which $\bar{\alpha}$ was identified with α^n), then it would follow that $\bar{\alpha}^{(m)} \vdash \alpha^{n-1}$ via (1, 2), a contradiction. Therefore, it is readily verified that

(4.16) there exists $\alpha \in V$, such that $\alpha_1 > \bar{\alpha}_1$, $\alpha_2 = \bar{\alpha}_2$, $\alpha_4 = \bar{\alpha}_4$, which can be identified with an element of a sequence of type (4.1), and for which the first alternative in (4.6) holds.

If V_2 is the range of the second component of imputations in V, then $\bar{\alpha}_2$ is an accumulation point in V_2. As $\bar{\alpha}$ was arbitrary, it follows that, for all imputations $\alpha \in V$ in some neighborhood of α^* and satisfying (4.15), α_2 is an accumulation point in V_2. This contradicts the hypothesis that V is denumerable.

Thus, $S = (1, 2)$ is impossible.

Discussion for $S = (1, 2, 4)$. This corresponds to case (4.13). The inequality (4.7) discards the possibility $\omega^{n^*} \to \alpha^*$ as $n \to \infty$. Then it is easily checked that

$$\omega_1^{n^*} \downarrow \alpha_1^*, \quad \omega_2^{n^*} \uparrow \alpha_2^* . \tag{4.17}$$

Moreover, $\omega_3^{n^*} < \alpha_3^*$, $\omega_4^{n^*} > \alpha_4^*$, and

(1, 3), (2, 3), and (3, 4) are strictly effective for (ω^{n^*}).

As $\omega_1^{n^*} + \omega_2^{n^*} \downarrow \alpha_1^* + \alpha_2^*$, it must be the case that $\omega_3^{n^*} + \omega_4^{n^*} \uparrow \alpha_3^* + \alpha_4^*$. By

Theorem 3.1, one of the following cases must hold, at least through a subsequence:

(4.18) $$\omega_3^{n*} \uparrow \omega_3^{**}, \quad \omega_4^{n*} \equiv \omega_4^{**},$$

or

(4.19) $$\omega_3^{n*} \equiv \omega_3^{**}, \quad \omega_4^{n*} \uparrow \omega_4^{**}.$$

As (ω^{n*}) contains a subsequence of type (4.1), with a possible permutation of subindexes, the former discussion shows that (4.18) is impossible and (4.19) must hold. Let the subsequence of (ω^{n*}) satisfying (4.17) and (4.19) be (σ^n). Then

$$\sigma_1^n \downarrow \alpha_1^*, \quad \sigma_2^n \uparrow \alpha_2^*, \quad \sigma_3^n \equiv \sigma_3^* < \alpha_3^*, \quad \sigma_4^n \uparrow \sigma_4^* > \alpha_4^*.$$

Now an argument similar to the one for the discussion of the case $S = (1, 2)$ can be made. Let $\bar{\alpha} \in V$ be an imputation satisfying (4.15). If $\bar{\alpha}_1$ is close enough to α_1^*, then $\bar{\alpha}$ can be considered an element, say α^n, of a sequence satisfying (4.1) and (4.13). To $\bar{\alpha}$ there must correspond a sequence (ω^{np}) as described before. If the second alternative of (4.6) applies to $\bar{\alpha}$, there will exist, as in (4.14), $\bar{\sigma} \in V$ such that $\bar{\sigma}_1 > \bar{\alpha}_1$, $\bar{\sigma}_2 = \bar{\alpha}_2$, and such that (1, 3), (2, 3), and (3, 4) are strictly effective for $\bar{\sigma}$. It is easily checked that $\bar{\sigma}_3 = \sigma_3^*$. Again $\bar{\sigma}$ can be considered an element of a sequence such as (σ^n). By recurrence, new imputations $\bar{\alpha}'$, $\bar{\sigma}'$, $\bar{\alpha}''$, $\bar{\sigma}''$, ..., $\bar{\alpha}^{(i)}$, $\bar{\sigma}^{(i)}$, ... can be obtained, satisfying

$$\bar{\sigma}_1^{(i)} > \bar{\alpha}_1^{(i)} > \bar{\sigma}_1^{(i-1)}, \quad \bar{\sigma}_2^{(i)} = \bar{\alpha}_2^{(i)} = \bar{\alpha}_2^{(i-1)},$$

provided that the second alternative in (4.6) holds for all of them. If this procedure could be carried on up to some $\alpha^{(m)}$ such that $\bar{\alpha}_1^{(m)} > \sigma_1^{n-1}$, then $\bar{\alpha}^{(m)} \vdash \sigma^{n-1}$ via (1, 2, 3), a contradiction.

Then we obtain a result as in (4.16), and, by a similar argument as before, V_2 is shown to contain an interval of the real line, a contradiction.

Therefore S can not be (1, 2, 4), and the proof of the

impossibility of (4.1) is complete.

PROOF FOR (4.2). Let (α^n) be a sequence in V satisfying (4.2). As V is denumerable, there exists a sequence $(\beta^n) \subset H - V$ such that

$$\beta_1^n \downarrow \alpha_1^*, \quad \beta_2^n \equiv \alpha_2^*, \quad \beta_3^n \uparrow \alpha_3^*, \quad \beta_4^n \equiv \alpha_4^*.$$

As in the discussion for case (4.1), there exist $(\omega^n) \subset V$ and a subset $S \subset I$ such that $\omega^n \underset{S}{\mapsto} \beta^n$ and either $\omega^n \to \omega^*$ or $\omega^n \equiv \omega^*$, at least through subsequences of (β^n) and (ω^n). It can be readily established that $S \ni 3$, for otherwise $\omega^n \underset{S}{\mapsto} \alpha^*$. Moreover $S \ni 1$ and

$$\omega_1^n \downarrow \alpha_1^*, \quad \omega_3^n \uparrow \alpha_3^*,$$

for otherwise $\omega^n \mapsto \alpha^{n'}$ for some n, n'. Since $\omega_1^n + \omega_3^n > \beta_1^n + \beta_3^n = \alpha_1^* + \alpha_3^*$, this yields $\omega_1^n + \omega_3^n \downarrow \alpha_1^* + \alpha_3^*$. Hence

$$\omega_2^n + \omega_4^n \uparrow \alpha_2^* + \alpha_4^*.$$

By Theorem 3.1,

$$\text{either} \quad \omega_2^n \uparrow \omega_2^*, \quad \omega_4^n \equiv \omega_4^* \quad \text{or} \quad \omega_2^n \equiv \omega_2^*, \quad \omega_4^n \uparrow \omega_4^*,$$

at least through a subsequence. Therefore (ω^n) contains a subsequence of type (4.1), with a possible permutation of subindexes, which was proved to be impossible. Thus, the impossibility of case (4.2) is proved.

PROOF FOR (4.3). The discussion of this case is completely parallel to the one for case (4.1). If $(\alpha^n) \subset V$ satisfies (4.3), a double sequence $(\beta^{np}) \subset H - V$ can be chosen such that, as $p \to \infty$,

$$\beta_1^{np} \uparrow \alpha_1^n, \quad \beta_2^{np} \downarrow \alpha_2^n, \quad \beta_3^{np} \equiv \alpha_3^n, \quad \beta_4^{np} \equiv \alpha_4^n .$$

There exist $(\omega^{np}) \subset V$ and $S \subset I$ such that, through subsequences of (β^{np}) and (ω^{np}), (4.4) and (4.5) hold. From a discussion for S similar to that

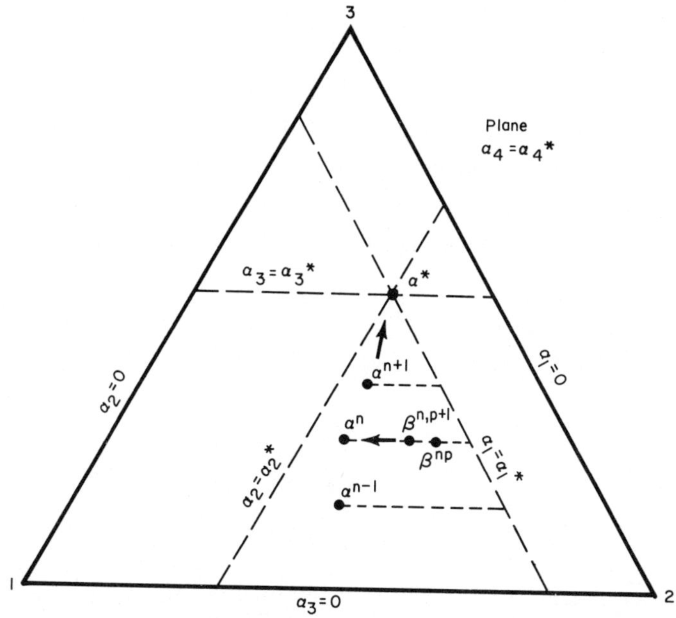

Fig. 2 — Sequences (α^n) and (β^{np}) in the proof for (4.3)

in case (4.1), there are only two possibilities for S which are not immediately discarded; (1, 3) and (1, 3, 4). The proof of their impossibility is practically the same as for (1, 2) and (1, 2, 4) in case (4.1), and will be therefore omitted.

The proof of Theorem 4.1 is thus complete.

§ 5. AN UPPER BOUND FOR THE NUMBER OF IMPUTATIONS IN A FINITE SOLUTION FOR $n = 4$

The purpose of this section is to prove the following theorem:

THEOREM 5.1. The number of imputations that a finite solution of a constant-sum four-person game can contain has an upper bound.

Suppose the theorem were false. According to Corollary 3.1, the number of planes parallel to the faces of the hypertetrahedron of imputations in which a particular solution is contained cannot exceed 32. Hence there would be games in which such a plane contains an aribtrarily large number of imputations of a solution. By considerations of symmetry, there would therefore be, for arbitrarily large m, constant-sum four-person games with finite solutions V in which at least one of these two possibilities holds:

(5.1) There are m imputations $\alpha^1, \alpha^2, \ldots, \alpha^m \in V$ such that $\alpha_1^i > \alpha_1^{i+1}$, $\alpha_2^i < \alpha_2^{i+1}$, $\alpha_3^i < \alpha_3^{i+1}$, $\alpha_4^i = \alpha_4^{i+1}$, for $i = 1, 2, \ldots, m-1$.

(5.2) There are m such imputations satisfying: $\alpha_1^i > \alpha_1^{i+1}$, $\alpha_2^i < \alpha_2^{i+1}$, $\alpha_3^i = \alpha_3^{i+1}$, $\alpha_4^i = \alpha_4^{i+1}$, for $i = 1, 2, \ldots, m-1$.

The proof of Theorem 5.1 will consist in proving the impossibility of cases (5.1) and (5.2).

PROOF FOR (5.1). First, notice that in case (5.1),

(5.3) $(1, 4)$ is strictly effective for α^i, $i = 2, \ldots, m$,

for otherwise $\alpha^{i+1} \vdash \alpha^i$ via $(2, 3)$. By a theorem proved by W. H. Mills [4], for each α^i there exists $S^i \subset I$ and $\omega^i \in V$ such that

(5.4) S^i is effective for ω^i and α^i, $S^i \ni 2$, $\alpha_2^i = \omega_2^i$, and $\omega_j^i > \alpha_j^i$ for $j \in S^i - (2)$.

By (5.3), the only possibilities for S^i are $(1, 2)$, $(1, 2, 3)$, $(1, 2, 4)$, $(2,4)$, and $(2, 3, 4)$. It follows that for an arbitrary large subset of (α^i), a constant subset S satisfies (5.4). The sequences (ω^i) and (α^i) can therefore be assumed chosen so that $S^i \equiv S$. Then $S \neq (2, 4)$, $(2, 3, 4)$ for otherwise $\omega^{i+1} \vdash \alpha^i$. Also $S \neq (1, 2, 3)$ for otherwise either $\omega^i \underset{S}{\vdash} \alpha^{i-1}$, $\alpha^{i-2} \vdash \omega^{i^S}$ via $(1, 4)$. There remain the possibilities $S = (1, 2)$

and $S = (1, 2, 4)$.

For $S = (1, 2)$ it must hold that

(5.5) $(1, 2)$ is strictly effective for (α^i).

For $S = (1, 2, 4)$ it must hold that

(5.6) $(2, 4)$ is strictly effective for (α^i),

for otherwise ω^i is dominated via $(1, 3)$.

By a similar discussion for the index 3, it follows that one of the following statements must be true:

(5.7) $(1, 3)$ is strictly effective for (α^i)

or

(5.8) $(3, 4)$ is strictly effective for (α^i).

As (5.5) is incompatible with (5.8) and (5.6) with (5.7), the previous discussion together with (5.3) yields two possible cases:

(5.9) $(1, 2)$, $(1, 3)$, $(1, 4)$ are strictly effective for (α^i),

(5.10) $(1, 4)$, $(2, 4)$, $(3, 4)$ are strictly effective for (α^i).

Discussion of case (5.9). By Mills' theorem, for each α^i there exists $S^i \subset I$ and $\omega^i \in V$ such that

(5.11) S^i is effective for ω^i and α^i, $S^i \ni 1$,

$\alpha^i_1 = \omega^i_1$, and $\omega^i_j > \alpha^i_j$ for $j \in S^i - (1)$.

It clearly follows that $\omega^i_4 \geq \alpha^i_4$, $i = 1, \ldots, m - 1$, for otherwise $\alpha^{i+1} \vdash \omega^i$ via $(1, 4)$. Therefore, either $\omega^i_2 < \alpha^i_2$ or $\omega^i_3 < \alpha^i_3$. It can be assumed, by symmetry, that $\omega_3^i < \alpha_3^i$ for an arbitrarily large subset of (α^i). The only possibilities for S^i are therefore (1, 2), (1, 4) and (1, 2, 4). It again can be assumed that $S^i \equiv S$, a constant subset of I. Then $S \neq (1, 4)$, for otherwise $\omega^i \mapsto \alpha^{i+1}$ via (1, 4). If $S = (1, 2)$, then

$$\omega_1^i = \alpha_1^i, \quad \omega_2^i > \alpha_2^i, \quad \omega_3^i < \alpha_3^i, \quad \omega_4^i = \alpha_4^i.$$

It must hold that $\omega_2^i \leq \alpha_2^{i+1}$, $\omega_3^i \geq \alpha_3^{i-1}$, for otherwise $\omega^i \mapsto \alpha^{i+1}$ via (1, 2) or $\alpha^{i-1} \mapsto \omega^i$ via (1, 3), respectively. Hence

$$\omega_1^i > \omega_1^{i+1}, \quad \omega_2^i < \omega_2^{i+1}, \quad \omega_3^i < \omega_3^{i+1}, \quad \omega_4^i = \omega_4^{i+1},$$

and, moreover,

$$(1, 2), (1, 3) \text{ and } (1, 4) \text{ are strictly effective for } (\omega^i).$$

By Mills' theorem, there exists $T^i \subset I$ and $\sigma^i \in V$ such that

$$T^i \text{ is effective for } \sigma^i \text{ and } \omega^i, \ T^i \ni 2,$$
$$\sigma_2^i = \omega_2^i, \text{ and } \sigma_j^i > \omega_j^i \text{ for } j \in T^i - (1).$$

The former discussion showed that T^i is either (1, 2) or (1, 2, 4). In either case, $\sigma^i \mapsto \alpha^i$ via T^i, a contradiction.

Finally, if $S = (1, 2, 4)$, then

$$\omega_1^i = \alpha_1^i, \quad \omega_2^i > \alpha_2^i, \quad \omega_3^i < \alpha_3^i, \quad \omega_4^i > \alpha_4^i;$$

also $\alpha_2^{i+1} \geq \omega_2^i$, for otherwise $\omega^i \mapsto \alpha^{i+1}$ via (1, 2, 4). Also,

$$(1, 3), (2, 3) \text{ are strictly effective for } (\omega^i),$$

for, otherwise, $\omega^i \mapsto \alpha^{i+1}$ via (2, 4) or $\omega^i \mapsto \alpha^i$ via (1, 4) respectively. Since $\omega_1^i > \omega_1^{i+1}$ and $\omega_2^i < \omega_2^{i+1}$, this implies that $\omega_3^i = \omega_3^{i+1}$. Then, clearly, $\alpha^{i-2} \mapsto \omega^i$ via (1, 3), a contradiction. Therefore, case (5.9) is impossible.

Discussion of case (5.10). Case (5.10) arises when only $S = (1, 2, 4)$ satisfies (5.4). Therefore, there exists $(\omega^i) \subset V$ such that

$$\omega_1^i > \alpha_1^i, \quad \omega_2^i = \alpha_2^i, \quad \omega_3^i < \alpha_3^i, \quad \omega_4^i > \alpha_4^i.$$

For (ω^i) the sets (1, 3), (2, 3) and (3, 4) are all strictly effective, for otherwise (respectively) $\omega^i \mapsto \alpha^i$ via (1, 4), or $\omega^i \mapsto \alpha^{i-1}$ via (2, 4), or (3, 4) would not be strictly effective for (α^i), contradicting (5.10). Since $\omega_1^i > \omega_1^{i+1}$ and $\omega_2^i < \omega_2^{i+1}$, this implies $\omega_3^i = \alpha_3^{i+1}$.

By the symmetry between the subindexes 2 and 3 in (5.1), a similar argument can be applied to yield another subset $(\sigma^i) \in V$, such that

$$\sigma_1^i > \alpha_1^i, \quad \sigma_2^i < \alpha_2^i, \quad \sigma_3^i = \alpha_3^i, \quad \sigma_4^i > \alpha_4^i,$$

and such that (1, 2), (2, 3) and (2, 4) are strictly effective for (σ^i), and $\sigma_1^i > \sigma_1^{i+1}$, $\sigma_2^i = \sigma_2^{i+1}$, $\sigma_3^i < \sigma_3^{i+1}$.

As $\omega_1^i > \sigma_1^{i+1}$ and $\omega_2^i > \sigma_2^{i+1}$, it follows that $\omega_4^i \leq \sigma_4^{i+1}$ for otherwise $\omega^i \mapsto \sigma^{i+1}$ via (1, 2, 4). Similarly, $\sigma_4^{i+1} \leq \omega_4^{i+2}$; hence $\omega_4^i \leq \omega_4^{i+2}$. Therefore, by considering subsets of (ω^i) and (σ^i) which can still be arbitrarily large, and by relabeling superscripts, it can be assumed that

$$\sigma_4^i \leq \sigma_4^{i+1}, \quad \omega_4^i \leq \omega_4^{i+1}. \tag{5.12}$$

There is an obvious similarity among (α^i), (ω^i), and (σ^i), except for the possibility of "=" sign in (5.12).

The same argument as before can be applied to (ω^i) and (σ^i). By Mills' theorem there exist $(\alpha'^i) \subset V$ and $T^i \subset I$ such that

$$T^i \text{ is effective for } (\alpha'^i) \text{ and } (\omega^i),\ T^i \ni 2,$$
$$\alpha_2'^i = \omega_2^i, \text{ and } \alpha_j'^i > \omega_j^i, \text{ for } j \in T^i - (2).$$

It can be shown as before that $T^i \equiv T = (1, 2, 3)$. Hence, it follows that

$$\alpha_1'^i > \omega_1^i > \alpha_1^i, \quad \alpha_2'^i = \omega_2^i = \alpha_2^i, \quad \alpha_3'^i > \omega_3^i, \quad \alpha_4'^i < \omega_4^i. \tag{5.13}$$

Similarly, by applying Mills' theorem to (σ^i), there exists (α''^i) satisfying

$$(5.14)\qquad \alpha_1''^i > \sigma_1^i > \alpha_1^i,\quad \alpha_2''^i > \sigma_2^i,\quad \alpha_3''^i = \sigma_3^i = \alpha_3^i,\quad \alpha_4''^i < \sigma_4^i.$$

It is easily checked that (1, 4), (2, 4) and (3, 4) are strictly effective for (α'^i) and (α''^i). Then (5.13) and (5.14) imply that

$$\alpha_2''^i < \alpha_2^i,\quad \alpha_3'^i < \alpha_3^i,\quad \alpha_4'^i = \alpha_4''^i = \alpha_4^i.$$

It can be assumed, by symmetry, that for arbitrarily large subsets of (α'^i) and (α''^i),

$$\alpha_1'^i \geq \alpha_1''^i.$$

Applying Mills' theorem to (α'^i) in the same way as before, a new subset (ω'^i) is finally obtained such that

$$\omega_1'^i > \alpha_1'^i \geq \alpha_1''^i,\quad \omega_2'^i = \alpha_2'^i = \alpha_2^i > \alpha_2''^i,\quad \omega_4'^i > \alpha_4'^i = \alpha_4''^i.$$

Therefore, $\omega'^i \vdash \alpha''^i$ via (1, 2, 4), which is a contradiction. This shows that case (5.10) can never occur, and completes the proof of the impossibility of (5.1).

PROOF FOR (5.2). Suppose there is an arbitrarily large set $(\alpha^i) \subset V$ satisfying (5.2). By Mills' theorem, for each α^i there exists $S^i \subset I$ and $\omega^i \in V$ fulfilling (5.11). It easily follows that the only possibilities for S^i are (1, 2), (1, 2, 3), (1, 2, 4). In any of those cases it must hold that

$$\omega_1^i = \alpha_1^i,\quad \alpha_2^i < \omega_2^i \leq \alpha_2^{i+1}.$$

As case (5.1) was proved to be impossible, this implies that for an arbitrary large subset of (ω^i)

$$\omega_3^i = \omega_3^{i+1}, \quad \omega_4^i = \omega_4^{i+1}, \text{ and } \alpha_3^i + \alpha_4^i > \omega_3^i + \omega_4^i.$$

Applying the same reasoning to (ω^i) and proceeding by recurrence, new subsets (ω'^i), (ω''^i), ..., $(\omega^{(j)i})$, ... can be obtained which satisfy:

$$\omega_1^{(j)i} = \omega_1^{(j+1)i}, \quad \omega_2^{(j)i} < \omega_2^{(j+1)i},$$

$$\omega_3^{(j)i} = \omega_3^{(j)i+1}, \quad \omega_4^{(j)i} = \omega_4^{(j)i+1}, \text{ and}$$

$$\omega_3^{(j)i} + \omega_4^{(j)i} > \omega_3^{(j+1)i} + \omega_4^{(j+1)i}.$$

For an arbitrarily large number of superscripts i, one of the following conditions must then hold:

$$\omega_3^{(j)i} > \omega_3^{(j)i} \text{ or } \omega_4^{(j)i} > \omega_4^{(j+1)i},$$

with a suitable relabeling of the superscripts.

If the diagonal subset $(\omega^{(i)i})$ is now considered, then

$$\omega_1^{(i)i} > \omega_1^{(i+1)i+1}, \quad \omega_2^{(i)i} < \omega_2^{(i+1)i+1}$$

and

$$\text{either } \omega_3^{(i)i} > \omega_3^{(i+1)i+1} \text{ or } \omega_4^{(i)i} > \omega_4^{(i+1)i+1}.$$

As the subset $(\omega^{(i)i})$ can be made arbitrarily large, and belongs to case (5.1) which was proved to be impossible, a contradiction arises. Hence (5.2) can never occur either, and the proof of Theorem 5.1 is complete.

§ 6. FINAL REMARKS

The method used in this paper, although successful in the discussion for $n = 4$, turns out to be cumbersome for games with more than four players. Nevertheless, there is some hope that the method permits an adequate generalization which might make it possible to attack the problems for general n.

The result obtained in Sec. 4 does not discard the possibility of the existence of a solution containing an accumulation point with a neighborhood in which there are only countably many points of the solution. Shapley [2] has shown an example in which such accumulation points exist for $n > 4$.

The upper bound that can be computed through the argument in Sec. 5 is far greater than the one conjectured by Shapley [2]. If it could be proved that for each imputation α belonging to a finite solution there is a subset $S \underset{\neq}{\subset} I$ which is exactly effective for α, the upper bound could be considerably lowered.

BIBLIOGRAPHY

[1] VON NEUMANN, J. and MORGENSTERN, O., Theory of Games and Economic Behavior, Princeton, 3rd ed. (1953).

[2] SHAPLEY, L. S., "A solution containing an arbitrary closed component," Annals of Mathematics Study No. 40, Princeton (1959), pp. 87-93.

[3] GILLIES, D. B., "Solutions to general non-zero-sum games," Annals of Mathematics Study No. 40, Princeton (1959), pp. 47-85.

[4] MILLS, W. H., "The four person game — finite solutions on the face of cube," Annals of Mathematics Study No. 40, Princeton (1959), pp. 125-143.

Alberto Raul Galmarino

Northeastern University
Boston, Massachusetts

THE DOUBLY DISCRIMINATORY SOLUTIONS OF FOUR PERSON CONSTANT-SUM GAMES*

Michael H. Hebert

§ 1. ABSTRACT

The concept of a discriminatory solution was first introduced by von Neumann and Morgenstern: a solution to a game is said to be discriminatory if there is one player discriminated against by the other players in the solution, in that they determine his payoff and that this is the same in all imputations of the solution. That is: in a discriminatory solution the component corresponding to the discriminated player is constant. A doubly discriminatory solution is a solution in which two players are discriminated against, i.e., a solution with two constant components.

In this work we shall determine the doubly discriminatory solutions of four-person constant-sum games.

It appears that only "half" of all four-person constant-sum games have doubly discriminatory solutions, that these solutions are easily defined through their constant components, and that in each case at least one of the discriminated players is excluded. The method of study consists in a discussion of the domination structure of the imputation space by means of a partition of this space which seems natural for this purpose.

---

* This work is a revised and shortened version of the author's "Die zweifach diskriminatorischen Lösungen der Vier-Personen-Konstantsommenspiele," prepared under the direction of Professor Ewald Burger, Cologne, and submitted to the University of Frankfurt am Main in July 1960.

Translated from the German by Guillermo Owen.

§ 2. PRELIMINARIES

1. Symbols and Definitions*

The set-theoretical symbols $\in$, $\subseteq$, $\cap$, $\cup$ and $-$ are used in the usual manner. $\{r, s, \ldots, u\}$ is the set consisting of elements $r, s, \ldots, u$. $\emptyset$ is the empty set. $|X|$ is the number of elements in the finite set X. $\{x | \cdots\}$ is the set of all x with the property designated by $\cdots$.

$I = \{1, 2, 3, 4\}$ is the set of the four players. The subsets of I are coalitions. A four-person constant-sum game is a function v by which a real number is made to correspond to each coalition of I, satisfying:

(1) $v(\emptyset) = 0$

(2) $v(R) + v(S) \leq v(R \cup S)$ for all $R, S \subseteq I$ with $R \cap S = \emptyset$

(3) $v(T) + v(I - T) = v(I)$ for all $T \subseteq I$ (constant-sum property)

(4) $v(U) = 0$ for $|U| = 1$, $v(I) = 1$ (0,1-normalization)

Hence we have: $v(Q) = 0$ for $|Q| < 2$, $0 \leq v(Q) \leq 1$ for $|Q| = 2$, $v(Q) = 1$ for $|Q| > 2$. For $v(\{i\})$, $v(\{i,j\})$, etc., we shall also write v_i, v_{ij}, etc., where i and j are players of I. W designates the set of all four-person constant-sum games.

An imputation x is a quadruple of real numbers satisfying

$$(5) \quad \begin{cases} x = (x_1, \ldots, x_4), \quad 0 \leq x_i \quad \text{for} \quad i = 1, \ldots, 4, \\ \sum_{i \in I} x_i = 1 \end{cases}$$

The space of imputations A is the set of all imputations. Given $v \in W$ the domination-relation, $\vdash$, is defined in A: $x \vdash y$ ("x dominates y") if and only if a coalition $S \neq \emptyset$ exists, satisfying

$$(6) \qquad y_i < x_i \qquad \text{for all} \quad i \in S$$

$$(7) \qquad \sum_{i \in S} x_i \leq v(S)$$

---

* As in L. S. Shapley, "Quota Solutions of n-Person Games," [4].

In this case S is called <u>effective</u>, and we say "x dominates y via S." Only two- and three-person coalitions can be effective.

For a subspace X of A, the set dom X, the <u>dominion</u> of X, is defined by

(8) $\quad \text{dom } X = \{y \mid y \in A, \text{ there exists } x \in X \text{ such that } x \vdash y\}.$

Moreover, for $S \subseteq I$ let the set S-dom X be that part of dom X which is dominated by elements of X via S. A <u>solution</u> of the game $v \in W$ is a subset $X \subseteq A$ with the properties

(9) $$X \cap \text{dom } X = \emptyset$$

(10) $$X \cup \text{dom } X = A.$$

A <u>doubly discriminatory</u> subset $X \subseteq A$ is a subset with the property:

(11) There exist two real numbers c_i and c_j with $i, j \in I$, $i \neq j$, with the properties: $0 \leq c_i$, $0 \leq c_j$, $c_i + c_j \leq 1$, and, for all $x \in X$, $x_i = c_i$, $x_j = c_j$.

We define $c = 1 - c_i - c_j$. For a doubly discriminatory subset X we also write $X(c_i, c_j)$. Thus i and j are called <u>discriminated players</u>; the other players, i.e., the non-discriminated players, we shall denote by m and n. We shall call a subset $X \subseteq A$ a doubly discriminatory solution if it is doubly discriminatory as well as a solution, i.e., if it has properties (9), (10) and (11).

2. The Problem

Let D be the set of all pairs (v, X) where v is a four-person constant-sum game and X a doubly discriminatory solution to v, i.e.,

$$(12) \qquad D = \left\{ (v,\ X) \;\middle|\; \begin{array}{l} v \in W \\ X \subseteq A \\ X \cap \text{dom } X = \emptyset \\ X \cup \text{dom } X = A \\ X = X(c_i,\ c_j),\ \text{doubly discriminatory} \end{array} \right\}$$

Hence D characterizes the set of all four-person constant-sum games that have doubly discriminatory solutions as well as the doubly discriminatory solutions of these games. The problem is, then, to express D explicitly. Starting from equation (12) we shall change, step by step, the five properties that characterize the elements of D, so that, finally, $(v,\ X) \in D$ will be characterized by new properties that let us know what class of games v are in W and — for a fixed v in this class — what class of subsets $X \subseteq A$ are doubly discriminatory solutions. The changes have to follow equivalently, i.e., we have to be careful that the "new" properties are consequences of the "old", and the "old" are consequences of the "new". The steps of the changes are shown in equations (12), (13), (15), (22), (25), (26), (27), (31), (32), and (35), in which the left side is always D, and, in (35), the right side is the desired explicit expression for D. This chain of equations may also serve as the shortest description of our manner of study.

3. Some Easy Steps

A subset $Y \subseteq A$ is called <u>domination-free</u> if and only if $Y \cap \text{dom } Y = \emptyset$.

THEOREM 1. The doubly discriminatory subsets of A are domination-free.

PROOF. Let Y be a doubly discriminatory subset of A. Then $z_i + z_j = c_i + c_j$ and $z_m + z_n = c$ for all $z \in Y$. Hence, for any pair $(x,\ y)$ with $x,\ y \in Y$, if $y_m < x_m$, then $x_n < y_n$. Hence there exists no pair of imputations in Y satisfying the domination relation (6) through a two- or three-person coalition. Therefore $Y \cap \text{dom } Y = \emptyset$.

Because of Theorem 1, the property $x \cap \text{dom } X = \emptyset$ is not necessary for the characterization of $v(X) \in D$ — it follows from the property "X is doubly discriminatory." Hence:

$$(13) \qquad D = \left\{ (v,X) \;\middle|\; \begin{array}{l} v \in W \\ X \subseteq A \\ X \cup \text{dom } X = A \\ X = X(c_i,\ c_j) \quad \text{doubly discriminatory} \end{array} \right\}$$

A doubly discriminatory subset $Y(c_i, c_j) \subseteq A$ will be called <u>maximal</u> if and only if it contains all imputations $z \in A$ for which $z_i = c_i$ and $z_j = c_j$.

THEOREM 2. Doubly discriminatory solutions of four-person constant-sum games are maximal.

PROOF. Suppose $(v, X) \in D$, $X = X(c_i, c_j)$ and X is not maximal. Then there is some $y \in A$ with the properties: $y \notin X$, $y_i = c_i$, $y_j = c_j$. Let $X' = X \cup \{y\}$. Clearly X' is doubly discriminatory, $X' = X'(c_i, c_j)$. By Theorem 1, X' is domination-free. Hence $y \notin X$, $y \notin \text{dom } X$, $y \in A$ and therefore $X \cup \text{dom } X \neq A$. But X is a solution and hence $X \cup \text{dom } X = A$. Hence the hypothesis is false and X is maximal.*

Before making use of Theorem 2, we shall define M:

$$(14) \qquad M = \left\{ L \;\middle|\; \begin{array}{l} L \subseteq A \\ L \;\; \text{is doubly discriminatory} \\ L \;\; \text{is maximal} \end{array} \right\}$$

If $(v, X) \in D$, then, by Theorem 2, $X \in M$. If $Y \in M$, and Y is a solution, then $Y \subseteq A$ and Y is doubly discriminatory. If $Z \in M$, then by Theorem 1 $Z \cap \text{dom } Z = \emptyset$ and, for $Z \cup \text{dom } Z = A$, we can write $A - Z = \text{dom } Z$. This means, for D,

---

* Note that in this proof of maximality the same idea is used as in the characterization by von Neumann and Morgenstern of solutions to two-person non-constant-sum games [6].

$$(15)\qquad D = \left\{ (v, X) \;\middle|\; \begin{array}{l} v \in W \\ X \in M \\ A - X = \text{dom } X \end{array} \right\}$$

Before we finish this section, we shall give, for future use, some properties of the elements of M. Let $L = L(c_i, c_j) \in M$. (a) We have $\ell_i = c_i$, $\ell_j = c_j$, $0 \leq \ell_m \leq c$, $0 \leq \ell_n \leq c$ for all $\ell \in L$; hence $\ell \in L$ is determined by giving one of the non-discriminated components. (b) For $p \in I$, let $\ell'_p = \max_{\ell \in L} \ell_p$. Since L is closed, ℓ'_p exists for all $p \in I$, and we have $\ell'_i = c_i$, $\ell'_j = c_j$, $\ell'_m = \ell'_n = c$. For $p \in \{m, n\}$ let $\ell'(p) = (\ell_i, \ell_j, \ell_m, \ell_n)$ with $\ell_i = c_i$, $\ell_j = c_j$, $\ell_p = c$, $\ell_q = 0$ for $p \neq q$, $q \in \{m, n\}$. Then $\ell'(m)$ and $\ell'(n)$ are the only imputations in L which are maximal in the components i, j, m, and i, j, n, respectively. Geometrically, $\ell'(m)$ and $\ell'(n)$ are the endpoints of a straight line which crosses the simplex A from side to side, parallel to one of the edges of the simplex.

§ 3. A NATURAL PARTITION OF THE SPACE OF IMPUTATIONS, RELATED TO THE QUESTION OF WHETHER A GIVEN SUBSET IS A SOLUTION

1. The partition

From now on, we shall understand by "coalitions" only two- and three-person coalitions. Given an imputation y and coalition T, we define Z(y, T):

$$(16)\qquad Z(y, T) = \{u \mid u \in A,\ y_p < u_p \text{ for all } p \in T\}.$$

Let X be a proper subset of A, let $y \in A - X$, and let S be a coalition. If $y \in S\text{-dom } X$, then, by (6) and (16), $X \cap Z(y, S)$ is not empty, and y and X determine an integer $r(y, X)$ which tells for how many two- and three-person coalitions R, T, ... the intersections $X \cap Z(y, R)$, $X \cap Z(y, T)$, ..., are non-empty. Using r, we partition $A - X$, into $E_0(X)$, $E_1(X)$, $E_2(X)$ and $E_3(X)$, as follows: for $i = 0, 1, 2$, $E_i(X)$ consists of all imputations in $A - X$ for which $r = i$, and $E_3(X)$ consists of all imputations in $A - X$ for which $r > 2$. With X, $E_0(X)$, ..., $E_3(X)$ we have the above-mentioned partition of A. We call it "natural", because it is simple and — as we shall see — it facilitates the decision of whether X is a solution. The partition does not depend on the choice of $v \in W$; it comes through pure geometrical methods. We make use of the fact that domination consists of two conditions, and use only its "geometrical" part, condition (6).

In the above formulation, the partition is not yet sufficient for our purpose. We can, however, sharpen it by the following observation: We obtain the same partition if, in its definition, we substitute the words "two-person coalition" for "coalition." For, if $x \in E_0(X)$, this means x is an imputation such that, for all coalitions Q, the intersection $X \cap Z(x, Q)$ is empty, and clearly it will be empty for all two-person coalitions. Conversely, if, for any two-person coalitions P the intersection $X \cap Z(w, P)$ is empty, then it must also be empty for three-person coalitions, i.e., $w \in E_0(X)$. If $x \in E_1(X)$, this means that, for x, there is exactly one coalition S such that $X \cap Z(x, S)$ is non-empty, and S must be a two-person coalition, for if S were a three-person coalition, then the intersections would also be non-empty for the three two-person subcoalitions of S. By the same argument, if $x \in E_2(X)$, then the two coalitions corresponding to x must be two-person coalitions. Finally, if $x \in E_3(X)$, this means that there are at least three coalitions so that the given Z-intersections are non-empty, and then there must be at least three such two-person coalitions. For, either the three coalitions are two-person coalitions, and our statement is true, or else at least one is a three-person coalition, and then the statement is also true, for this three-person coalition contains three two-person coalitions, whose Z-intersections are non-empty. Thus we define, finally, for $X \subseteq A$

$$(17) \qquad E_0(X) = \left\{ x \;\middle|\; \begin{array}{l} x \in A - X \\ X \cap Z(x, Q) = \emptyset \quad \text{for all} \quad Q \subseteq I,\ |Q| = 2 \end{array} \right\}$$

$$(18) \qquad E_1(X) = \left\{ x \;\middle|\; \begin{array}{l} x \in A - X \\ X \cap Z(x, R) \neq \emptyset \quad \text{for some} \quad R \subseteq I,\ |R| = 2 \\ X \cap Z(x, Q) = \emptyset \quad \text{for all} \quad Q \subseteq I,\ |Q| = 2,\ Q \neq R \end{array} \right\}$$

$$(19)\qquad E_2(X) = \left\{ x \;\middle|\; \begin{array}{l} x \in A - X \\ X \cap Z(x, R) \neq \emptyset,\ X \cap Z(x, S) \neq \emptyset, \text{ for some } R, \\ \quad S \subseteq I,\ R \neq S,\ |R| = |S| = 2 \\ X \cap Z(x, Q) = \emptyset \text{ for all } Q \subseteq I,\ |Q| = 2,\ Q \neq R, S \end{array} \right\}$$

$$(20)\qquad E_3(X) = \left\{ x \;\middle|\; \begin{array}{l} x \in A - X \\ X \cap Z(x, R) \neq \emptyset, \\ X \cap Z(x, S) \neq \emptyset,\ X \cap Z(x, T) \neq \emptyset \text{ for some } R, S, \\ \quad T \subseteq I,\ R \neq S \neq T \neq R, \end{array} \right\}$$

Then $E_0(X), \ldots, E_3(X)$ satisfy

$$(21)\qquad \left\{ \begin{array}{l} A - X = E_0(X) \cup E_1(X) \cup E_2(X) \cup E_3(X) \\ \emptyset = E_p(X) \cap E_q(X) \quad \text{for } p \neq q;\ p, q = 0, 1, 2, 3. \end{array} \right\}$$

We have to study the solution properties of maximal doubly discriminatory subsets of A, that is, of the elements of M, and shall apply our partition now to these sets. Using (17) through (21) we obtain for D:

$$(22)\qquad D = \left\{ (v, L) \;\middle|\; \begin{array}{l} v \in W \\ L = L(c_i, c_j) \in M \\ E_0(L) \cup E_1(L) \cup E_2(L) \cup E_3(L) = \text{dom } L \end{array} \right\}$$

Without knowing the form of the subsets $E_0(L), \ldots, E_3(L)$, we can make statements (23) and (24); they are true for each $L \in M$:

$$(23)\qquad E_0(L) \cap \text{dom } L = \emptyset$$

$$(24)\qquad E_3(L) \subseteq \text{dom } L.$$

PROOF OF (23). Given a pair (x, ℓ), if $x \in E_0(L)$ and $\ell \mapsto x$, $\ell \in L$, this contradicts the domination relation (6).

It follows from (23) that L can only be a solution if $E_0(L)$ is empty. But if $E_0(L)$ is empty then $E_1(L) \cup E_2(L) \cup E_3(L) = A - L$, and we obtain

$$(25)\qquad D = \left\{ (v, L) \;\middle|\; \begin{array}{l} v \in W \\ L = L(c_i, c_j) \in M \\ E_0(L) = \emptyset \\ E_1(L) \cup E_2(L) \cup E_3(L) = \text{dom } L \end{array} \right\}$$

PROOF OF (24). Let $L = L(c_i, c_j) \in M$ be fixed. It is enough to prove that for any imputation $x \in E_3(L)$ there exists an imputation $\ell \in L$ which has three components greater than those of x, because, if U is a three-person coalition, then $v(U) = 1$ by (3) and (4), and hence $\Sigma_{u \in U} \ell_u \leq v(U)$. If ℓ exceeds x in the U-components, then $\ell \vdash x$ via U. Assume then that $x \in E_3(L)$. Let R, S and T be the three two-person coalitions which correspond to x by (20).

Case 1. Each of the three coalitions R, S and T contains at least one discriminated player. Then their union $R \cup S \cup T$ contains both discriminated players, and at least one non-discriminated player. That is, $\{i, j, p\} \subseteq R \cup S \cup T$, with $p \in \{m, n\}$. By (20) there are imputations in L which exceed x in the R-components [S-components] [T-components]. Then, for all $q \in \{i, j, p\}$, $x_q < \ell'_q$. Hence x is exceeded by $\ell'(p)$ in the three components i, j, and p.

Case 2. Among the three coalitions R, S and T there is one which contains no discriminated players. Let, say, $R = \{m, n\}$. By (20), there is an imputation $\ell \in L$ which exceeds x in the R-components. As $R \cup S \cup T$ contains at least one player other than m and n, and hence a discriminated player, say i, we have $x_i < \ell_i$, so that x is exceeded by ℓ in the three components i, m, and n.

If $E_1(L) \cup E_2(L) \cup E_3(L) = \text{dom } L$, then $E_1(L) \cup E_2(L) \subseteq \text{dom } L$. If, on the other hand, $E_1(L) \cup E_2(L) \subseteq \text{dom } L$, then, by (21), (23) and (24), $E_1(L) \cup E_2(L) \cup E_3(L) = \text{dom } L$. This means, for D,

$$(26)\qquad D = \left\{ (v, L) \;\middle|\; \begin{array}{l} v \in W \\ L = L(c_i, c_j) \in M \\ E_0(L) = \emptyset \\ E_1(L) \cup E_2(L) \subseteq \text{dom } L \end{array} \right\}$$

For a subset of A to lie in the dominion of another subset of A, it is sufficient that it be empty. Hence for (26) we can write:

$$D = \left\{ (v, L) \;\middle|\; \begin{array}{l} v \in W \\ L = L(c_i, c_j) \in M \\ E_0(L) = \emptyset \\ E_1(L) = \emptyset \text{ or } [E_1(L) \neq \emptyset \text{ and } E_1(L) \subseteq \text{dom } L] \\ E_2(L) = \emptyset \text{ or } [E_2(L) \neq \emptyset \text{ and } E_2(L) \subseteq \text{dom } L] \end{array} \right\}. \tag{27}$$

Now we shall look for necessary and sufficient conditions that $E_0(L) = \emptyset$, $E_1(L) = \emptyset$, $E_2(L) = \emptyset$, $E_1(L) \subseteq \text{dom } L$, and $E_2(L) \subseteq \text{dom } L$. For this purpose we shall calculate $E_0(L)$, $E_1(L)$, and $E_2(L)$. This actually happens with the following Theorem 3. In what follows, let $L = L(c_i, c_j) \in M$ be arbitrary but fixed.

2. Determination of the Parts of the Partition

THEOREM 3. Given an imputation $x \in A - L$ and a two-person coalition S, there is an imputation $\ell \in L$ which exceeds x in the S-components if and only if $x_m + x_n < c$ in case $S = \{m, n\}$, or $x_p < \ell'_p$ for all $p \in S$, when $S \neq \{m, n\}$.

PROOF. Let $S = \{m, n\}$. If $\ell \in L$, $x_m < \ell_m$, $x_n < \ell_n$, then $x_m + x_n < \ell_m + \ell_n = c$. Conversely, if $x_m + x_n < c$ let $z = (z_1, \ldots, z_n)$ with $z_i = c_i$, $z_j = c_j$, $2(z_m - x_m) = 2(z_n - x_n) = c - (x_m + x_n)$. Then z is an imputation, for all its components lie between 0 and 1, and their sum equals 1. Also, $z \in L$, because $z_i = c_i$, $z_j = c_j$, and $L = L(c_i, c_j)$ is maximal. Since, by hypothesis $0 < c - (x_m + x_n)$, it follows that $x_m < z_m$, $x_n < z_n$.

Let $S \neq \{m, n\}$. Suppose $\ell \in L$, $x_p < \ell_p$ for all $p \in S$. Because $\ell_q \leq \ell'_q$ for all $q \in I$ and $\ell \in L$ it follows that $x_p < \ell'_p$ for all $p \in S$. Conversely, suppose $x_p < \ell'_p$ for all $p \in S$. As S does not contain both non-discriminated players, there is a non-discriminated player $p \in \{m, n\}$ such that $S \subseteq I - \{p\}$. Then, for $q \in \{m, n\}$, $q \neq p$, $\ell'(q)$ satisfies our theorem, because the S-components are included among the

$\{i, j, q\}$-components.

Determination of $E_0(L)$. By (17) and Theorem 3, $E_0(L)$ consists of all imputations $x \in A - L$ which possess the following properties:

1. $c < x_m + x_n$.
2. $\ell'_p \leq x_p$ or $\ell'_q \leq x_q$ for all $\{p, q\} \neq \{m, n\}$.

In particular, it follows from the second property that $c_i \leq x_i$ or $c_j \leq x_j$, because $\{i, j\} \neq \{m, n\}$. It is not possible that $c_i \leq x_i$ simultaneously with $c_j \leq x_j$, for then, if equality holds in both cases, x lies in L, whereas, if strict inequality holds in either case, then, because of the first property, the sum of the components of x exceeds 1 in contradiction to (5). Hence we have either $c_i \leq x_i$ or $c_j \leq x_j$, i.e., either $x_i < c_i$ or $x_j < c_j$. If, say, $x_i < c_i$, it follows from the second property that $\ell'_m \leq x_m$ and $\ell'_n \leq x_n$, since we know that either $\ell'_p \leq x_p$ or $\ell'_q < x_q$ for $\{p, q\} = \{i, m\}$ and $\{p, q\} = \{i, n\}$. The same will follow from $x_j < c_j$. Hence $E_0(L)$ consists of two disjoint parts, $E_0(L, i)$ and $E_0(L, j)$, whose union equals $E_0(L)$. Thus, for $p = i, j$:

$$E_0(L, p) = \left\{ x \;\middle|\; \begin{array}{ll} x \in A & \\ x_k \leq \ell'_k & \text{for } k = p \\ \ell'_k \leq x_k & \text{for } k \neq p \end{array} \right\} \tag{28}$$

Determination of $E_1(L)$. Let $E_1(L, p, q)$ be the set of all $x \in E_1(L)$ such that $L \cap Z(x, \{p, q\})$ is non-empty. Then $E_1(L)$ consists of the six disjoint sets $E_1(L, p, q)$ with $\{p, q\} \subseteq I$, so that $E_1(L) = \bigcup_{\{p,q\}} E_1(L, p, q)$, where $\{p, q\}$ varies over all two-person coalitions.

Let $\{m, n\} = \{p, q\}$. Then, by Theorem 3, $E_1(L, p, q)$ consists of all imputations $x \in A - L$ with the following properties:

1. $x_m + x_n < c$, from which follows that $x_m < c$ and $x_n < c$, since x_m and x_n are nonnegative, and
2. $c_i \leq x_i$, $c_j \leq x_j$, from which it follows that $x_m + x_n < c$.

Let $\{p, q\} \neq \{m, n\}$. Then by Theorem 3 $E_1(L, p, q)$ consists of all imputations $x \in A - L$ with the following properties:

1. $x_k < \ell'_k$ for all $k \in \{p, q\}$ and
2. $\ell'_k \leq x_k$ for all $k \notin \{p, q\}$.

Hence, for all $\{p, q\} \subseteq I$,

$$\text{(29)} \qquad E_1(L, p, q) = \left\{ x \;\middle|\; \begin{array}{l} x \in A - L \\ x_k < \ell'_k \quad \text{for} \quad k \in \{p, q\} \\ \ell'_k \leq x_k \quad \text{for} \quad k \notin \{p, q\} \end{array} \right\}$$

Determination of $E_2(L)$. Let $x \in E_2(L)$ and let R and S be those two two-person coalitions which, by (19), correspond to x. Then the pair (R, S) has the following four properties:

(a) $R \neq S$.

(b) $R \cup S \neq I$; for otherwise we would have $x_p < \ell'_p$ for all $p \in I$, and hence, by Theorem 3, there is an imputation $\ell \in L$ which exceeds x in the U-components for each of the five two-person coalitions U distinct from $\{m, n\}$. But we have assumed that only R and S have these properties with respect to x.

(c) $R \cap S = \{p\}$, $p \in \{i, j\}$. It follows from (a) and (b) that $R \cap S$ has one element. Suppose, say, that $R = \{m, n\}$, and $S = \{i, m\}$. Then, by Theorem 3, $U = \{i, n\}$ would be a third coalition for which an $\ell \in L$ would exist which exceeded x in the U-components. Suppose on the other hand that, say, $R = \{i, m\}$ and $S = \{j, m\}$. Then, by Theorem 3, $U = \{i, j\}$ would be a third coalition for which an $\ell \in L$ would exist exceeding x in the U-components.

(d) $R \neq \{i, j\} \neq S$. If, say, $R = \{i, j\}$ and $S = \{i, m\}$, then by Theorem 3, $U = \{j, m\}$ would be a third coalition for which an $\ell \in L$ exists exceeding x in the U-components.

The only pairs of two-person coalitions satisfying (a), ..., (d) are: $\{i, m\}$, $\{i, n\}$ and $\{j, m\}$, $\{j, n\}$. Hence $E_2(L)$ consists of two disjoint parts $E_2(L, i)$ and $E_2(L, j)$ whose union is $E_2(L)$. Hence, for $p = i, j$, and using Theorem 3,

$$\text{(30)} \qquad E_2(L, p) = \left\{ x \;\middle|\; \begin{array}{l} x \in A \\ x_k < \ell'_k \quad \text{for} \quad k \in \{m, n, p\} \\ \ell'_k \leq x_k \quad \text{for} \quad k \notin \{m, n, p\} \\ c \leq x_m + x_n \end{array} \right\}$$

With this, the sets $E_0(L)$, $E_1(L)$ and $E_2(L)$ are determined. For D this means:

$$D = \left\{ (v, L) \;\middle|\; \begin{array}{l} v \in W \\ L = L(c_i, c_j) \in M \\ E_0(L, p) = \emptyset \quad \text{for} \quad p = i, j \\ E_1(L, p, q) = \emptyset \quad \text{or} \quad [E_1(L, p, q) \neq \emptyset \quad \text{and} \\ \quad E_1(L, p, q) \subseteq \text{dom } L] \quad \text{for all} \quad \{p, q\} \subseteq I \\ E_2(L, p) = \emptyset \quad \text{or} \quad [E_2(L, p) \neq \emptyset \quad \text{and} \\ \quad E_2(L, p) \subseteq \text{dom } L] \quad \text{for} \quad p = i, j \end{array} \right\} \tag{31}$$

Now, for each of the sets $E_0(L, k)$, $E_1(L, p, q)$, $E_2(L, k)$, with $k = i, j$ and for all $\{p, q\} \subseteq I$, we shall search for necessary and sufficient conditions that this set be non-empty.

3. Conditions that the Parts of the Partition be Non-empty

$E_0(L, i) \neq \emptyset$ if and only if $0 < c_i$, $c \leq c_i$.

PROOF. Suppose $E_0(L, i) \neq \emptyset$. Then there exists an imputation x with the property that $x_i < c_i$, $c_j \leq x_j$, $c \leq x_m$ and $c \leq x_n$. From the first of these inequalities it follows that $0 < c_i$, and from the others it follows that $c_j + 2c \leq 1$, which is equivalent to $c \leq c_i$. Suppose $0 < c_i$, $c \leq c_i$. Let $x = (x_i, \ldots, x_n)$ with $x_i = 0$, $x_j = 1 - 2c$, $x_m = x_n = c$. Then x is an imputation, for all its components lie between 0 and 1 and their sum equals 1. But then $x \in E_0(L, i)$, because $x_i < c_i$, $c_j \leq x_j$, $x_m = x_n = c$.

$E_0(L, j) \neq \emptyset$ if and only if $0 < c_j$, $c \leq c_j$.

PROOF. Precisely the same as for $E_0(L, i)$, changing players i and j.

$E_1(L, i, j) \neq \emptyset$ if and only if $0 < c_i$, $0 < c_j$, and $2c \leq 1$.

PROOF. Suppose $E_1(L, i, j) \neq \emptyset$. Then there exists an imputation $x \in A - L$ with the property that $x_i < c_i$, $x_j < c_j$, $c \leq x_m$ and $c \leq x_n$. It follows that $0 < c_i$, $0 < c_j$, and $2c \leq x_m + x_n$. As $x_m + x_n \leq 1$, it follows that $0 < c_i$, $0 < c_j$, and $2c \leq x_m + x_n$. As $x_m + x_n \leq 1$, it follows that $2c \leq 1$. Suppose $0 < c_i$, $0 < c_j$, and $2c \leq 1$. Let $x = (x_i, \ldots, x_n)$ with $x_i = x_j = 0$, $2x_m = 2x_n = 1$. Then x is an imputation, since all its components lie between zero and 1 and their sum is 1. But then $x \in E_1(L, i, j)$ because $x_i < c_i$, $x_j < c_j$ and $c \leq x_m = x_n$.

$$E_1(L, m, n) \neq \emptyset \text{ if and only if } 0 < c.$$

PROOF. Suppose $E_1(L, m, n) \neq \emptyset$. Then there exists an imputation $x \in A - L$ with the property that $c_i \leq x_i$, $c_j \leq x_j$, $x_m < c$ and $x_n < c$. It follows that $0 < c$. Suppose $0 < c$. Let $x = (x_i, \ldots, x_n)$ with $x_i = c_i + c/2$, $x_j = c_j + c/2$, $x_m = x_n = 0$. Then x is an imputation because $x_i = c_i + c/2 = (1 + c_i + c_j)/2 \leq 1$ as $c_i - c_j \leq 1$; similarly $x_j \leq 1$, and all components lie between 0 and 1 and their sum equals 1. But then $x \in E_1(L, m, n)$ because $x \in A - L$ (since $x_i \neq c_i$) and $c_i < x_i$, $c_j < x_j$, $x_m < c$, $x_n < c$.

$$E_1(L, i, m) \neq \emptyset \text{ if and only if } 0 < c_i \text{ and } 0 < c.$$

PROOF. Suppose $E_1(L, i, m) \neq \emptyset$. Then there exists an imputation $x \in A - L$ with the property that $x_i < c_i$, $c_j \leq x_j$, $x_m < c$, and $c \leq x_n$. It follows that $0 < c_i$ and $0 < c$. Suppose $0 < c_i$ and $0 < c$. Let $x = (x_i, \ldots, x_n)$ and with $x_i = x_m = 0$, $2x_j = 2c_j + c_i$ and $2x_n = 2c + c_i$. Then x is an imputation, because all its components lie between 0 and 1 and their sum equals 1. But then $x \in E_1(L, i, m)$ because $x_i < c_i$, $x_m < c$, $c_j < x_j$ and $c < x_n$.

$$E_1(L, i, n) \neq \emptyset \text{ if and only if } 0 < c_i \text{ and } 0 < c.$$

$$E_1(L, j, m) \neq \emptyset \text{ if and only if } 0 < c_j \text{ and } 0 < c.$$

$$E_1(L, j, n) \neq \emptyset \text{ if and only if } 0 < c_j \text{ and } 0 < c.$$

PROOF. The proofs follow directly from that for $E_1(L, i, m)$ by proper changes of the players.

$$E_2(L, i) \neq \emptyset \text{ if and only if } 0 < c_i \text{ and } 0 < c.$$

PROOF. Suppose $E_2(L, i) \neq \emptyset$. Then there exists an imputation x with the property that $x_i < c_i$, $c_j \leq x_j$, $x_m < c$, $x_n < c$, and $c \leq x_m + x_n$. It follows that $0 < c_i$ and $0 < c$. Suppose $0 < c_i$ and $0 < c$. Let $x = (x_i, \ldots, x_n)$ with $x_i = 0$, $2x_m = 2x_n = c$, and $x_j = c_i + c_j$. Then x is an imputation, because all its components lie between 0 and 1 and their sum equals 1. But then $x \in E_2(L, i)$ because $x_i < c_i$, $c_j < x_j$, $x_m < c$, $x_n < c$ and $x_m + x_n = c$.

$$E_2(L, j) \neq \emptyset \text{ if and only if } 0 < c_j \text{ and } 0 < c.$$

PROOF. The proof follows directly from that for $E_2(L, i)$, interchanging players i and j.

Now we can also give necessary and sufficient conditions that the sets $E_0(L, i), \ldots, E_2(L, j)$ be empty:

$$E_0(L) = \emptyset \text{ if and only if } c_i < c \text{ and } c_j < c.$$

For if $0 = c_i = c_j$, then $c = 1$ and $c_i < c$, $c_j < c$; if $0 < c_i$, $c_j < c$, then $0 < c$, and hence $c_i < c$; if $0 < c_j$, $c_i < c$, then similarly $c_j < c$.

$E_1(L, i, j) = \emptyset$ if and only if $c_i = 0$ or $c_j = 0$ or $1 < 2c$.

$E_1(L, m, n) = \emptyset$ if and only if $c = 0$.

$E_1(L, i, m) = \emptyset$ if and only if $E_1(L, i, n) = \emptyset$, if and only if $E_2(L, i) = \emptyset$, if and only if $c_i = 0$ or $c = 0$.

$E_1(L, j, m) = \emptyset$ if and only if $E_1(L, j, n) = \emptyset$ if and only if $E_2(L, j) = \emptyset$ if and only if $c_j = 0$ or $c = 0$.

4. Domination Conditions in the Parts of the Partition

We look now for necessary and sufficient conditions that for all $\{p, q\} \subseteq I$ and all $k \in \{i, j\}$, the sets $E_1(L, p, q)$ and $E_2(L, k)$ lie in dom L, under the hypothesis that they are non-empty and that $E_0(L)$ is empty.

$$E_1(L, i, j) \subseteq \text{dom } L \text{ if and only if } v_{mn} \leq c.$$

PROOF. Suppose $E_1(L, i, j) \subseteq \text{dom } L$. Then, for each imputation $x \in E_1(L, i, j)$ there is an imputation $\ell \in L$ such that $\ell \mapsto x$ via $\{i, j\}$. Hence, by (7), $\ell_i + \ell_j < v_{ij}$. The sum of the $\{i, j\}$ components is constant in L, and equal to $c_i + c_j$, so that $c_i + c_j \leq v_{ij}$; by (3) and the definition of c, this is equivalent to $v_{mn} \leq c$. Suppose $v_{mn} \leq c$. Then, as above, $c_i + c_j \leq v_{ij}$ and $\ell_i + \ell_j \leq v_{ij}$ for all $\ell \in L$. By (18), for all $x \in E_1(L, i, j)$, $x_i < c_i$ and $x_j < c_j$. Then, for all pairs (ℓ, x) with $\ell \in L$, and $x \in E_1(L, i, j)$, we have: $x_i < \ell_i$, $x_j < \ell_j$, and $\ell_i + \ell_j \leq v_{ij}$. I.e., each imputation in L dominates each imputation in $E_1(L, i, j)$. It follows that $E_1(L, i, j) \subseteq \text{dom } L$.

$$E_1(L, m, n) \subseteq \text{dom } L \text{ if and only if } c \leq v_{mn}.$$

PROOF. Let $E_1(L, m, n) \subseteq \text{dom } L$. Then for each imputation $x \in E_1(L, m, n)$ there is an imputation $\ell \in L$ such that $\ell \mapsto x$ via $\{m, n\}$. Hence, by (7): $\ell_m + \ell_n \leq v_{mn}$. The sum of the $\{m, n\}$-components is constant and equal to c in L, and it follows that $c \leq v_{mn}$. Suppose $c \leq v_{mn}$. Then as above $\ell_m + \ell_n \leq v_{mn}$ for all $\ell \in L$. By (18), for each imputation $x \in E_1(L, m, n)$ there is an imputation $\ell \in L$ such that $x_m < \ell_m$ and $x_n < \ell_n$. As, moreover, $\ell_m + \ell_n \leq v_{mn}$, there is to each imputation $x \in E_1(L, m, n)$ an imputation $\ell \in L$ with $\ell \mapsto x$ via $\{m, n\}$. I.e., $E_1(L, m, n) \subseteq \text{dom } L$.

$$E_1(L, i, m) \subseteq \text{dom } L \text{ if and only if } 2c_i < v_{im}.$$

PROOF. Suppose $E_1(L, i, m) \subseteq \text{dom } L$. Let $z = (z_i, \ldots, z_n)$ with $z_i = 0$, $z_j = c_j$, $z_m = c_i$ and $z_n = c$. Then z is an imputation, because all its components lie between 0 and 1 and their sum equals 1. z is not in L, because $z_i = 0$ and $c_i \neq 0$ since $E_1(L, i, m)$ is non-empty. But then, by (29), $z \in E_1(L, i, m)$ because $z \in A - L$, $z_i < c_i$, $z_j = c_j$, $z_n = c$ and $z_m < c$, since, as by hypothesis $E_0(L, i)$ is empty, $c_i < c$. By (18) there exists an imputation $\ell \in L$ such that $z_i < \ell_i$, $z_m < \ell_m$ and $\ell_i + \ell_m \leq v_{im}$. It follows, as $z_m = c_i = \ell_i$, that $2c_i < v_{im}$. Suppose $2c_i < v_{im}$. Suppose moreover, that for the imputations $x \in E_1(L, i, m)$ and $w \in L$, $w \mapsto x$ via $\{i, m\}$. If $\bar{x}$ is also an imputation in $E_1(L, i, m)$ and $\bar{x}_m \leq x_m$, then we have also $w \mapsto \bar{x}$ because $\bar{x}_i < c_i$ by (18), $\bar{x}_m \leq x_m < w_m$, and $w_i + w_m \leq v_{im}$ by supposition. This proves that, if there is an imputation $E_1(L, i, m)$

with a maximal m^{th} component, then for $E_1(L, i, m)$ to lie in dom L it is sufficient that this imputation do so. Clearly, for the imputation z given above, $z_m = \max_{y \in E_1(L,i,m)} y_m$ because the $\{i, j, n\}$-components of z are minimal in $E_1(L, i, m)$. Let the imputation $\ell \in L$ be given by $\ell_m = \min\{v_{im} - c_i, c\}$. We shall prove that $\ell \vdash z$ via $\{i, m\}$; it will follow that $E_1(L, i, m) \subseteq$ dom L.

To prove the assertion, we observe first that $z_i < \ell_i$, because $z_i = 0$, $\ell_i = c_i$ and $0 < c_i$, since $E_1(L, i, m)$ is non-empty. Also $z_m < \ell_m$, because if (1) $v_{im} - c_i < c$, then $z_m < \ell_m$ since $z_m = c_i$, $\ell_m = v_{im} - c_i$ and $c_i < v_{im} - c_i$ by hypothesis. If (2) $c \leq v_{im} - c_i$, then $z_m < \ell_m$, since $z_m = c_i$, $\ell_m = c$, and $c_i < c$ as $E_0(L, i)$ is empty. Finally, $\ell_i + \ell_m \leq v_{im}$, because if (1) $v_{im} - c_i < c$, then $\ell_i + \ell_m = v_{im}$, and if (2) $c < v_{im} - c_i$, then $\ell_i + \ell_m = c_i + c \leq v_{im}$. Hence we have $z_i < \ell_i$, $z_m < \ell_m$, and $\ell_i + \ell_m \leq v_{im}$, so $\ell \vdash z$ via $\{i, m\}$.

$E_1(L, i, n) \subseteq$ dom L if and only if $2c_i < v_{in}$.

$E_1(L, j, m) \subseteq$ dom L if and only if $2c_j < v_{jm}$.

$E_1(L, j, n) \subseteq$ dom L if and only if $2c_j < v_{jn}$.

PROOFS. The proofs follow directly from that for $E_1(L, i, m)$ through proper permutations of the players.

We must find conditions also that $E_2(L, i)$ and $E_2(L, j)$ lie in dom L. We have shown that, if $E_2(L, i)$ is non-empty, then so are $E_1(L, i, m)$ and $E_1(L, i, n)$, and conversely, and that for $(v, L) \in D$ we have $2c_i < v_{im}$, $2c_i < v_{in}$. These two inequalities can also be used in looking for the desired conditions on $E_2(L, i)$, since we are only interested in pairs (v, L) with $(v, L) \in D$. Similarly, in dealing with $E_2(L, j)$ we shall assume the inequalities $2c_j < v_{jm}$ and $2c_j < v_{jn}$.

$E_2(L, i) \subseteq$ dom L if and only if $1 - c_j < v_{im} + v_{in} - 2c_i$.

PROOF. Suppose that $E_2(L, i) \subseteq$ dom L.

Case 1. $c \leq v_{im} - c_i$. Together with $2c_i < v_{in}$, it follows that $1 - c_j = c + c_i < v_{im} + v_{in} - 2c_i$.

Case 2. $v_{im} - c_i < c$. Let $u = (u_1, \ldots, u_n)$ with $u_i = 0$, $u_j = c_j$, $u_m = v_{im} - c_i$ and $u_n = 1 - c_j - (v_{im} - c_i)$. Then u is an imputation because $u_i = 0$, $0 \leq u_j \leq 1$ since $0 \leq c_j \leq 1$; $0 \leq u_m \leq 1$ since $0 < c_i < v_{im} - c_i \leq 1$; $0 < u_n$, since $v_{im} - (v_{im} - c_i) < c + c_i - (v_{im} - c_i) = 1 - c_j - (v_{im} - c_i)$ by hypothesis and $0 < c_i = (v_{im} - c_i)$, also by hypothesis; the sum of the components of u equals 1; hence also, $u_n \leq 1$. Thus we have proved that $u \in A$. But then $u \in E_2(L, i)$, because $u_i < c_i$ since $0 = u_i$ and $0 < c_i$ by hypothesis; $u_j = c_j$; $u_m < c$ since $v_{im} - c_i < c$ by hypothesis; $u_n < c$ because $1 - c_j - (v_{im} - c_i) = c - (v_{im} - 2c_i) < c$ since $2c_i < v_{im}$ by hypothesis; and $c < 1 - c_j = u_m + u_n$. Thus we have proved that $u \in E_2(L, i)$. It follows, by (19) and by the hypothesis $E_2(L, i) \subseteq \text{dom } L$ that $u \in R\text{-dom } L$ or $u \in S\text{-dom } L$ with $R = \{i, m\}$ and $S = \{i, n\}$. If $u \in R\text{-dom } L$, then there is some $\ell \in L$ such that $u_m < \ell_m$; however, $u_m = v_{im} - c_i$, hence $v_{im} - c_i < \ell_m$ and therefore $v_{im} < \ell_{im} + \ell_m$; this means $u \notin R\text{-dom } L$. Thus we must have $u \in S\text{-dom } (L)$, $S = \{i, n\}$. Hence there is some $w \in L$ such that $w \vdash u$ via S, i.e., $u_i < w_i$, $u_n < w_n$, and $w_i + w_n \leq v_{in}$. It follows that $u_n < w_n \leq v_{in} - w_i$, and therefore $1 - c_j - (v_{im} - c_i) < v_{in} - c_i$, i.e., $1 - c_j < v_{im} - v_{in} - 2c_i$.

Suppose now that $1 - c_j < v_{im} - v_{in} - 2c_i$. We prove that, for every imputation $x \in E_2(L, i)$, there exists an imputation $\ell \in L$ such that $\ell \vdash x$ via $\{i, m)$, or an imputation $\bar{\ell} \in L$ such that $\bar{\ell} \vdash x$ via $\{i, n\}$. Suppose, then, that $x \in E_2(L, i)$.

Case 1. There exist an imputation $w \in L$ such that $x_i < w_i$, $x_m < w_m$ and $w_i + w_m \leq v_{im}$. Then w satisfies our claim, because $w \vdash x$ via $\{i, m\}$.

Case 2. There is no $y \in L$ such that $x_i < y_i$, $x_m < y_m$ and $y_i + y_m \leq v_{im}$; i.e., for all imputations $z \in L$ such that $x_i < z_i$ and $x_m < z_m$, we have $v_{im} < z_i + z_m$, i.e., $v_{im} - c_i < z_m$. By (19), however, there is a $z \in L$ such that $x_i < z_i$ and $x_m < z_m$. Because $z_m \leq c$ it follows that $v_{im} - c_i < c$. We claim that $v_{im} - c_i \leq x_m$. To prove this, suppose $x_m < v_{im} - c_i$. Then, because $0 < v_{im} - c_i < c$, there exists $\ell \in L$ such that $\ell_m = v_{im} - c_i$. Then $x_i < \ell_i$ because $x_i < c_i$ by (30) and $c_i = \ell_i$. And, $x_m < \ell_m$, because $x_m < v_{im} - c_i$ by hypothesis and $v_{im} - c_i = \ell_m$. And $\ell_i + \ell_m = v_{im}$. Hence $\ell \in L$, $x_i < \ell_i$, $x_m < \ell_m$ and $\ell_i + \ell_m = v_{im}$, which contradicts the hypothesis that for all $\ell \in L$ with $x_i < \ell_i$ and $x_m < \ell_m$,

$v_{im} < \ell_i + \ell_m$. Hence we must have $v_{im} - c_i \leq x_m$.

From this inequality, together with our hypothesis $1 - c_j < v_{im} + v_{in} - 2c_i$, we obtain the inequality $1 - c_j < x_m + v_{in} - c_i$. We replace the "1" on the left side of this inequality with the sum of the components of x, subtract x_m from both sides, and obtain $x_i + x_j + x_n - c_j < v_{in} - c_i$. By (30), $x_j - c_j$ is nonnegative and hence $x_i + x_n < v_{in} - c_i$. Suppose $u \in L$, with $u_n = \min\{v_{in} - c_i, c\}$. Suppose then that $c < v_{in} - c_i$. Then $x_i < u_i$ on the same grounds as above, and $x_n < u_n$ because by (30) $x_n < c$ and $u_n = c$. Also, $u_i + u_n < v_{in}$, because $u_i + u_n = c_i + c < v_{in}$ since $c < v_{in} - c_i$. Hence $u \mapsto x$ via $\{i, n\}$. Now suppose $v_{in} - c_i \leq c$. Then $x_i < u_i$ on the same grounds as above, and $x_n < u_n$ because $x_n \leq x_n + x_i < v_{in} - c_i = u_n$. Also $u_i + u_n = v_{in}$. Hence $u \mapsto x$ via $\{i, n\}$. As this proof holds for an arbitrary $x \in E_2(L, i)$, it follows that $E_2(L, i) \subseteq \text{dom } L$.

$$E_2(L, j) \subseteq \text{dom } L \text{ if and only if } 1 - c_i < v_{jm} + v_{jn} - 2c_j.$$

PROOF. The proof follows directly from the proof for $E_2(L, i)$ by interchanging players i and j.

We now take the conditions we have obtained, and list them, labeling them with A, ..., Q:

A	$c_i < c$ and $c_j < c$
B	$c_i = 0$
C	$c_j = 0$
D	$1 < 2c$
E	$c = 0$
F	$0 < c_i$
G	$0 < c_j$
H	$2c \leq 1$
I	$0 < c$
J	$v_{mn} \leq c$
K	$c \leq v_{mn}$
L	$2c_i < v_{im}$

$\underline{M}$ $\quad 2c_i < v_{in}$

$\underline{N}$ $\quad 2c_j < v_{jm}$

$\underline{O}$ $\quad 2c_j < v_{jn}$

$\underline{P}$ $\quad 1 - c_j < v_{im} + v_{in} - 2c_i$

$\underline{Q}$ $\quad 1 - c_i < v_{jm} + v_{jn} - 2c_j$

Let us look once again at equation (31). Our purpose was to replace the last three conditions of (31) with equivalent expressions which contain only c_i, c_j, and v. We have accomplished this: by using $\underline{A}, \ldots, \underline{Q}$ we can write these three expressions equivalently in the desired way. For example, for the expression "$E_1(L, i, m) = \emptyset$ or $[E_1(L, i, m) \neq \emptyset$ and $E_1(L, i, m) \subseteq \text{dom } L]$" we can insert the expression "$c_i = 0$ or $c = 0$ or $[0 < c_i$ and $0 < c$ and $2c_i < v_{im}]$". For this latter expression, since $\underline{F}$ and $\underline{I}$ are the contradictory opposites of $\underline{B}$ and $\underline{E}$, we can write, more briefly, "$B \times E \times L$" where the cross "$\times$" is taken to mean disjunction. We obtain, for D:

$$(32) \qquad D = \left\{ (v, L) \;\middle|\; \begin{array}{l} v \in W \\ L = L(c_i, c_j) \in M \\ \underline{A} \\ \underline{B} \times \underline{C} \times \underline{D} \times \underline{J} \\ \underline{E} \times \underline{K} \\ \underline{B} \times \underline{E} \times \underline{L} \\ \underline{B} \times \underline{E} \times \underline{M} \\ \underline{C} \times \underline{E} \times \underline{N} \\ \underline{C} \times \underline{E} \times \underline{O} \\ \underline{B} \times \underline{E} \times \underline{P} \\ \underline{C} \times \underline{E} \times \underline{Q} \end{array} \right\}$$

Here the given conditions bring us to the end of our search for the sets $E_0(L), \ldots, E_2(L, j)$. Here the possibilities offered by our "natural" partition of the complement of L are brought out completely. Now, in the next section, we can find the desired solutions easily.

§ 4. DOUBLY DISCRIMINATORY SOLUTIONS

We derive now all complete systems of conditions from the conditions A, ..., Q, for which — if $v \in W$ and $L \in M$ — L is a solution of v. For this purpose, we shall call a system "complete" if it contains at least one condition proper to each of the sets $E_0(L)$, ..., $E_2(L, j)$, i.e., if it contains at least one element of the disjunction for each of the last nine parts of Equation (32). We shall not yet discuss whether the complete systems we obtain are consistent, i.e., whether any two conditions in a system contradict each other, for the answer to this question will appear below without special care. However, we shall not worry about complete systems that are obviously inconsistent.

Each system S contains the condition A: $c_i < c$ and $c_j < c$. Hence no system will contain condition E, because A implies $0 < c$. Therefore every system must contain condition K: $c \leq v_{mn}$.

First Case: Conditions B and C are fulfilled: $c_i = c_j = 0$. Then the system (A, B, C, K) is obviously complete. But A follows from B and C, so that we obtain $S_1 = (B, C, K)$ as a complete system.

Second Case: Condition B is fulfilled, but condition C is not, i.e., conditions B and G are fulfilled: $0 = c_i$, $0 < c_j$. A system S which contains conditions A, B, G, K must, if it is to be complete contain also conditions N, O, and Q. We shall prove now that conditions A, N, and O follow from conditions B, G, K and Q, i.e., that the system $S_2 = (B, G, K, Q)$ is already complete. From B and Q follows: $1 < v_{jm} + v_{jn} - 2c_j$. As $v_{jm} \leq 1$ and $v_{jn} \leq 1$ it follows that $0 < v_{jn} - 2c_j$ and $0 \; v_{jm} - 2c_j$; these are conditions N and O. From this same inequality $1 < v_{jm} + v_{jn} - 2c_j$ follows: $2c_j < c_{jn} + v_{jn} - 1$. It follows that $2c_j < 1$, because $v_{jm} + v_{jn} - 1 \leq 1$. Because of B, $C = 1 - c_j$. From $2c_j < 1$ it follows that $c_j < 1 - c_j$, and hence $e_j < C$. Hence $c_i < c$, $c_j < c$; that is condition A.

Third Case: Condition C is fulfilled and B is not fulfilled. This case is treated similarly to the second case, but interchanging players i and j. We obtain the complete system $S_3 = (C, F, K, P)$.

Fourth Case: Neither condition B nor condition C is fulfilled. A system which contains neither of these conditions must contain conditions

$\underline{P}$ and $\underline{Q}$ to be complete. However, it is then inconsistent. For, added together, conditions $\underline{P}$ and $\underline{Q}$ give: $c_i < -c_j$! .

The systems S_1, S_2, and S_3 are clearly consistent. Hence S_1, S_2 and S_3 exhaust the class of complete, consistent systems. S_1 says: $c_i = c_j = v_{ij} = 0$. S_2 says: $0 = c_i$, $0 < c_j$, $v_{ij} \leq c_j$ and $2c_j < v_{jm} + v_{jn} - 1$. S_3 says: $0 = c_j$, $0 < c_i$, $v_{ij} \leq c_i$ and $2c_i < v_{im} + v_{in} - 1$. But S_1 says nothing about $v_{jm} + v_{jn} - 1$. The case $0 < v_{jm} + v_{jn} - 1$, when appearing under S_1, may be transferred to S_2, and the case $v_{jm} + v_{jn} - 1 < 0$ (or, equivalently, $0 < v_{im} + v_{in} - 1$) may, similarly, be transferred to S_3, and we obtain the three systems T_1, T_2, and T_3:

$$(33)\qquad \begin{cases} T_1 : & c_i = c_j = v_{ij} = v_{jm} + v_{jn} - 1 = 0 \\ T_2 : & c_i = 0,\ 2v_{ij} \leq 2c_j < v_{jm} + v_{jn} - 1 \\ T_3 : & c_j = 0,\ 2v_{ij} \leq v_{im} + v_{in} - 1 \end{cases}$$

The systems T_1, T_2, T_3 tell us which games $v \in W$ possess doubly discriminatory solutions with players i and j discriminated; these are clearly the games possessing property A_1, A_2, or A_3:

$$(34)\qquad \begin{cases} A_1 : & v_{ij} = v_{jm} + v_{jn} - 1 = 0 \\ A_2 : & 2v_{ij} < v_{jm} + v_{jn} - 1 \\ A_3 : & 2v_{ij} < v_{im} + v_{in} - 1\ . \end{cases}$$

The last link in our chain of Equations (12), (13), (15), (22), (25), (26), (27), (31), (32), the explicit expression of D, is now:

$$(35)\qquad D = \left\{ (v, L) \;\middle|\; \begin{array}{l} v \in W \\ A_1 \text{ or } A_2 \text{ or } A_3 \\ L = L(c_i, c_j) \in M \\ \text{if } A_1, \text{ then } c_i = c_j = 0 \\ \text{if } A_2, \text{ then } c_i = 0,\ 2v_{ij} \leq 2c_j < v_{jm} + v_{jn} - 1 \\ \text{if } A_3, \text{ then } c_j = 0,\ 2v_{ij} < 2c_i < v_{im} + v_{in} - 1 \end{array} \right\}$$

We can formulate our results in the following

> THEOREM. A four-person constant-sum game v has a doubly discriminatory solution with discriminated players i and j if and only if A_1: $v_{ij} = v_{jm} + v_{jn} - 1 = 0$ or A_2: $2v_{ij} < v_{jm} + v_{jn} - 1$ or A_3: $2v_{ij} < v_{im} + v_{in} - 1$ holds. The solutions of v which discriminate against i and j are then the subsets $L(c_i, c_j)$ of the space of imputations, with i^{th} and j^{th} components equal to c_i and c_j, respectively, and which are maximal in the space of imputations. In case A_1 we have $c_i = c_j = 0$; in case A_2 we have $2v_{ij} \leq 2c_j < v_{jm} + v_{jn} - 1$ and $c_i = 0$; in case A_3 we have $2v_{ij} \leq 2c_i < v_{im} + v_{in} - 1$ and $c_j = 0$. The game v has no other doubly discriminatory solutions with players i and j discriminated.

We shall take a look now at the domain of existence of doubly discriminatory solutions. A game $v \in W$ is completely determined by the values of three two-person coalitions, as long as no two of them are complementary. Because of the "constant-sum property" (3), the values of the other three two-person coalitions are thereby determined and the values for coalitions with more or less than two persons are fixed by (3) and (4). Thus we can map the games $v \in W$ biuniquely and in a nautral manner onto the unit cube Q with corners $(0, 0, 0)$, $(0, 0, 1)$, ..., $(1, 1, 1)$; as has been shown by von Neumann and Morgenstern [6]. Figure 1 shows this cube. We take from [6] the choice of coordinate axes and the labeling of the corners.

The games in W with property A_1 for an arbitrary pair (i, j) are represented by the edges of the tetrahedron with vertices I, V, VI and VII. The games corresponding to property A_2 or A_3 for a given pair (i, j) are represented by the points of Q which are always on the far side of the planes "$2v_{ij} = v_{km} + v_{kn} - 1$", with $k = i, j$, as viewed from the mid-point of Q (excluding the plane). By the construction of these planes we obtain the polyhedron shown in Figure 2. On the faces of the tetrahedron pyramids are constructed with the points II', III', IV' and VIII' as vertex. For example, II' is the point of intersection of the planes:

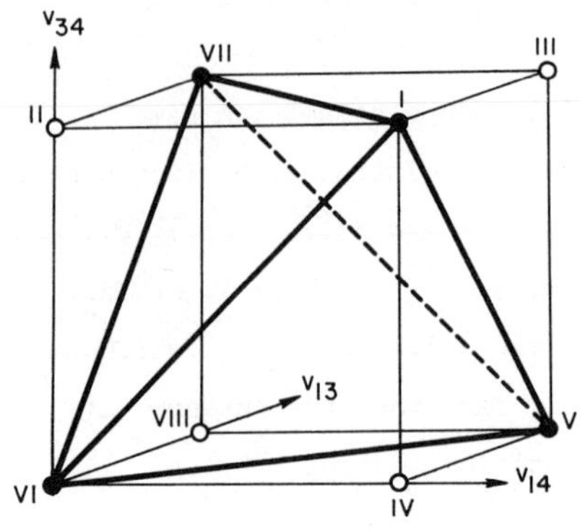

Fig. 1

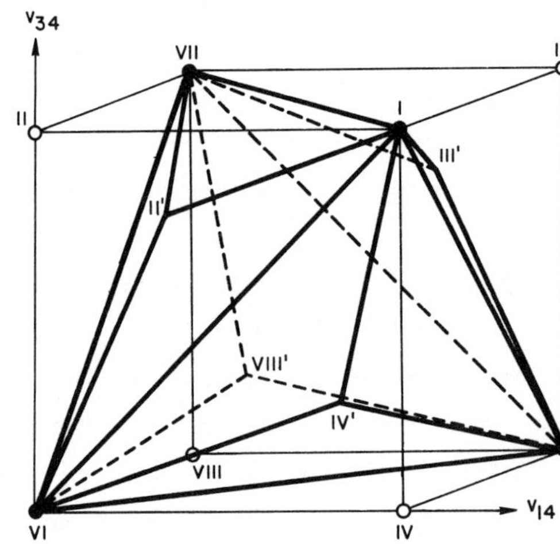

Fig. 2

$v_{23} + v_{24} - 1 = 2v_{12}$, $2v_{13} = v_{23} + v_{34} - 1$, and $2v_{14} = v_{24} + v_{34} - 1$. The volume of the polyhedron is half the volume of the cube. In this sense, half of all four-person constant-sum games contains doubly discriminatory solutions.

The domain of existence of doubly discriminatory solutions consists, then, of the exterior of the tetrahedron enlarged by the pyramids, together with the edges of the tetrahedron. We remark that in every doubly discriminatory solution, at least one of the discriminated players receives 0, i.e., he is excluded.

§ 5. COMPARATIVE LITERATURE

In [5], L. S. Shapley characterizes all games with solutions that exclude one or more players, i.e., discriminate against them completely. We find the following theorem: ([5], p. 5, Theorem 3): "A necessary and sufficient condition that the game v have a solution which excludes the set S is that the fraction v_{I-S} be solvable, and either winning or effective." Here, by the "fraction" v_R is meant the function defined on the subsets of $R \subseteq I$, such that $v_R(T) = v(T)$ for $T \subseteq R$, $T \neq R$, and $v_R(T) = 1$ for $T = R$. Hence the fraction v_R of v is an r-person game. It is called "winning" if $v(R) = 1$ and "effective" if the core* of v_R contains no interior points. Now, if $v \in W$, $\{p, q\}, \subseteq I$, and $R = I - \{p, q\}$, then v_R is a two-person game. Its space of imputations is equal to its core; hence it has interior points and is not effective. Hence only those four-person constant-sum games for which R is winning will have solutions which exclude the players in $I - R$. Returning to our treatment in § 4, this means that the four-person constant-sum games which are represented by the boundary points of the cube are precisely those having solutions which exclude two of the players; specifically, the boundary points with $v_{pq} = 1$ are those in which the players in $I - \{p, q\}$ can be excluded.

In [3], W. H. Mills has given all solutions for those four-person constant-sum games represented by the points on the edges of the cube. We have, above, determined all four-person constant-sum games which have doubly discriminatory solutions and all doubly discriminatory solutions to these games. We shall show below, that our "edge solutions" agree with Mills' doubly discriminatory solutions.

We shall now use the notation in [3], staying, however, in 0, 1-normalization. In [3], upper-case latin letters represent imputations, and the corresponding lower-case letters their components. Greek and upper-case script letters represent sets of imputations. The games $v_{13} = v_{14} = 1$, $v_{12} = t$ with $0 \leq t \leq 1$ are considered. These are the games on the edge III-IV in Figures 1 and 2. It is shown ([3], p. 368) that through a permutation of the players these results can be transferred from one edge to the

---

* The core of a game v with player set I is the set of all imputations x such that $v(S) \leq \Sigma_{i \in S} x_i$ for all $S \subseteq I$.

other edges.

For a game on this edge, Mills gives a complete list of the solutions ([3], p. 376):

(a) The trivial solutions of Theorem 2.

(b) The solutions for which $1 - t \le j_3 + k_4$. These solutions consist of the monotone curve L and the set $\mathfrak{M}$. If $1 - t < j_3 + k_4$, then $\mathfrak{M} = M(\bar{\alpha}, C)$. If $1 - t = j_3 + k_4$, then $\mathfrak{M} = \{J, K\}$.

(c) The solutions for which $j_3 + k_4 < 1 - t$ and $j_3 < c_3$. They consist of the monotone curve $\mathcal{C}$ and of the line segment $\mathcal{B}'$.

(d) The solutions for which $j_3 + k_4 < 1 - t$ and $j_3 < c_3$. If $\bar{\alpha} < t - 2c_2 + c_4$, they consist of the curve $\mathcal{C}$ and the line segment $\mathcal{B}'$. If $t - 2c_2 + c_4 \le \bar{\alpha} \le t - c_2 + c_4$, they consist of the monotone curve $\mathcal{C}$, of the line segment $\mathcal{B}'$, and of the point D''.

(e) The solutions which can be obtained from those in (c) and (d) by interchanging players 3 and 4.

We give the definition of the above-mentioned terms: A solution ν is "trivial" if player 1 is discriminated. In the following, let ν be non-trivial. C is the imputation in ν with $c_1 = \min_{P\in\nu} p_1$; we have $c_1 = 0$ by Lemma 1. $\bar{\alpha} = \max_{P\in\nu} p_1$. $M(\bar{\alpha}, C) = \{P|P$ is an imputation, $p_1 = \bar{\alpha}$, $p_i = c_i$ for at least one value of $i\}$; we have $M(\bar{\alpha}, C) \ne \emptyset$. There is exactly one imputation J with $j_3 = \max_{P\in M(\bar{\alpha},C)} p_3$, and there is exactly one imputation K with $k_4 = \max_{P\in M(\bar{\alpha},C)} p_4$. $\mathcal{C}$ is a continuous curve from C to D, monotone in the sense that, if both P and Q are on $\mathcal{C}$ and $q_1 < p_1$, then $p_i \le q_i$ for $i = 2, 3, 4$. D is an imputation: $d_2 = \max_{P\in\mathfrak{U}'} p_2$, $d_3 = \max_{P\in\mathfrak{V}'} p_3$, $d_4 = \max_{P\in\mathfrak{V}'} p_4$, $d_1 = 1 - d_2 - d_3 - d_4$, where $\mathfrak{U}' = \mathfrak{U}^* \cap \mathfrak{V}'$, $\mathfrak{U}^* = \mathcal{A}^* \cap \nu$, $\mathcal{A}^* = \{P|p_1 + p_2 \le t\}$, $\mathfrak{V}' = \{P|p_1 = \alpha,\ P \in \nu\}$, and $\alpha = \max_{P\in\mathfrak{U}^*} p_1$. $\mathcal{B}' = \{B|\alpha \le b_1,\ b_2 = p_2,\ b_3 = j_3\}$. (See [3], pp. 368-375.) We do not give the definition of D'', because it will not appear in the study below.

<u>Case (a)</u>: Theorem 2 ([3], p. 370) states that the solutions which discriminate player 1 are of the form ν_c, and ν_c is a solution if and only

if $0 \le c < \max(\frac{1}{2}, \frac{1+t}{3})$. Then c is a real number such that $0 \le c \le 1$ and $\nu_c = \{P | p_1 = c,\ p_2 = 0\}$ if $t \le c$, and $\nu_c = \{P | p_1 = c,\ p_1 + p_2 \le t\}$ if $c < t$, where in our formulation we avoid the distinction of cases which appears in [3]. Here only those pairs of discriminated players are considered which contain player 1.

Suppose $t \le c$. All solutions of this case are of the form $\{P | p_1 = c,\ p_2 = 0\}$ and discriminate players 1 and 2. Pairs (t, c) with $t < \frac{1}{2}$ and $c < \frac{1}{2}$ — and no others — satisfy, simultaneously the inequalities $t < c$ and $0 \le c < \max(\frac{1}{2}, \frac{1+t}{3})$ and describe, in this way, these solutions and the corresponding games.

Suppose $c < t$. Then $\nu_c = \{P | p_1 = c,\ p_2 \le t - c\}$. Clearly $(c, 0, 1-c, 0) \in \nu_c$, $(c, t-c, 1-t, 0) \in \nu_c$, and $(c, t-c, 0, 1-t) \in \nu_c$. Hence no player except player 1 is discriminated. This gives:

(a') The games $v_{13} = v_{14} = 1$, $v_{12} < \frac{1}{2}$ — and no others — have doubly discriminatory solutions which discriminate player 1. These solutions are of the form $\{P | p_1 = c,\ p_2 = 0\}$ with $v_{12} \le c < \frac{1}{2}$.

The solutions of cases (b) through (e) are clearly not trivial and hence do not discriminate player 1. Unlike case (a), they are given implicitly, and dealing with them therefore presents great difficulties. In the future, let ν — if nothing is said to the contrary — be an arbitrary non-trivial solution. Then α, $\bar{\alpha}$, C, $M(\bar{\alpha}, C)$, J and K are well-defined and depend only on ν.

<u>Case (b)</u>. Let $\nu = \mathfrak{C} \cup \mathfrak{M}$ and let ν be doubly discriminatory. Let $1 - t < j_3 + k_4$ and $\mathfrak{M} = M(\bar{\alpha}, C)$. Two components (other than the first) are constant in ν, and hence also in $M(\bar{\alpha}, C)$; hence $M(\bar{\alpha}, C)$ has only one element and $J = K = M$, where M is the one imputation in $M(\bar{\alpha}, C)$: $M = (\bar{\alpha}, m_2, m_3, m_4)$. We have $\bar{\alpha} + m_k \le 1$ for $k = 2, 3, 4$ and $1 \le \bar{\alpha} + c_k$ for $k \in \{3, 4\}$ by Lemma 8 ([3], p. 372). Hence $m_k \le 1 - \bar{\alpha} \le c_k$ for $k = 3, 4$. Suppose player 3 is discriminated. Then $m_3 = 1 - \bar{\alpha}$ and $M = (\bar{\alpha}, 0, 1 - \bar{\alpha}, 0)$. Let player 2 also be discriminated. Then $1 - t < j_3 + k_4 = m_3 + m_4 = 1 - \bar{\alpha}$. Hence $\bar{\alpha} < t$ and therefore $\bar{\alpha} < 1$. Let $N = (1, 0, 0, 0)$. $N \notin R\text{-dom}\ \nu$ for $1 \in R$, because $\bar{\alpha} < n_1$. $N \notin S\text{-dom}\ \nu$ for $2 \in S$, because $p_2 = 0$ for all $P \in \nu$. $N \notin T\text{-dom}\ \nu$ for $T = \{3, 4\}$

is not effective, since, for all $p \in \nu$, $v_{34} = 1 - t < 1 - \bar{\alpha} = p_3 \leq p_3 + p_4$. Hence the assumption that player 2 was also discriminated is false. Suppose that, in addition to player 3, player 4 is also discriminated. Then $p_3 = 1 - \bar{\alpha}$, $p_4 = 0$ for all $p \in \nu$, and $N \notin \text{dom}\ \nu$. Hence it is also false to assume that both players 3 and 4 are discriminated. Because of the symmetry of players 3 and 4, it is impossible to have ν doubly discriminatory in the case $1 - t < j_3 + k_4$.

Suppose $1 - t = j_3 + k_4$ and therefore $\mathfrak{M} = \{J, K\}$. Once again two players other than player 1 are discriminated, and $M(\bar{\alpha}, C)$ has only the one element $J = K = M$. $\mathcal{C}$ is closed, hence $M \in \mathcal{L}$ and $M = D$. Let player 3 be discriminated. Then, once again, $M = (\bar{\alpha}, 0, 1 - \bar{\alpha}, 0)$. $1 - t = j_3 + k_4 = m_3 + m_4 = 1 - \bar{\alpha}$; hence $\bar{\alpha} = t$ and $M = (t, 0, 1 - t, 0)$.

Suppose $t < 1$. Then $N \notin R\text{-dom}\ \nu$ if $1 \in R$. Let player 2 also be discriminated. Then $N \notin S\text{-dom}\ \nu$ for $2 \in S$, and $N \notin T\text{-dom}\ \nu$ for $T = \{3, 4\}$, because M is the only imputation for which $\{3, 4\}$ is effective and $m_2 = 0$. Hence if player 3 is discriminated and $t < 1$, player 2 is not discriminated. Suppose that, in addition to player 3, player 4 is also discriminated. Then $N \notin R\text{-dom}\ \nu$ for $\{1, 4\} \cap R \neq \emptyset$, and $N \notin S\text{-dom}\ \nu$ for $S = \{2, 3\}$, because v_{23} is not effective on the edge we are considering. Hence we must have $t = 1$ because of the symmetry among players 3 and 4, and then $\nu = \{P | p_i = p_j = 0\}$ with $\{i, j\} = \{2, 3\}, \{2, 4\}, \{3, 4\}$.

Suppose $t = 1$, $\nu = \{P | p_2 = p_3 = 0\}$. Then $C = (0, 0, 0, 1)$; $\mathfrak{A}^*$ is the set of all imputations; $\mathfrak{U}^* = \nu$; $\alpha = 1$; $\mathfrak{V}' = \{(1, 0, 0, 0)\}$; $\mathfrak{U}' = \mathfrak{V}'$; $d_2 = 0 = d_3 = d_4$; $D = (1, 0, 0, 0)$; $\bar{\alpha} = 1$; $M(\bar{\alpha}, C) = \{(1, 0, 0, 0\}$; $J = K = D$; $\mathcal{C} = \nu$; $\nu = \mathcal{C} \cup \{J, K\}$; $j_3 + k_4 = 1 - t$. Hence ν is a solution in the sense of (b). If $t = 1$, then players 2, 3, and 4 have symmetric roles. Hence we have

> (b') In case (b), only the game $v_{12} = v_{13} = v_{14} = 1$ has doubly discriminatory solutions. More precisely, this game has the three doubly discriminatory solutions $\{P | p_i = p_j = 0\}$ for $\{i, j\} = \{2, 3\}, \{3, 4\}, \{2, 4\}$, and no others.

<u>Case (c)</u>. Suppose $\nu = \mathcal{C} \cup \mathcal{B}'$, $j_3 + k_4 < 1 - t$, $j_3 = c_3$, and let ν be doubly discriminatory. We claim: if $\mathcal{B}'$ has only one element, then

$\mathcal{B}' = \{(1 - c_3, 0, c_3, 0)\}$ and $\alpha = 1 - c_3$. Proof: α, j_2, j_3 are components of an imputation, since $0 \leq j_4 = 1 - j_1 - j_2 - j_3 = 1 - \bar{\alpha} - j_2 - j_3 \leq 1 - \alpha - j_2 - j_3 \leq 1$, and hence $(\alpha, j_2, j_3, 1 - \alpha - j_2 - j_3) \in \mathcal{B}'$. Suppose $\mathcal{B}'$ has only one element, B. Then $B = (\alpha, j_2, j_3, 1 - \alpha - j_2 - j_3)$. But $j_2 + j_3 \leq 1$; hence $(1 - j_2 - j_3, j_2, j_3, 0)$ is an imputation $\alpha < \bar{\alpha} = 1 - j_2 - j_3 - j_4 \leq 1 - j_2 - j_3$. Hence $\alpha = 1 - j_2 - j_3 = \bar{\alpha}$ and $B = (\alpha, j_2, j_3, 0)$. By hypothesis, $j_3 = c_3$. $j_2 = 1 - \bar{\alpha} - c_3 \leq 0$, since $1 - c_3 \leq \alpha$ by Lemma 8 ([3], p. 372). Hence $j_2 = 0$, $B = (1 - c_3, 0, c_3, 0)$ and $\bar{\alpha} = (1 - c_3)$, and our claim is proved.

Suppose player 4 is discriminated in ν. Then $\mathcal{B}'$ has the one element $B = (1 - c_3, 0, c_3, 0)$. $p_4 = 0$ for all $P \in \nu$. If R-dom $\nu \neq \emptyset$, then $4 \notin R$. $R \neq \{2, 3\}$, because $\{2, 3\}$ is not effective. Hence $1 \in R$ and $(1, 0, 0, 0) \in \nu$. $1 = \bar{\alpha} = \alpha \leq t$. Hence $t = 1$. By hypothesis, however, $j_3 + k_4 < 1 - t$, i.e., $t < 1$. Hence player 4 cannot be discriminated in ν. It follows that players 2 and 3 are discriminated in ν. $p_2 = j_2$ and $p_3 = c_3$ for all $P \in \nu$. Hence $\nu = \{P | p_2 = j_2, p_3 = c_3\}$, because doubly discriminatory solutions are maximal. Once again, $j_2 = 0$ and therefore $c_2 = 0$, and $\nu = \{P | p_2 = 0, p_3 = c_3\}$. Suppose $J' = (\bar{\alpha}, 0, 1 - \bar{\alpha}, 0)$, $K' = (\bar{\alpha}, 0, 0, 1 - \bar{\alpha})$. Then $j_3' = \max_{Q \in \{P | p_1 = \bar{\alpha}\}} q_3$ and $k_4' = \max_{Q \in \{p_1 = \bar{\alpha}\}} q_4$. Hence $J' = J$, $K' = K$ and $j_3 = k_4$. Besides, $j_3 + k_4 = 2c_3$.

Suppose $\nu = \{P | p_2 = 0, p_3 = c_3\}$, $2c_3 < 1 - t$, $t < 1$. Then $\min_{P \in \nu} p_1 = 0$; $\mathfrak{U}^* = \mathfrak{A}^* \cap \nu \{P | p_2 = 0, p_3 = c_3, p_1 < 1 - t\}$; $\alpha = 1 - t$; $\mathfrak{V}' = \{P | p_1 = \alpha, p \in \nu\} = \{(1 - t, 0, c_3, t - c_3)\}$; $\mathfrak{U}' = \mathfrak{V}'$; $d_2 = 0$, $d_3 = c_3$, $d_4 = t - c_3$, $d_1 = 1 - t = \alpha$, $D = (\alpha, 0, c_3, t - c_3)$. Hence ν consists of the line running from $C = (0, 0, c_3, 1 - c_3)$ to D and of the set $\{B | \alpha \leq b_1, b_2 = 0, b_3 = c_3\}$. Then $\bar{\alpha} = 1 - c_3$; $J = (1 - c_3, 0, c_3, 0)$; $K = (1 - c_3, 0, 0, c_3)$. Hence $j_2 = 0$, $j_3 = k_4 = c_3$ and $j_3 + k_4 < 1 - t$ because $2c_3 < 1 - t$. Hence ν is a solution in the sense of case (c). We have:

(c') In case (c), only the games $v_{13} = v_{14} = 1$, $v_{12} < 1$ have doubly discriminatory solutions. These games have doubly discriminatory solutions $\{P | p_2 = 0, p_3 = c\}$ with $2c < v_{34}$, and no others.

Case (d). Suppose $\nu = \mathcal{C} \cup \mathcal{B}'$, $j_3 + k_4 < 1 - t$, $j_3 < c_3$ and ν is doubly discriminatory. Then $\alpha \leq \overline{\alpha} = j$, $J \in B' \subseteq \nu$. Hence players 2 and 4 are discriminated and $\mathcal{B}'$ has only one element, B. As in case (c), $b_1 = \overline{\alpha} = \alpha$, $b_4 = 0$. Hence $C = (0, j_2, c_3, 0)$, $J = (\overline{\alpha}, j_2, j_3, 0)$. If R-dom $\nu \neq 0$, then $1 \in R$. Hence $(1, 0, 0, 0) \in \nu$ and $t = 1$. This contradicts $j_3 + k_4 < 1 - t$. Hence this case has no doubly discriminatory solutions.

Case (e).

(e') In case (e), only the games $v_{13} = v_{14} = 1$, $v_{12} < 1$, have doubly discriminatory solutions. These games have doubly discriminatory solutions $\{P | p_2 = 0, p_4 = c\}$ with $2c < v_{34}$, and no others.

We determine now from our data in § 4 the edge solutions. In case A_1, exactly three of the equations $v_{ij} = v_{jm} + v_{jn} - 1 = 0$ can be satisfied with $v_{13} = v_{14} = 1$, namely: $v_{24} = v_{23} + v_{12} - 1 = 0$, $v_{23} = v_{34} + v_{13} - 1 = 0$, and $v_{34} = v_{24} + v_{14} - 1 = 0$. Hence, necessarily, $v_{34} = 0$. In case A_1, hence, we have the game $v_{13} = v_{14} = 1$, $v_{34} = 0$, and, for the solutions, the constants:

$$(36) \qquad \left\{ \begin{array}{l} v_{34} = c_2 = c_3 = 0 \\ v_{34} = c_2 = c_4 = 0 \\ v_{34} = c_3 = c_4 = 0 \,. \end{array} \right.$$

This corresponds to (b').

In case A_2 [resp. A_3], exactly three of the inequalities of the form $2v_{ij} < v_{jm} + v_{jn} - 1$ can be satisfied with: $v_{13} = v_{14} = 1$, namely: $2v_{12} < 1$ for $i = 2$, $j = 1$; $0 < v_{34}$ for $i = 2$, $j = 3$; $0 < v_{34}$ for $i = 2$, $j = 4$. Hence we obtain the edge solutions:

$$(37) \qquad \left\{ \begin{array}{ll} v_{12} \leq c_1 < 1/2, & c_2 = 0 \\ 0 \leq c_3 < v_{34}/2, & c_2 = 0 \\ 0 \leq c_4 < v_{34}/2, & c_2 = 0 \,. \end{array} \right.$$

This corresponds to (a'), (c') and (e'). Hence our edge solutions agree completely with Mills' doubly discriminatory solutions.

BIBLIOGRAPHY

[1] BURGER, E., Einführung in die Theorie der Spiele, Berlin 1959.

[2] LUCE, R. D. and RAIFFA, H., Games and Decisions, New York 1957.

[3] MILLS, W. H., "The four person game — edge of the cube," Annals of Mathematics, 59 (1954), pp. 367-378.

[4] SHAPLEY, L. S., "Quota solutions of n-person games," Annals of Mathematics Study No. 28 (Princeton, 1953), pp. 343-359.

[5] SHAPLEY, L. S., "Notes on n-person games — VI: On solutions that exclude one or more players," The RAND Corporation, Research Memorandum RM-2533, March 3, 1960.

[6] VON NEUMANN, J. and MORGENSTERN, O., Theory of Games and Economic Behavior, Princeton 1944, 3rd ed. 1953.

Michael H. Hebert

Leverkusen, Germany

THREE-PERSON COOPERATIVE GAMES WITHOUT SIDE PAYMENTS

R. E. Stearns

§ 1. INTRODUCTION

A good deal of insight about the classical n-person game theory of von Neumann and Morgenstern can be obtained by studying the three-person case. An exhaustive study of the solutions to these three-person games is contained in [1] (pps. 282-288, 403-419, and 550-555) and this provides a rich source of examples. The object of this paper is to make the same sort of study for three-person cooperative games without side payments, as defined by Aumann and Peleg [2]. The results in this paper should be sufficient to enable someone to find all the solutions to a particular three person (Aumann-Peleg) game. There will always be at least one solution.

The attractions of this model are discussed in detail in [2] and [3], so we shall only mention them briefly. The first is that it is a generalization of the classical theory. This is because the rules of the game might allow unrestricted side payments. The second is that it is based upon an axiomatic treatment of the characteristic function which is broad enough to include several notions of effectiveness. We shall illustrate this point after making some definitions. The formal theory may be compared with the development of the classical solution concept from the notion of a superadditive set function.

The main results of this paper are contained in Sections 6 and 7. The definitions needed to understand these results are in Sections 2 and 4. It may be helpful at some point for the reader to take an advance look at the figures, which are explained at the beginning of Section 7. In way of a summary, a method of determining all solutions is described at the end of Section 7. Finally, in Section 8 these results are related to extended games.

Many of the proofs in Sections 5, 6, and 7 are too long to be included in full, and many of the fine details are indicated without proof. The techniques used are given, however, and the interested reader should be able to supply the details. Two-dimensional diagrams similar to the figures are most helpful in this respect.

Section 3 is self-contained, and is of some interest in itself, as it applies to n-person games, $n \geq 3$, and more generally to compact convex polyhedrons.

§ 2. NOTATION AND DEFINITIONS

Let N be the set of players and n the number of players. Let E^N be an n-dimensional Euclidean space whose coordinates are indexed by N. The points in E^N are called <u>payoff vectors</u>, but we shall generally use the less descriptive word <u>point</u>. The components of these vectors will be indexed with superscripts; e.g., $x \in E^N$, $x = (x^1, x^2, \ldots, x^n)$. We write $x >_S y$ to represent $x^i > y^i$ for all $i \in S$. Similarly, we define $x \geq_S y$ and $x =_S y$. When $S = N$, we write simply $x > y$ or $x \geq y$.

When referring to three-person games, we let $N = \{1, 2, 3\}$. <u>When the symbols i, j and k are used to represent elements of this set, it is assumed that they represent distinct elements.</u> We abbreviate $v(\{i, j\})$ with $v\{ij\}$.

DEFINITION 2.1. An <u>n-person characteristic function</u> is a set N with n members, together with a function v that carries each subset S of N into a subset $v(S)$ of E^N so that:

(1) $v(S)$ is closed;

(2) $v(S)$ is comprehensive; i.e., $x \in v(S)$, $y \in E^N$, and $y \leq_S x$ imply $y \in v(S)$;

(3) $v(S)$ is convex;

(4) v is super-additive; i.e., if S and T are disjoint subsets of N, then $v(S \cup T) \supset v(S) \cap v(T)$;

(5) $v(\emptyset) = E^N$.

Intuitively, v is a function that maps a coalition S onto the set of payoff vectors for which S is "effective." Two notions of effectiveness that are known to determine such a characteristic function ([2] and [3]) are defined as follows:

(1) A coalition S is said to be <u>α-effective</u> for $x \in E^N$ if and only if there exists a (correlated mixed) strategy for S that assures to each member i of S a payoff of at least x^i against any strategy of $N - S$.

(2) A coalition S is <u>β-effective</u> for $x \in E^N$ if and only if, for each strategy of $N - S$, there is a strategy for S that yields to each member i of S a payoff of at least x^i.

β-effectiveness seems appropriate for repeated play situations because of the Blackwell approachability-excludability theory [4] which says that S is β-effective for x if and only if S can ensure that, as the game is repeated in time, the average payoff to every i of S approaches a value greater or equal to x^i. The relation of β-effectiveness to the concepts associated with a supergame is discussed in detail in Section 10 of [3]. These notions of effectiveness are given here to provide motivation and are not used in the sequel.

DEFINITION 2.2. An <u>n-person game</u> is an n-person characteristic function (N, v) together with a compact convex polyhedral subset H of E^N such that a payoff vector x is in $v(N)$ if and only if there is a payoff vector y in H such that $y \geq x$.

Intuitively, H is the convex hull of the finite set of possible outcomes of the game. By cooperating, the set of all players can achieve any point in this set or any convex combination of these points. In the definition of a game given in [2], the condition on H is that $H \subset v(N)$. This is a weaker condition than 2.2 and is the actual condition used in this paper. Thus the results here hold for either definition. 2.2 is the definition from [3].

To see that the classical theory is a special case of this theory, let $\bar{v}$ be the von Neumann-Morgenstern characteristic function and $\bar{A}$ the set of individually rational points for which N is effective. Take $v(S) = \{x \in E^N \mid \Sigma_S x^i \leq \bar{v}(S)\}$ and $H = \bar{A}$, and the classical game fits the definition from [2].

Definition 2.2 is actually the definition of an "ordinary game," as given in [3], whereas the term "game" is applied in [3] to what is called an "extended game" in this paper (Definition 8.1), and in [2]. In the classical theory, the extended games play an important role in the composition theory, whereas they have no part in the composition of games without side payments (see [2]). In the extended classical theory, the A-stable sets are simply related to the $\bar{A}$-stable sets, but we shall see in Section 8 that this relationship does not extend to the new theory for $n \geq 4$. Thus, at the present time, the ordinary game seems to be the more important concept.

DEFINITION 2.3. Suppose $y \in v(S)$. If $y >_S x$, we say that y <u>dominates</u> x <u>via</u> S and write $y \text{ Dom}_S\, x$. We write y Dom x to indicate $y \text{ Dom}_S\, x$ for some S such that $y \in v(S)$ and Dom K to indicate the set of x in E^N such that y Dom x for some y in K.

DEFINITION 2.4. If $K \subset P \subset E^N$, K is said to be <u>P-stable</u> if and only if $K = P - \text{Dom } K$. The set $P - \text{Dom } P$ is called the <u>P-core</u>. Clearly if P is compact, so are the P-stable sets and the P-core.

It is easy to show that there is a number v_i (real, ∞, or $-\infty$) such that $v(\{i\}) = \{x \in E^N \mid x^i \leq v_i\}$. This leads to the following:

DEFINITION 2.5. A point x is called <u>individually rational</u> if $x^i \geq v_i$ for all $i \in N$. The set of points in H that are individually rational is denoted by $\bar{A}$ and is also a polyhedron. The point x is called <u>group rational</u> if there are no $y \in H$ such that $y > x$. The set of points in $\bar{A}$ which are group rational is denoted by A, and the A-core is denoted by C.

It is known that the A-core is equal to the $\bar{A}$-core. The proof of this will be given informally in the next section. (See also [3].)

THEOREM 2.6 ([2] and [6]). A set V is A-stable if and only if it is $\bar{A}$-stable.

PROOF. If $x \in H$, let $f(x)$ be a group rational point in H such that $f(x) > x$ or $f(x) = x$. Clearly if $x \in \bar{A}$, then $f(x) \in A$. If $x \in \bar{A} - A$, then x is dominated by a member of the A-stable set V, either by $f(x)$ via N or by the point in V which dominates $f(x)$. Conversely, x cannot be in any $\bar{A}$-stable set V because then $f(x)$ would not be in either V (because $f(x)\ \mathrm{Dom}_N\ x$) or in Dom V (because $y\ \mathrm{Dom}_S\ f(x)$ implies $y\ \mathrm{Dom}_S\ x$).

DEFINITION 2.7. A set V is a solution to the game $G = (N, v, H)$ if and only if V is A-stable.

§ 3. A MAPPING THEOREM

The central theorem (3.3) of this section is a mapping theorem which enables us to come to grips with some special non-convex subsets of $\bar{A}$ (such as the set A). After the theorem, we shall discuss three useful partial orderings that satisfy the conditions of this theorem. Considerable use will be made of the polyhedral nature of H.

DEFINITION 3.1. Let R be a partial ordering on the points of E^N, and let $K = \{(x \times y) \in E^N \times E^N \mid xRy\}$. R is said to be closed if and only if K is closed, and R is convex if and only if K is convex. If $H \subset E^N$, let $A(R, H)$ be the set of points in H that are maximal in R; i.e., $x \in H$, $y \in A(R, H)$, and xRy imply $x = y$.

LEMMA 3.2. If $H \subset E^N$ is a compact convex polyhedron and R is a convex partial ordering, then $A(R, H)$ is the union of closed faces of H.

PROOF. If $x \in A(R, H)$ is a vertex, then z is a closed face; otherwise z must be in the relative interior of some closed face F. If F is not contained in $A(R, H)$, there are points x and y in H such that $x \in F$, yRx, and $y \neq x$. We may write $z = \pi x + (1 - \pi)z'$ for some $z' \in F$ and $0 < \pi \leq 1$. But since $z'Rz'$ and yRx, $[\pi y + (1 - \pi)z']Rz$ by convexity, contradicting $z \in A(R, H)$.

THEOREM 3.3. If H is a compact convex polyhedral subset of E^N, if R is a closed convex partial ordering, if x and y are points in H, and if L_{xy} is the line segment between x and y, then there exists a continuous mapping f of L_{xy} into $A(R, H)$ such that $f(z)Rz$ for all z in L_{xy}.

PROOF. For each closed face F of $A(R, H)$, let

$$C_{xy}^{F} = \left\{z \in L_{xy} \mid \text{ there is a } \bar{z} \in F \text{ such that } \bar{z}Rz\right\}.$$

Gillies' Lemma 7 (p. 55 of [7]) can be used to show that each z in L_{xy} belongs to some C_{xy}^{F}. To see that it also belongs to a nontrivial C_{xy}^{F} (when $x \neq y$), consider a sequence of distinct points in L_{xy} converging to z. An infinite number of these must belong to the same C_{xy}^{F} which is, therefore, nontrivial and which contains z because R is closed.

Let $K(x, y)$ be the number of nontrivial C_{xy}^{F}. If $K(x, y) = 0$, then $x = y$, and clearly there is a map of $\{x\}$ into $A(R, H)$ satisfying the theorem. Now assume that the theorem is true whenever $K(x, y) \leq m$, and suppose that $K(a, b) = m + 1$ for some a and b in H. Let F' be a face such that $C_{ab}^{F'}$ is nontrivial and contains a. $C_{ab}^{F'}$ is a closed set because R is closed. Let c be the point in $C_{ab}^{F'}$ closest to b. Let $\bar{a}$ and $\bar{c}$ be points in F' such that $\bar{a}Ra$ and $\bar{c}Rc$. Let $f(\pi a + (1 - \pi)c) = \pi\bar{a} + (1 - \pi)\bar{c}$ for $1 \geq \pi \geq 0$. Then $f(z)Rz$ by convexity, and we are finished if $b = c$. If $b \neq c$, let $g(\pi b + (1 - \pi)c) = \pi b + (1 - \pi)\bar{c}$.

Clearly $K(c, b) \leq m$ since $C^{F'}_{cb}$ is trivial, and $K(\overline{c}, b) \leq K(c, b)$ because $g(z)Rz$ by convexity and hence g^{-1} maps nontrivial C^{F}_{cb} onto nontrivial C^{F}_{cb}. By the induction hypothesis, there is a mapping h of $L_{\overline{c}b}$ into $A(R, H)$ satisfying the theorem. Since $h(g(c)) = h(\overline{c}) = \overline{c} = f(c)$, f can be extended over L_{ab} by setting $f(z) = h(g(z))$ for $z \in L_{cb}$. Since $h(g(z))Rg(z)Rz$, we have $f(x)Rz$, and the theorem is proved.

The relation " $\geq$ " is obviously a closed convex partial ordering. We write $|x - y| < \varepsilon$ to mean $|x^i - y^i| < \varepsilon$ for all $i \in N$. An important consequence of the previous theorem is the following:

LEMMA 3.4. If a is a point in $A^* = A(\geq, \overline{A})$, if b is a point in $\overline{A}$, and if ε is a positive number, then there exists a point c in A^* such that $|c - a| < \varepsilon$ and, for any $S \subset N$,

(i) $b \geq_S a$ implies $c \geq_S a$,

(ii) $b >_S a$ implies $c >_S a$,

(iii) $b \operatorname{Dom}_S a$ implies $c \operatorname{Dom}_S a$,

(iv) $a \notin v(S)$ implies $c \notin v(S)$, and

(v) $a \geq_S b$ implies $c \geq_S b$.

PROOF. By 3.3, there is a continuous mapping f of L_{ab} into A^* such that $f(z) \geq z$. Since $a \in A^*$, $f(a) = a$. For any $z \in L_{ab}$ such that $z \neq a$, $b \geq_S a$ implies $a \leq_S z \leq f(z)$ and $b >_S a$ implies $a <_S z \leq f(z)$, and hence (i) and (ii) hold if we take $c = f(z)$. Similarly, (v) holds. Because f is continuous, we can ensure $|c - a| < \varepsilon$ and (iv): $f(z) \in E^N - v(S)$, (an open set) whenever $a \notin v(S)$, by taking z close enough to a. The additional constraint $\varepsilon < \min(b^i - a^i)$ for $b^i > a^i$ ensures that $b \operatorname{Dom}_S a$ implies $b \operatorname{Dom}_S c$, and hence (iii).

In order to work with the set A, we make the following:

DEFINITION 3.5. If m is a number greater than one and x and y are in E^N, we write $xR_m y$ if and only if $m\cdot(x^i - y^i) \geq (x^j - y^j)$ for all i and j in N.

Clearly xR_my implies $x > y$ or $x = y$, and R_m is a closed convex partial ordering, the set K of 3.1 being the convex hull of n^2 linear inequalities.

LEMMA 3.6 (Aumann [3]). There is a large number M such that $A(R_M,\bar{A}) = A$.

PROOF. Clearly $A \subset A(R_m,\bar{A})$ for all m. For each open face F of $\bar{A}$ such that there is an $x \in F$ and $y \in \bar{A}$ such that $y > x$, there is a constant M_F (for example $M_F = \mathrm{Max}_{N\times N}(y^j - x^j)/(y^i - x^i))$ such that $F \cap A(R_m,\bar{A}) = \emptyset$ for all $m \geq M_F$. (Remember $A(R_m,\bar{A})$ is a union of closed faces.) Choose $M \geq M_F$ for all open F of $\bar{A}$, a finite set.

LEMMA 3.7. If a is a point in A, b is a point in $\bar{A}$, and ϵ is a positive number, then there exists a point c in A such that $|c - a| < \epsilon$ and, for any $S \subset N$,

(i) $b \geq_S a$ implies $c \geq_S a$,

(ii) $b >_S a$ implies $c >_S a$,

(iii) $b \ \mathrm{Dom}_S \ a$ implies $c \ \mathrm{Dom}_S \ a$,

(iv) $a \notin v(S)$ implies $c \notin v(S)$, and

(v) $a \geq_S b$ implies $c \geq_S b$.

PROOF. The proof of 3.4 applies word for word, using A for A^* and R_M for $\geq$.

This lemma provides a useful tool for proving theorems about A. For example, part (iii) proves that $A \cap \mathrm{Dom}\, A = A \cap \mathrm{Dom}\, \bar{A}$. Hence the equality of the A-core and the $\bar{A}$-core follows from $\bar{A} - A \subset \mathrm{Dom}_N\, A$.

We wish to give one final relation.

DEFINITION 3.8. If S and $T = N - S$ are nonempty subsets of N and m is a number greater than zero, then we write xR_m^Sy if and only if $x \geq y$ and $m \sum_T (x^j - y^j) \geq (x^i - y^i)$ for each i in S.

R_m^S is seen to be a closed convex partial ordering such that xR_m^Sy and $x =_T y$ imply $x = y$.

LEMMA 3.9. Let S be a set such that $\emptyset \neq S \neq N$ and let $B = \{x \in A \mid y \in A$ and $y \geq x$ imply $y =_{N-S} x\}$. For all M large enough, $A(R_M^S, \bar{A}) = B$.

PROOF. Each undesirable open face of $\bar{A}$ is excluded by choosing M larger than some M_F.

Note: In [5], two relations are used based on the equations $(x^i - y^i) \geq -M(x^j - y^j)$ and $M(x^j - y^j) \geq (x^i - y^i)$ for $i \in N$ and $j \in S$. For large M, these relations and 3.3 enable one to find arcs along special edges of A^* and A.

§ 4. SPECIAL DEFINITIONS

This section gives all the definitions used in the theorems of Sections 6 and 7 which characterize the solutions to three-person games. These definitions are special, in that most are meaningful only for three-person games.

DEFINITION 4.1. Let $A^* = A(\geq, \bar{A}) = \{x \in \bar{A} \mid y \in \bar{A}$ and $y \geq x$ imply $y = x\}$; the set of points maximal with respect to "$\geq$". Let $A^i = A(R_M^{\{i\}}, \bar{A}) = \{x \in \bar{A} \mid y \in \bar{A}$ and $y \geq x$ imply $y =_{\{jk\}} x\}$ where M is defined in 3.9.

DEFINITION 4.2. For $S \subset N$, let $W(S) = \{x \in A \cap v(S) \mid$ there do not exist any $y \in \bar{A}$ such that $y \operatorname{Dom}_S x\}$.

DEFINITION 4.3. Let $b_j^i = \text{l.u.b. } x^i$ for $x \in v\{ij\} \cap \bar{A}$. If $v\{ij\} \cap \bar{A} = \emptyset$, we take $b_j^i = -\infty$.

DEFINITION 4.4. Let $Q_i = \left\{z \in A^i \cap v\{ij\} \cap v\{ik\} \mid z^i = b_j^i \text{ and } z^i = b_k^i\right\}$.

DEFINITION 4.5. Let $Z = A^* \cap v\{1,2\} \cap v\{1,3\} \cap v\{2,3\}$.

DEFINITION 4.6. $B \subset Z$ shall be called <u>class II</u> if and only if there is an $i \in N$ such that Q_i is nontrivial, Q_i contains B, and $Q_i \cap (B \cup C) \neq \emptyset$; otherwise B shall be called <u>class I</u>. (Recall that C is the A-core.)

DEFINITION 4.7. The point a in E^N is said to <u>define a triangle</u> if and only if, for each $i \in N$, there is a $p_i \in A^*$ such that $p_i =_{\{jk\}} a$ and $p_i^i \leq a^i$. We write $T_a = \{p_1, p_2, p_3\}$.

DEFINITION 4.8. Suppose E is a compact subset of A^i such that x and y in E imply $x^i = y^i$, and suppose e_j is the point in E with maximal k^{th} coordinate and e_k the point with maximal j^{th}. Then define $t_i(E) = \{x \in A^* \mid x^j = e_k^j \text{ and } x^k = e_j^k\}$. (Because of 4.1, $t_i(E)$ can have at most one point.)

DEFINITION 4.9. If p is a point in A and i a player in N, let $D_i(p) = \{x \in A \mid x \geq_{\{jk\}} p \text{ and } x^i \leq p^i\}$.

DEFINITION 4.10. If E is as in 4.8, let $D_i(E) = \{x \in A \mid x^j \geq e_k^j \text{ and } x^k \geq e_j^k\}$.

§ 5. SPECIAL LEMMAS

In this section we present the basic lemmas required for the later theorems. Except for 5.1, it is assumed that we have a three-person game.

LEMMA 5.1. The sets A, A^*, A^i, and $w(S)$ for $S \subset N$ are compact.

PROOF. A, A^*, and A^i are compact by 3.2, 3.6, and 3.9. A sequence of points in $w(S)$ must converge to a point in $A \cap v(S)$ and this limit cannot be dominated via S because part of the sequence would also be dominated.

LEMMA 5.2. If x is a point in $A \cap v\{ij\}$ and y is a point in $w\{ij\}$ such that $x^i = y^i$ and $x^j > y^j$, then x is in $w\{ij\}$ and $x^i = y^i = b^i_j$.

PROOF. Suppose z is a point in $\bar{A} \cap v\{ij\}$ such that $z^i > x^i$. Then there is a point $\bar{z}$ in L_{xz} near x such that $\bar{z} >_{\{ij\}} y$. But $\bar{z} \in v\{ij\}$ because $v\{ij\}$ is convex and this contradicts $y \in w\{ij\}$. The result follows.

LEMMA 5.3. If x_1 and x_2 in A^* (or in A) and x_3 in $\bar{A}$ are points such that $x^j_3 \geq x^j_1 \geq x^j_2$ and $x^k_3 \geq x^k_2 \geq x^k_1$, then there exists a point z in A^* (in A) such that $z^j = x^j_1$ and $z^k = x^k_2$.

PROOF. If $x^j_1 = x^j_2$ or $x^k_1 = x^k_2$, the lemma is immediate with $z = x_2$ or $z = x_1$. Otherwise, let y be a point in $B = \{p \in A^* \mid p^j \geq x^j_1$ and $p^k \geq x^k_2\}$ with maximal i^{th} coordinate. (This set contains a point $c \geq x_3$ and is compact.) Suppose $y^i \geq \min(x^i_1, x^i_2)$. Let $\pi_1 = (y^k - x^k_2)/(x^k_2 - x^k_1)$, $\pi_2 = (y^j - x^j_1)/(x^j_1 - x^j_2)$, $\pi_3 = (1 + \pi_1 + \pi_2)^{-1}$, and $z = \pi_1\pi_3x_1 + \pi_2\pi_3x_2 + \pi_3y$. Clearly $z \in \bar{A}$ (by convexity), $z^j = x^j_1$, $z^k = x^k_2$, and $z^i \geq \min(x^i_1, x^i_2)$. Thus $z \geq x_1$ or $z \geq x_2$ and z is in A^* or A depending on whether x_1 and x_2 are in A^* or A. Finally, if $y^i < \min(x^i_1, x^i_2)$, we must have $y^j = x^j_1$ for otherwise 3.4 (or 3.7) using $a = y$, $b = x_2$ and $\epsilon < y^j - x^j_1$ gives a point c in B with $c^j > x^j_1$, $c^k \geq x^k_2$ (by 3.4(v) with $S = \{k\}$) and $c^i > y^i$ (by 3.4(ii) with $S = \{i\}$), contrary to the choice of y. Similarly $y^k = x^k_2$; and set $y = z$.

LEMMA 5.4. Suppose the set E of 4.8 is nontrivial. If $D_i(E) \neq \emptyset$ (see 4.10), then there is a point p in $t_i(E)$ such that $D_i(E) = D_i(p)$.

PROOF. Let x_3 be a point in $D_i(E)$. Equality $e^j_j = e^j_k$ obviously implies that E is trivial; therefore, $x^j_3 \geq e^j_k > e^j_j$. The point $e_j \in A^i$

must be in A^* for if $y \in A^*$, $y \geq e_j$, and $y^i > e_j^i$, then 3.4 gives a point $c \in A^*$ with $c^j > e_j^j$ and $c \geq e_j$ (use $a = y$, $b = x_3$, $\epsilon < y^i - e_j^i$). Similarly for e_k, and p exists by 5.3 with $x_1 = e_k$ and $x_2 = e_j$.

Next we give two lemmas which deal with the problem of existence. The details of the proof of 5.6 are especially uninteresting and some of the smaller steps are indicated without proof.

LEMMA 5.5. Suppose p is a point in $A^i \cap v\{ij\} \cap v\{ik\}$ such that $D_i(p) \cap v\{jk\} \subset w\{jk\}$, and let J be the set of all $x_i \in A$ such that:

(i) there exist points x_j and x_k in $D_i(p) \cap v\{ij\} \cap v\{ik\}$ such that $x_j^k = x_i^k$, $x_k^j = x_i^j$, and $x_j^i = x_k^i$;

(ii) $x_j \in w\{ik\}$ or $x_j^j = p^j$; and

(iii) $x_k \in w\{ij\}$ or $x_k^k = p^k$.

Then $W = J \cup (C \cap D_i(p))$ is a $D_i(p)$-stable set.

PROOF. (1) Let J_j be the set of $x_i \in A$ satisfying conditions (i) and (ii), and let J_k be the set satisfying (i) and (iii). We shall show that each x_i in $v\{ij\} \cap v\{ik\} \cap D_i(p)$ is also in $J_j \cup J_k$. Let x_j and x_k be points satisfying (i) which have maximal i^{th} coordinate. This is possible since $x_j = x_i$ and $x_k = x_i$ satisfy (i) and any limits of such points again satisfy (i). Now suppose that neither (ii) nor (iii) are satisfied. Then by 3.7, there is a point $y_j \in D_i(p)$ such that $y_j \operatorname{Dom}_{\{ik\}} x_j$ (take $\epsilon < x_j^j - p^j$) and a point $y_k \in D_i(p)$ such that $y_k \operatorname{Dom}_{\{ij\}} x_k$. By symmetry we may assume $y_j^i \geq y_k^i$. By 5.3 (with $x_1 = y_k$, $x_2 = x_i$, $x_3 = y_j$, and i and j interchanged), there is a point z_j with $z^k = x_i^k$ and $z^i = y_k^i$. Again by 5.3, there is a z_k with $z_k^j = x^j$ and $z_k^i = z_j^i > x_k^i$. The points z_j and z_k are $\{ij\}$ and $\{ik\}$ comprehended by y_k and y_j, and it is seen that $z_j^j \geq y_j^j$ and $z_k^k \geq y_k^k$ because y_j and y_k are in A. Thus, z_j and z_k contradict the choice of x_j and x_k, and (1) is proved.

(2) We wish to show $J = J_j \cap J_k$. Let x_i be in J_j because of y_j and y_k, and in J_k because of z_j and z_k. If $y_j^i = z_k^i$, take $x_j = y_j$ and $x_k = z_k$. Otherwise, we may assume $y_j^i > z_k^i$ and take $x_j = y_j$ and $x_k = y_k$; y_k being in $w\{ij\}$ by 5.2.

(3) If x is not in C, we may assume without loss of generality that it is in $v\{ij\} - v\{ik\}$ or in J_j. Let $a \in w\{ij\}$ be a point such that $a \operatorname{Dom}_{\{ij\}} x$ and let $f: L_{ax} \to A$ be a continuous map such that $f(y)R_M y$ for $y \in L_{ax}$ (see 3.6 and 3.3). Let b be the first point in $f(L_{ax})$, as y runs from x to a, such that $f(y) \in w\{ij\}$. If $b \in D_i(p)$, let $b = c$; otherwise, let c be the first point in $f(L_{ax})$ such that $f^k(y) = p^k$. If $c \notin v\{ik\}$, then $c^k > p^k$ (otherwise $p \operatorname{Comp}_{\{ik\}} c$); hence $c \in C$ and $c \operatorname{Dom}_{\{ik\}} x$. Otherwise, $c \in J_k$ and one sees that, because J_k and J_j are closed, that some $d \in f(L_{ax})$ between c and x must be in $J_k \cap J_j = J$ and $d \operatorname{Dom}_{\{ij\}} x$.

(4) It remains to be shown that W is self-consistent. If it is not, some $x_i \in J$ is dominated by a point $y \in W$, say via $\{ij\}$. We cannot have $y \in C$, for then y, not dominating x_k via $\{ij\}$ nor being dominated by x_j via $\{ik\}$, must satisfy $y^i = x_k^i$; but then $y^i = b_j^i$ by 4.2 $(y^j > x_i^j = x_k^j)$ and $y = x_k = p \in v\{ij\}$, a contradiction. If $y \in J$, and y_j and y_k are associated points, we cannot have $y_k^i > x_k^i$ because then $y_k \operatorname{Dom}_{\{ij\}} x_k$, cannot have $y_k^i < x_k^i$ because then $x_j \operatorname{Dom}_{\{ik\}} y_j$, and cannot have $y_k^i = x_k^i$ because then $x_k^i = b_j^i$ by 5.2, contradicting $p \in v\{ij\}$. Thus (4) and the lemma are proved.

LEMMA 5.6. If G is a game such that $t_\ell(Q_\ell) \cap Z \neq \emptyset$ for all nontrivial Q_ℓ; then either there is a triangle T_a such that $p_i \in w\{jk\}$, $p_k \in w\{ij\}$, and $p_j \in w\{ik\}$, or else $Z \subset \operatorname{Dom} C$ and either $v\{ij\} \cap v\{ik\} \cap A^i \subset \operatorname{Dom} C$ or there is a point p_i in $A^i \cap C \cap v\{ij\} \cap v\{ik\}$ such that $v\{ij\} \cap v\{ik\} \cap A^i \subset D_i(p_i) \cup \operatorname{Dom} C$.

PROOF. (1) Let B be the set of points in A^i which are not dominated by $\bar{A}$ via $\{ij\}$ or $\{ik\}$. Let J be the set of points x_i in A^i such that there exist points x_j and x_k in A^i satisfying $x_j^i = x_k^i \geq x_i^i$, $x_j^k = x_i^k$, $x_k^j = x_i^j$, $x_j \in w\{ik\}$, and $x_k \in w\{ij\}$. Let $W = J \cup B$. W is obviously compact and, applying the method of the previous proof with $R_M^{\{i\}}$ (see 3.9) instead of R_M, we see that each point of $A^i - W$ is dominated by W via $\{ij\}$ or $\{ik\}$.

(2) Next we wish to show that, if $W \cap v\{jk\} \neq \emptyset$, there is a point

y in $W \cap w\{jk\}$. Let Y be the set of points in $W \cap v\{jk\}$ with minimal i^{th} coordinate. If Y contains a point in $A^i - A^*$, then we let this point be y for clearly $y \in w\{jk\}$. Hence we may assume $Y \subset A^*$.

Now assume that Y contains a single point y. If y is not in $w\{jk\}$, we take a sequence of points in $A^* \cap v\{jk\}$ larger than y over $\{jk\}$ which converges to y, and consider the corresponding sequence of points $z(m)$ in W which dominate these points via $\{ij\}$ or $\{ik\}$. It is easily verified that $z^i(m) < y^i$ for each m. (It is necessary to observe that, if $y \in J$ and x is another point in J, then $x^j > y^j$ and $y^k > x^k$, or $x^k > y^k$ and $y^j > x^j$, imply, because of the associated points, that Q_i is nontrivial and there is a point p in $W \cap v\{jk\} \cap t_i(Q_i)$ with $p^i < y^i$, contradicting the choice of y.) Thus the $z(m)$ must converge to y, and some of them must be in $v\{jk\}$, contradicting the choice of y.

If Y contains several points, the same argument works with the proper choice of y. It is seen that these points are contained in B, and we may choose y to be either the point in $Y \cap w\{ij\}$ with maximal k^{th} coordinate, the point in $Y \cap w\{ik\}$ with maximal j^{th} coordinate, or any point not in $w\{ij\} \cup w\{ik\}$. If y is the point with maximal k^{th} coordinate, the only difficult case is when there is a point $z \in B$ with $z^i = y^i$ and $z^k > y^k$ and a point x such that $x \operatorname{Dom}_{\{jk\}} y$; but this is impossible because a point with larger k^{th} coordinate can then be found in $Y \cap w\{ij\}$ by applying 5.3 to z, points of A in an image $f(L_{xz})$, and y.

(3) Suppose that $y \in J$. Let y_j and y_k be the associated points. If y_j and y_k are in A^*, let y_i be a point in $t_i(\{y_j, y_k\})$ (nonempty by 5.4); clearly $y_i =_{\{jk\}} y$ and $y_i \in w\{jk\}$. Thus $T_a = \{y_i, y_j, y_k\}$ where $a = (y_j^i, y^j, y^k)$ satisfies the lemma. If either y_j or y_k is not in A^*, it is seen that $y_j = y_k \geq y$. We then let p be the point in $A^i \cap v\{ij\} \cap v\{ik\}$ greater than y_j with largest i^{th} coordinate. Clearly $p \in C \cap w\{jk\}$. If $p \in A^*$, then T_p satisfies the lemma. Otherwise we carry out (4) on p.

(4) Now we suppose that either p is defined as in (3) or else $y \in B-J$, in which case we let $p = y$. The first step is to prove that $x \in v\{ij\} \cap v\{ik\} \cap A^i$ and $x^i > p^i$ imply $x \in \operatorname{Dom} C$. Clearly $p \geq_{\{jk\}} x$ and unless $x^j = p^j$ or $x^k = p^k$, $p \operatorname{Dom}_{\{jk\}} x$. (Of course $p \in C$.) Suppose $x^j = p^j$. Clearly $p^j = b_i^j$ and $x \in A^*$. If also $p^j = b_k^j$, then x and $\overline{p}$

are in nontrivial Q_j where $\overline{p}$ is the point in A^* greater or equal to p, and the point in $t_j(Q_j)$ $\{ik\}$-comprehends $\overline{p}$ contrary to the choice of p. If $p^j \neq b_k^j$, then the technique developed in (5) (applied to $\{jk\}$ instead of $\{ij\}$) proves $x \in \text{Dom } C$ via $\{jk\}$.

(5) Suppose $x \in A^i \cap (v\{ij\} - w\{ij\})$ is a point such that $x^i = p^i$. There must be a point $b \in w\{ij\} \cap A^i$ such that $b \text{ Dom}_{\{ij\}} x$. For each integer m, there must be a point $z(m)$ in $\left\{z \in A^i \cap w\{ij\} \mid z^i \geq p^i + \left(\frac{1}{2}\right)^m (b^i - p^i)\right\}$ such that the ith coordinate is minimized, the jth maximized, and $z(m) \text{ Dom}_{\{ij\}} x$. Let $z = \lim z(m)$. If $z^i > p^i$, z cannot be dominated by a point c via $\{jk\}$ because we could assume that $c^i > z^i$ and $c \in A^i$, use an image $f(L_{xc})$ in A^i to find a point s in $v\{ij\}$ such that $s^j = z^j$ and $s^i < z^i$, and then find a $z(m) \geq_{\{ij\}} s$ such that $z^i(m) < z^i$ contradicting $z = \lim z(m)$. Thus $z \in C$ and $x \in \text{Dom } C$. If $z^i = x^i$, then an examination of various cases proves that some $z(m)$ must be in C. Again we have shown $x \in \text{Dom } C$.

(6) Suppose that x is a point in $A^i \cap v\{ij\} \cap v\{ik\}$ such that $x^i < p^i$. We shall show that either $x \in \text{Dom } C$, or $x \in J$, or x is dominated by a point c in J such that $c \in v\{jk\}$ implies $c \in w\{jk\}$. We use an argument similar to that of the previous lemma, with $R_M^{\{i\}}$ instead of R_M. When applying step (3) of 5.5, we observe from (5) that a point z such that $z \text{ Dom}_{\{ij\}} x$ and $z^i = p^i$ must be dominated by C via $\{ij\}$, or must itself be in C, or must be in J_k. Thus if x is not in Dom C, our construction leads to a point c in J which dominates x and is such that $c^i \leq p^i$, and (6) follows.

(7) The lemma now follows. If there is no T_a, we look for a point maximal in $J \cap v\{ij\} \cap v\{ik\} \cap A^i - \text{Dom } C$. If this set is nonempty, there must be such points and they are not $\{jk\}$-dominated. There can only be one such point p_i for otherwise Q_i is nontrivial, etc. It is easily seen that p_i must be equal to its associated points and is hence in C. By (6), it must also be the case that $v\{ij\} \cap v\{ik\} \cap A^i \subset \text{Dom } C \cup D_i(p_i)$. Now $Z \cap C$ is empty (any point in this set defines a triangle) and so $Z \cap D_i(p_i)$ $(p_i \in C)$ is also empty; hence $Z \subset \text{Dom } C$ (remember $Z \subset v\{ij\} \cap v\{ik\} \cap A^*$). Q.E.D.

§ 6. PARTS OF THE SOLUTION

We wish to consider various parts of a solution separately. The heart of our classification comes from the following:

THEOREM 6.1. If V is a solution to the game G, then $V \cap Z$ is one of the following:

(i) the empty set;

(ii) a nontrivial set E such that for some $i \in N$, x and y in E imply $x^i = y^i$;

(iii) a triangle T_a;

(iv) a class II set.

In fact, $V \cap Z$ must fit one of the first three cases, but we wish to treat class II sets separately.

PROOF. It is easy to verify that if the points of $V \cap Z$ do not have a common value for some coordinate i, then $V \cap Z$ has three points p_1, p_2, p_3 with $p_j^i = p_k^i$ for all i; for otherwise the set would be inconsistent. If $p_i^i > p_j^i$ for some i, then $p_i^i > p_j^i$ for all i because each p_i, being in A^*, must then have some coordinate larger than p_j and p_k. Applying 3.4 with $a = p_i$ and $b = \frac{1}{2} p_j + \frac{1}{2} p_k$, we obtain a point $c \in A^*$ which $\{jk\}$-dominates p_i and is such that $c \geq_{\{ik\}} p_j$. But c cannot be in V or dominated by V, a contradiction.

Before making a detailed examination of the types of solutions suggested by this theorem, we wish to consider other parts of a solution.

THEOREM 6.2. If V is an A-stable set and p is a point in $V \cap v\{ij\} \cap v\{ik\}$, and if $x \in V$ and $x^i = p^i$ imply $x \leq_{\{jk\}} p$; then $V \cap D_i(p)$ is $D_i(p)$-stable.

PROOF. Let b be a point in $D_i(p) - V$ and let $c \in V$ be a point such that c Dom b. This cannot be via $\{jk\}$ since then $c \text{ Dom}_{\{jk\}} p$. If via $\{ij\}$, we cannot have $c^i > p^i$ for then $c \text{ Dom}_{\{ij\}} p$; nor can we have $c^i = p^i$ for then $c^j \leq p^j \leq b^j$ by hypothesis. If $c^i < p^i$, then $c^k \geq p^k$

to avoid $p \text{ Dom}_{\{ik\}} c$ and so $c \in D_i(p)$. Similarly for $\{ik\}$, and the result follows.

THEOREM 6.3. Let p be a point in $A^i \cap v\{ij\} \cap v\{ik\}$ such that $D_i(p) \cap v\{jk\} \subset w\{jk\}$. Then a necessary and sufficient condition that $W \subset D_i(p)$ be $D_i(p)$-stable is that p be in W and W be a maximal connected self-consistent subset of $D_i(p) \cap [C \cup (A^i \cap v\{ij\} \cap v\{ik\})]$.

PROOF. Necessity. (1) Of course $p \in W$ because p is in the $D_i(p)$-core. Suppose x is a point in $(W - v\{ij\}) \cap (v\{ij\} - w\{ij\})$. Then $x^i \leq p^i$, and hence $x^k > p^k$; for otherwise $p \text{ Comp}_{\{ik\}} x$. Applying 3.7 with $\epsilon < p^k - x^k$, we find a point c in $D_i(p)$ such that $c \text{ Dom}_{\{ij\}} x$ and $c \notin v\{ik\}$. But obviously c is not in W nor in Dom W since either x or p would be dominated by W, a contradiction. Now suppose that x is in $W \cap w\{ij\} - v\{ik\}$ and x is dominated by $\bar{A}$ via $\{jk\}$. Since we cannot have $p \text{ Comp}_{\{ik\}} x$ or $x \text{ Dom}_{\{jk\}} p$, we must have $x^k > p^k$, $x^j = p^j$, and (since $p \in A^i$) $x^i < p^i$. By 5.2, $p^j = b^j_i$. Apply 3.7 to obtain a point c such that $c \text{ Dom}_{\{jk\}} x$, $c \notin v\{ik\}$, and (since $c^j > b^j_i$) $c \notin v\{ij\}$. Then $c \in D_i(p)$ cannot be in either W or Dom W, a contradiction. Using symmetries, the only remaining possibility is $x \in (v\{ij\} \cap v\{ik\}) \cup C$.

Now suppose that x is in $(W - C) \cap (A - A^i)$ and there is a point y in A such that $y \geq x$ and $y^j > x^j$. Then x cannot be $\{ik\}$-dominated because of $x \in A$ and 3.7, and therefore x must be $\{ij\}$-dominated. By 3.7, there is a point z in A such that $z \text{ Dom}_{\{ij\}} x$ and close enough to x so that $z \in A - A^i$ (A^i is compact by 5.1). It is easily verified that $z \in D_i(p)$ and is undominated via $\{ik\}$. But z cannot be in W or Dom W, a contradiction. Therefore $x \in C \cup (A^i \cap v\{ij\} \cap v\{ik\})$.

(2) Next we will show that if a and b are distinct points in W such that $a^i = b^i$, then a and b lie in the same component of $C \cap D_i(p)$. Let f be a continuous mapping of L_{ab} into A such that $f(x) \, R_M \, x$ for $x \in L_{ab}$. Since a and b are distinct elements of A, we may assume without loss of generality that $a^j \geq b^j$ and $b^k \geq a^k$ ($b >_{\{jk\}} a$

is impossible because of $a \in A$, $p^i > a^i$, and 3.7). If $z \in f(L_{ab})$, then $z \geq_{\{ij\}} b$, $z \geq_{\{ik\}} a$, and $z \geq_{\{jk\}} p$; therefore $z \in W$. Furthermore, if a and b are in C, so is $f(L_{ab})$. Suppose $b \in v\{ik\}$. If $a^j = b^j$ or $a^k = b^k$, a or b is in $A - A^i$ and a and b are in C. If $a^k < b^k$, there is a point $y \in \bar{A}$ such that $y \operatorname{Dom}_{\{ik\}} a$ (or else $a^i = b^i_k$ by 5.2, which is impossible because of p) and, because of 3.7 with $\epsilon < a^j - b^j$, we may assume $y \in D_i(p)$ and $y \geq_{\{ij\}} b$. But y can be in neither W or Dom W, hence $b \in C$, and (2) is proved by symmetry.

(3) Suppose that W, compact by 2.4, is the union of disjoint closed sets B and B'. Let $a \in B$ and $b \in B'$ be two points which minimize the function $|x^i - y^i|$ for $x \in B$ and $y \in B'$. By (2), $a^i \neq b^i$, and we may assume $b^i > a^i$. Apply 3.4 with $\epsilon < b^i - a^i$ to obtain a point $c \in D_i(p)$ with $b^i > c^i > a^i$. Since c is not in B or B', there is a point $d \in W$ such that d Dom c, and we may assume without loss of generality that it is via $\{ij\}$. We clearly have $b^j < c^j < d^j \leq a^j$. Now $d^i > b^i$ is out because then $d \operatorname{Dom}_{\{ij\}} b$; $d^i = b^i$ is also out because then (1) $b \in C$ and (5.2) $b^i = b^i_j$, a contradiction. Therefore W is connected. It is obviously maximal.

<u>Sufficiency</u>. The details of this proof are too long, dirty, and uninteresting to be included here, so we give only a macroscopic sketch. First one shows that $W \cap \operatorname{Dom} C = \emptyset$. This is because that, as a point varies over connected W from a point with i^{th} coordinate equal to that of a point $c \in C$ to a point $x \in W \cap v\{ij\} \cap v\{ik\}$, it passes through a point $z \in C \cap v\{ij\} \cap v\{ik\}$, and this "protects" x from C and makes x Dom c impossible. Thus, as one proceeds "down" W from p (in the sense that the ith coordinate decreases), one must include all components of the core in order for W to be maximal. Finally, one cannot get stuck at a point $y \in v\{ij\} \cap v\{ik\}$ because W could always be extended with the $D_i(y)$-stable set of 5.5, which is necessarily connected. Thus W inevitably proceeds downward, taking in the core, and dominating everything. Q.E.D.

Thus $D_i(p)$-stable sets are easy to construct when p is not dominated via $\{jk\}$. One starts at p and constructs a curve in $v\{ij\} \cap v\{ik\} \cap D_i(p)$ that is monotone decreasing in x^i and monotone non-decreasing in x^j and x^k. If one reaches the core, add that component of C and look for another point of $v\{ij\} \cap v\{ik\} \cap A^i$ in that component, from which to continue the curve. When the process is finished, the set is constructed.

§ 7. CLASSIFICATION OF SOLUTIONS

In this section, we consider the possible sets $V \cap Z$ given in 6.1 and give necessary and sufficient conditions for such a set to be part of a solution. Solutions of each type are illustrated in the figures.

Explanation of the Figures. The figures represent the sets A^* for various games. We imagine that the set A^* has been flattened into a plane so that points on the same horizontal line have the same first coordinate, those on the same 60-degree line the same second coordinate, and those on the same 120-degree line the same third coordinate. The diagrams have guide lines such that points in a line parallel to the ith guide line have constant i^{th} coordinate. The sets $w\{ij\}$ are indicated with dashed lines and are labeled only in Fig. 1, it being clear which set is which. Sometimes, as in Fig. 1, part of these dashed lines may be covered by boldface lines. The set $v\{2,3\}$ is the set of points above the line $w\{2,3\}$; $v\{1,2\}$ is to the left of $w\{1,2\}$; and $v\{1,3\}$ to the right of $w\{1,3\}$. The one-dimensional parts of the solutions are indicated by boldface lines and the two-dimensional parts, which always belong to the core, by shading. One should remember that the sets A^* and $A^* \cap v\{ij\}$ are not convex, and are not expected to be convex when flattened out.

The big advantage of this type of diagram, in addition to fitting a sheet of paper, is that the set of points dominated by another point lie in the same wedge-shaped regions as in the familiar classical diagrams (e.g., [1], Fig. 72). Thus it is easy to check for domination and stability. Strictly speaking, the diagrams show only the part of the solution in A^*, but these sets are related to the full solution V by the following result from [5]:* V is A-stable and $V^* = A^* \cap V$ if and only if V^* is

---

* To prove $V^* = A^* \cap V$, the most difficult case is where there is a point $a \in V$, a point $b \in V^*$, such that $b \operatorname{Dom}_{\{ij\}} a$, and a point $y \in A^*$ such that $y \geq a$ and $y^j > a^j$. One uses 3.3 and the partial ordering based on $(x^i - y^i) \geq -M(x^j - y^j)$ for $i \in N$ and $j \in \{ik\}$ to find a point $c \in w\{ij\}$ along the edge of A^* with points maximal in $\{ik\}$. One can show $c \operatorname{Dom}_{\{ij\}} b$, $c \geq_{\{jk\}} a$, and $c \in V^*$. Furthermore, $a \in C$ implies $c \in C^*$. The biggest step in proving $V = A - \operatorname{Dom}(V^* \cup C)$ is to show that, for each $a \in A - \operatorname{Dom}(V^* \cup C)$, there is a point $x \in V^* \cup C$ such that $x \geq a$ and $a \in v\{ij\}$ implies $x \in v\{ij\}$ for all i and j.

A^*-stable and $V = A - \mathrm{Dom}(C \cup V^*)$. (It is also true that $C^* = C \cap A^*$.) Thus the set V^* has to be A^*-stable and we shall assume that $A = A^*$ for the games in question and hence $V^* = V$.

It should also be noted that in general the set A^* cannot be flattened into a plane. Thus, most of the diagrams do not represent bona fide games, but are two-dimensional pseudo-games endowed with the properties essential for this theory, i.e., 2.1.1, 2.1.2, 3.4, and 5.2. Nevertheless, there are genuine games which are topologically similar to these; games with disconnected cores, etc.

We now take up the four cases of 6.1.

THEOREM 7.1. Suppose $B \subset Z$ is a class II set because of Q_i (see 4.6). Then G has a solution V such that $V \cap Z = B$ if and only if

(i) $B = Q_i \cap Z$;

(ii) $t(Q_i) \cap v\{jk\} = \emptyset$.

The possible V are precisely those sets of the form $C \cup E \cup W$ where $E = \left\{x \in A \mid x^i = b^i_j = b^i_k \text{ and } y \in A \text{ and } y^i = x^i \text{ imply } y^j \leq x^j \text{ or } y^k \leq x^k\right\}$ and W is a $D_i(Q_i)$-stable set (see 5.4 and 6.3).

This theorem is illustrated by Figures 1 and 2. In Figure 1, $\emptyset$ is a class II set and $Z = \emptyset$. In Figure 2, both Q_1 and Z have two components and $t(Q_1) = \emptyset$. The solution in Figure 2 is unique.

PROOF. First we shall show that $Q_i \subset V$. Let x be a point in $Q_i \cap (B \cup C)$ (4.6), y be any point in $Q_i - V$, and z a point in V such that z Dom y. This must be via $\{jk\}$ because $y^i = b^i_j = b^i_k$. Now x is in V because it is in $B \cup C$, y is in A^* because of z, and $z^i < x^i = y^i$ because $y \in A^*$. By symmetry, we may assume $x^j > y^j$ and $x^k < y^k$ rather than vice versa (x and y are in A^i). To avoid x Dom z or z Dom x, and because $z^k > y^k > x^k$, we see that $z^j = x^j > y^j$, x is in A^*, and there is a point c in A^* such that $c^j = x^j$ and $c^k = y^k$ (by 5.3). Now c must be in Z, being comprehended by x, y, and z, and c cannot be in V since it

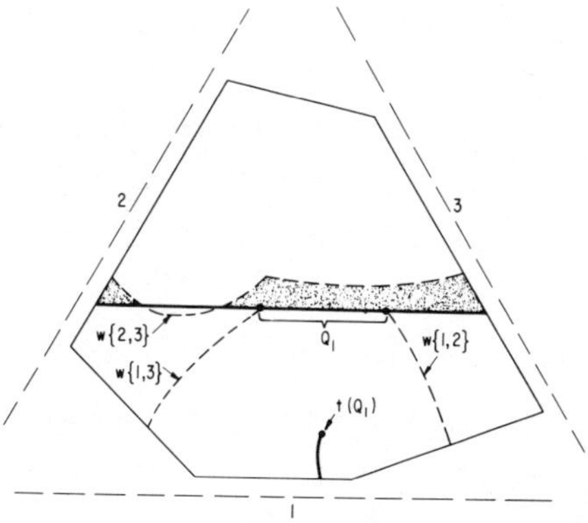

Fig. 1

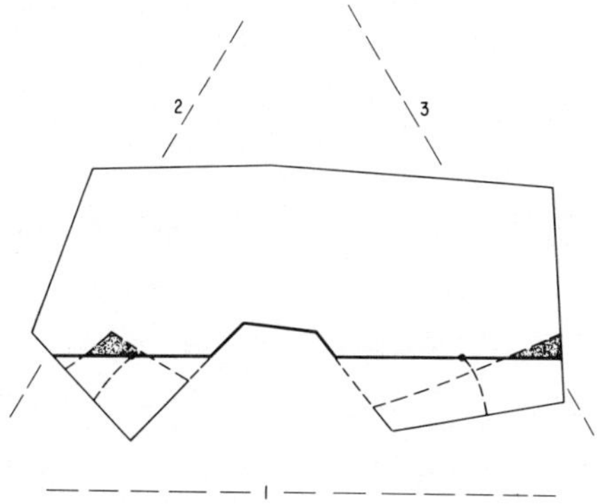

Fig. 2

is not in Q_i. Since $c \geq_{\{ij\}} z$ and $c \geq_{\{jk\}} x$, c must be dominated by a point $b \in V$ via $\{ij\}$. But then either $z \text{ Dom}_{\{jk\}} b$, or $b \text{ Comp}_{\{ik\}} z$ and $z \in Z$, a contradiction in either case, and (i) follows.

Letting q_k and q_j be the points in Q_i with maximal j^{th} and maximal k^{th} coordinates respectively, it is easy to verify that $t(Q_i) \subset V$. Letting p be a point in $t(Q_i)$ (if any), $p \notin Z$ because $p \in V$ and we see that $q_k \text{ Comp}_{\{ij\}} p$ and $q_j \text{ Comp}_{\{ik\}} p$ imply $p \notin v\{jk\}$. (Note that $p \neq q_i \neq q_j$ because Q_i is nontrivial.) Therefore (ii) holds.

It is easy to verify that $D_i(Q_i)$, $D_j(q_j)$, and $D_k(q_k)$ satisfy the conditions of 5.2. The points of E lie in either the $D_j(q_j)$-core or the $D_k(q_k)$-core and must be in V. Also C and W must be in V and since these dominate everything, $V = C \cup E \cup W$. If conditions (i) and (ii) are satisfied, it is easy to verify that any such set is a solution. The existence of a $D_i(Q_i)$-stable set is guaranteed by 5.4 and 5.5. The theorem is proved.

LEMMA 7.2. Suppose that $\emptyset$ is class I and $Z \subset \text{Dom } C$. Then for each $i \in N$, either (1) $v\{ij\} \cap v\{ik\} \cap A^i \subset \text{Dom } C$ or (2) there is a point $p_i \in C \cap v\{ij\} \cap v\{ik\} \cap A^i$ such that $v\{ij\} \cap v\{ik\} \cap A^i \subset \text{Dom } C \cup D_i(p_i)$.

PROOF. This will follow from 5.6 if we can show that $t(Q_\ell) \cap Z \neq \emptyset$ for each nontrivial Q_ℓ. Now $Q_\ell \cap C = \emptyset$ by 4.6 and hence $Q_\ell \subset Z$. Therefore, for a point x in Q_ℓ, there is a point c in C such that $c \operatorname{Dom}_S x$ where $S = N - \{\ell\}$. Since c is undominated by Q_ℓ, $c \in D_i(Q_i)$ and 5.4 applies; $t_\ell(Q_\ell)$ is nonempty and is contained in Z. Q.E.D.

THEOREM 7.3. If $\emptyset$ is class I, then G has a solution V such that $V \cap Z = \emptyset$ if and only if $Z \subset \operatorname{Dom} C$. The solutions all have the form $C \cup W_1 \cup W_2 \cup W_3$ where $W_i = \emptyset$ or W_i is a $D_i(p_1)$-stable set depending on whether case (1) or case (2) of 7.2 holds.

Figures 3 and 4 illustrate this theorem. In Figure 3, $Z = \emptyset$. Notice that the core is not connected. In Figure 4, $Z \neq \emptyset$ and there is no point p_1.

PROOF. (1) First we will show that $V \cap Z = \emptyset$ implies that each point in Z is dominated by V via only one two element set. Suppose to the contrary that $z \in Z$, $a \in V$, and $b \in V$ are points such that $a \operatorname{Dom}_{\{ij\}} z$ and $b \operatorname{Dom}_{\{ik\}} z$. We may assume that a has been chosen so that $x \in A$ and $x =_{\{ij\}} a$ imply $x^k \leq a^k$; for if $x^k > a^k$, then $x^k \in v\{ij\} \cap V$ and x could have been chosen instead of a. Similarly we may assume $x \in A$ and $x =_{\{ik\}} b$ imply $x^j \leq b^j$. Now since $z \in A^*$, we have $b^k > z^k > a^k$ and $a^j > z^j > b^j$. But then $a^i = b^i$ because a and b do not dominate each other. One can now verify that a and b are in A^*. Therefore by 5.3, there is a point $y \in A^*$ with $y^i = b^i$ and $y^j = z^j$, and this point is in Z because it is comprehended by a, b, and z. Letting p be a point in V dominating y, it is seen that, in order for V to be self-consistent, we have $p^j = a^j$, $p^k = b^k$, and $p \in v\{jk\}$. We thus may assume that p is in $t_i(\{a,b\})$ and $p \in Z$ (p comprehended by a and b), but this contradicts $V \cap Z = \emptyset$.

(2) Next we show that any nontrivial Q_i is connected. If Q_i is the union of nonempty disjoint closed subsets B and B', let $a \in B$ and $b \in B'$ be points which minimize the function $|x^j - y^j|$ for $x \in B$ and $y \in B'$. We may assume that $a^j > b^j$ ($a^j = b^j$ and $a^i = b^i$ would imply

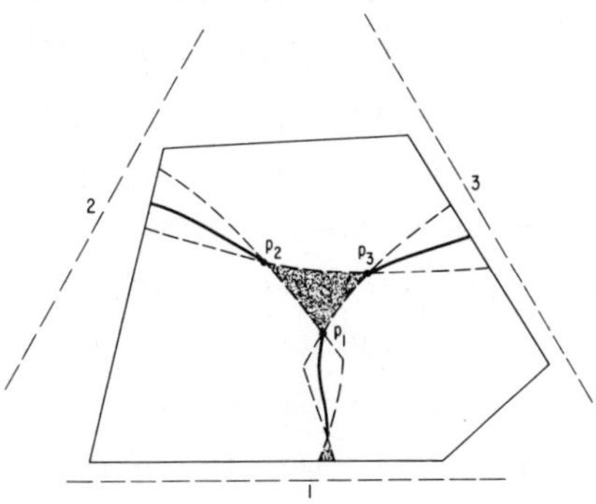

Fig. 3

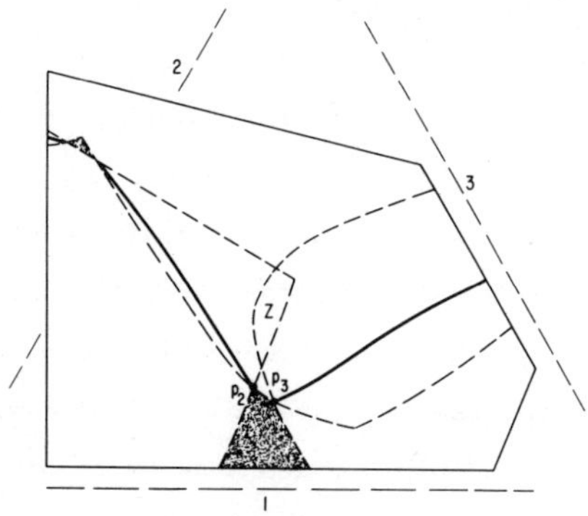

Fig. 4

$a = b$). Since $\emptyset$ is class I, $a \notin C$ and there is a point $y \in \bar{A}$ such that $y \text{ Dom}_{\{jk\}} a$. Now a is in A^* because of y. Because of 3.4, we may assume that y is in A^* and $a^j - y^j < a^j - b^j$; hence $a^j > y^j > b^j$. By 5.3 (using $x_1 = y$, $x_2 = a$, $x_3 = b$, $i = k$) there is a point $c \in Q_i$ with $a^j > c^j > b^j$, a contradiction.

(3) We wish to show that if Q_i is nontrivial then $t_i(Q_i) \cap v\{jk\} \neq \emptyset$. Let q_j be the point in Q_i with maximal k^{th} coordinate, q_k the point with maximal j^{th} coordinate, and z the point in $B = \{x \in A^* \cap v\{jk\} \mid x^j \leq q_k^j \text{ and } x^k \leq q_j^k\}$ with minimal i^{th} coordinate. Of course $B \neq \emptyset$ and $z^i \leq q_j^i$ because $Q_i \subset Z$ ($\emptyset$ is class I). Using 5.3, it is clear that $a = (q_j^i, z^j, z^k)$ defines a triangle $T_a = (z, z_j, z_k)$. Using 5.3 and the images $f(L_{zz_j})$, $f(L_{zz_k})$, and $f(L_{z_j z_k})$, we see that $B' = \{x \in A^* \mid x \leq a\}$ is a connected subset of Z. Because of (1), points of B' must be dominated via a single coalition and this coalition must be $\{jk\}$ because $B' \cap w\{ij\} \cap w\{ik\} \neq \emptyset$.

Let $y \in A^*$ be a point such that $y \text{ Dom}_{\{jk\}} z$. If $y^i \geq q_k^j$ and $y^k \geq q_j^k$, we are done by 5.4. Otherwise, we can assume by symmetry that $q_j >_{\{ik\}} y$ and (since $y^i \leq q_j^i$ contradicts choice of z) $y >_{\{jk\}} q_k$. An application of 5.3 (with $q_j = x_3$) gives a point $b \in A^*$ with $b^i = z^i$ and $b^k = y^k$. A second application (with $y = x_3$) gives a point $c \in A^*$ with $q_j^k > c^k = b^k = y^k > z^k$ and $c^j = z^j \leq q_k^j$. Thus $c \in B$ and $c^i < z^i$, contradicting the choice of z.

(4) Because of (3), 5.6 applies. We cannot have the case of a

triangle T_a because $\{x \in A^* \mid x \leq a\}$ is a connected subset of Z (use argument in (3)) and all its points cannot be dominated by the same coalition as implied by (1) ($p_i \in w\{ij\}$, etc.). Thus $Z \subset \text{Dom } C$.

(5) Going the other way, $Z \subset \text{Dom } C$ implies 7.2 and it is seen that the points p_i must be in all solutions. It can be shown that 6.2 applies to these points. Therefore, the solution must contain a set of the form $C \cup W_1 \cup W_2 \cup W_3$. It remains to be shown that any such set is a solution. Because the p_i are in C, it is easy to see that this set is self-consistent.

(6) One can show by considering several cases that if a point $x \in A$ is not in $B_i = A^i \cap v\{ij\} \cap v\{ik\}$ for any i, nor in C; then x is dominated by points in A via exactly one two-element coalition S, say $S = \{ij\}$. We find a point $y \in w\{ij\}$ such that $y \text{ Dom}_{\{ij\}} x$ and consider the image $f(L_{xy})$, where $f(z)R_m z$ for $z \in L_{xy}$. Let c be the first point in $f(L_{xy})$ going from x to y such that $c \in B_1 \cup B_2 \cup B_3 \cup C$. (The first point in $w\{ij\}$ must be in C if there is no other choice of c.) The only two-element set by which c is dominated is seen to be $\{ij\}$ and we observe that $c \text{ Dom}_{\{ij\}} x$. Thus c or the point in $W_1 \cup W_2 \cup W_3 \cup C$ dominating c dominates x and $W_1 \cup W_2 \cup W_3 \cup C$ is a solution. The existence of such a set is guaranteed by 5.5 and the theorem is proved.

THEOREM 7.4. Suppose E is a nontrivial class I subset of Z such that x and y in E implies $x^i = y^i$. then G has a solution V such that $V \cap Z = E$ if and only if:

(i) E is compact;

(ii) E is connected;

(iii) $e_j \in w\{ik\}$ and $e_k \in w\{ij\}$ (see 4.8);

(iv) $t_i(E) \cap v\{jk\} = \emptyset$;

(v) if x in $v\{ik\} \cap A$ is a point such that $x^i > e^i_j$ and $x^k = e^k_j$, then $e^k_j \neq b^k_j$ (and also for j and k interchanged).

The possible V are precisely those sets of the form $E \cup W_1 \cup W_2 \cup W_3$ where W_j is a $D_j(e_j)$-stable set such that $W_j \cap Z = \{e_j\}$, W_k is a $D_k(e_k)$-stable set such that $W_k \cap Z = \{e_k\}$, and W_i is a $D_i(E)$-stable set.

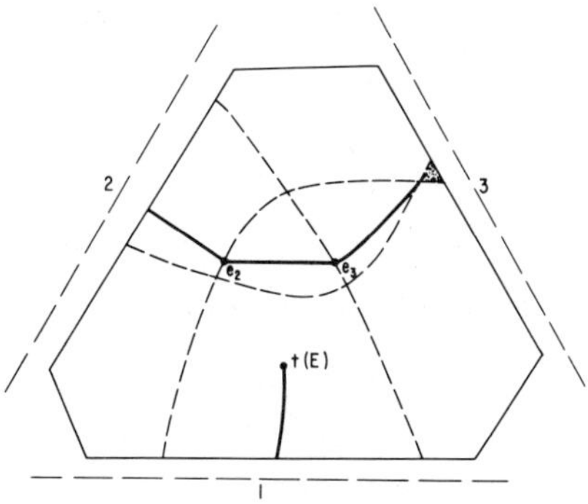

Fig. 5

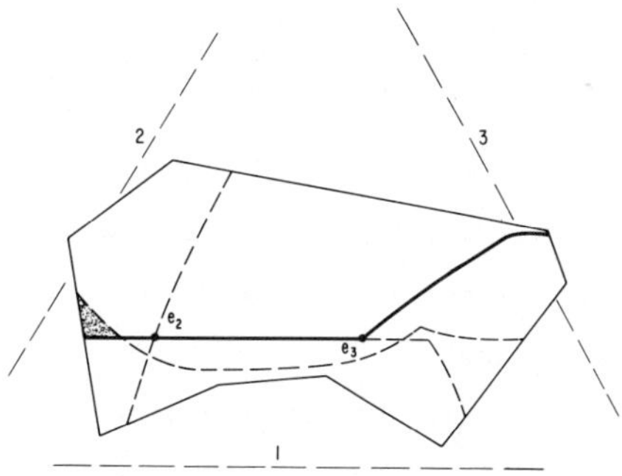

Fig. 6

Solutions of this type are shown in Figures 5 and 6. In Figure 6 it is interesting to note that there are some points in $Z - E$ with the same first coordinate as the points in E.

PROOF. (1) E is obviously compact (2.4). Suppose that E is the union of two disjoint closed sets B and B'. Let $a \in B$ and $b \in B'$ be two points which minimize the function $|x^j - y^j|$ for $x \in B$ and $y \in B'$. We may assume $a^j > b^j$ and $b^k > a^k$. We cannot have both $a \in w\{ik\}$ and $b \in w\{ij\}$ because then $a^i = b^i = b^i_j = b^i_k$ by 5.2, and $E \subset Q_i$, contradicting E is class I.

Assume that $a \notin w\{ik\}$. For each integer m, let $c(m)$ be a point in A^* such that $c(m)\ \mathrm{Dom}_{\{ik\}}\ a$ and such that $|c(m) - a| \leq (1/2)^m$, and let $d(m)$ be a point in V such that $d(m)\ \mathrm{Dom}\ c(m)$. This must be via $\{ij\}$ or $\{jk\}$, and a check of the various cases indicates that we must have $d(m) \geq_{\{ik\}} a$ and $c^j(m) < d^j(m) \leq a^j$; hence the $d(m)$ approach a and $c(0)\ \mathrm{Dom}_{\{ik\}}\ d$, where $d = d(\overline{m})$ for some large $\overline{m}$. Since $d\ \mathrm{Dom}_{\{ik\}}\ a$ and $d\ \mathrm{Dom}_{\{ij\}}\ b$ are impossible, we must have $d^i = a^i$ and $d \in v\{jk\}$. Now $a\ \mathrm{Comp}_{\{ij\}}\ d$, so if d is in A^*, then d is in Z and we have a contradiction; for then d is in E, but cannot be in either B or B'. Suppose $p \in \overline{A}$ is a point such that $p \geq d$. Then $p^j = d^j$, for otherwise we could use $c(0)$ and 3.7 to prove $d \notin A$. Then also $p^k = d^k$, for otherwise a and 3.7 could be used to find a point $\bar{p} \in A$ so that $\bar{p} \geq d$ and $\bar{p}^j > d^j$. Now $d^k > c^k(\bar{m}) > a^k$, so if $p^i > d$, there would be a point $r \in L_{pd}$ near d such that $r \in V$ and $r \; \mathrm{Dom}_{\{ik\}} \; a$, which is impossible. Therefore $p^i = d^i$, $p = d$, $d \in A^*$, and E is connected.

(2) Assuming $e_j \notin w\{ik\}$, the argument of the previous paragraph applies (setting $e_j = a$), with slight modification, and leads to a point $d \in E$ with $d^k > e_j^k$, contradicting the definition of e_j; therefore $e_j \in w(ik)$.

(3) One may verify that $A - \mathrm{Dom}\, E = D_j(e_j) \cup D_k(e_k) \cup D_i(E) \cup E$. The points in $D_j(e_j)$ are seen to undominated by points in $V - W_j$, where $W_j = V \cap D_j(e_j)$. Thus W_j must be $D_j(e_j)$-stable. Now (v) must hold, because otherwise x would be in the $D_j(e_j)$-core and in V, but $x \; \mathrm{Dom}_{\{ik\}} \; e_k$ and $e_k \in V$.

Next we wish to show that any $D_j(e_k)$-stable set W_j such that $W_j \cap Z = \{e_j\}$ does not dominate points in $D_k(e_k)$ or $D_i(E)$. Suppose $p \in W_j$ does dominate points in $D_k(e_k) \cup D_i(E)$. One sees that $p \in v\{ik\}$. Suppose $p^i = e_j^i$. Then $p^k > e_j^k$ and p is not in C because a repeated application of 5.2 proves $E \subset Q_i$, contradicting E is class I. Thus $p \in A^j \cap v\{ij\} \cap v\{jk\}$, by 6.3. Let y be the point in A^* with $y =_{\{ik\}} p$ and $y^j \geq p^j$. By 6.3, either $y \in C$ or $y \in Z$, a contradiction in either case. Now suppose $p^k = e_j^k$. Then $p^i > e_j^i$ and p is not in C because (v) is violated with $x = p$. The argument finishes as in the previous case. The above remarks apply equally well to $D_k(e_k)$.

(4) We now see that $W_i = D_i(E) \cap V$ must be $D_i(E)$-stable. The set $t_i(E)$ cannot be $\{jk\}$-dominated by W_i because E would be also. Thus, $t_i(E) \subset W_i$ and (iv) must hold since $t_i(E) \cap Z = \emptyset$, any $p \in t_i(E)$ being automatically in $A^* \cap v\{ij\} \cap v\{ik\}$. On the other hand, (iv) guarantees that arbitrary $D_i(E)$-stable sets do not dominate other parts of the solution. Thus all sets of the indicated form are solutions.

(5) It remains to be shown that there exist solutions of the indicated form. This is done using 5.4 and 5.5. It is not difficult to show that the W_j and W_k constructed by 5.5 satisfy $W_j \cap Z = \{e_j\}$ and $W_k \cap Z = \{e_k\}$. Q.E.D.

Note. The set of real numbers π for which $E(\pi) = \{x \in A^* \cap v\{ij\} \cap v\{ik\} \mid x^i = \pi\}$ satisfies 7.4 is a connected set; and is bounded below

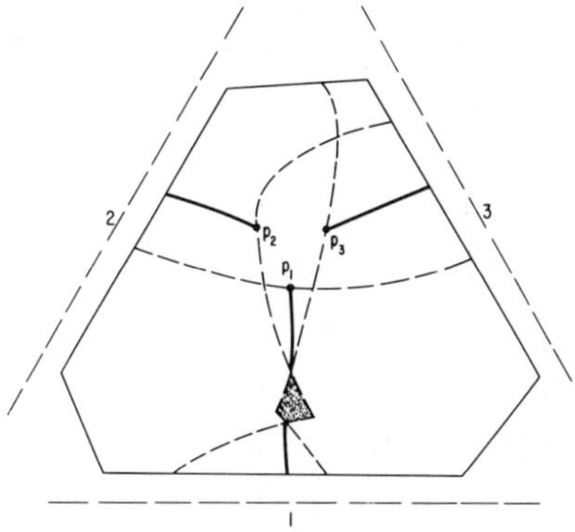

Fig. 7

because of conditions (ii), (iii), (v), and $E(\pi) \subset Z$, and bounded above because of condition (iv). In cases as in Fig. 6, (iv) prevents E from being too short and (ii) and $E \subset Z$ prevent it from being too long.

THEOREM 7.5. Suppose that $T_a = \{p_1, p_2, p_3\}$ is a class I subset of Z. Then G has a solution V such that $V \cap Z = T_a$ if and only if:

(i) $p_i \in w\{jk\}$ for $i \in N$;

(ii) if x in $v\{ik\} \cap A$ is a point such that $x^i > p^i_j$ and $x^k = p^k_j$, then $x^k \neq b^k_j$ (for all permutations of i, j, and k).

The possible V are precisely those sets of the form $W_1 \cup W_2 \cup W_3$ where W_i is a $D_i(p_i)$-stable set such that $W_i \cap Z = \{p_i\}$.

Figure 7 illustrates this theorem.

PROOF. (1) To show (i), apply argument (2) of 7.4 with $e_j = p_j$.

(2) Suppose that y is a point in $V \cap v\{ij\} \cap v\{ik\}$ such that $y^i = p^i_i$, $y^j > a^j$ $(= p^j_i)$, and hence $y^k < a^k$. Clearly $p_i = a$, for otherwise $p_j \operatorname{Dom}_{\{ik\}} y$. Therefore $y \in w\{ij\}$ by 5.2. If also $y \in w\{ik\}$, then $b^i_j = b^i_k$, and one can verify that Q_i is a nontrivial set containing a,

contradicting T_a is class I. Thus $y \notin w\{ik\}$, and we let $c(m)$ be a sequence of points in A such that $c(m) \text{ Dom}_{\{ik\}} y$ and $d(m) \in V$ a sequence such that $d(m) \text{ Dom } c(m)$. By the familiar argument of 7.4(1), a contradiction is achieved. We conclude that 6.2 applies to p_i, and it follows that V must be of the prescribed form.

(3) Condition (ii) was proved in (2) when $p_i = a$. Otherwise, the failure of (ii) implies that x is in the $D_j(p_j)$-core and $x \in V$, a contradiction because $x \text{ Dom}_{\{ik\}} p_k$.

(4) If (i) and (ii) hold, 5.5 is seen to define sets W_i so that $W_i \cap Z = \{p_i\}$. Any set V of the form $W_1 \cup W_2 \cup W_3$ is self-consistent by the argument in (5) of 7.4. We see that any such V is a solution. Q.E.D.

The classification of solutions according to 6.1 has been completed and all that remains is to prove the existence theorem. A proof of the zero-sum case can be found in a paper by Peleg [6].

THEOREM 7.6. Every three-person game has a solution.

PROOF. If $t_i(Q_i) \cap v\{jk\} = \emptyset$ for some nontrivial Q_i, then $B = Q_i \cap Z$ is seen to saitsfy 7.1. Otherwise 5.6 holds. If $Z \subset \text{Dom } C$, then we obtain a solution by the proof of 7.4. Finally suppose that we have the triangle T_a from 5.6. We can clearly choose a so that $T_b \subset Z$ and $b \geq a$ implies $b = a$. It is impossible for T_a to be class I, for if Q_i is nontrivial and contains T_a, then the triangle $\{q_j, q_k, p\} = T_b$, where $b = (b^i_j, p^j, p^k)$ for $p \in t_i(Q_i)$, is contained in Z and satisfies (i) of 7.5; but $b \geq a$ contradicting the choice of a. If (ii) of 7.5 fails, then Q_k is seen to be nontrivial and again we are led to a larger b. Thus T_a must satisfy 7.5 and the game has a solution. Q.E.D.

<u>Method of determining all solutions</u>: (1) First check to see if there is a nontrivial Q_i such that $t_i(Q_i) \cap v\{jk\} = \emptyset$. There can be at most one such Q_i. If a Q_i does meet this condition, then the game has solutions derived from 7.1 with $B = Q_i \cap Z$. If $D_i(Q_i) = \emptyset$, we must take $W = \emptyset$; but if $D_i(Q_i) \neq \emptyset$, there is a point p in $t_i(Q_i)$ by 5.4 and W can be any set described in 6.3. The construction of such sets is discussed after 6.3. Any other solutions to this game will have the form of 7.4. If

there are solutions of this form, $E = Q_i$ will be contained in Z, satisfy the five conditions of 7.4 and be a limit of possible choices of E. The W_i are, of course, constructed by 6.3.

(2) If the solutions are not given by (1), then check to see if $Z \subset \text{Dom } C$. If so, then $\emptyset$ is class I and the solutions are given by 7.3. The W_i are constructed by 6.3.

(3) If the solutions are not given by (1) or (2), then there must be a triangle T_a which satisfies the conditions of 7.5. In rare cases, there may be many such triangles, all with a common point. In addition to solutions based on 7.5, the game will usually have solutions based on 7.4 for each i in N and the least upper bound for the possible values of the ith coordinate for points in various choices of E will be a^i. As always, the W_i are constructed by 6.3.

§ 8. EXTENDED GAMES

We shall explain how the preceding results may be carried over to the extended case.

DEFINITION 8.1. An <u>extended</u> <u>n-person</u> <u>game</u> is an n-person characteristic function (N, v) together with a compact convex polyhedral subset H of E^N.

We let $\bar{C}$ be the $\bar{A}$ core which is in general larger than the A-core.

THEOREM 8.2. If G is an extended three-person game and W is an A-stable set, then $W \cup \bar{C}$ is an $\bar{A}$-stable set. If V is $\bar{A}$-stable, then $V \cap A$ is A-stable.

The most difficult part in the proof as given in [5] is to show that, if $a \in A$ and $b \in \bar{C}$ are points such that $b \text{ Dom } a$, then there is a point $c \in C$ such $c \text{ Dom } a$. In the worst case, c is found on the edge of A using a relation based on the inequalities $M(x^j - y^j) \geq (x^i - y^i)$ for $i \in N$ and $j \in S$.

In the classical theory, 8.2 is true for n-person games. In the present theory, 8.2 does not generalize, even to the four person case, as is seen by the following example:

$$N = \{1, 2, 3, 4\}$$

$$H = \text{convex hull of} \begin{cases} p_1 = (10, 0, 0, 5) \\ p_2 = (0, 10, 0, 0) \\ p_3 = (0, 0, 10, 0) \\ p_4 = (4, -10, -10, 4) \end{cases}$$

$$v\{1, 2, 4\} = v\{1, 3, 4\} = \{x \mid x^1 \leq 6,\ x^4 \leq 3\}$$

$$v\{1, 4\} = \{x \mid 3x^1 + 2x^4 \leq 20\}$$

$$v(\emptyset) = E^N, \text{ and } v(S) = \emptyset \text{ for other } S \subset N.$$

One easily verifies that $\overline{A} = H$ and A is the convex hull of p_1, p_2 and p_3. The only $\overline{A}$-stable set is $\overline{C}$ and the only A-stable set is $C \cup \{p\}$ where $p = (2, 4, 4, 1)$. But $p_4 \in \overline{C}$ and $p_4 \operatorname{Dom}_{\{1,4\}} p$.

BIBLIOGRAPHY

[1] von NEUMANN, J. and MORGENSTERN, O., Theory of Games and Economic Behavior, Princeton , 1944, 3rd ed., 1953.

[2] AUMANN, R. J. and PELEG, B., "von Neumann-Morgenstern solutions to cooperative games without side payments," Bull. Amer. Math. Soc.,66, (1960), pp. 173-179.

[3] AUMANN, R. J., "The core of a cooperative game without side payments," Trans. Amer. Math. Soc. 98, (1961), pp. 539-552.

[4] BLACKWELL, D., "An analog of the minimax theorem for vector payoffs," Pacific J. Math. 6, (1956), pp. 1-8.

[5] STEARNS, R. E., Three Person Cooperative Games Without Side Payments, Ph.D. Thesis, Princeton (1961).

[6] PELEG, B., "Solutions to cooperative games without side payments," Trans. Amer. Math. Soc.

[7] GILLIES, D. B., "Solutions to general non-zero sum games," Annals of Mathematics Study No. 44, (Princeton 1959), pp. 47-85.

R. E. Stearns

General Electric Research Laboratory
Schenectady, New York

SOME THOUGHTS ON THE THEORY OF COOPERATIVE GAMES*

Gerd Jentzsch
Edited by Robert J. Aumann

INTRODUCTION

The author of this paper, Gerd Jentzsch, died while still a young man on March 26, 1959. This is apparently his only publication. Judging from its originality and all-around brilliance, his death was a loss of the first magnitude to game theory.

The von Neumann-Morgenstern (N-M) theory of n-person games [4] is concerned with cooperative games in which side payments are permitted and utility is "unrestrictedly transferable" — that is, each player's utility for money is linear in money.** Jentzsch's investigations grew out of an attempt to generalize the N-M theory either by eliminating the requirement that the utility functions*** be linear, or more generally, by eliminating side payments altogether. He notices at the outset that the notion of "effectiveness" — which is crucial in the N-M theory — does not generalize in a straightforward manner. In the classical theory, a coalition K is effective for a payoff vector f if, roughly speaking, the coalition can assure itself of getting at least f. An equivalent definition of effectiveness is that the <u>opposition</u> — the complement of K — cannot prevent K from obtaining at least f. But when utilities are nonlinear in money or side payments are forbidden, these two definitions of effectiveness are in general no longer equivalent — in Jentzsch's terminology, the game need not be "clear" (Example 4). Jentzsch addresses himself to the task of broadening the class of games considered by von Neumann and Morgenstern, while still retaining the clearness property.

---

* Translated from the German by Translavic Associates, for The RAND Corporation.

** See R. D. Luce and H. Raiffa, <u>Games and Decisions</u>, p. 168.

*** By the phrase "utility function" we shall henceforth <u>always</u> mean "utility of money as a function of money."

The chief result is Theorem 21. Rather than stating it here in its most general form, we will describe its application to games with side payments ("money games" for short) in which the utility functions need not be linear. The problem that Jentzsch considers is, what kinds of utility functions of the players will always lead to clear games (as linear utility functions do)? More precisely, what conditions, when placed on the utility functions of the players, will ensure that all money games in which these players participate are clear? The answer is that each coalition must have a kind of "social utility function" for money. For example, this involves the demand that \$50 be indifferent — from the point of view of the coalition as a whole — to some probability combination of 0 dollars and \$100 (though not necessarily the 1/2-1/2 combination). The sums of money involved (\$50, \$0, \$100) are not given to the individual players, but to the coalition as a whole for distribution among its members. "Indifferent" has a very precise meaning here: The two sets of (utility) payoff vectors that can result from the two possibilities must coincide.

The existence of such a "social utility function" is a considerable restriction. Jentzsch remarks without proof that Bernoullian (i.e., logarithmic) individual utility functions lead to a social utility function (Examples 11, 16) and that other individual utility functions that lead to a social utility function can be obtained as solutions of a third-order differential equation with one parameter (which he does not specify). These questions must certainly be investigated further. But on the whole, Jentzsch's result shows that clearness is the exception rather than the rule — that games with nonlinear utility functions or without side payments cannot be "expected" to be clear.

The difference between the two kinds of effectiveness was appreciated by others, working independently of Jentzsch, as far back as 1957 — which is probably the approximate date of Jentzsch's investigations.* It was explicitly mentioned by Aumann and Peleg [2], who used the names α- and β-effectiveness for the two kinds. A survey of the whole field of cooperative games without side payments is given in [1], which has a bibliography of 15 items; but of this work, Jentzsch knew only of the pioneering investigation of Shapley and Shubik [5]. This is

* In fact, the idea is related to Blackwell's approachability-excludability theory [3] which appeared already in 1956.

another example of the known phenomenon of the intrinsic "ripeness" of a scientific concept — leading to simultaneous independent discovery by widely separated investigators. It should be emphasized, though, that it is only in the basic recognition of the difference between α- and β-effectiveness that Jentzsch's work overlaps that of others; the main result of this paper has not been found by anybody else, and appears here for the first time. Indeed other workers have approached the subject from a somewhat different viewpoint — they have tried to "live with" the difference, whereas Jentzsch characterized the conditions under which it could be eliminated (see [1]).

An attempt has been made to keep editorial comment separate from Jentzsch's original text. All the footnotes are the editor's, as are the two "Editor's Notes." The long formal proofs given by Jentzsch for Theorems 10 and 13 have been replaced by short intuitive sketches. There has been some rearranging of the material, and the more straightforward proofs have been left out. Those are all the changes.

Since Jentzsch is interested only in the question of effectiveness, he fixes once and for all a coalition K, and considers only the joint strategies of the coalition, the joint strategies of the opposition, and the payoff to the coalition. The resulting formal object is called a "K-game," and this is the object of investigation throughout.* The individual strategies of members of the coalition and of the opposition, and the payoff to the opposition, are of no interest in this context, and are therefore suppressed in the formal model.

R. J. A.

The Hebrew University
Jerusalem

* It is formally identical to Blackwell's "game with vector payoffs" [3].

§ 1. K-GAMES

1.1. Definitions

DEFINITION 1. $\Gamma = \Gamma(S, T, r)$ is called a K-game if K is a nonempty finite set, $E = E(K) = \{f = (f(k))_{k \in K} | f(k)$ real for $k \in K\}$, S is a nonempty set, T is a nonempty set, and r is a mapping of $S \times T$ into E.

We call K "the coalition", its elements "coalition members", or simply "players." Members of E are called "coalition payoff vectors", or simply "payoff vectors." If the coalition members k anticipate the utilities f(k), then we say that the coalition anticipates the payoff vector f ($f = (f(k))_{k \in K}$). We call S "the set of strategies of the coalition," T "the set of strategies of the opposition," and r "the payoff function." In a K-game the coalition chooses one of its strategies $s \in S$, and the opposition independently chooses one of its strategies $t \in T$. The payoff vector r(s, t) results.

DEFINITION 2. Let Γ and Γ' be two K-games, $\Gamma = \Gamma(S, T, r)$, $\Gamma' = \Gamma(S', T', r')$. We say Γ is "isomorphic" to Γ' if there is a one-to-one mapping σ of S onto S' and a one-to-one mapping τ of T onto T' such that for all $s \in S$, $t \in T$, $s' \in S'$, $t' \in T'$, $r(s, t) = r'(\sigma(s), \tau(t))$, $r'(s', t') = r(\sigma^{-1}(s'), \tau^{-1}(t'))$.

Isomorphic copies of K-games differ only in notation.

DEFINITION 3. Let $f \in E(K)$, $h \in E(K)$. We write

$$f \leq h, \text{ if } f(k) \leq h(k) \text{ for all } k \in K$$
$$f \geq h, \text{ if } f(k) \geq h(k) \text{ for all } k \in K.$$

If the coalition has a choice between payoff vectors f and h, and if $f \leq h$, $f \neq h$, then it will prefer h; h obviously guarantees to each coalition member a utility which is no less, and possibly greater than that received under f. Let us assume that the interests of the opposition are directly opposed to those of the coalition. If the opposition can determine whether the coalition receives f or h, and if $f \leq h$, $f \neq h$, then the opposition will see to it that the coalition only receives f.

DEFINITION 4. Let $\Gamma = \Gamma(S, T, r)$ be a K-game, $S' \subset S$, $T' \subset T$. S' is called "complete" in Γ, if for every $s \in S$ there is an $s' \in S'$ such that $r(s', t) \geq r(s, t)$ for all $t \in T$. T' is called "complete" in Γ, if for every $t \in T$ there is a $t' \in T'$ such that $r(s, t') \leq r(s, t)$ for all $s \in S$.

DEFINITION 5. Let Γ and Γ' be two K-games, $\Gamma = \Gamma(S, T, r)$, $\Gamma' = \Gamma(S', T', r')$. Γ' is called a "deflation" of Γ, if $S' \subset S$, $T' \subset T$, $r'(s, t) = r(s, t)$ for all $s \in S'$, $t \in T'$, and S' and T' are complete in Γ.

The coalition attempts to obtain the best possible payoff vectors and the opposition tries to counteract this. Nothing essential is altered in the possibilities open to either one, if they restrict their strategies to choices from complete subsets of their strategy sets. They can thus change from a K-game to one of its deflations.

DEFINITION 6. Two K-games Γ and Γ' are called "equivalent" if there is a natural number N and a chain of K-games $\Gamma = \Gamma_0, \ldots, \Gamma_N = \Gamma'$ such that for all $n = 1, \ldots, N$,

Γ_n is a deflation of Γ_{n-1},

Γ_n is an isomorph of Γ_{n-1}, or

Γ_{n-1} is a deflation of Γ_n.

DEFINITION 7. Every K-game $\Gamma(S, T, r)$ has "characteristic sets" $V(s)$, V, U, $U(t)$, where $s \in S$, $t \in T$, defined by

$$V(s) = \{f \in E(K) \mid \text{for all } t \in T \;\; f \leq r(s, t)\},$$

$$U(t) = \{f \in E(K) \mid \text{there is an } s \in S \text{ such that } f \leq r(s, t)\},$$

$$V = \bigcup_{s \in S} V(s), \qquad U = \bigcap_{t \in T} U(t).$$

These sets may be interpreted in the following manner: $V(s)$ is the set of payoff vectors which the coalition can enforce by using strategy s. V is the set of payoff vectors which the coalition can enforce.

$U(t)$ is the set of payoff vectors which the opposition cannot prevent by using strategy t (for if the coalition discovers the "correct" strategy s, then $f \leq r(s, t)$). U is the set of payoff vectors which the opposition cannot prevent.

THEOREM 1. If U and V are characteristic sets of a K-game (according to Definition 7), then $V \subset U$.

The opposition cannot prevent any payoff vector that the coalition can enforce.

PROOF. If $f \in V$, then there is an $s \in S$ such that $f \in V(s)$. Then for every t, $f \leq r(s, t)$. Therefore $f \in U(t)$ for all $t \in T$. So $f \in U$. f was arbitrary, so $V \subset U$.

DEFINITION 8. Let Γ be a K-game, U and V characteristic sets of Γ. Γ is called "clear" if $V = U$.

In a clear K-game the coalition can enforce precisely those payoff vectors that the opposition cannot prevent.

DEFINITION 9. Let $\Gamma(S, T, r)$ be a K-game with the characteristic sets V, U, $U(t^0)$, where $t^0 \in T$. t^0 is called "optimal" and $\Gamma(S, T, r)$ "classical" if $V = U(t^0)$.

By using an optimal strategy the opposition can prevent every payoff vector that the coalition cannot enforce. In a classical K-game the opposition has an optimal strategy.

THEOREM 2. Every classical K-game is clear.

PROOF. According to Theorem 1, $V \subset U$. According to Definition 7, $U(t^0) \supset U$. From $V = U(t^0)$ it therefore follows that $V = U$.

THEOREM 3. The properties "clear" and "classical" are invariant under equivalence (Definition 6).

The proof is straightforward, and is omitted.

The existence of nonclear K-games is shown by the following examples.

EXAMPLE 1. Let

$$K = \{1\},\ S = \{0,\ 1\},\ T = \{0,\ 1\},$$

$$r(0,\ 0) = r(1,\ 1) = (1),\ r(0,\ 1) = r(1,\ 0) = (-1),$$

$$\Gamma_1 = \Gamma(S,\ T,\ r).$$

The characteristic sets of Γ_1 are

$$V(0) = V(1) = V = \{f \in E(K)\ |\ f(1) \leq -1\},$$

$$U(0) = U(1) = U = \{f \in E(K)\ |\ f(1) \leq +1\},$$

Γ_1 is not clear.

Player 1 (as a one-man coalition) plays Matching Pennies against the opposition. Since only pure strategies are permitted, he cannot prevent the loss of his penny; i.e., he cannot "guarantee" more than the loss of the penny. On the other hand, the opposition cannot prevent player 1 from winning the penny.

EXAMPLE 2. (See Fig. 1.) Let

$$K = \{1, 2\}, \ S = \{0, 1\}, \ T = \{0, 1\},$$

$$r(0, 0) = (2, -1), \ r(0, 1) = (-2, 1)$$

$$r(1, 0) = (1, -2) \quad r(1, 1) = (-1, 2)$$

$$\Gamma_2 = \Gamma(S.T.r).$$

The characteristic sets of Γ_2 are

$$V(0) = \{f \mid f \leq (-2, -1)\},$$

$$V(1) = \{f \mid f \leq (-1, -2)\},$$

$$V = V(0) \cup V(1),$$

$$U(0) = \{f \mid f \leq (2, -1)\},$$

$$U(1) = \{f \mid f \leq (-1, 2)\},$$

$$U = \{f \mid f \leq (-1, -1)\}.$$

There is no answer to the question of how a player (or a coalition) should act in a nonclear situation such as that in Matching Pennies. J. von Neumann circumvented these difficulties in the case of a two-person zero sum game by introducing mixed strategies. We will now see if something similar is also possible for K-games.

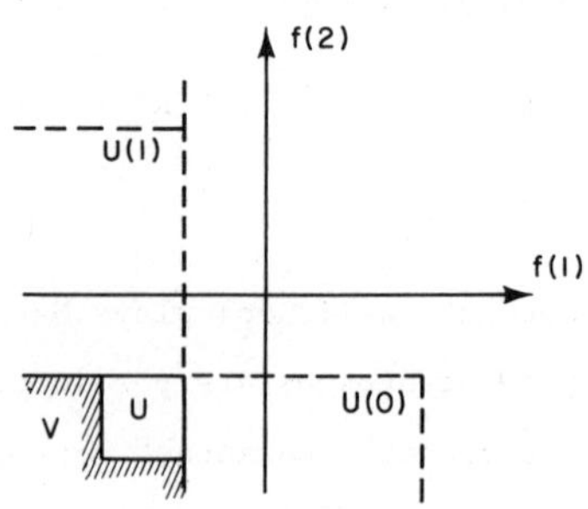

Fig. 1

1.2. Convex Cooperative Games

DEFINITION 10. Addition and scalar multiplication on $E = E(K)$ are defined as is usual for vectors, i.e., $(f + h)(k) = f(k) + h(k)$ and $(cf)(k) = cf(k)$.

DEFINITION 11. Let $\Gamma = \Gamma(S, T, r)$, $\Gamma' = \Gamma(S', T', r')$ be two K-games. Γ' is called a "mixed extension of Γ" if S' is the set of all finite probability distributions over S, T' is the set of all finite probability distributions over T and $r'(s', t')$ is the expectation of the random variable $r(s, t)$ when s and t are distributed according to s' and t' respectively. Intuitively, Γ' is obtained from Γ by permitting the use of mixed strategies.

DEFINITION 12. A K-game $\Gamma = \Gamma(S, T, r)$ is called "convex" is for every pair $s' \in S$, $s'' \in S$ and for every probability p, $0 < p < 1$, there is an $s \in S$ such that $r(s, t) = (1-p)r(s', t) + pr(s'', t)$ for all $t \in T$, and if for every pair $t' \in T$, $t'' \in T$ and for every p, $0 < p < 1$, there is a $t \in T$ such that $r(s, t) = (1-p)r(s, t') + pr(s, t'')$ for all $s \in S$.

THEOREM 4. Every convex K-game is equivalent to its mixed extension.

The proof is straightforward, and is omitted.

EXAMPLE 3. The mixed extension of "matching pennies" (Example 1) is classical with the optimal strategy of the opposition $t^0 = 1/2$. Matching Pennies thus becomes classical by the introduction of mixed strategies.*

---

* In fact, if Γ is any K-game where K has only one member and S and T are both finite, then the mixed extension of Γ is classical. Indeed, such a game is essentially a finite two-person zero-sum game, and its classicality is equivalent to the minimax theorem.

The next example shows that there are convex nonclear K-games.

EXAMPLE 4. (See Fig. 2.) Let

$$K = \{1, 2\}, \; S = \{s \mid 0 \le s \le 1\}, \; T = \{t \mid 0 \le t \le 1\},$$

$$r(s, t) = (1-s)(1-t)(2, -1) + (1-s)t(-2, 1) + s(1-t)(1, -2) + st(-1, 2),$$

$$\Gamma_4 = \Gamma(S, T, r).$$

Γ_4 is convex. It is the mixed extension of Γ_2. Its characteristic sets are

$$V(s) = \{f \mid f \le (s-2, -1-s)\};$$

$$V = \{f \mid f \le (-1, -1), \; f(1) + f(2) \le -3\},$$

$$U(t) = \{f \mid f \le (2-4t, 2t-1)\}, \text{ if } t \le 1/2,$$

$$U(t) = \{f \mid f \le (1-2t, 4t-2)\}, \text{ if } t \ge 1/2,$$

$$U = U(0) \cap U(1) = \{f \mid f \le (-1, -1)\}.$$

Γ_4 is not clear. The opposition cannot prevent the payoff vector $(-1, -1)$, and the coalition cannot compel it.

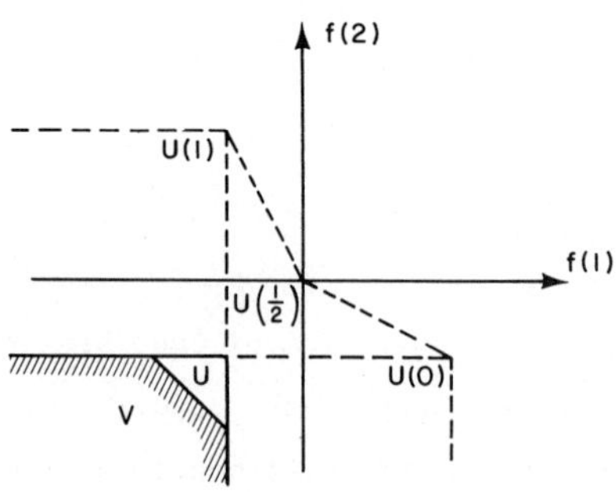

Fig. 2

1.3. Perfect Information

We have seen that in K-games the admission of mixed strategies was not sufficient to "clarify matters." Now we will show in two examples that perfect information does not play the same role in cooperative games that it does in two-person zero-sum games.

EXAMPLE 5. Let

$K = \{1\}$, $S = \{0, 1\}$, $T = \{00, 01, 10, 11\}$,
$r(0, 00) = r(0, 01) = (2, -1)$, $r(0, 10) = r(0, 11) = (-2, 1)$,
$r(1, 00) = r(1, 10) = (1, -2)$, $r(1, 01) = r(1, 11) = (-1, 2)$,
$\Gamma_5 = \Gamma(S, T, r)$.

Γ_5 can be represented by the "tree" in Fig. 3. "K" indicates a move by the coalition and "0" indicates a move by the opposition. Γ_5 is of perfect

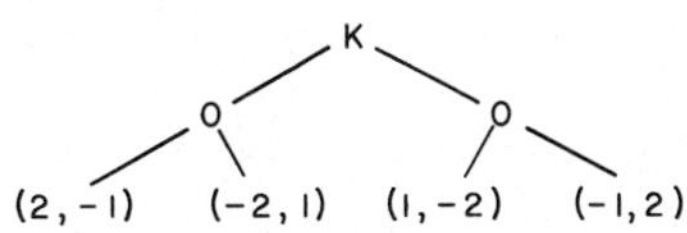

Fig. 3

information in the sense of von Neumann and Morgenstern [4]. The characteristic sets of Γ_5 are:

$$V(0) = \{f \mid f \leq (-2, -1)\},$$
$$V(1) = \{f \mid f \leq (-1, -2)\},$$
$$V = V(0) \cup V(1),$$
$$U(00) = \{f \mid f \leq (2, -1)\},$$
$$U(01) = \{f \mid f \leq (2, -1) \text{ or } f \leq (-1, 2)\},$$
$$U(10) = \{f \mid f \leq (-2, 1) \text{ or } f \leq (1, -2)\},$$
$$U(11) = \{f \mid f \leq (-1, 2)\},$$
$$U = U(00) \cap U(10) \cap U(11).$$

We have $U = V$, so Γ_5 is clear. But it is not classical; the opposition has no optimal strategy.

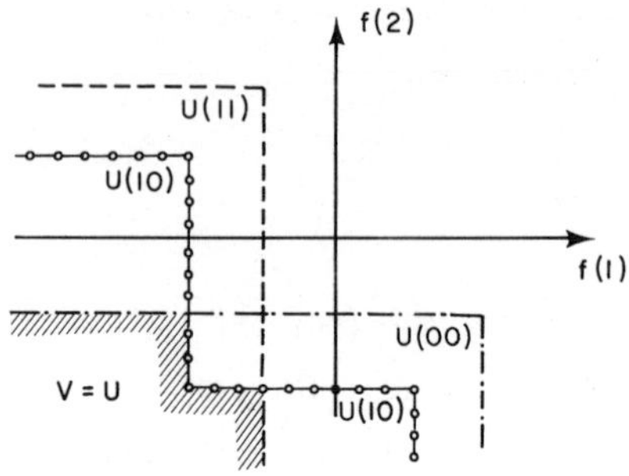

Fig. 4

EXAMPLE 6. Let Γ_6 be the mixed extension of Γ_5 (Example 5). Then it can be verified that $(-1, -1)$ is in U but not in V; indeed if $f \in V$ and $f(1) = f(2)$, then $f \leq (-1.5, -1.5)$. Hence Γ_6 is not clear.

THEOREM 5. A K-game of perfect information is not necessarily classical. There are nonclear K-games with perfect information. There are clear nonclassical K-games. The mixed extension of a clear K-game is not necessarily clear.

PROOF. Examples 5 and 6.

In von Neumann and Morgenstern the distinction between clear and nonclear, and between classical and nonclassical games did not arise. In their theory it was assumed that the coalition was concerned only with the sum of the payoffs of the coalition members, and that the coalition could distribute this sum arbitrarily. How this should be done in particular (especially when the "payoffs" are really "expected payoffs") was not mentioned. In the next section we will investigate several examples of side payments in K-games.

1.4. Side Payments

In the examples studied so far the coalition members were not allowed to make side payments to each other. Now we wish to allow the coalition members to make side payments to each other in one form or another.

EXAMPLE 7. Let

$$K = \{1, 2\}, \ S = \{s | s \text{ real}\}, \ T = \{t | 0 \le t \le 1\},$$

$$r(s, t) = (1-t)(1-s, s) + t(-s, 1+s),$$

$$\Gamma_7 = \Gamma(S, T, r).$$

Coalition member 1 pays to member 2 the amount of s dollars (he receives $-s$ dollars from member 2 in case $s < 0$). The opposition does not know this. It must pay one dollar to player 1 or 2, and may determine the recipient by a chance mechanism. The coalition members are guided by the expected payoff.

The characteristic sets of Γ_7 are

$$V(s) = \{f | f \le (-s, s)\},$$

$$V = \{f | f(1) + f(2) \le 0\},$$

$$U(t) = \{f | f(1) + f(2) \le 1\},$$

$$U = \{f | f(1) + f(2) \le 1\}.$$

Γ_7 is not clear.

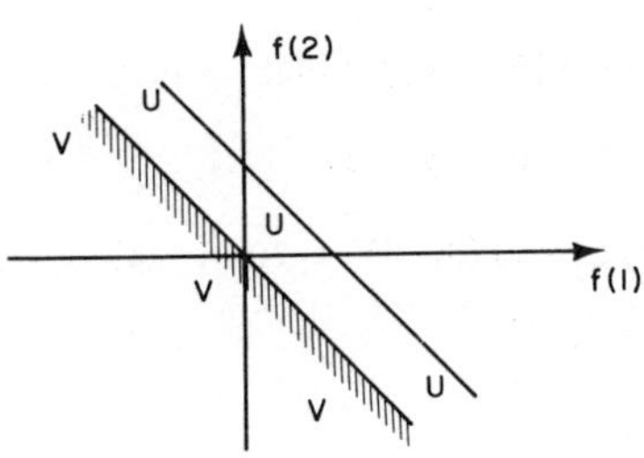

Fig. 5

EXAMPLE 8. Let

$$K = \{1, 2\},\ S = \{s = (s', s'')|s' \text{ real}, s'' \text{ real}\},$$

$$T = \{t|0 \leq t \leq 1\},\ r(s, t) = (1-t)(1-s', s') + t(-s'', 1+s''),$$

$$\Gamma_8 = \Gamma(S, T, r).$$

Here again the opposition pays one dollar to one of the coalition members, and may choose the recipient by a chance mechanism. The coalition members make the following agreement: player 1 will pay s' dollars in case he receives the dollar from the opposition; he will pay s'' dollars in case the dollar goes to player 2. The opposition does not know of this agreement.

The characteristic sets of Γ_8 are:

$$V(s) = \{f|f \leq (\text{Min}\{1-s', -s''\}, \text{Min}\{s', 1+s''\})\},$$

$$V = \{f|f(1) + f(2) \leq 1\},$$

$$U(t) = U = V.$$

Γ_8 is classical; every strategy of the opposition is optimal.

Examples 7 and 8 show that in general it is not the same whether the side-payments must be carried out before beginning the "actual play" (von Neumann and Morgenstern [4] only discuss examples in which this is the case) or whether the side-payments may have strings attached.

EXAMPLE 9. Let

$$K = \{1, 2\},\ S = \{s|0 \leq s \leq 1000\},$$

$$T = \{t|0 \leq t \leq 1\},\ 0 < p < 1,$$

$$r(s, t) = (1-t)((1-p)(0, 0) + p(-s, s)) + t(-s, s),$$

$$\Gamma_9 = \Gamma(S, T, r).$$

The characteristic sets of Γ_9 are:

$$V(s) = \{f \mid f \leq (-s,\ ps)\},$$

$$V = \{f \mid f \leq (0,\ 1000p) \text{ and } pf(1) + f(2) \leq 0\},$$

$$U(t) = \{f \mid f \leq (0,\ 1000\ ((1-t)p + t)) \text{ and } f(1) + f(2) \leq 0\},$$

$$U = U(0) = \{f \mid f \leq (0,\ 1000p) \text{ and } f(1) + f(2) \leq 0\}.$$

Γ_9 is not clear. To be sure, every payoff vector that can be at all prevented by the opposition, can already be prevented by using the strategy $t = 0$; but $t = 0$ is not optimal. Figure 6 illustrates the case $p = 1/2$.

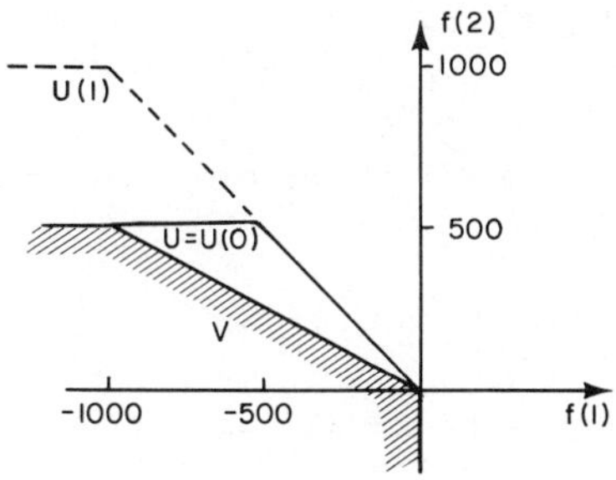

Fig. 6

Γ_9 can be interpreted as follows: the coalition members have no ready cash at the outset. However, player 1 has prospects of a bank loan in the amount of $1000. He promises s dollars to his coalition partner in case he should receive the loan. The bank makes inquiries about player 1 from the opposition. If the opposition gives a favorable report, then player 1 will get the credit; otherwise he will get it only with probability p.

The coalition knows the probability p. However it only will learn indirectly if the opposition has given an unfavorable report, and then only when credit is refused player 1. It is again assumed that the coalition members are guided by the expected payoff (player 1 naturally must subtract the bank debt from his account). Our example has been kept very simple. For instance, we have neglected interest. The availability of cash plays a large role in economic theory, as well as in practice. Our example shows that one cannot count on "clear" cooperative games as models for situations in which the availability of cash plays a role.

§ 2. K-GAMES WHOSE PAYOFFS ARE CATALOGUES

2.1. Catalogues. K-Games Whose Payoffs Are Catalogues

DEFINITION 13. Let K be a coalition, $E = E(K)$. A subset F of E is called a "K-catalogue" or simply a "catalogue" if

(K1) F is nonempty,

(K2) F is convex,

(K3) $f \leq f'$ and $f' \in F$ imply $f \in F$.

(K3) says that F contains all the vectors that it dominates.*

EXAMPLE 10. For each real y, let

$$F(y) = \{f \in E(K) \mid \Sigma_{k \in K} f(k) \leq y\}.$$

We call $F(y)$ a "von Neumann K-catalogue."

DEFINITION 14. If

$$F_n \subset E,\ p_n \geq 0,\ F_n \neq \emptyset \text{ for } n = 1, \ldots, N,\ \Sigma_{n=1}^{N} p_n = 1,$$

then let

$$\Sigma_{n=1}^{N} p_n F_n = \{f \in E \mid f = \Sigma_{n=1}^{N} p_n f_n,\ f_n \in F_n \text{ for } n = 1, \ldots, N\}.$$

THEOREM 6. If for $n = 1, \ldots, N$, F_n is a K-catalogue, and if

$$p_n \geq 0,\ \Sigma_{n=1}^{N} p_n = 1,\ \Sigma_{n=1}^{N} p_n F_n = F,$$

then F is likewise a K-catalogue.

* Stearns [6] calls this "comprehensiveness."

In words: a convex combination of K-catalogues is a K-catalogue. Intuitively, if the coalition can with probability p_n select an arbitrary payoff vector from the catalogue F_n, then it can certainly select an arbitrary payoff vector from the catalogue F.

The proof is straightforward and is omitted.

DEFINITION 15. We call Γ a "K-game whose payoffs are catalogues", or a "K-game for catalogues", if M, N, L, W are nonempty finite sets,

$$z = (z_\ell)_{\ell \in L},\ z_\ell \geq 0 \quad \text{for} \quad \ell \in L,\ \Sigma_{\ell \in L}\, z_\ell = 1.$$

d is a mapping of the cartesian product $M \times N \times L$ into W (so that $d(m, n, \ell) \in W$ for all $m \in M$, $n \in N$, $\ell \in L$), F_w is a K-catalogue for all $w \in W$,

$$X = \{x | x = (x_m)_{m \in M},\ x_m \geq 0 \quad \text{for} \quad m \in M,\ \Sigma_{m \in M}\, x_m = 1\},$$

$$S = \{s | s = (x, (f_{mw})_{m \in M, w \in W}),\ x \in X,\ f_{mw} \in F_w \quad \text{for} \quad m \in M,\ w \in W\},$$

$$T = \{t | t = (t_n)_{n \in N},\ t_n \geq 0 \quad \text{for} \quad n \in N,\ \Sigma_{n \in N}\, t_n = 1\},$$

$$r(s, t) = \Sigma_{m \in M} \Sigma_{n \in N} \Sigma_{\ell \in L}\, x_m t_n z_\ell f_{m, d(m,n,\ell)} \quad \text{and}$$

$$\Gamma = \Gamma(S, T, r).$$

We then also write $\Gamma = \Gamma(M, N, L, W, z, d, (F_w)_{w \in W})$. In a K-game for catalogues, the coalition selects its pure fighting strategies $m \in M$ and its distribution methods* $(f_{mw})_{w \in W}$ with probabilities x_m, where $f_{mw} \in F_w$ for $w \in W$. The opposition independently chooses one of its pure strategies n with probability t_n. Chance selects one of its "strategies" ℓ with probability z_ℓ. If the fighting strategy m and strategies n and ℓ are used, a situation w results in which the coalition is entitled to choose one payoff vector from the catalogue F_w. According to plan it selects f_{mw}. Thus the vector $f_{m,d(m,n,\ell)}$ results with probability $x_m t_n z_\ell$.

* A "strategy" in the sense of Definition 1 is a pair consisting of a "fighting strategy" and a "distribution method."

THEOREM 7. K-games for catalogues are convex.

The proof is straightforward, and is omitted.

Editor's Note: K-games for catalogues are a natural generalization of K-games with unrestricted side payments. In the latter, strategies (more precisely, "fighting strategies") x and t are chosen by the coalition and opposition respectively; the resulting total payoff $v(x, t)$ can then be divided in an arbitrary way by the coalition. Conceivably, the coalition could even throw away part of the total payoff. This means that the actual result of the strategies (x, t) is the von Neumann K-catalogue

$$F = \{f: \Sigma_{k\in K} f(k) \leq v(x, t)\}$$

from which the coalition may choose any payoff vector it pleases (see Fig. 7).

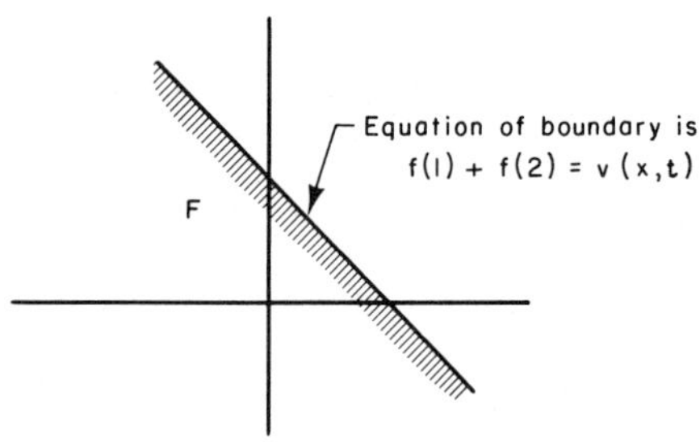

Fig. 7

In a K-game for catalogues, the payoff to a strategy pair (s, t) is not necessarily a von Neumann catalogue, but may be an arbitrary catalogue. This could result from a situation in which side payments are permitted but somehow restricted, or side payments are unrestricted in monetary terms but utility is not linear in money (see Example 11 below).

Briefly: A K-game for catalogues is a finite two-person game in matrix form, in which the entries of the matrix are K-catalogues rather than numbers. The two "players" of the game are the coalition and the opposition. After a play is completed, the coalition picks a payoff vector from the catalogue that resulted from the play. Provision is also made for chance moves.

EXAMPLE 11. If

$$y > 0, B(y) = \{f \in E(K) | \Sigma_{k \in K} \exp(f(k)) \leq y\},$$

then we call $B(y)$ "Bernoullian."

Explanation: An amount y is divided among the coalition members. Following Bernoulli, the members have logarithmic utility functions. Assuming an appropriate choice of scale, coalition member k therefore gets the utility $\log(y(k))$ from the portion $y(k) > 0$. From $\Sigma_{k \in K} y(k) \leq y$ and $f(k) = \log(y(k))$ it follows that $\Sigma_{k \in K} \exp(f(k)) \leq y$. $B(y)$ is convex. Intuitively, $B(y)$ is the set of payoff vectors that are accessible to the coalition if it has the amount y to divide.

2.2. K-Games For Catalogues From $\mathfrak{F}$

In the following, K will be an arbitrary but fixed coalition. When sets of catalogues are mentioned, this will mean that the catalogues are all K-catalogues for the same K.

DEFINITION 16. If $\mathfrak{F}$ is a nonempty set of catalogues, and $\Gamma = \Gamma(M, N, L, W, z, d, (F_w)_{w \in W})$ is a K-game for catalogues which has the property that $F_w \in \mathfrak{F}$ for all $w \in W$, then we call Γ a K-game "for catalogues from $\mathfrak{F}$."

EXAMPLE 12. Let $\mathfrak{F}_1$ be the set of all von Neumann catalogues $F(y)$. We call a K-game for catalogues from $\mathfrak{F}_1$ a "von Neumann K-game."

We will see later that von Neumann K-games are always classical. In order to demonstrate this and more general results, we need some more definitions and preliminary theorems.

DEFINITION 17. A set $\mathfrak{F}$ of catalogues is called "clear" if every K-game for catalogues from $\mathfrak{F}$ is clear. A set $\mathfrak{F}$ of catalogues is called "classical" if every K-game for catalogues from $\mathfrak{F}$ is classical.

THEOREM 8. Every classical set of catalogues is clear.

PROOF. Follows from Theorem 2.

EXAMPLE 13. We stated above that the set $\mathfrak{F}_1$ of von Neumann catalogues is classical, and that this would be demonstrated in the sequel. It can be shown that the set $\mathfrak{F}_2$ of Bernoullian catalogues B(y) (Example 11) is also classical.

DEFINITION 18. If $\mathfrak{F}$ is a nonempty set of catalogues, then

$$CH(\mathfrak{F}) = \{F \subset E(K) \mid F = \Sigma_{n=1}^{N} p_n F_n,\ p_n \geq 0,\ F_n \in F,\ \Sigma_{n=1}^{N} p_n = 1\}.$$

We call $CH(\mathfrak{F})$ the "convex hull" of $\mathfrak{F}$.

THEOREM 9. $CH(\mathfrak{F})$ consists of catalogues.

PROOF. Theorem 6.

THEOREM 10. If $\mathfrak{F}$ is a set of catalogues, Γ a K-game for catalogues from $CH(\mathfrak{F})$, then there is a K-game for catalogues from $\mathfrak{F}$ which is equivalent to Γ.

The idea of the proof is as follows: each strategy triple (opposition, coalition, chance) in Γ yields a catalogue $\mathfrak{F}$ in $CH(\mathfrak{F})$. The catalogue $\mathfrak{F}$ is a probability combination of catalogues $\mathfrak{F}_n$ in $\mathfrak{F}$, say with probabilities p_n. If instead of awarding $\mathfrak{F}$ for the given strategy triple, we allow chance to choose $\mathfrak{F}_n$ with probability p_n and then award $\mathfrak{F}_n$, we get the desired K-game for catalogues from $\mathfrak{F}$ that is equivalent to Γ. Details of the proof are left to the reader.

THEOREM 11. If a set $\mathfrak{F}$ of catalogues is clear, then $CH(\mathfrak{F})$ is also clear. If a set $\mathfrak{F}$ of catalogues is classical, then $CH(\mathfrak{F})$ is also classical.

PROOF. Follows from Theorems 3 and 10.

DEFINITION 19. Let $\mathfrak{F}$ be a nonempty finite set of catalogues, and let

$$F_0 = \bigcap_{F \in \mathfrak{F}} F,$$

$$F_1 \quad \text{the convex hull of} \quad \bigcup_{F \in \mathfrak{F}} F,$$

$$F_p = (1-p)F_0 + pF_1 \quad \text{for} \quad 0 \leq p \leq 1.$$

Then $\mathfrak{F}$ is called r.o. ("real-ordered", i.e., ordered like the reals) if for every $F \in \mathfrak{F}$ there is a p such that $0 \leq p \leq 1$ and

(r.o.1) $$F = F_p,$$

and if for every $f \in F_p$ there is an $f_0 \in F_0$ and an $f_1 \in F_1$ such that

(r.o.2) $$f = (1-p)f_0 + pf_1 \quad \text{and} \quad f_0 \leq f \leq f_1.$$

A set of catalogues is called "r.o." if every finite nonempty subset of catalogues is r.o.

Editor's Note. That (r.o.2) does not follow from (r.o.1) can be seen by the example sketched in Fig. 8. If we define $\mathfrak{F} = \{F_p = (1-p)F_0 + pF_1 | 0 \leq p \leq 1\}$, then

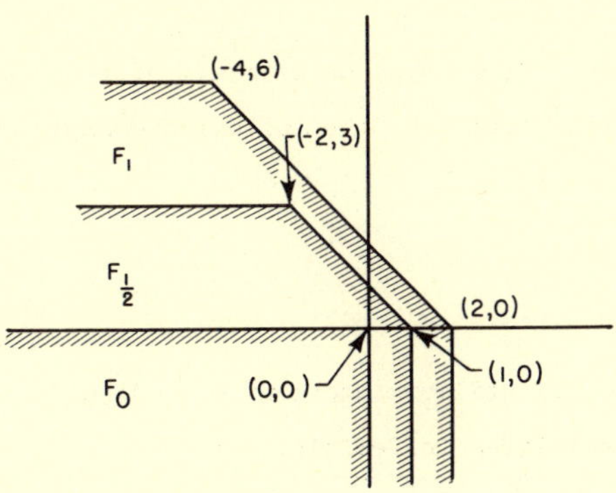

Fig. 8

$\mathfrak{F}$ satisfies (r.o.1) but not (r.o.2); indeed $(-2, 3) \in F_{1/2}$, but there do not exist $f_0 \in F_0$ and $f_1 \in F_1$ such that $f_0 \leq (-2, 3) \leq f_1$ and $f = 1/2\ f_0 + 1/2\ f_1$.

THEOREM 12. $\mathfrak{F}_1$, the set of von Neumann catalogues (see Example 10), is r.o.

This is geometrically evident. The formal proof is left to the reader.

THEOREM 13. Every r.o. set of catalogues is classical.

The idea of the proof is as follows: Because $\mathfrak{F}$ is r.o., it is possible to associate with each F in $\mathfrak{F}$ a real "index"* α such that $(1-p)F_\alpha + pF_\beta = F_{(1-p)\alpha+p\beta}$, and $F_\alpha \supset F_\beta$ if and only if $\alpha \geq \beta$. The coalition is therefore interested in playing in such a way as to maximize the expected value of the index and the opposition in minimizing it. Thus we have essentially a finite** two-person 0-sum game, and such games are classical.*** The details of the proof are left to the reader.

COROLLARY. The set of von Neumann catalogues is classical. Every von Neumann K-game is classical.

PROOF. Theorems 12 and 13.

Theorem 13 does not provide us with all of the classical sets of catalogues. This is illustrated by the following example.

---

* This is the "social utility function" mentioned in the introduction.

** Finiteness is specified in Definition 15.

*** See footnote to Example 3.

EXAMPLE 14. Let

$$K = \{1, 2\},$$

$$F = \{f \mid f(1) + f(2) \leq 1,\ f(1) \leq 1,\ f(2) < 1\},$$

$$H = \{f \mid f(1) + f(2) \leq 1,\ f(1) < 1,\ f(2) \leq 1\},$$

$$\mathfrak{F}_3 = \{F, H\}.$$

$\mathfrak{F}_3$ is obviously not r.o. ((r.o.1) is false). It is, however, classical; indeed any mixed strategy of the opposition that utilizes every pure strategy with positive probability is optimal.

The converse of Theorem 8 is also false. There are clear sets of catalogues that are not classical.

EXAMPLE 15.

$$K = \{1, 2\},$$

$$\mathfrak{F}_4 = \{\{f \mid f(1) \leq 0\},\ \{f \mid f(2) \leq 0\}\}.$$

$\mathfrak{F}_4$ is clear but not classical. Four cases can arise:

$$V = U = \{f \mid f \leq (0, 0)\},$$

$$V = U = \{f \mid f(1) \leq 0\},$$

$$V = U = \{f \mid f(2) \leq 0\},$$

$$V = U = E(\{1, 2\}).$$

Example 15 shows that the definition of catalogue admits quite pathological cases. In the following section we will introduce the concept of a "regular" catalogue. The converses of Theorems 8 and 13 are valid for regular catalogues.

EXAMPLE 16. It can be shown directly that the set $\mathfrak{F}_2$ of Bernoullian catalogues is r.o., from which it follows by Theorem 13 that it is classical (as remarked in Example 13). In addition to the Bernoulli utility functions there is a family of other utility functions (obtained as solutions of a differential equation of third order with one parameter) that generate r.o. and therefore classical sets of catalogues. We will not deal with them here.*

§ 3. K-GAMES FOR REGULAR CATALOGUES

3.1. Regular Catalogues

Inner products, the norm, and limits in E are defined as usual; that is, $f\cdot h = \Sigma_{k\in K} f(k)h(k)$, $|f| = \sqrt{f\cdot f}$, and $\lim f_n = f$ if and only if $\lim f_n(k) = f(k)$ for all k (or equivalently, $\lim |f - f_n| = 0$). Furthermore, we agree that $f > 0$ shall mean $f(k) > 0$ for all $k \in K$.

DEFINITION 20. Let $u \in E$, $u > 0$ and $F \subset E$, $F \neq \emptyset$. Then we define $u(F) = \sup_{f\in F} u\cdot f$, and $F/u = \{f \in F | u\cdot f = u(F)\}$.

DEFINITION 21. A catalogue F is called "regular" if for all $u \in E$ such that $u > 0$ we have

(K4) $u(F) < \infty$,

(K5) F is closed.

EXAMPLE 17. If $f_n \in E$ for $n = 1, \ldots, N$, and $F(f_1, \ldots, f_N) = \{f \in E|$ there are $p_1, \ldots, p_N$ such that $p_n \geq 0$, $\Sigma p_n = 1$, and $\Sigma p_n f_n = f\}$, then we say that F is "generated" by the payoff vectors $f_1, \ldots, f_n$.

EXAMPLE 18. If F is a catalogue, $f \in E$, and for all $f' \in F$ we have $f' \leq f$, then we say F is "bounded" (short for "bounded from above"). Catalogues generated by a finite number of payoff vectors are bounded. Bounded closed catalogues are regular. Catalogues generated by a finite

---

* On the other side of the ledger, it is easy to give examples of utility functions that generate sets of catalogues that are neither r.o. nor clear nor classical.

number of payoff vectors are regular.*

LEMMA 1. If F is a regular catalogue, and if (f_n), $f_n \in F$, $n = 1, 2, \ldots,$ is an arbitrary sequence of payoff vectors in F, then at least one of the following two cases applies:

Case 1: For some subsequence (f_{nr}) of (f_n), $\lim f_{nr} = f \in F$.

Case 2: For some subsequence (f_{nr}), $\lim |f_{nr}| = +\infty$, $\lim |f_{nr}|^{-1} f_{nr} = e \in E$, and $|e| = 1$, $e \leq 0$.

PROOF. The sequence (f_n) is either bounded or unbounded (in norm).

Case 1: If the sequence (f_n) is bounded then it contains a convergent subsequence; this converges to an element of F, because F is closed.

Case 2: If the sequence (f_n) is unbounded then it contains a subsequence whose sequence of norms tends to infinity. If we divide every element of this sequence by its norm, we will obtain a sequence of unit vectors. This is bounded and therefore contains a convergent subsequence. The set of unit vectors is closed; the unit vectors therefore converge to a unit vector. Now it remains to be shown that for this unit vector e we have $e \leq 0$, i.e., $e(k) \leq 0$ for all $k \in K$.

Let us assume there is a $k'' \in K$ such that $e(k'') > 0$. Define $u \in E$ by the expression

$$u(k') = 1 \quad \text{for} \quad k' \neq k'', \; u(k'') = (e(k''))^{-1}(|K| + 1),$$

where $|K|$ is the number of coalition partners. Then

---

* Geometrically, a regular catalogue F is one which has a supporting hyperplane in each strictly positive direction. That is, for each $u > 0$ there is hyperplane perpendicular to u that supports F. von Neumann catalogues are _not_ regular. Bernoullian catalogues _are_ regular. In general, catalogues generated by utility functions v_k of the coalition members are regular if $v_k(y)/y \to 0$ as $y \to \infty$.

$$u \cdot e = u(k'')e(k'') + \Sigma_{k' \neq k''} u(k')e(k')$$

and because $e(k') \geq -1$, $u \cdot e \geq (|K| + 1) - (|K| - 1) = 2$. Because $\lim |f_{nr}|^{-1} f_{nr} = e$, for sufficiently large n_r we have $u \cdot (|f_{nr}|^{-1} f_{nr}) \geq 1$. Then it follows that $u \cdot f_{nr} \geq |f_{nr}|$ for sufficiently large n_r and since $(f_{nr}) \subset F$, $\lim |f_{nr}| = +\infty$, it follows that $u(F) = +\infty$. But since $u > 0$ this contradicts the definition of a regular catalogue. This completes the proof.

THEOREM 14. If F is a regular catalogue, $u \in E$, and $u > 0$, then F/u is a nonempty, bounded, closed and convex set.

PROOF. There is a sequence of payoff vectors f_n such that $f_n \in F$ and $\lim u \cdot f_n = u(F)$. Let us apply Lemma 1 to this sequence. Suppose Case 2 applies; we then have $\lim |f_{nr}|^{-1} = 0$, $\lim u \cdot f_{nr} = u(F) < \infty$, $\lim |f_{nr}|^{-1} f_{nr} = e$ with $|e| = 1$ and $e \leq 0$. Then $\lim |f_{nr}|^{-1}\, u \cdot f_{nr} = 0$ and $\lim u \cdot (|f_{nr}|^{-1} f_{nr}) = u \cdot e < 0$, a contradiction. Therefore, Case 1 must apply: $\lim f_{nr} = f \in F$. For this f, $u \cdot f = \lim u \cdot f_{nr} = u(F)$, therefore $f \in F/u$; so F/u is nonempty. Next, F/u must also be bounded; otherwise there would be a sequence f_n in F/u which contains no convergent subsequence. Lemma 1 applied to this sequence would result in $u \cdot f_{nr} = u(F)$, $\lim |f_{nr}|^{-1} = 0$, therefore $\lim |f_{nr}|^{-1}\, u \cdot f_{nr} = 0$ and $\lim u \cdot (|f_{nr}|^{-1} f_{nr}) = u \cdot e < 0$, a contradiction. Finally, as the intersection of the convex and closed set F and the convex and closed hyperplane $\{f | u \cdot f = u(F)\}$, F/u is closed and convex.

LEMMA 2. If F is a regular catalogue, $h \in E$, and $u \cdot h \leq u(F)$ for all $u \in E$ such that $u > 0$, then $h \in F$.

PROOF. Since F is nonempty and closed (K5), there is an $f \in F$ such that

(1) $$(h-f) \cdot (h-f) \leq (h-f') \cdot (h-f') \quad \text{for all } f' \in F.$$

F is convex. Therefore for the same f, the following also holds:

(2) $$(h-f)\cdot f = \sup_{f'\in F}(h-f)\cdot f'.$$

Because of (K3), $f \leq h$, i.e.,

(3) $$(h-f) \geq 0;$$

for otherwise we could define f' by $f'(k) = \min\{f(k), h(k)\}$. Then we would have $f \in F$ and $f' \leq f$, and so according to (K3), $f' \in F$. Hence

$$(h-f')\cdot(h-f') = (h-f)\cdot(h-f) - \Sigma_{f(k)>h(k)}(f(k) - h(k))^2,$$

contradicting (1); this establishes (3). Next, select an arbitrary but fixed $u \in E$ with $u > 0$. For real $c > 0$, set $u_c = h - f + cu$. Certainly $u_c \in E$ and $u_c > 0$, therefore $h\cdot u_c \leq u_c(F)$, by hypothesis. Therefore

$$(h-f)\cdot h + cu\cdot h \leq \sup_{f'\in F}((h-f)\cdot f' + cu\cdot f') \leq \sup_{f'\in F}(h-f)\cdot f' + c\sup_{f''\in F}u\cdot f''.$$

Application of (2) and Definition 20 yields

$$(h-f)\cdot h + cu\cdot h \leq (h-f)\cdot f + cu(F), \text{ i.e.,}$$
$$(h-f)\cdot(h-f) \leq c(u(F) - u\cdot h) \quad \text{for all} \quad c > 0.$$

Since $u(F) - u\cdot h$ is finite, this is only possible if $(h-f)\cdot(h-f) = 0$; therefore $h = f$. This proves $h \in F$.

THEOREM 15. If F and H are two regular catalogues, $0 \leq p \leq 1$, $G = (1-p)F + pH$, then G is a regular catalogue.

PROOF. It follows from Theorem 6 that G is a catalogue. We must establish the fact that G has the properties (K4) and (K5) in Definition 21.

Let $u \in E$, $u > 0$. Then $u(G) = (1-p)u(F) + pu(H)$. Therefore G has the property (K4). Now we must show that G is closed (K5). For

$p = 0$ $(G = F)$ and $p = 1$ $(G = H)$, there is nothing to prove. So let $0 < p < 1$. If $g_n \in G$ for $n = 1, 2, \ldots,$ and $\lim g_n = g$, then there are sequences $f_n \in F$, $h_n \in H$, $n = 1, 2, \ldots$ such that for all n, $g_n = (1-p)f_n + ph_n$. We apply Lemma 1 to the sequence (f_n).

Case 1: $\lim f_{nr} = f \in F$. Because $h_{nr} = p^{-1}(g_{nr} - (1-p)f_{nr})$, there is an $h \in H$ such that $\lim h_{nr} = h = p^{-1}(g - (1-p)f)$; therefore $g = (1-p)f + ph$, $g \in G$.

Case 2 cannot occur. For if

$$\lim |f_{nr}| = +\infty, \ \lim |f_{nr}|^{-1}f_{nr} = e \le 0,$$

then we would have

$$\lim |h_{nr}| = +\infty \ \text{ and } \ \lim |h_{nr}|^{-1}h_{nr} = -e \ge 0.$$

(p is different from zero and one and the sequence (g_{nr}) is convergent, therefore bounded.) However, Lemma 1 applied to the sequence (h_{nr}) would give the result that $\lim |h_{nr}|^{-1}h_{nr} \le 0$, since (h_{nr}) can contain no convergent subsequence. Since the requirements $|e| = 1$, $e \le 0$, $-e \le 0$ are irreconcilable, the proof is complete.

THEOREM 16. If $\mathfrak{F}$ is a set of regular catalogues, then $CH(\mathfrak{F})$ is also a set of regular catalogues.

PROOF. Follows from Theorem 15.

LEMMA 3. If F and H are two regular catalogues, and if for $g \in E$ we set $P(g) = \{p \mid 0 \le p \le 1,$ $g = (1-p)f + ph,\ f \in F,\ h \in H\}$, then $P(g)$ is convex and closed.

PROOF. Convexity is straightforward. To prove closedness, let $p_n \in P(g)$ for $n = 1, 2, \ldots,$ $\lim p_n = p$. There are sequences $f_n \in F$, $h_n \in H$ such that for $n = 1, 2, \ldots$

(1) $$g = (1-p_n)f_n + p_n h_n.$$

Let us apply Lemma 1 to the sequence f_n.

(2) $$\text{Case 1:} \quad \lim f_{nr} = f, \; f \in F.$$

We again apply Lemma 1 to the appropriate sequence $\{h_{nr}\}$.

Case 1.1. The sequence $\{h_{nr}\}$ has a convergent subsequence. Then we have

$$\lim p_{nr\mu} = p, \; \lim f_{nr\mu} = f, \; \lim h_{nr\mu} = h \in H;$$

passing to the limit, we obtain $g = (1-p)f + ph$, $f \in F$, $h \in H$, and so $p \in P(g)$.

Case 1.2. $\lim f_{nr\mu} = f \in F$, $\lim |h_{nr\mu}| = +\infty$, $\lim |h_{nr\mu}|^{-1} h_{nr\mu} = e \leq 0$, $|e| = 1$. Because of (1) and (2), this is only possible if $\lim p_n = p = 0$. If we show that $g \in F$, then we will have shown that $p \in P(g)$. Now because of (1) and $\lim p_{nr\mu} = 0$ and (2),

$$g - f = \lim (p_{nr\mu} \, h_{nr\mu}) = \lim (p_{nr\mu}|h_{nr\mu}|)(|h_{nr\mu}|^{-1} h_{nr\mu}) = |g - f| e \leq 0.$$

Therefore $g \leq f$, $f \in F$; so according to (K3), $g \in F$.

Case 2. $\lim |f_{nr}| = +\infty$, $\lim |f_{nr}|^{-1} f_{nr} = e$, $|e| = 1$, $e \leq 0$. Again we will apply Lemma 1 to the corresponding sequence $\{h_{nr}\}$.

Case 2.1. A subsequence of h_{nr} converges. This is symmetric to Case 1.2; one proves that $p = 1$ and $g \in H$, from which it follows that $p \in P(g)$.

Case 2.2. $\lim |f_{nr\mu}| = \infty$, $\lim |h_{nr\mu}| = \infty$, $\lim |f_{nr\mu}|^{-1} f_{nr\mu} = e_f$, $\lim |h_{nr\mu}|^{-1} h_{nr\mu} = e_h$, $e_f \leq 0$, $e_h \leq 0$. If for $n > n_0$ it were always true that $0 < p_n < 1$, then it would be true for $n_{r\mu} > n_0$ that $|g - f_{nr\mu}|^{-1} (g - f_{nr\mu}) = |h_{nr\mu} - g|^{-1} (h_{nr\mu} - g)$. In the limit we would have $e_h = -e_f$, and this leads to the contradiction $|e_f| = 1$, $e_f \geq 0$, $e_f \leq 0$. It follows that $p_n = 0$ for infinitely many n, or $p_n = 1$ for infinitely many n; then, however, $\lim p_n = 0$, or, respectively, $\lim p_n = 1$. In each case, though, $p_n = p$ for certain n. Therefore $p \in P(g)$, and the proof is complete.

Note: By repeated application of Lemma 1, we have used both the fact that the catalogues F and H are closed, as well as their property (K4). Lemma 3 is not valid for closed nonregular catalogues.

3.2. Clear Sets of Regular Catalogues

THEOREM 17. If $\mathfrak{F}$ is a clear set of regular catalogues and if F and H are in $\mathfrak{F}$, then $F \cup H$ is convex.

PROOF. For $g \in E$ let $P(g)$ be as in Lemma 3. $P(g) \neq \emptyset$ if and only if g is an element of the convex hull of $F \cup H$. Further, $0 \in P(g)$ if and only if $g \in F$, and $1 \in P(g)$ if and only if $g \in H$. We must show that $0 \in P(g)$ or $1 \in P(g)$ if $P(g) \neq \emptyset$. Let $P(g) \neq \emptyset$. According to Lemma 3, $P(g)$ is convex and closed. Therefore, there are probabilities p', p'', $0 \leq p' \leq p'' \leq 1$ such that $P(g) = \{p \mid p' \leq p \leq p''\}$. Let $c = \text{Min}\,\{p', 1-p''\}$. If $c = 0$, then $p' = 0$ or $p'' = 1$, and $0 \in P(g)$ or $1 \in P(g)$. Therefore it is sufficient to show that $c = 0$.

As an aid let us define a K-game Γ_g for catalogues from $CH(\mathfrak{F})$.

$$\Gamma_g = \Gamma(M,\ N,\ L,\ W,\ z,\ d,\ (F_w)_{w \in W}),\ \text{where}$$
$$M = \{0,\ 1\},\ N = \{0,\ 1\},\ L = \{0\},\ W = \{00,\ 01,\ 10,\ 11\},$$
$$d(m,\ n,\ 0) = mn,$$
$$F_{00} = (1-p' + c)F + (p' - c)H,\ F_{01} = (1-p'')F + p''H$$
$$F_{10} = (1-p')F + p'H,\ F_{11} = (1-p'' - c)F + (p'' + c)H$$

Then

$$X = \{(x_0,\ x_1) \mid x_0 \geq 0,\ x_1 \geq 0,\ x_0 + x_1 = 1\},$$
$$T = \{t = (t_0,\ t_1) \mid t_0 \geq 0,\ t_1 \geq 0,\ t_0 + t_1 = 1\},$$
$$S = \{s = (x_0,\ x_1,\ (f_{mw})_{m \in M, w \in W}),\ (x_0,\ x_1) \in X,\ f_{mw} \in F_w\},$$

and if we simply write f_{mn} in place of $f_{m,\ d(m,n,0)} = f_{m,mn}$, then

$$r(s,\ t) = x_0t_0f_{00} + x_0t_1f_{01} + x_1t_0f_{10} + x_1t_1f_{11}.$$

Choose $t \in T$ arbitrarily, $t = (t_0, t_1)$. We will now define $s \in S$ so that $x_0 = t_1$, $x_1 = t_0$,

$$f_{00} = (1-p' + c)f + (p' - c)h, \quad f_{01} = (1-p'')f + p''h,$$
$$f_{10} = (1-p')f + p'h, \quad f_{11} = (1-p'' - c)f + (p'' + c)h,$$

where $f \in F$, $h \in H$. Then for this t and the s we have just defined, we have $r(s, t) = (1-p)f + ph$, where $p = t_0p' + t_1p''$. Because of the convexity of $P(g)$, we have $p \in P(g)$. We can therefore select $f \in F$ and $h \in H$ such that $(1-p)f + ph = g$. Thus for every $t \in T$ there is an $s \in S$ such that $r(s, t) = g$. Hence $g \in U$, where U is as defined in Definition 7.

$\mathfrak{F}$ was assumed to be clear. According to Theorem 11, $CH(\mathfrak{F})$ is also clear. Γ_g is a K-game for catalogues from $CH(\mathfrak{F})$. Therefore $V = U$ must be true, and hence $g \in V$. So there is an $s \in S$ such that for all $t \in T$ we have $r(s, t) \geq g$. For this s there are two possibilities: 1. $x_0 = 1$, 2. $x_0 < 1$.

1. $x_0 = 1$, $x_1 = 0$. For $t' = (1, 0)$, we must have $r(s, t') \geq g$. But this means $f_{00} \geq g$. So there must be an $f \in F$ and an $h \in H$ such that $f_{00} = (1-p' + c)f + (p' - c)h \geq g$. Because of the property (K3) of the catalogues F and H, there are even $f \in F$ and $h \in H$ such that $(1 - p' + c)f + (p' - c)h = g$. Therefore $(p' - c) \in P(g) = \{p | p' \leq p \leq p''\}$. Because $c \geq 0$, this is only possible if $c = 0$.

2. $x_0 < 1$. For $t'' = (0, 1)$, we have $r(s, t'') \geq g$. So $r(s, t'') = x_0f_{01} + x_1f_{11} \geq g$. Hence there must be elements $f_0 \in F$, $f_1 \in F$, $h_0 \in H$, $h_1 \in H$ such that $x_0(1-p'')f_0 + x_1(1-p'' - c)f_1 + x_0p''h_0 + x_1(p'' + c)h_1 \geq g$. Because of the convexity of F and H, then, there are also elements $f \in F$ and $h \in H$ such that $(1-p'' - x_1c)f + (p'' + c)h \geq g$. As above, we conclude from this that $(p'' + x_1c) \in P(g)$, from which because of $x_1 > 0$ it follows that $c = 0$. Thus it has been shown in each case that we must have $c = 0$.

THEOREM 18. If $\mathfrak{F}$ is a clear set of regular catalogues, $F \in \mathfrak{F}$, $H \in \mathfrak{F}$, $f' \in F$, $h' \in H$, $0 < p < 1$, $g = (1-p)f' + ph'$, then there are elements $f \in F$, $h \in H$ such that $(1-p)f + ph = g$ and moreover, either $f \leq g \leq h$ or $h \leq g \leq f$.

PROOF. By the given assumptions, g is an element of the convex hull of $F \cup H$. According to Theorem 17, $F \cup H$ is itself convex, therefore $g \in F \cup H$. There are three possible cases: (1) $g \in F$ and $g \in H$, (2) $g \in F$ and $g \notin H$, (3) $g \notin F$ and $g \in H$.

1. If $g \in F$ and $g \in H$, then we can select $f = g = h$.

2. This case is symmetric to the third case.

3. Assume $g \notin F$, $g \in H$. We define a K-game with catalogues from $\mathfrak{F}$ as follows:

$$\Gamma'_g = \Gamma(M, N, L, W, z, d, (F_w)_{w\in W}) \text{ with } M = \{0\}, N = \{0, 1\},$$
$$L = \{0, 1\}, W = \{0, 1\}, z = (z_0, z_1) = (1-p, p),$$
$$d(0, 0, 0) = 0, d(0, n, \ell) = 1 \text{ for } (n, \ell) \neq (0, 0),$$
$$F_0 = F, F_1 = H.$$

In simplified notation we have the following:

$$S = \{s | s = (f, h), f \in F, h \in H\},$$
$$T = \{t | t = (t_0, t_1), t_0 \geq 0, t_1 \geq 0, t_0 + t_1 = 1\},$$
$$r(s, t) = t_0(1-p)f + (t_0p + t_1)h.$$

Choose $t \in T$ arbitrarily, $t = (t_0, t_1)$. Because $p > 0$ and $t \in t$, $(t_0p + t_1) > 0$. Because $h' \in H$ and $g \in H$, since H is convex, $(t_0p + t)^{-1}(t_0ph' + t_1g) = h'' \in H$. If we set $s = (f', h'')$, then $s \in S$ and

$$r(s, t) = t_0(1-p)f' + (t_0p + t_1)(t_0p + t_1)^{-1} (t_0ph' + t_1g)$$
$$= t_0((1-p)f' + ph') + t_1g.$$

It was assumed that $g = (1-p)f' + ph'$, therefore $r(s, t) = g$. So for every $t \in T$ there is an $s \in S$ such that $r(s, t) = g$. Therefore $g \in U$ (where U is as in Definition 7).

$\mathfrak{F}$ was assumed to be clear, therefore the K-game Γ'_g for catalogues from $\mathfrak{F}$ must be clear. Since $g \in U$ it follows that $g \in V$, so $g \in V(s)$ for some $s \in S$. So there must be an $\tilde{f} \in F$ and an $h \in H$ such that for all $t \in T$, we have $r((\tilde{f}, h), t) \geq g$. In particular this inequality must hold for $t = (0, 1)$ and for $t = (1, 0)$. The following inequalities result: $h \geq g$ and $(1-p)\tilde{f} + ph \geq g$. If we set $f = (1-p)^{-1}(g-ph)$, then because of $\tilde{f} \geq f$ and (K3) for F, we deduce $f \in F$. So $h \geq g \geq f$ and $(1-p)f + ph = g$.

LEMMA 4. If $\mathfrak{F}$ is a clear set of regular catalogues, $F \in \mathfrak{F}$, $H \in \mathfrak{F}$, $u \in E(K)$, $u > 0$, $u(F) \leq u(H)$, then there is an $f \in F/u$ and an $h \in H/u$ such that $f \leq h$.

PROOF. By Theorem 14, F/u and H/u are nonempty. Therefore there are $f' \in F/u$ and $h' \in H/u$. We will select a p, $0 < p < 1$, set $g = (1-p)f' + ph'$ and apply Theorem 18. As a result we obtain an $f \in F$ and an $h \in H$ such that $g = (1-p)f + ph$ and $f \leq g \leq h$ or $h \leq g \leq f$. Since $f' \in F/u$, $h' \in H/u$, therefore $(1-p)u \cdot f + pu \cdot h = u \cdot g = (1-p)u(F) + pu(H)$. Because of $0 < p < 1$ and the defintion of $u(F)$ and $u(H)$, this is only possible if $u \cdot f = u(F)$ and $u(H) = u \cdot h$. Therefore $f \in F/u$, $h \in H/u$.

If we had $f \geq h$ and $f \neq h$, then we would have $u \cdot f > u \cdot h$, which is not so because $u(F) \leq u(H)$. Therefore $f \leq h$.

THEOREM 19. If $\mathfrak{F}$ is a clear set of regular catalogues, $F \in \mathfrak{F}$, $H \in \mathfrak{F}$, and if for some $\underline{u} \in E$ such that $\underline{u} > 0$ we have $\underline{u}(F) < \underline{u}(H)$, then for all $u \in E$ such that $u > 0$, we likewise have $u(F) < u(H)$.

PROOF. It is clearly sufficient to prove the assertion for all $u \in E$ such that $u > \underline{u}$. So let $\underline{u} < u$, $u \in E$. According to Lemma 4 there are payoff vectors $f_0 \in F/\underline{u}$, $h_0 \in F/\underline{u}$ such that $f_0 \leq h_0$. Likewise there is an $f' \in F/u$ and an $h' \in H/u$ such that $f' \leq h'$ or $h' \leq f'$. We will now select a natural number $N \geq 1$ such that $N(\underline{u}(h) - \underline{u}(F)) \geq (u-\underline{u}) \cdot (f' - f_0)$. For $n = 0, \ldots, N$ we set $u = N^{-1}(N-n)\underline{u} + N^{-1}nu$. Then $u_0 = \underline{u}$, $u_N = u$ and for $n < N$, $(u_{n+1} - u_n) = N^{-1}(u - \underline{u})$. For $n = 0, \ldots, N$, according to Lemma 4 there are elements $f_n \in F/u_n$ and $h_n \in H/u_n$ such that $f_n \leq h_n$ or $h_n \leq f_n$; choose $f_N = f'$, $h_N = h'$. We have $u_0(F) < u_0(H)$.

We now prove that if $u_n(F) < u_n(H)$ for $n = 0, \ldots, m-1$, $1 \leq m \leq N$, then $u_m(F) < u_m(H)$. By complete induction we will then obtain

$u_N(F) < u_N(H)$, and therefore, as asserted, $u(F) < u(H)$.

Let $0 < m \leq N$, $u_n(F) < u_n(H)$ for $n = 0, \ldots, m-1$. Then $f_n \leq h_n$ for $n = 0, \ldots, m-1$. We set $d_n = (u_{n+1}(H) - u_{n+1}(F)) - (u_n(H) - u_n(F))$ for $n = 0, \ldots, N-1$. Then $u_m(H) - u_m(F) = u_0(H) - u_0(F) + \Sigma_{n=1}^{m-1} d_n$. Now $d_n = (u_{n+1} - u_n)\cdot(h_n - f_n) + u_{n+1}\cdot(h_{n+1} - h_n) - u_{n+1}\cdot(f_{n+1} - f_n)$. By induction hypothesis, $h_n - f_n \geq 0$ for $n = 0, \ldots, m-1$. Moreover, it was assumed that $u > \underline{u}$, therefore $u_{n+1} - u_n > 0$. Consequently it follows that $(u_{n+1} - u_n)\cdot(h_n - f_n) > 0$ for $n = 0, \ldots, m-1$. Because $h_{n+1} \in H/u_{n+1}$ and $h_n \in H$, we have $u_{n+1}\cdot(h_{n+1} - h_n) \geq 0$ for $n = 0, \ldots, N-1$. Similarly, $u_{n+1}\cdot(f_{n+1} - f_n) \geq 0$ and $u_n\cdot(f_{n+1} - f_n) \leq 0$, and a fortiori $(u_{n+1} - u_n)\cdot(f_{n+1} - f_n) \geq u_{n+1}\cdot(f_{n+1} - f_n) \geq 0$ for $n = 0, \ldots, N-1$. We arrive at the estimate

$$u_m(H) - u_m(F) > u_0(H) - u_0(F) - \Sigma_{n=0}^{N-1} N^{-1} (u-\underline{u})\cdot(f_{n+1} - f_n)$$

$$= \underline{u}(H) - \underline{u}(F) - N^{-1} (u-\underline{u})\cdot(f' - f_0).$$

However, N was so defined that the right side is greater than zero; so $u_m(H) > u_m(F)$. Thus the general inductive step has been proved. Hence $u(H) > u(F)$, and the proof is complete.

COROLLARY 1. If $\mathfrak{F}$ is a clear set of regular catalogues, $F \in \mathfrak{F}$, $H \in \mathfrak{F}$, and if for some $u \in E$ such that $u > 0$ we have $u(F) = u(H)$, then $F = H$.

PROOF. By Theorem 19, $u(F) = u(H)$ for all $u \in E$ with $u > 0$. For all $f \in F$, $u\cdot f \leq u(H)$ for all $u > 0$; for all $h \in H$, $u\cdot h \leq u(F)$ for all $u > 0$. Applying Lemma 2 twice, we obtain $F \subset H$ and $H \subset F$, therefore $F = H$.

COROLLARY 2. Every clear set of regular catalogues is completely ordered by inclusion.

PROOF. If F and H are two catalogues of this set, and $u > 0$ is chosen arbitrarily, then $u(F) \leq u(H)$ or $u(H) \leq u(F)$. Theorem 19 in combination with Lemma 2 then yields $F \subset H$ or $H \subset F$.

THEOREM 20. Every clear set of regular catalogues is r.o.

PROOF. Let $\mathfrak{F}$ be an arbitrary clear set of regular catalogues. Then every nonempty finite subset of $\mathfrak{F}$ is also clear. Now let $\mathfrak{F}$ be a nonempty finite set of regular catalogues, F_1 = convex hull of $\bigcup_{F\in\mathfrak{F}} F$, $F_0 = \bigcap_{F\in\mathfrak{F}} F$ and $F_p = (1-p)F_0 + pF_1$ for $0 \leq p \leq 1$. $\mathfrak{F}$ is finite and, according to Corollary 2 of Theorem 19, ordered by inclusion. Therefore the intersection F_0 and the union of the catalogues in $\mathfrak{F}$ are themselves in $\mathfrak{F}$. The union is therefore convex, so F_1 is contained in $\mathfrak{F}$. Thus F_p is contained in $CH(\mathfrak{F})$ for all p, $0 \leq p \leq 1$. $CH(\mathfrak{F})$ is a clear set of regular catalogues (Theorems 11 and 16). Now let $F \in \mathfrak{F}$; then $F \in CH(\mathfrak{F})$. Choose $u \in E$ with $u > 0$. Because $F_0 \subset F \subset F_1$, $u(F_0) \leq u(F) \leq u(F_1)$. Therefore there is a p with $0 \leq p \leq 1$ such that $u(F) = (1-p)u(F_0) + pu(F_1)$. Now $u(F_p) = (1-p)u(F_0) + pu(F_1) = u(F)$. F and F_p both belong to the clear set of regular catalogues $CH(\mathfrak{F})$; so by Corollary 1 of Theorem 19, they are equal, i.e., $F = F_p$. This is condition (r.o.1).

Condition (r.o.2) follows from Theorem 18 for $0 < p < 1$; for if $f_0 \in F_0$ and $f_1 \in F_1$ are such that $f = (1-p)f_0 + pf_1$ and $f_0 \geq f \geq f_1$, then because $F_0 \subset F_1$ we have $f_0 \in F_1$; and by property (K3) of the catalogue F_0, $f_1 \in F_0$. When $p = 0$ or 1, condition (r.o.2) is easily verified.

If $\mathfrak{F}$ is an arbitrary clear set of regular catalogues, then according to what was just shown, every nonempty finite subset of $\mathfrak{F}$ is r.o. Then by Definition 19, $\mathfrak{F}$ is r.o.

THEOREM 21. For sets $\mathfrak{F}$ of regular catalogues, the properties "$\mathfrak{F}$ is clear", "$\mathfrak{F}$ is r.o.", and "$\mathfrak{F}$ is classical" are equivalent.

PROOF. If $\mathfrak{F}$ is a set of regular catalogues, then the following holds: If $\mathfrak{F}$ is clear, then $\mathfrak{F}$ is r.o. (Theorem 20). If $\mathfrak{F}$ is r.o, then $\mathfrak{F}$ is classical (Theorem 13). If $\mathfrak{F}$ is classical, then $\mathfrak{F}$ is clear (Theorem 8).

It can be shown that the set of Bernoullian catalogues is r.o., and therefore classical.

BIBLIOGRAPHY

[1] AUMANN, R. J., "Cooperative games without side payments," in Recent Advances in Game Theory, papers delivered at a meeting of the Princeton University Conference, October 4-6, 1961, privately printed for members of the conference, pp. 83-99.

[2] AUMANN, R. J. and PELEG, B., "von Neumann-Morgenstern solutions to cooperative games without side payments," Bull. Amer. Math. Soc. 66, 1960, pp. 173-179.

[3] BLACKWELL, D., "An analog of the minimax theorem for vector payoffs," Pac. J. of Math. 6, 1956, pp. 1-8.

[4] VON NEUMANN, J. and MORGENSTERN, O., Theory of Games and Economic Behavior, Princeton University Press, 1944.

[5] SHAPLEY, L. S. and SHUBIK, M., "Solutions of n-person games with ordinal utilities," (abstract) Econometrica 21, 1953, p. 348.

[6] STEARNS, R., "Three-person Cooperative Games Without Side Payments." This Study.

Gerd Jentzsch

Institute for Applied Economic Research
Tübingen

THE BARGAINING SET FOR COOPERATIVE GAMES*

Robert J. Aumann and Michael Maschler

§ 1. INTRODUCTION

This paper grew out of an attempt to translate into mathematical formulas what people may argue when faced with a cooperative n-person game described by a characteristic function.

The basic difficulty in n-person game theory is due to the lack of a clear meaning as to what is the purpose of the game. Certainly, the purpose is not just to get the maximum amount of profits, because if every player will demand the maximum he can get in a coalition, no agreement will be reached. Thus, one decides that the purpose of the game is to reach some kind of stability, to which the players would or should agree if they want any agreement to be enforced. This stability should reflect in some sense the power of each player, which results from the rules of the game.

In this paper, we assume that all the players can "bargain" together, with perfect communication, and settle at a "stable" outcome which is based on the "threats" and "counter threats" that they possess. The set of all the stable outcomes, called the <u>bargaining set</u>,** is defined in Section 2 and some of its properties are discussed. In particular, this set can always be determined by solving systems of algebraic linear inequalities.

The bargaining sets for the 2- and 3-person games are fully described (Sections 3, 4, 5) and some cases of 4-person games are treated, in which not all the coalitions are permissible (Sections 6, 8).

---

* This research was sponsored in part by the Office of Naval Research under Nonr-1858(16), and in part by the Carnegie Corporation of New York.

** The definition of the bargaining set appeared in [1].

Some counter examples for various conjectures, as well as existence theorems, are treated in Section 7, and possible modifications are suggested in Sections 9 and 10. In Section 11 we discuss some similarities and deviations between our theory and other solution concepts, such as W. Vickrey's concept of self-policing patterns [8] and D. Luce's concept of ψ-stability [3]. We also outline how our theory can be modified if the game is given by Thrall's characteristic function [7] or by the Aumann-Peleg characteristic function for cooperative games without side payments.

We conclude by pointing out that "central parts" of the von Neumann-Morgenstern solution for some games appear also in the bargaining set. The reason for this phenomenon is obscure to us.

§ 2. THE BARGAINING SET

We shall consider an n-person cooperative game Γ, described by its characteristic function. More precisely, a set $N = \{1, 2, \ldots, n\}$ of n players is given, together with a collection $\{B\}$ of non-empty subsets B of N, called <u>permissible coalitions</u>. For each B, $B \in \{B\}$, a number $v(B)$ is given and it is called <u>the value of the coalition</u> B.

For the sake of simplicity, we shall assume throughout this paper that all 1-person coalitions are in $\{B\}$ and have a zero value, i.e.,

$$i \in \{B\}, \quad v(i) = 0. \tag{2.1.a}$$

In addition, we shall also assume that

$$v(B) \geq 0, \quad B \in \{B\}. \tag{2.1.b}$$

It will turn out later that no essential change will occur if we add to $\{B\}$ all other, non-permissible coalitions, assigning them the value zero.

A <u>payoff configuration</u> (p.c.) will now be defined as an expression of the form

$$(x; \mathcal{B}) \equiv (x_1, x_2, \ldots, x_n; B_1, B_2, \ldots, B_m), \tag{2.2}$$

where $B_1, B_2, \ldots, B_m$ are mutually disjoint sets of $\{B\}$ whose union is N ; i.e.,

$$(2.3) \qquad B_j \cap B_k = \phi, \quad j \neq k; \quad \bigcup_{j=1}^{m} B_j = N;$$

and the x_i are real numbers which satisfy

$$(2.4) \qquad \sum_{i \in B_j} x_i = v(B_j); \quad j = 1, 2, \ldots, m.$$

A p.c. is therefore a representation of a possible outcome of the game, in which the players divide themselves into the coalitions $B_1, B_2, \ldots, B_m$, each coalition shares its value among its members, and each player receives the amount x_i, $i = 1, 2, \ldots, n$.

When people are faced with such a game, each one trying to get as high an amount as he thinks he can get, it is reasonable to expect that some of the p.c.'s will never form. E.g., one does not expect that a p.c. will occur with $x_{i_0} < 0$, since the player i_0 alone can secure more by playing as a 1-person coalition. We are willing to make a strong assumption, namely, that the outcome (2.2) will be a <u>coalitionally rational p.c.</u> (c.r.p.c.),* i.e., for each B, $B \in \{B\}$, $B \subset B_j$, $j = 1, 2, \ldots, m$,

$$(2.5) \qquad \sum_{i \in B} x_i \geq v(B).$$

Thus, we assume that a coalition will not form if some of its members can obtain more by themselves forming a permissible coalition.

The assumption of coalition rationality differs from the assumption of belonging to the core by the restricting condition $B \subset B_j$. This restriction avoids some of the difficulties which arise when dealing with games whose core is empty. (See R. D. Luce and H. Raiffa [3].)

In itself, the coalition rationality assumption is a very strong one, as it forces the game to be essentially superadditive within those coalitions which are actually formed. Indeed, a coalition whose value is less than the sum of the values of disjoint subcoalitions cannot occur in any c.r.p.c., and can as well be declared non-permissible or have its value be

---

* In R. J. Aumann and M. Maschler [1], a c.r.p.c. is called a p.c.

replaced by zero. Moreover, this assumption is open to the same theoretical objections which are discussed at length in R. D. Luce and H. Raiffa [3]. As a matter of fact, our theory can be developed without the coalition rationality assumption, as indicated in Section 10. Nevertheless, as we are only interested in "stable" outcomes, we feel it instructive to make this assumption.

Several phenomena can be observed when watching people who are confronted with a game such as described above. Usually, negotiations start, each one tries to get at least as much as he expects, and at the same time there is an attempt to enter into a "safe" coalition. This latter factor applies, in particular, to those coalitions which are planned to operate for a long period. The search for "safety" gives rise to feelings of sympathy and antipathy which play an important role in the final decisions. Guarantees of all kinds are demanded, contracts are signed, etc. *If people do not feel safe enough, they often do not enter a coalition even if they can win more in it*.

The demand for safety is usually considered legitimate and a sound way to convince the partners to get a smaller amount of profit *in order that no one in the coalition will feel deprived*. There is a desire for "fair play", which can be achieved in various ways. Often it is accepted that "if all things are equal" it is "fair" to divide the profits equally. Sometimes people share the profits according to some fixed ratio established by other precedents, etc.

If "all things" are *not* equal, people will still be happy with their coalition if they agree that the "stronger" partners will get more. Thus, during the negotiations, prior to the coalition formation, each player will try to convince his partners that in some sense he is strong. This he can try in various ways, among which an important factor is his ability to show that he has other, perhaps better, alternatives. His partners, besides pointing out their own alternatives, may argue in return that even without his help they can perhaps keep their proposed shares. Thus, a negotiation quite often takes the form of a sequence of "threats" and "counter threats," or "objections" against "counter objections." It is this principle that we shall try to formulate mathematically. It seems that a certain kind of stability is reached if all objections can be answered by counter objections.

Perhaps it is not enough that any objection by one person could be met. It is possible that a subset of the players of a coalition unite, during the negotiation period, and threaten another subset. If we insist on a strong stability, we have to take care also of such threats. This, in fact, will be done.

To be sure, there are other means used during the bargaining period, such as threats based on the so-called "interpersonal comparison of utilities," sanctions in other games, propaganda, etc. These will be ignored in this paper.

The following example will illustrate our purpose. Let $n = 3$, $v(1) = v(2) = v(3) = v(123) = 0$, $v(12) = 100$, $v(13) = 100$, $v(23) = 50$. Consider the p.c.

(2.6) $$(80, 20, 0; 12, 3).$$

Now, player 2 can object by pointing out that in the p.c.

(2.7) $$(0, 21, 29; 1, 23)$$

he and player 3 get more. Player 1 has no counter objection because he cannot keep his 80 while offering player 3 at least 29. Thus, (2.6) is unstable. On the other hand,

(2.8) $$(75, 25, 0; 12, 3)$$

is stable. An objection of player 2, e.g.,

(2.9) $$(0, 26, 24; 1, 23),$$

can be met by a counter objection

(2.10) $$(75, 0, 25; 13, 2);$$

or an objection of player 1, e.g.,

(2.11) $(76,\ 0,\ 24;\ 13,\ 2)$

can be met by the counter objection

(2.12) $(0,\ 25,\ 25;\ 1,\ 23)$.

In these counter objections, the threatened player can keep his share and offer his partners at least what the player who objects offered. It will turn out that the only stable p.c.'s in this game are

(2.13) $(0,0,0;\ 1,2,3)$; $(75,25,0;\ 12,3)$; $(75,0,25;\ 13,2)$; $(0,25,25;\ 1,23)$.

Let it be said at once that our paper was largely motivated by the fact that most of our friends, to whom this game was presented, started their considerations from the p.c.'s (2.8) and (2.10). We tried to find what characterizes these p.c.'s and how they can be generalized to more complicated cases.

Let Γ be a game, as described above. Let K be a non-empty subset of the set of players N. A player i will be called a <u>partner of</u> K <u>in a p.c.</u> $(x;\mathcal{B})$, if he is a member of a coalition in $\mathcal{B}$ which intersects K. The set $P[K;\ (x;\mathcal{B})]$ of all the partners of K in $(x;\mathcal{B})$ is, therefore,

(2.14) $$P[K;\ (x;\mathcal{B})] \equiv \{i \mid i \in B_j,\ B_j \cap K \neq \phi\}.$$

Note that $K \subset P[K;\ (x;\mathcal{B})]$; i.e., each member of K is also a partner of K, contrary to everyday usage. K needs only the consent of its partners in order to get its part of x.

DEFINITION 2.1. Let $(x;\mathcal{B})$ be a coalitionally rational payoff configuration (2.2), (2.5) for a game Γ. Let K and L be non-empty disjoint subsets of a coalition B_j which appears in $(x;\mathcal{B})$. An <u>objection</u> of K against L in $(x;\mathcal{B})$ will be a c.r.p.c.

(2.15) $$(\mathscr{Y}; \mathscr{C}) \equiv (y_1, y_2, \ldots, y_n; C_1, C_2, \ldots, C_\ell)$$

for which

(2.16) $$P[K; (\mathscr{Y}; \mathscr{C})] \cap L = \phi,$$

(2.17) $$y_i > x_i \quad \text{for all} \quad i, i \in K,$$

(2.18) $$y_i \geq x_i \quad \text{for all} \quad i, i \in P[K; (\mathscr{Y}; \mathscr{C})].$$

Verbally, in their objection, players K claim that, without the aid of players L ((2.16)), they can get more in another c.r.p.c. ((2.17)), and the new situation is reasonable because their new partners do not get less than what they got in the previous p.c. ((2.18)).

DEFINITION 2.2. Let $(x; \mathscr{B})$ be a coalitionally rational payoff configuration (2.2), (2.5) in a game Γ, and let $(\mathscr{Y}; \mathscr{C})$ be an objection of a set K against a set L in $(x; \mathscr{B})$, $K, L \subset B_j$. A <u>counter objection</u> of L against K is a c.r.p.c.

(2.19) $$(\mathscr{Z}; \mathscr{D}) \equiv (z_1, z_2, \ldots, z_n; D_1, D_2, \ldots, D_k)$$

for which

(2.20) $$P[L; (\mathscr{Z}; \mathscr{D})] \not\supset K,$$

(2.21) $$z_i \geq x_i \quad \text{for all} \quad i, i \in P[L; (\mathscr{Z}; \mathscr{D})],$$

(2.22) $$z_i \geq y_i \quad \text{for all} \quad i, i \in P[L; (\mathscr{Z}; \mathscr{D})] \cap P[K; (\mathscr{Y}; \mathscr{C})].$$

Verbally, in their counter objection, players L claim that they can hold their original properties ((2.21)), promise their partners at least their original share ((2.21)), and if they need partners of K in his objection, they can give them not less than what they were offered in the objection ((2.22)). Sometimes, the members of L have to use the tactics of "divide and rule" by using members of K as partners, but they may not use all members of K ((2.20)).

DEFINITION 2.3. A c.r.p.c. $(x; \mathcal{B})$ is called stable if for each objection of a K against an L in $(x; \mathcal{B})$ there is a counter objection of L against K. The bargaining set $\mathfrak{M}$ of a game Γ is the set of all stable c.r.p.c.'s.

The feeling of "safety" suggested by this definition lies in the assurance that all threats within a coalition can be met. It may be felt, perhaps, that there is a lack of symmetry when comparing (2.16) and (2.20), but the situation is not symmetric in the first place. An objection (2.15) can serve, in general, as an objection of K (or another "K") against various groups "L", each one of which has to have a counter objection.

To be sure, even if there is a desire for a stability as demanded in the p.c.'s of the bargaining set, this does not mean that the outcome will belong to $\mathfrak{M}$. A player, e.g., may agree to sacrifice some of his profits in order to make sure that he enters a coalition. Other factors, mentioned above, may also cause deviations from $\mathfrak{M}$. However, if the demand for stability is strong enough, we hope that the outcome will not be too far from $\mathfrak{M}$; in this sense the theory has a normative aspect. Moreover, as the number of the players increases, there arise many possible threats, and, using the concepts involved in the definition, one may compute and show the players where they are "safe" and what threats they do possess. This is another normative aspect.*

The bargaining set is never empty. Indeed, (0, 0, ..., 0; 1, 2, ..., n) always belongs to $\mathfrak{M}$.

In a coalition of zero value, any objection (if there is one) can be countered by the other players playing as 1-person coalitions.

A dummy always gets zero in a c.r.p.c., therefore he cannot belong to any objecting K. On the other hand, he can always keep his zero by

---

* Results close to the bargaining set have been observed in an experimental study [4].

playing alone. He can be of no use for any objection or counter objection, since the same can be effected without his help. Thus, a dummy has no essential effect on $\mathfrak{M}$.

The definition of $\mathfrak{M}$ does not use "interpersonal comparisons of utilities" and it is independent of the names of the players.

THEOREM 2.1. The bargaining set $\mathfrak{M}$ of a game Γ can be represented as the set of solutions of a conjunctive-disjunctive* system of linear inequalities involving the x_i as unknowns. It is, therefore, a union of a finite number of polyhedral convex sets in the n-space with coordinates $(x_1, x_2, \ldots, x_n)$.

PROOF.** In any finite expression with coordinates which has the form of quantifiers followed by linear inequalities connected by the words "or" and "and", the free variables — if such exist — which satisfy the expression are those and only those which satisfy a certain disjunctive-conjunctive system of linear inequalities. This is a known theorem in logic, but for the sake of completeness we sketch the proof. It is sufficient to prove the theorem when there is only one quantifier. Moreover, we may assume that this quantifier is $\exists$, because $\forall = \sim \exists \sim$. The theorem now follows from the fact that the projection of a polyhedron is a polyhedron.

§ 3. THE TWO-PERSON GAME

The bargaining set $\mathfrak{M}$ for the game:

$$v(1) = v(2) = 0 \qquad v(12) = a \geq 0, \tag{3.1}$$

consists of all possible c.r.p.c.'s; i.e.,

---

* I.e., a system of linear inequalities connected by the words "or" and "and".

** We are indebted to Professor M. Rabin and Professor A. Robinson for pointing this out.

$$(3.2) \quad \begin{cases} (0,\ 0;\ 1,\ 2) & \\ (x_1,\ x_2;\ 12) & x_1 + x_2 = a,\ \ x_1 \geq 0,\ \ x_2 \geq 0. \end{cases}$$

Indeed, there are no possible objections.

§ 4. THE 3-PERSON GAME. PERMISSIBLE COALITIONS OF LESS THAN THREE PLAYERS

In this section we shall study the game:

$$(4.1) \quad v(1) = v(2) = v(3) = 0;\ v(12) = a;\ v(23) = b;\ v(13) = c;\ a,\ b,\ c \geq 0.$$

THEOREM 4.1. In the game (4.1), essentially two cases arise.

Case A. If a, b, c satisfy the "triangle inequality"

$$(4.2) \quad a \leq b + c,\ b \leq a + c,\ c \leq a + b,$$

then the bargaining set $\mathfrak{M}$ is:

$$(4.3) \quad \begin{cases} (\ 0\ ,\ 0\ ,\ 0\ ;\ 1,\ 2,\ 3) \\ (\frac{a+c-b}{2},\ \frac{a+b-c}{2},\ 0\ ;\ 12,\ 3\ ) \\ (\frac{a+c-b}{2},\ 0\ ,\ \frac{c+b-a}{2};\ 13,\ 2\ ) \\ (\ 0\ ,\ \frac{a+b-c}{2},\ \frac{c+b-a}{2};\ 1,\ 23\ ) \end{cases}$$

Case B. If, e.g., $a > b + c$, then the bargaining set $\mathfrak{M}$ is:

$$(4.4) \quad \begin{cases} (\ 0\ ,\ 0\ ,\ 0\ ;\ 1,\ 2,\ 3) \\ (c \leq x_1 \leq a-b,\ a - x_1,\ 0\ ;\ 12,\ 3\ ) \\ (\ c\ ,\ 0\ ,\ 0\ ;\ 13,\ 2\ ) \\ (\ 0\ ,\ b\ ,\ 0\ ;\ 1,\ 23\ ) \end{cases}$$

Before proving this theorem, we shall give some illustrations which will throw some light on the nature of the bargaining sets.

Example 1. Let $a = 100$, $b = 100$, $c = 50$. The triangle inequality is satisfied, and therefore $\mathfrak{M}$ is discrete: $\{(0, 0, 0; 1, 2, 3),$ $(25, 75, 0; 12, 3)$ $(25, 0, 25; 13, 2)$ $(0, 75, 25; 1, 23)\}$.

One can approach this solution also by the following intuitive argument: Suppose that player 1 receives α, then player 2 gets $100 - \alpha$ and he is thus willing to pay player 3 at most $100 - (100 - \alpha) = \alpha$. Thus, player 3 will be willing to pay player 1 at most $50 - \alpha$. If $50 - \alpha > \alpha$, then player 1 will prefer to join player 3. This will cause player 2 to agree to get less. If $50 - \alpha < \alpha$, player 2 will demand more as he will get more from player 3 if player 1 insists on getting α. Thus an equilibrium will be reached only if $\alpha = 50 - \alpha$, in which case $\alpha = 25$.

Example 2. The above argument fails in the case $a = 20$, $b = 30$, $c = 100$. Here one obtains $\alpha = 45$ in which case player 2 will lose money. This he can avoid by playing alone. Our bargaining set is no longer discrete: $\{(0, 0, 0; 1, 2, 3)$ $(20 \leq x_1 \leq 70, 0, 100 - x_1; 13, 2)$ $(20, 0, 0; 12, 3)$ $(0, 0, 30; 1, 23)\}$. One can reason as follows: Player 1, being in the coalition 13, will not be satisfied in getting less than 20, since otherwise he will do better by joining player 2. Similarly, player 3 will demand at least 30. Fortunately, both demands can be satisfied, and player 2 cannot cause any harm since he is a weak player.

Example 3. Let $a = 100$, $b = 100$, $c = 0$. We observe that the bargaining set is again discrete: $\{(0, 0, 0; 1, 2, 3)$ $(0, 100, 0; 12, 3)$ $(0, 0, 0; 13, 2)$ $(0, 100, 0; 1, 23)\}$. This solution reflects the character of an "unrestricted competition" in our bargaining set. Indeed, player 2 can practically receive the amount 100 because whatever the positive demand of player 1 will be, player 3 will be "satisfied" in getting less, and vice versa. One observes that our theory does not take into account the psychological threat that player 2 may also "lose" his profit 100, and probably will therefore be willing to pay some amount in order to be in a coalition with player 1 or with player 3. In practical situations several side conditions may come into consideration such as: (1) It may be customary not to enter a coalition unless a certain minimum amount or percentage of profit is guaranteed in advance. (2) A "Cartel" agreement is decided between player 1 and player 3, in which both of them declare not to enter a coalition with player 2 without getting at least a certain amount of profit. (3) There is enforced a "Cartel" or an "Anti-Cartel" law in the country. (4) It is known that in order to ensure a certain profit, one is willing to give up a certain amount or percentage in order to "push" an equilibrium situation to his side.

Proof of Theorem 4.1. Certainly, $(0, 0, 0; 1, 2, 3) \in \mathfrak{M}$.

Next, let us examine under what circumstances a payoff configuration $(x_1, x_2, 0; 12, 3)$ can belong to the bargaining set. It should be coalitionally rational, and therefore it must satisfy

$$x_1 \geq 0, \quad x_2 \geq 0; \quad x_1 + x_2 = v(12). \tag{4.5}$$

LEMMA 1. A necessary and sufficient condition that player 1 has no objection is:

$$x_1 \geq v(13). \tag{4.6}$$

PROOF. Indeed, if $x_1 \geq v(13)$, then player 1 has no objection either by playing alone (see (4.5)) or by participating in the coalition 13. If $x_1 < v(13)$, then player 1 can suggest the objection

$$\left(\frac{v(13) + x_1}{2}, \quad 0, \quad \frac{v(13) - x_1}{2}; \quad 13, \ 2\right). \tag{4.7}$$

This is a coalitionally rational payoff configuration.

LEMMA 2. A necessary and sufficient condition that player 1 has an objection and to each such objection player 2 has a counter objection, is:

$$x_1 < v(13), \tag{4.8}$$

$$x_1 - x_2 \geq v(13) - v(23) \quad \text{or} \quad x_2 = 0. \tag{4.9}$$

PROOF. Indeed, if (4.8) and (4.9) hold, then, by Lemma 1, player 1 has an objection. This can only be (see (4.5)) of the form

(4.10) $$(x_1 + \varepsilon,\quad 0,\quad v(13) - x_1 - \varepsilon;\quad 13,\ 2),$$

where ε is a sufficiently small positive number. If $x_2 = 0$, then (4.10) is itself also a counter objection; otherwise,

(4.11) $$(0,\quad v(23) - v(13) + x_1 + \varepsilon,\quad v(13) - x_1 - \varepsilon;\ 1,\ 23)$$

is a possible counter objection. By (4.8), player 2 will now receive even more than x_2. If (4.8) does not hold, then there is no objection for player 1, by Lemma 1. If (4.8) holds, but

(4.12) $$x_2 > 0 \quad \text{and} \quad x_1 - x_2 < v(13) - v(23),$$

then player 1 can object by (4.10), choosing ε so small that $v(23) - v(13) + x_1 + \varepsilon < x_2$. Now, player 2 does not have any counter objection, either by playing alone or by forming a coalition with player 3.

Summing up, and making the necessary permutations, we obtain:

LEMMA 3. A necessary and sufficient condition that a payoff configuration $(x_1, x_2, 0;\ 12,\ 3)$ will belong to the bargaining set $\mathfrak{M}$, is that x_1 and x_2 will satisfy (4.5) as well as at least one of the following columns:

(4.13)

$x_1 \geq v(13)$	$x_1 < v(13)$	$x_1 < v(13)$
	$x_2 = 0$	$x_1 - x_2 \geq v(13) - v(23)$

and also at least one of the following columns:

(4.14)

$x_2 \geq v(23)$	$x_2 < v(23)$	$x_2 < v(23)$
	$x_1 = 0$	$x_2 - x_1 \geq v(23) - v(13)$

Taking into account that $x_1 + x_2 = a$, these inequalities reduce to

$$(4.15)\qquad \begin{array}{c|c|c} 0 \le x_1 \le a & x_1 < c & 0 \le x_1 \le a \\ c \le x_1 & x_1 = a & \frac{a+c-b}{2} \le x_1 < c \end{array}$$

$$(4.16)\qquad \begin{array}{c|c|c} x_1 \le a - b & a - b < x_1 & a - b < x_1 \le \frac{a+c-b}{2} \\ & x_1 = 0 & \end{array}$$

We now use the assumption $a, b, c \ge 0$, and the inequalities (4.15), (4.16). A detailed calculation yields the following results:

Case A. If a, b, c satisfy the "triangle inequalities" (4.2), then

$$x_1 = \frac{a + c - b}{2} \tag{4.17}$$

is the only solution.

Case B. If $a > b + c$, then each x_1 satisfying

$$c \le x_1 \le a - b, \tag{4.18}$$

is a solution, and there are no other solutions.

Case C. If $b > a + c$, then $x_1 = 0$ is the only solution.

Case D. If $c > a + b$, then $x_1 = a$ is the only solution.

These are the only possible cases, they exclude each other, and therefore the proof of Theorem 4.1 has been completed.

§ 5. THE GENERAL 3-PERSON GAME

Let us add the coalition 123 with its value $v(123) = d \ge 0$ to the game treated in Section 4. This coalition will have no effect on the previous p.c.'s of the bargaining set. Indeed, this coalition cannot be used for objections and counter objections, because it contains all the players L and K. Thus, it only remains to find out under what condition does a p.c. $(x_1, x_2, x_3; 123)$ belong to the new bargaining set.

As it should be coalitionally rational, it is necessary that x_1, x_2, x_3 satisfy:

(5.1) $x_1, x_2, x_3 \geq 0;\ x_1 + x_2 \geq a,\ x_2 + x_3 \geq b,\ x_1 + x_3 \geq c;\ x_1 + x_2 + x_3 = d.$

On the other hand, if (5.1) is satisfied, there can be no objection and hence this pair belongs to $\mathfrak{M}$.

In order that the inequalities (5.1) have at least one solution, it is necessary and sufficient that

(5.2) $$d \geq a,\ b,\ c, \quad d \geq \frac{a + b + c}{2}\ .$$

We have thus proved:

THEOREM 5.1. In the 3-person game for which

$$v(1) = v(2) = v(3) = 0,\ v(12) = a,\ v(23) = b,$$
$$v(13) = c,\ v(123) = d,\ a,\ b,\ c,\ d \geq 0,$$

the bargaining set $\mathfrak{M}$ consists of the p.c.'s given by Theorem 4.1, and also of the p.c.'s $(x_1, x_2, x_3; 123)$ which satisfy (5.1). The latter p.c.'s exist if and only if (5.2) is satisfied.

§ 6. THE 4-PERSON GAME. COALITIONS OF 1 PERSON AND 3 PERSONS

Consider the 4-person game, in which the permissible coalitions are all the single-person and the three-person coalitions. Let their values be

(6.1) $$\begin{cases} v(1) = v(2) = v(3) = v(4) = 0, & v(123) = a, \quad v(124) = b, \\ v(134) = c, \quad v(234) = d, & a,\ b,\ c,\ d \geq 0. \end{cases}$$

Evidently $(0, 0, 0, 0; 1, 2, 3, 4)$ belongs to the bargaining set $\mathfrak{M}$. Similar considerations to those which were used in Section 4 lead to the inequalities which are listed in Appendix 1. These inequalities express a necessary and sufficient condition in order that the payoff configuration $(x_1, x_2, x_3; 123, 4)$ belongs to the bargaining set.

We omit the calculations, which are somewhat lengthy but easy, and state the results. There are essentially four different cases:

<u>Case A</u>. If

$$(6.2)\qquad \begin{cases} 2a \leq b + c + d, \\ 2b \leq a + c + d, \\ 2c \leq a + b + d, \\ 2d \leq a + b + c, \end{cases}$$

then the bargaining set is

$$(6.3)\qquad \begin{cases} \left(0 \quad 0 \quad 0 \quad 0 \quad ;\ 1,2,3,4\right) \\ \left(\frac{a+b+c-2d}{3} \quad \frac{a+b+d-2c}{3} \quad \frac{a+c+d-2b}{3} \quad 0 \quad ;\ 123,\ 4\right) \\ \left(\frac{a+b+c-2d}{3} \quad \frac{a+b+d-2c}{3} \quad 0 \quad \frac{b+c+d-2a}{3};\ 124,\ 3\right) \\ \left(\frac{a+b+c-2d}{3} \quad 0 \quad \frac{a+c+d-2b}{3} \quad \frac{b+c+d-2a}{3};\ 134,\ 2\right) \\ \left(0 \quad \frac{a+b+d-2c}{3} \quad \frac{a+c+d-2b}{3} \quad \frac{b+c+d-2a}{3};\ 234,\ 1\right) \end{cases}$$

<u>Case B</u>. If

$$(6.4)\qquad \begin{cases} 2a > b + c + d, & \\ 2b \leq a + c + d, & b \leq c + d, \\ 2c \leq a + b + d, & c \leq b + d, \\ 2d \leq a + b + c, & d \leq b + c, \end{cases}$$

then the bargaining set is:

$$(6.5)\quad \begin{cases} (\;0 & 0 & 0 & 0;\; 1,2,3,4\;) \\ (\;x_1 & a - x_1 - x_3 & x_3 & 0;\; 123,\; 4\;) \\ (\frac{b + c - d}{2} & \frac{b + d - c}{2} & 0 & 0;\; 124,\; 3\;) \\ (\frac{b + c - d}{2} & 0 & \frac{c + d - b}{2} & 0;\; 134,\; 2\;) \\ (\;0 & \frac{b + d - c}{2} & \frac{c + d - b}{2} & 0;\; 234,\; 1\;)\;. \end{cases}$$

Here, x_1 and x_3 satisfy the inequalities

$$(6.6)\qquad 0 \leq x_1 \leq a - d, \quad 0 \leq x_3 \leq a - b, \quad c \leq x_1 + x_3 \leq a.$$

<u>Case C</u>. If

$$(6.7)\qquad \begin{cases} 2a > b + c + d, & \\ 2b \leq a + c + d, & b > c + d, \\ 2c \leq a + b + d, & \\ 2d \leq a + b + c, & \end{cases}$$

then the bargaining set is:

$$(6.8)\quad \begin{cases} (\;0 & 0 & 0 & 0;\; 1,2,3,4) \\ (\;x_1 & a - x_1 - x_3 & x_3 & 0;\; 123,\; 4\;) \\ (c \leq \xi_1 \leq b - d & b - \xi_1 & 0 & 0;\; 124,\; 3\;) \\ (\;c & 0 & 0 & 0;\; 134,\; 2\;) \\ (\;0 & d & 0 & 0;\; 234,\; 1\;) \end{cases}$$

Here, x_1 and x_3 satisfy the inequalities (6.6).

<u>Case D</u>. If

$$(6.9)\qquad \begin{cases} 2a > b + c + d, \\ 2b > a + c + d, \\ a \geq b, \end{cases}$$

the bargaining set is the same as in Case C.

Only Case A is completely discrete; all other cases contain the continuum (6.6). Equations (6.2) can be considered as a generalization of the triangle inequalities. In fact, it follows from (6.2) that any three of the numbers a, b, c, d satisfy the triangle inequalities. Moreover, an equality $a = b + c$, for example, can occur only if $a = d$. The converse does not hold. (E.g., $a = 8$, $b = c = d = 5$.)

It is possible to approach the bargaining set in Case A as follows: If players 1 and 2 get α and β, respectively, then player 3 gets $a - \alpha - \beta$ in the coalition 123. With these values, player 4 will get $b - \alpha - \beta$ in the coalition 124, $c - a + \beta$ in the coalition 134, and $d - a + \alpha$ in the coalition 234. In order that no coalition can exert threats on others, it is necessary and sufficient that

$$b - \alpha - \beta = c - a + \beta = d - a + \alpha. \tag{6.10}$$

Hence

$$\alpha = \frac{a + b + c - 2d}{3}, \quad \beta = \frac{a + b + d - 2c}{3}. \tag{6.11}$$

If

$$2a \geq b + c + d, \quad a \geq b, c, d, \tag{6.12}$$

then the coalition 123 is strong and player 4 cannot get more than zero. If we decide to omit him from the game, and look at each of the remaining two persons in a coalition which contained him as a new 2-person coalition which has the same value as before, we get a 3-person game in which $v(123) = a$, $v(12) = b$, $v(23) = d$, $v(13) = c$, $v(1) = v(2) = v(3) = 0$.

A comparison with the previous two sections shows that the bargaining set of the new game is essentially the same as the game treated in Cases B, C, D, and Case A, if equality holds in the first relation of (6.12).

(Note that each of the systems (6.4) and (6.7) implies $a > b, c, d$, (6.9) implies $a \geq b$, $a > c, d$, and (6.7) as well as (6.9) implies $b > c + d$.)

We can therefore conclude:

THEOREM 6.1. A 4-person game, in which the permissible coalitions are all the 1-person and 3-person coalitions always has a bargaining set in which all possible partitions into coalitions appear. The set is discrete if and only if (6.2) is satisfied. If (6.12) holds, the bargaining set is essentially the same as the one of a full 3-person game obtained from the original one by deleting a player who belongs to the complement of a maximal valued coalition. This player always get 0.

REMARK 1. The same situation occurs in a 3-person game in which the only non-permissible coalition is the 3-person coalition. If the triangle inequality does not hold, then one coalition is strong enough to reduce the game to a 2-person game with essentially the same bargaining set. The weak player gets 0.

REMARK 2. The conditions (6.12) are necessary and sufficient for the existence of a c.r.p.c. $(x_1, x_2, x_3, 0; 123, 4)$ such that

$$x_1, x_2, x_3 \geq 0, \quad x_1 + x_2 \geq b, \quad x_1 + x_3 \geq c, \quad x_2 + x_3 \geq d. \tag{6.13}$$

Obviously, such p.c.'s cannot be objected against. However, in any c.r.p.c, in which player 4 is in a 3-person coalition and receives more than 0, there exists an objection against player 4 which cannot be countered. Thus, the coalition 123 "dictates" everything; this is why player 4 cannot claim more than 0.

REMARK 3. The following *ad hoc* rule serves for the discrete case: The value of each coalition is equally divided among its members. If a person enters a coalition he gets the sum of "his shares" minus the sum of the "shares" which his partners get from coalitions which do not include him. For example: The first player's shares are $\frac{a}{3}, \frac{b}{3}, \frac{c}{3}$. If he is entering the coalition 123, his partners have their shares $\frac{d}{3}, \frac{d}{3}$ from the coalition 234, which is that coalition that does not contain player 1; therefore, this player gets

$$\frac{a}{3} + \frac{b}{3} + \frac{c}{3} - \frac{d}{3} - \frac{d}{3} \tag{6.14}$$

if he enters the coalition 123.

The same rule applies also to the 3-person game, with 1- and 2-person coalitions, in the discrete case.

REMARK 4. The discrete case exhibits a game with a nonnegative "3-quota." Each player always receives his "quota" in the bargaining set if he succeeds in becoming a member of a 3-person coalition. Quota games are treated in [5], and the results of this section are further generalized in [6].

§ 7. EXISTENCE THEOREMS. COUNTER EXAMPLES

DEFINITION 7.1. A permissible coalition in a game Γ will be called effective, if it is possible to divide its value among its members in such a way that no permissible sub-coalition can alone make more.

Conditions (5.2), for example, is a necessary and sufficient condition that the coalition 123 will be effective in the game treated in Section 5.

Clearly, we can assume that all subsets of N are permissible coalitions and that those having a positive value are effective, since we are dealing only with c.r.p.c.'s. The zero-valued coalitions will be called trivial coalitions.

The first question which may arise is whether each partition of the players, in which the only trivial coalitions are 1-person coalitions, is represented in $\mathfrak{M}$. The answer is no.

Example 7.1. n = 5, the non-trivial coalitions are 12, 35, 134, 2345, with values:

$$v(12) = 10, \quad v(35) = 85, \quad v(134) = 148, \quad v(2345) = 160. \tag{7.1}$$

Consider the coalitionally rational payoff configuration

(7.2) $(\alpha, \beta, 0, 0, 0; 12, 3, 4, 5),$

where, of course, $0 \leq \alpha \leq 10$, $\alpha + \beta = 10$. Now, player 1 can object by

(7.3) $(11, 0, 29, 108, 0; 134, 2, 5).$

This objection is justified — i.e., player 2 has no counter objection — if $\alpha < 10$. Indeed, any attempt of player 2 to keep his positive share β will end with a coalitionally non-rational p.c. Thus, (7.2) can belong to $\mathfrak{M}$ only if $\alpha = 10$, $\beta = 0$. But this case is also ruled out, since now player 2 has a justified objection: (0, 1, 100, 44, 15; 1, 2345).

Let Γ be a game, some of the values of the coalitions of which are positive. Is it possible that no p.c. belongs to $\mathfrak{M}$ unless all the players get zero? In other words — is it possible that, in spite of some coalitions having a positive value, it would be worthless to enter into such coalitions, if one insists on the stability demanded by the definition of $\mathfrak{M}$? This, in fact, may happen as the following example shows:

Example 7.2.*

$$(7.4)\quad \begin{cases} v(12b) = 1, & b = 3, 4, 5, 6. \\ v(1ab) = 1, & a = 3, 4; \quad b = 5, 6. \\ v(2pq) = 1, & p = 3, \quad q = 4 \quad \text{or} \quad p = 5, \quad q = 6. \\ v(3456) = 1, & \\ v(B) = 1, & B \text{ contains at least one of the above-mentioned coalitions.} \\ v(B) = 0, & \text{otherwise} \end{cases}$$

It is a long but easy computation to verify that for this game $(x; \mathcal{B}) \in \mathfrak{M}$ implies $x_i = 0$, $i = 1, 2, \ldots, 6$.

The following theorem might be helpful in gaining some more insight into the nature of the bargaining set $\mathfrak{M}$.

---

* This game was given by J. von Neumann and O. Morgenstern [9], pp. 467-469, as an example of a simple game which is not a weighted majority game and for which no main simple solution exists.

THEOREM 7.1. Let Γ be an n-person game, in which 12 is a permissible coalition. Let $\mathcal{B}^o \equiv 12, B_2, \ldots, B_m$ be a fixed partition. Let $(x; \mathcal{B}^o) \equiv (x_1, x_2, \ldots, x_n; \mathcal{B}^o)$ be a c.r.p.c. and let J be the set of all the numbers σ_1, $0 \leq \sigma_1 \leq v(12)$, such that player 1 has a justified objection* against player 2, in $(\sigma_1, v(12) - \sigma_1, x_3, x_4, \ldots, x_n; \mathcal{B}^o)$; then J is an open set relative to the closed interval $[0, v(12)]$.

PROOF. If

$$(x_1, x_2, \ldots, x_n;\ 12, B_2, \ldots, B_m) \tag{7.5}$$

is a coalitionally rational payoff configuration, then so is also

$$(x_1 + \epsilon, x_2 - \epsilon, x_3, \ldots, x_n;\ 12, B_2, \ldots, B_m), \tag{7.6}$$

provided that $-x_1 \leq \epsilon \leq v(12) - x_1$.

If $x_1 \in J$, then $\delta \equiv v(12) - x_1 > 0$, since otherwise player 2 can counter object by playing alone.

Let $(y; \mathcal{C})$ be an objection of player 1 against player 2; then $y_1 > x_1$. Let z_2 be the maximum that player 2 can get by joining a coalition such that his partners (if such exist) get what they are supposed to get in a counter objection. Obviously such a maximum exists, and $z_2 < x_2$, because $x_1 \in J$. Choose ϵ such that

$$-x_1 \leq \epsilon < \min(\delta, y_1 - x_1, x_2 - z_2); \tag{7.7}$$

then $x_1 + \epsilon$ will also belong to J. Indeed, (7.6) will be coalitionally rational; $(y; \mathcal{C})$ will remain an objection which is justified.

Thus, if $x_1 \in J$, then so are all the points on the interval $[0, x_1 + \epsilon]$.

---

* By the term "a justified objection" we mean an objection which has no counter objection.

THEOREM 7.2. Let Γ be an n-person game, in which 12 is a permissible coalition and all the permissible coalitions are 1, 2 and 3-person coalitions; then, if $(x; \mathcal{B}^o)$ is a c.r.p.c., there exists a c.r.p.c.

$$(7.8) \qquad (\xi_1, \xi_2, x_3, x_4, \ldots, x_n; \; 12, B_2, \ldots, B_m)$$

such that neither player 1 nor player 2 has any justified objection. Here $\mathcal{B}^o \equiv 12, B_2, \ldots, B_m$.

PROOF. We proved in Theorem 7.1 that the numbers x_1, for which player 1 has a justified objection, form an open set T_1 with respect to $[0, v(12)]$. Similarly, the numbers x_1 for which player 2 has a justified objection form an open set T_2 with respect to the same interval ($x_3, \ldots, x_n$ remain fixed). We shall show that T_1 and T_2 are disjoint, from which it will follow that there is a point ξ_1 in $[0, v(12)]$, which is neither in T_1, nor in T_2, and therefore (7.8) will satisfy the requirements. (None of the sets is the closed interval because $v(12) \notin T_1$, $0 \notin T_2$.)

Indeed, suppose that

$$(7.9) \qquad (\sigma_1, \sigma_2, x_3, \ldots, x_n; \; 12, B_2, \ldots, B_m)$$

is a c.r.p.c. in which both players have justified objections. Player 1, in his objection, must join a coalition C which contains more than one person and does not contain player 2. Similarly, player 2 must join, in his objection, a coalition D which consists of more than one person and does not contain player 1. If $C \cap D = \phi$, then player 2's objection can serve as a counter objection for player 1's objection, the latter being therefore not justified. If $C \cap D \equiv E = \phi$, then E contains one or two members. Without loss of generality, we can assume that the total amount that the players in E got in player 2's objection was not less than what they got from player 1's objection. If E contains one member, player 2's objection is a counter objection to player 1's objection. If E contains two members, this is not always true, because in order to counter object, player 2 has to modify, perhaps, his payments to the members of E. By doing so, a payoff configuration may result, which is not coalitionally rational; i.e., one player in E and player 2 can now make more by together forming a coalition. But if this is the case, then this coalition can serve in a counter objection; e.g., by player 2 taking σ_2 for himself and letting the other player get the rest.

REMARK. The theorem fails to hold if we remove the restriction on the the number of the players in the permissible coalitions. A counter example is provided in Example 7.1.

Application. Suppose that each one of two men got a license to build a gasoline station. Each one considers the possibility of taking at most two partners. They expect various profits from the corresponding possible coalitions. The other partners do not have licenses. Of course, the two men consider also their joint coalition. Under these assumptions, Theorem 7.2 says that the coalition of the two licensees is represented in the bargaining set.

§ 8. THE 4-PERSON GAME IN WHICH ONLY 1- AND 2-PERSON COALITIONS ARE PERMISSIBLE

The inequalities which determine under what condition is $(x_1, x_2, x_3, x_4; 12, 34)$ in $\mathfrak{M}$, for the game:

$$(8.1)\begin{cases} v(1) = v(2) = v(3) = v(4) = 0, \ v(12) = a, \ v(23) = b, \ v(34) = c, \\ v(13) = d, \ v(24) = e, \ v(14) = f, \ a, b, c, d, e, f \geq 0, \end{cases}$$

are given in Appendix 2.

Theorem 7.2 ensures that any partition which contains only one 2-person coalition is represented in $\mathfrak{M}$. We shall now study the case of partitions into two couples. Our object is to prove that some such partitions appear in $\mathfrak{M}$. It turns out that this can be proved even if we limit ourselves to maximal partitions, i.e., to those partitions in which the sum of the values of the coalitions is maximal. This restriction helps us by reducing the number of inequalities which need to be examined.

THEOREM 8.1. Let Γ be the game (8.1), where

$$a + c \geq d + e, \quad a + c \geq b + f; \tag{8.2}$$

then there always exists a p.c. $(x_1, x_2, x_3, x_4; 12, 34)$ in the bargaining set $\mathfrak{M}$.

PROOF. We omit the calculations, but state the various cases.

Case A. If

$$a \leq b + d, \quad b \leq a + d, \quad d \leq a + b, \quad 2c \geq b + d - a, \tag{8.3}$$

then

$$\left(\frac{a + d - b}{2}, \frac{a + b - d}{2}, \frac{b + d - a}{2}, \frac{2c + a - b - d}{2}; \quad 12, 34\right) \in \mathfrak{M}. \tag{8.4}$$

If

$$a \leq b + d, \quad b \leq a + d, \quad d \leq a + b, \quad 2c < b + d - a, \tag{8.5}$$

then

$$\left(\frac{a + d - b}{2}, \frac{a + b - d}{2}, c, 0; \quad 12, 34\right) \in \mathfrak{M}. \tag{8.6}$$

Case B. If

$$a > b + d, \quad c > d + f, \tag{8.7}$$

then

$$(d, a - d, 0, c; \quad 12, 34) \in \mathfrak{M}. \tag{8.8}$$

Case C. If

$$a > b + d, \quad f > c + d, \quad b + c \geq e, \tag{8.9}$$

then

$$(8.10)\qquad (f - c,\ \ a + c - f,\ \ 0,\ c;\ \ 12,\ 34) \in \mathfrak{M}\ .$$

(Indeed, (8.9) and (8.2) imply $c \geq e - b \geq e - (a + c - f)$, hence $a + 2c \geq e + f$. The rest follows directly.) If

$$(8.11)\qquad a > b + d,\quad f > c + d,\quad e > b + c,$$

then

$$(8.12)\qquad \left(\frac{a + f - e}{2},\ \ \frac{a + e - f}{2},\ \ 0,\ c;\ \ 12,\ 34\right) \in \mathfrak{M}\ .$$

(Indeed, (8.2) and (8.11) imply $2d + e \leq d + a + c < a + f$. Also $2b + f \leq b + a + c < a + e$.)

<u>Case D</u>. If

$$(8.13)\qquad a > b + d,\quad d > f + c,\quad b + c \geq e,$$

then

$$(8.14)\qquad (a - b,\ b,\ 0,\ c;\ \ 12,\ 34) \in \mathfrak{M}\ .$$

If

$$(8.15)\qquad a > b + d,\quad d > f + c,\quad e > b + c,$$

then

$$(8.16)\qquad (a + c - e,\ \ e - c,\ 0,\ c;\ \ 12,\ 34) \in \mathfrak{M}\ .$$

<u>Case E</u>. If

$$(8.17)\qquad d > a + b,\quad d > c + f,$$

then

$$(8.18) \qquad (a,\ 0,\ d - a,\ \ c + a - d;\ \ 12, 34) \in \mathfrak{M}\ .$$

All other cases are either not maximal partitions, or they can be reduced to these cases by permuting the players:

$$(8.19) \qquad 1 \longleftrightarrow 3, \quad 2 \longleftrightarrow 4.$$

§ 9. THE RESTRICTED BARGAINING SET

In a given game there are in general many stable p.c.'s. Though we do not possess a criterion for choosing between them, there are cases in which it is clear that some p.c.'s in $\mathfrak{M}$ are "better" than others. We therefore suggest that the latter should be deleted from $\mathfrak{M}$, thus giving rise to the restricted bargaining set $\mathfrak{M}^*$.

A p.c. $(x; \mathcal{B})$ in $\mathfrak{M}$ should be deleted if one of the following cases occurs:

(i) There exists in $\mathfrak{M}$ a p.c. $(x^*; \mathcal{B}^*)$ with

$$(9.1) \qquad x_i^* > x_i; \quad i = 1, 2, \ldots, n.$$

(ii) There exists in $\mathfrak{M}$ a p.c. $(x^{**}; \mathcal{B}^{**})$, where the coalitions in $\mathcal{B}^{**}$ are unions of coalitions in $\mathcal{B}$, such that

$$(9.2) \qquad x_i^{**} > x_i$$

for all the players i which belong to a union of more than one coalition of $\mathcal{B}$ and

$$(9.3) \qquad x_i^{**} \geq x_i$$

for all the other players.

One sees that in the examples given in the previous sections, only those coalitions which have relatively big values (if such exist), will appear in $\mathfrak{M}^*$.

§ 10. POSSIBLE MODIFICATIONS

Inasmuch as our theory tries to cope with "reality," it is flexible enough to allow for some modifications.

For instance, if players are faced with the game treated in Example 7.2, they may claim that the demand for stability is too strong. They would rather relax this demand and still gain something from the game.

One can then offer them the following definition of a bargaining set $\mathfrak{M}_1$:

> DEFINITION 10.1. A c.r.p.c. $(x; \mathcal{B})$ belongs to the bargaining set $\mathfrak{M}_1$, if for any objection K against L, there is somebody in L who can counter object.

According to this definition, each player in a coalition B_j which contains K, who does not belong to the partners of K, is required to be a able to counter object; but several such players may perhaps be unable to protect their shares simultaneously. Clearly, the resulting bargaining set $\mathfrak{M}_1$, <u>includes</u> $\mathfrak{M}$, since the number of sets which is required to counter object is reduced. In this case, e.g., the players of the game treated in Example 7.2 may agree to

$$(10.1) \qquad (\tfrac{1}{3}, \tfrac{1}{3}, \tfrac{1}{3}, 0, 0, 0; \; 123, 4, 5, 6)$$

which belongs to $\mathfrak{M}_1$.

In some other real-life cases, one can estimate and tell in advance which coalitions may object and which coalitions may counter object. This leads to various bargaining sets and brings us to the circle of ideas surrounding ψ-stability. (See Luce and Raiffa [3], pp. 163-168, 174-176, 220-236.)

One may limit K to be always one-person and L to be the remaining members of the coalition, except for K's partners. This type of stability of one against the rest, which generates a bargaining set $\mathfrak{M}_2$, is still different from the stability demanded in $\mathfrak{M}$ as the following example shows:

Example 10.1. Consider the game:

$$n = 5,\ v(i) = 0,\ v(123) = 30,\ v(24) = v(35) = 50,\ v(1245) = v(1345) = 60. \tag{10.2}$$

Let

$$(x; \mathcal{B}) \equiv (10, 10, 10, 0, 0; 123, 4, 5). \tag{10.3}$$

If $K = 1$, 2, or 3, then the remaining players which belong to the same coalition and are not among his partners can always counter object; but the objection for $K = 23$,

$$(0, 11, 11, 39, 39;\ 1, 24, 35), \tag{10.4}$$

has no counter objection because player 1 cannot keep his profits. Thus $(x; \mathcal{B}) \in \mathfrak{M}_2$ but $\notin \mathfrak{M}$.

It is easy to show that $\mathfrak{M} \subset \mathfrak{M}_2$.

Sometimes people would like to feel safe not only within their coalitions but also from "outside" threats. It may happen, e.g., that several players from various coalitions will threaten together other people from these coalitions. A reasonable way to cope with this strong demand for stability would be to allow K and L to belong to several coalitions, provided that K and L are required to intersect the same coalitions. This will bring us to a bargaining set $\mathfrak{M}_o$ which is included in $\mathfrak{M}$. Let us remark that $\mathfrak{M}_o = \mathfrak{M}$ for the 2- and 3-person games, as well as for the 4-person game with only 1-, 3-, and 4-person coalitions permissible. If $n = 4$, where 1- and 2-person coalitions are permissible, one has to replace the inequalities of Appendix 2 by those listed in Appendix 3. Fortunately, these inequalities are satisfied in all the examples given in Section 8, and therefore Theorem 8.1 is valid if one replaces $\mathfrak{M}$ by $\mathfrak{M}_o$.

Finally, we would like to question the assumption of coalitional rationality. If we drop this condition, we may arrive at negative values in the bargaining set, but this does not have to bother us, since (0, 0, ..., 0; 1, 2, ..., n) will certainly remain in the bargaining set, and therefore we can demand that the restricted bargaining set will contain only individually rational p.c.'s. However, we shall show in Example 10.2 that the resulting restricted bargaining set may still contain non-coalitionally rational p.c.'s.

Example 10.2. Let Γ be the game

$$v(i) = 0,\ v(12) = v(45) = v(46) = v(56) = v(123) = 30,\ v(34) = 10. \tag{10.5}$$

In this game, the non-coalitionally rational p.c.

$$(10,\ 10,\ 10,\ 0,\ 15,\ 15;\ \ 123,\ 4,\ 56) \tag{10.6}$$

is stable if one drops the condition of coalitional rationality. In fact, it then belongs to the restricted bargaining set, since otherwise there exists a p.c. $(x; \mathcal{B})$ in the bargaining set with

$$\sum_{i=1}^{6} x_i > 60. \tag{10.7}$$

This can only occur if the coalition 34 is formed. Since, in addition, $x_3 \geq 10$, $x_4 \geq 0$, player 3 gets 10. This is impossible because in this case, player 4 has a justified objection, due to the fact that $x_1 + x_2 = 30$.

We have thus shown that the restricted bargaining set may contain a non-coalitionally rational p.c., if this condition is dropped from the definition of the bargaining set.

§ 11. CONCLUDING REMARKS

Perhaps, the nearest to our theory is W. Vickrey's concept of self-policing patterns [8]. His objections — called "heretical imputations" — are similar to ours; however, his counter objections — named "penalizing policing imputations" — are quite different.

Both the heretical and the penalizing policing imputations are in Vickrey's case imputations, whereas this is not the case in our theory. His penalizing policing imputation insists that at least one member of the "heretical coalition" is punished, whereas we only demand that the set L will be able to hold on to its property. However, the main difference lies, perhaps, in the fact that Vickrey is looking for a set of imputations — "self-policing patterns" — which are stable as a whole,* while our bargaining set consists of payoff configurations, each one of which is stable in itself.

It has been pointed out in Section 10 that if a lack of communication is known to exist, one can incorporate ideas from ψ-stability theory into ours. Both theories stress the dependence of an outcome on the coalition structure which actually forms. However, ψ-stability theory requires the payoffs to be imputations,** whereas we require that the outcome satisfies (2.4). (See, e.g., R. D. Luce and H. Raiffa [3], p. 222.) The coalition rationality requirement (2.5) is a special case of a ψ-stability requirement, if one requires that all the subsets of the coalitions in the coalition structure τ are values of $\psi(\tau)$. A similar requirement appears also in Milnor's class L of reasonable outcomes ([3], pp. 240-242).

In many practical situations, the characteristic function is not the best way to describe a game. It would rather be better to apply the "Thrall characteristic function" (see R. M. Thrall [7]), which associates with each coalition-structure a value for each coalition appearing in that structure. One can try to define the concepts of objections and counter objections, for such cases, and it is possible to do so in various ways.

It is also possible to apply the notions described in this paper to the Aumann-Peleg characteristic function for cooperative games without side payments [2], essentially without change.

Finally, we should like to point out that our theory gives in many cases answers similar to those appearing in classical theories. Thus, e.g., the bargaining set in the discrete case of the 3-person non-zero-sum game*** consists essentially of the "central" three points of the non-discriminatory von Neumann-Morgenstern solution (see [9], pp. 550-554), but does not contain the additional "wiggles" that occur in their solution. The bargaining set

---

* In particular, he looks for "strong solutions."

** Or at least feasible individually rational n-tuples ([3], p. 226).

*** Clearly, one has to modify the characteristic function in the obvious way so as to get superadditivity.

for the non-discrete case is essentially the core. This suggests a pattern in which the bargaining set forms the "central" or "intuitive" part of a von Neumann-Morgenstern solution, whereas the "complications" disappear.

APPENDIX 1

Let Γ be a 4-person game, the coalitions of which, and their values, are given by (6.1). In order that the pair $(x_1, x_2, x_3, 0; 123, 4)$ belongs to the bargaining set $\mathcal{M}$, it is necessary and sufficient that

$$0 \le x_1, \quad 0 \le x_3, \quad x_1 + x_3 \le a, \; x_2 = a - x_1 - x_3, \tag{A1.1}$$

and that at least one inequality (or equality) in each of the following lines should be satisfied.

$x_1 + x_3 \ge c$	$2x_1 + x_3 \ge a + c - d$	$x_1 + x_3 = a$
$x_3 \le a - b$	$x_3 - x_1 \le d - b$	$x_3 = 0$
$x_1 \le a - d$	$2x_1 + x_3 \le a + c - d$	$x_1 = 0$
$x_3 \le a - b$	$x_1 + 2x_3 \le a + c - b$	$x_3 = 0$
$x_1 + x_3 \ge c$	$x_1 + 2x_3 \ge a + c - b$	$x_1 + x_3 = a$
$x_1 \le a - d$	$x_3 - x_1 \ge d - b$	$x_1 = 0$

APPENDIX 2

Let Γ be a 4-person game, the coalitions of which, and their values, are given in (8.1). In order that the pair $(x_1, x_2, x_3, x_4; 12, 34)$ belong to the bargaining set $\mathcal{M}$, it is necessary and sufficient that

$$0 \le x_1 \le a, \quad 0 \le x_3 \le c, \quad x_1 + x_2 = a, \quad x_3 + x_4 = c, \tag{A2.1}$$

and that at least one inequality (or equality) in each line be satisfied.

If the partition 12, 34 is maximal (in the sense of (8.2)), the last column can be omitted.

$x_1 = a$	$x_1 + x_3 \geq d$	$2x_1 \geq a + d - b$	$x_1 + x_3 \geq a + c - e$
$x_1 = a$	$x_1 - x_3 \geq f - c$	$2x_1 \geq a + f - e$	$x_1 - x_3 \geq a - b$
$x_1 = 0$	$x_1 - x_3 \leq a - b$	$2x_1 \leq a + d - b$	$x_1 - x_3 \leq f - c$
$x_1 = 0$	$x_1 + x_3 \leq a + c - e$	$2x_1 \leq a + f - e$	$x_1 + x_3 \leq d$
$x_3 = c$	$x_1 + x_3 \geq d$	$2x_3 \geq c + d - f$	$x_1 + x_3 \geq a + c - e$
$x_3 = c$	$x_1 - x_3 \leq a - b$	$2x_3 \geq c + b - e$	$x_1 - x_3 \leq f - c$
$x_3 = 0$	$x_1 - x_3 \geq f - c$	$2x_3 \leq c + d - f$	$x_1 - x_3 \geq a - b$
$x_3 = 0$	$x_1 + x_3 \leq a + c - e$	$2x_3 \leq b + c - e$	$x_1 + x_3 \leq d$

APPENDIX 3

The following inequalities replace those given in Appendix 2, if one desires that $(x_1, x_2, x_3, x_4;\ 12, 34)$ shall belong to $\mathfrak{M}_o$. (See Section 10.) Again, at least one inequality in each line should be satisfied, as well as those given in (A2.1).

$x_1 + x_3 \geq d$	$x_1 = a$ $2x_3 \geq c + d - f$	$x_3 = c$ $2x_1 \geq a + d - b$	$x_1 + x_3 \geq a + c - e$
$x_1 - x_3 \geq f - c$	$x_1 = a$ $2x_3 \leq c + d - f$	$x_3 = 0$ $2x_1 \geq a + f - e$	$x_3 - x_1 \leq b - a$
$x_1 - x_3 \leq a - b$	$x_1 = 0$ $2x_3 \geq b + c - e$	$x_3 = c$ $2x_1 \leq a + d - b$	$x_1 - x_3 \leq f - c$
$x_1 + x_3 \leq a + c - e$	$x_1 = 0$ $2x_3 \leq b + c - e$	$x_3 = 0$ $2x_1 \leq a + f - e$	$x_1 + x_3 \leq d$

BIBLIOGRAPHY

[1] AUMANN, R. J. and MASCHLER, M., "An equilibrium theory for n-person cooperative games," American Math. Soc. Notices, Vol. 8 (1961), p. 261.

[2] AUMANN, R. J. and PELEG, B., "von Neumann-Morgenstern solutions to cooperative games without side payments," Bull. Amer. Math. Soc. 66 (1960), pp. 173-179.

[3] LUCE, R. D. and RAIFFA, H., Games and Decisions, John Wiley and Sons, New York (1957).

[4] MASCHLER, M., "An experiment on n-person cooperative games," Recent Advances in Game Theory, Proceedings of a Princeton University Conference, October 1961, privately printed for members of the conference (1962), pp. 49-56.

[5] MASCHLER, M., "Stable payoff configurations for quota games." (This Study.)

[6] MASCHLER, M., "n-Person games with only 1, n–1, and n-person permissible coalitions," J. Math. Analysis and Applications, Vol. 6 (1963), pp. 230-256.

[7] THRALL, R. M., "Generalized characteristic functions for n-person games," Recent Advances in Game Theory, Proceedings of a Princeton University Conference, October 1961, privately printed for members of the conference (1962), pp. 157-160.

[8] VICKREY, W., "Self-policing properties of certain imputation sets," Contributions to the Theory of Games, Vol. IV, Annals of Mathematics Studies, No. 40, Princeton University Press, Princeton (1959), pp. 213-246.

[9] VON NEUMANN, J. and MORGENSTERN, O., Theory of Games and Economic Behavior, 2nd ed., Princeton University Press, Princeton (1947).

Robert J. Aumann
Michael Maschler

Princeton University
The Hebrew University of Jerusalem

STABLE PAYOFF CONFIGURATIONS FOR QUOTA GAMES*

Michael Maschler

§ 1. INTRODUCTION

Cooperative n-person <u>quota</u> <u>games</u> are formed by assigning each player i a real number ω_i — called his quota — having the property that

$$(1.1) \qquad \omega_i + \omega_j \geq 0, \quad i \neq j, \quad i, j = 1, 2, \ldots, n,$$

and defining the characteristic function for 1- and 2-person coalitions to be

$$(1.2) \qquad v(i) = 0, \quad v(ij) = \omega_i + \omega_j, \quad i \neq j, \ i, j = 1, 2, \ldots, n.$$

Other coalitions are not permissible.

If all other subsets of the players were permissible coalitions as well, if the characteristic function were superadditive, and if

$$(1.3) \qquad v(12 \ldots n) = \omega_1 + \omega_2 + \ldots \omega_n,$$

one would obtain Shapley's quota games (see [5]).**

In this paper we study the properties of some "stable payoff configurations" (= stable outcomes) for quota games, as well as for Shapley's quota games. The definition of "stable payoff configurations" which we use here is one of several definitions which are proposed by R. J. Aumann and M. Maschler

---

* This research was sponsored in part by the Office of Naval Research, Nonr-1858(16) and in part by the Carnegie Corporation of New York.

Part of the content of this paper was presented at a conference on Recent Advances in Game Theory, held at Princeton, October 4-6, 1961.

** Except that Shapley avoids the normalization $v(i) = 0, \ i = 1, 2, \ldots, n$.

[1]. More precisely, we use Definitions 2.1, 2.2 and 2.3 of [1], except that instead of requiring that K and L be subsets of a coalition B_j in $(x; \mathcal{B})$ we ask that they be (non-empty and disjoint) sets of players which intersect the same coalitions in $\mathcal{B}$. In other words, the sets K and L are are required to satisfy

$$K \cap L = \emptyset, \quad K, L \neq \emptyset \tag{1.4}$$

$$K \cap B_j \neq \emptyset \Longleftrightarrow L \cap B_j \neq \emptyset, \; B_j \in \mathcal{B}. \tag{1.5}$$

The resulting <u>bargaining set</u> is denoted by $\mathfrak{M}_0$, as in [1], and a <u>stable payoff configuration</u> is defined here to be an element of $\mathfrak{M}_0$. In all other respects we use the terminology of [1].

We are mainly interested in the cases where not more than one 1-person coalition occurs. These are the most "practical" cases, because otherwise, two players can join and, in general, better themselves.

In Section 2 we characterize the stable payoff configurations of n-person quota games, $n \geq 3$, in the case where the players are partitioned into a maximal number of 2-person coalitions. We prove that these stable payoff configurations are those and only those in which each player gets <u>essentially</u> his quota. The word "essentially" is used since a player in a 1-person coalition must get 0, and so must a player whose quota is negative, while his associate (if any) then gets the value of the coalition. (The latter case is true only for $n \geq 4$ and $n \neq 5$.)

The various definitions for "stable payoff configurations," proposed in [1], correspond to various "levels" of stability that might be desired by the players. The definition which is chosen in this paper is the one which leads to the smallest class of stable payoff configurations. Intuitively, it applies to the case where the players wish "maximum" stability. This strong desire for stability may sometimes cause the players not to agree to participate in a coalition in a certain proposed coalition-structure. Indeed, we give such examples of Shapley's quota games (Section 3). One example shows that an imputation may be stable in one coalition-structure but not stable in another. Other examples show that a quota split is not always stable and sometimes the stable outcomes differ from any imputation which appears in the main solution given by Shapley in [5].

We choose to start with quota games, in order to show the basic properties of the stable payoff configurations. Essentially the same results have been obtained by B. Peleg for m-quota games (see [4]). For cases in which some players have negative quotas and $m(m + 1) > n$, the description of the stable payoff configurations is incomplete both here and in Peleg's paper. The only case which is settled is when $m = n-1$ (see [3]). The results for Shapley's quota games are similar, but only under some additional restrictions (Section 3). Some "artificial" complications may arise, due to the fact that some k-person coalitions, $2 < k < n$, may possess enough power to "block" a quota split.

I am grateful to Dr. H. Kesten for discussions that stimulated this paper.

§ 2. QUOTA GAMES

Following L. S. Shapley [5], we shall make the following:

DEFINITION 2.1. An n-person quota game is an n-person game in which the set of permissible coalitions $\{B\}$ consists of the 1- and 2-person coalitions, and its characteristic function satisfies

(2.1) $v(i) = 0, \; i \in B, \; v(B) \geq 0$ for each permissible coalition B,

(2.2) $v(ik) = \omega_i + \omega_k, \quad i, k = 1, 2, \ldots, n, \quad i \neq k.$

Here ω_i, $i = 1, 2, \ldots, n$, are fixed real numbers. The number ω_i, $i = 1, 2, \ldots, n$, is called the quota of player i, and the vector $\omega \equiv (\omega_1, \omega_2, \ldots, \omega_n)$ will be referred to as the quota of the game.

This definition is somewhat different from the one given by Shapley, and we shall therefore state some of his theorems modified to our case (Lemmas 2.1 - 2.4). Shapley's quota games will be treated in Section 3.

LEMMA 2.1. An n-person game is always* a quota game for $n = 1, 2, 3$.

LEMMA 2.2. The quota of a game, if such exists, is unique for $n > 2$.

LEMMA 2.3. At most one player has a negative quota.

The proofs of these Lemmas are obvious. A player with a negative quota will be called a _weak player_.

LEMMA 2.4. A game in which the set of permissible coalitions consists of the 1- and 2-person coalitions is a quota game if and only if

$$v(ij) + v(k\ell) = v(ik) + v(j\ell), \tag{2.3}$$

for all distinct i, j, k, ℓ.

PROOF. The necessity is immediate. Suppose that (2.3) holds. There is nothing to prove if $n = 1, 2, 3$. If $n \geq 4$, then

$$\omega_i \equiv \frac{1}{2}[v(ij) + v(ik) - v(jk)] \qquad (i, j, k \text{ distinct}) \tag{2.4}$$

is independent of j and k, and can serve as a quota for player i.

REMARK. The conditions (2.3) are equivalent to the conditions:

$$v(ij) + v(12) = v(1j) + v(2i), \tag{2.5}$$

for all distinct $1, 2, i, j$. These latter conditions are somewhat easier to check.

---

* Provided that the 3-person coalition is not permissible if $n = 3$.

THEOREM 2.1. Let Γ be an n-person quota game which contains no weak player. Let n be even, $n \geq 4$, and let $\mathcal{B}^*$ be a partition of all the players into pairs. Then, a payoff configuration (p.c.) $(x; \mathcal{B}^*)$ is stable if and only if each player gets his quota.

PROOF. Without loss of generality, we can assume that

$$\mathcal{B}^* = 12,\ 34,\ \ldots,\ (n-1)n. \tag{2.6}$$

Obviously,

$$(\omega_1, \omega_2, \ldots, \omega_n; \mathcal{B}^*) \in \mathfrak{M}_o, \tag{2.7}$$

since it is coalitionally rational (c.r.) and no player can object.

Suppose now that

$$(x; \mathcal{B}^*) \equiv (x_1, x_2, \ldots, x_n; \mathcal{B}^*) \in \mathfrak{M}_o. \tag{2.8}$$

Let K^* be the set of those players who get in $(x; \mathcal{B}^*)$ less than their quota. Certainly, the members of K^* belong to different coalitions, and their associates get more than their quotas. Let L^* be the set of these associates, then* $|K^*| = |L^*|$.

Case A. $|K^*| = 1$. We can assume that $1 \in K^*$. Since $n \geq 4$, player 1 can object against player 2, e.g., by joining player 3 in a quota split. Player 2 has no counter objection, either alone, because $x_2 > 0$, or by joining another player, because he received more than his quota in $(x; \mathcal{B}^*)$ and all other players except player 1 received their quotas in $(x; \mathcal{B}^*)$. This contradicts the assumption (2.8).

Case B. $|K^*| > 1$. We choose as the objecting K any fixed set of** $2\left[\frac{|K|}{2}\right]$ players from K^*, and let L be the set of their associates in

---

* We denote by $|P|$ the number of elements in a set P.

** $[a]$ denotes the largest integer not greater than a.

$(x; \mathcal{B}^*)$, $|K| = |L| \geq 2$. An objection $(y; \mathcal{C})$ of K against L in $(x; \mathcal{B}^*)$ is possible in which the players K form pairs with a quota split. Here, $P[K; (y; \mathcal{C})] = K$. In order to counter object, each player of L has to join an associate who will "agree" to accept less than his quota, because the members of L got more than their quotas in $(x; \mathcal{B}^*)$. As only 2-person coalitions are permissible, the number of these associates must be equal to $|L|$, i.e., greater than or equal to 2. But there is at most one player who may agree, namely the player in $K^* - K$ (if such exists), because any other player got his quota, either in $(x; \mathcal{B}^*)$ or as a member of $K = P[K; (y; \mathcal{C})]$ in $(y; \mathcal{C})$. Therefore, no counter objection is possible, and again the assumption (2.8) is contradicted. Therefore, $|K^*| = 0$, and $(x; \mathcal{B}^*)$ is the p.c. (2.7).

REMARK. If $n = 2$, any p.c. of the form

$$(x_1, x_2; 12), \quad x_1 \geq 0, \quad x_2 \geq 0, \quad x_1 + x_2 = v(12), \tag{2.9}$$

belongs to $\mathfrak{M}_o$. We can claim that even in this case each player received his quota, because we can define the quota to be $\omega_1 = x_1$, $\omega_2 = x_2$.

THEOREM 2.2. Let Γ be an n-person quota game, which contains no weak player. Let n be odd, $n \geq 1$, and let $\mathcal{B}^*$ be a partition of all the players but one into pairs. Then, a p.c. $(x; \mathcal{B}^*)$ is stable if and only if each player gets his quota, except the one who is left alone and therefore gets 0.

PROOF. The theorem is trivial for $n = 1$. Let $n \geq 3$, and assume that

$$\mathcal{B}^* = 12,\ 34,\ \ldots,\ (n-2)(n-1),\ n. \tag{2.10}$$

Certainly,

$$(\omega_1, \omega_2, \ldots, \omega_{n-1}, 0; \mathcal{B}^*) \in \mathfrak{M}_o. \tag{2.11}$$

Indeed, only one kind of objection is possible, in which a player i, $i = 1, 2, \ldots, n-1$, objects against his associate, player j, by joining player n and offering him some amount which is less than ω_n. (Even this is impossible if $\omega_n = 0$.) But player j can counter object by offering the same player n a quota split.

Suppose that

$$(2.12) \qquad (x; \mathcal{B}^*) \equiv (x_1, x_2, \ldots, x_{n-1}, 0; \mathcal{B}^*) \in \mathfrak{M}_o.$$

Let K^* be the set of all the players, except player n, who get in $(x; \mathcal{B}^*)$ less than their quotas. Again, the members of K^* belong to different coalitions and their associates get more than their quota. If L^* denotes the set of these associates, then $|K^*| = |L^*|$.

Case A. $|K^*| = 1$. Player i, who belongs to K^*, can object against his associate, player j, by joining player n with a quota split. Player j has no counter objection, either as a 1-person coalition, because $x_j > \omega_j \geq 0$, or by joining any 2-person coalition, because each player received at least his quota either in $(x; \mathcal{B}^*)$ or as player i's partner in player i's objection. Thus (2.12) is contradicted.

Case B. $|K^*| \geq 2$. Let the objecting K be any fixed $2\left[\frac{|K^*| + 1}{2}\right] - 1$ players from K^*, and let L be the set of their associates in $(x; \mathcal{B}^*)$, $|K| = |L| \geq 1$. K can object against L by allowing one member to join player n and the other members of K to form pairs with quota splits. At most one player in K^* is not in K, and therefore L cannot counter object, unless $|L| = 1$; in which case $|K^*| = 2$. In this case, however, K^* can object against L^* by forming a coalition with a quota split, and L^*, which has two members, cannot counter object, because only player n may perhaps "agree" to accept less than his quota. Thus, again (2.12) is contradicted. Therefore $|K^*| = 0$, and $(x; \mathcal{B}^*)$ is the p.c. (2.11).

THEOREM 2.3. Let Γ be an n-person game which contains one weak player. Let n be even, $n \geq 4$, and let $\mathcal{B}^*$ be a partition of all the players into pairs. Then, a p.c. $(x; \mathcal{B}^*)$ is stable if and only if each player gets his quota, except the weak player who gets zero and his associate, who gets the value of their coalition.*

PROOF. Without loss of generality, we can assume that player 1 is the weak player, and that

(2.13) $$\mathcal{B}^* = 12,\ 34,\ \ldots,\ (n-1)n.$$

We shall prove that

(2.14) $$(0,\ v(12),\ \omega_3,\ \ldots,\ \omega_n;\ \mathcal{B}^*) \in \mathcal{M}_o.$$

Indeed, in any possible objection, player 2 must join a player i, $i \geq 3$, because player 2 is the only player who gets in $(x; \mathcal{B}^*)$ less than his quota. Such an objection can be regarded either as an objection of player 2 against player 1, or as an objection of player i against his associate, player j, or as an objection of the players 2, i against the players 1, j. As player 2 cannot get more than his quota in the objection, any c.r.p.c., in which player 1 is a 1-person coalition, and players 2, j join in a quota split,** is a counter objection for all the three cases.

Suppose that

(2.15) $$(x; \mathcal{B}^*) \equiv (x_1,\ x_2,\ \ldots,\ x_n;\ \mathcal{B}^*) \in \mathcal{M}_o.$$

Let K^* be the set of players who get in $(x; \mathcal{B}^*)$ less than their quota. Certainly, player 2 belongs to K^*, because $x_1 \geq 0$.

If $|K^*| \geq 2$, let an objecting K be K^*, if $|K^*|$ is even, and $K^* - \{2\}$, if $|K^*|$ is odd. The players of K can object by forming pairs with a quota split. They so object against their associates in $\mathcal{B}^*$, who form a set L. Again, each player in L gets in $(x; \mathcal{B}^*)$ more than his quota, and $|L| \geq 2$. Only one player at most (player 1, if he is in L and if $x_1 = 0$) can counter object by playing alone; therefore, L cannot

* See the Remark following Theorem 2.1 for the case $n = 2$.

** The last coalition is not necessary if only player 2 objects.

counter object. This is a contradiction to the assumption (2.15). Thus, $|K^*| = 1$. Also $x_1 = 0$, because, otherwise, player 2 can object against player 1 by joining player 3 with a quota split, and player 1 has no counter objection. We have proved that $(x; \mathcal{B}^*)$ is the p.c. (2.14).

THEOREM 2.4. Let Γ be an n-person quota game which contains a weak player. Let n be odd, $n \geq 7$. Let $\mathcal{B}^*$ be a partition of all the players, but one, into pairs. Then, a p.c. $(x; \mathcal{B}^*)$ is stable if and only if the weak player gets 0, his associate, if such exists, gets the value of the coalition, and the player who is left alone gets 0, while all the other players get their quotas.

PROOF. Without loss of generality, we can assume that player 1 is the weak player. There are essentially two possible partitions:

(2.16) $$\mathcal{B}^* = \mathcal{B}^{(1)} \equiv 1,\ 23,\ 45,\ \ldots,\ (n-1)n,$$

(2.17) $$\mathcal{B}^{**} = \mathcal{B}^{(2)} \equiv 12,\ 34,\ \ldots,\ (n-2)(n-1),\ n.$$

Certainly,

(2.18) $$(0,\ \omega_2,\ \omega_3,\ \ldots,\ \omega_n;\ \mathcal{B}^{(1)}) \in \mathfrak{M}_o,$$

because no one can object. We shall prove that also

(2.19) $$(0,\ v(12),\ \omega_3,\ \ldots,\ \omega_{n-1},\ 0;\ \mathcal{B}^{(2)}) \in \mathfrak{M}_o.$$

Indeed, any objection must contain player 2, or player n, or both, as partners of the objecting K; because these are the only players who might get in (2.19) less than their quotas. Let player j and player m be the associates in $\mathcal{B}^{(2)}$ of players i and k, respectively. We have to consider only the following cases:

(i) Player 2 joins a player i, $i = 3, 4, \ldots, n$. Here $K = \{2\}$ or $K = \{i\}$ (for $i \neq n$), or $K = \{2, i\}$ (for $i \neq n$).

(ii) Player n joins a player k, $k = 1, 3, 4, 5, \ldots, n-1$. Here $K = \{k\}$.

(iii) Player 2 joins player i and player n joins player k, players $1, 2, i, j, k, m, n$ distinct. Here $K = \{2, k\}$ or $K = \{i, k\}$ or $K = \{2, i, k\}$.

All these K's determine the L's, and all the objections can be countered as follows:

(i) Player 1 acts as a 1-person coalition, and if $i \neq n$, player j joins player n in a quota split.

(ii) Player m joins player n in a quota split.

(iii) Player m joins player n in a quota split, player j joins player 2 in a quota split,* and player 1 acts as a 1-person coalition.**

Suppose that

$$(2.20) \qquad (x; \mathcal{B}^{(1)}) \equiv (0, x_2, x_3, \ldots, x_n; \mathcal{B}^{(1)}) \in \mathfrak{M}_o.$$

Let K^* be the set of those players who get less than their quotas in $(x; \mathcal{B}^{(1)})$. Certainly $1 \notin K^*$, and therefore $|K^*| = |L^*|$, where L^* is the set of the associates of the members of K^*.

Case A. If $|K^*| = 1$, there exists a player who has received his quota, because $n \geq 5$.*** The player in K^* can therefore object by joining him in a quota split, and the player in L^* has no counter-objection. This case contradicts the assumption (2.20) and is therefore ruled out.

Case B. If $|K^*| \geq 2$, the players of K^*, with the possible exception of at most one of them, can object by forming pairs with a quota split. Their associates cannot counter object because there are at least two of them, and there is at most one player available who may "agree" to accept less than his quota. This, again, contradicts assumption (2.20). Therefore,

---

* This is not necessary if $K = (2, k)$.

** This is not necessary if $K = (i, k)$. Similar remarks apply to case (i)

*** In this case only the restriction $n \geq 5$ is used.

$|K^*| = 0$, and $(x; \mathcal{B}^{(1)})$ is the p. c. (2.18).

Suppose that

$$(x; \mathcal{B}^{(2)}) \equiv (x_1, x_2, \ldots, x_{n-1}, 0; \mathcal{B}^{(2)}) \in \mathfrak{M}_o. \tag{2.21}$$

Let K^* be the players, different from player n, who get in $(x; \mathcal{B}^{(2)})$ less than their quotas. Certainly, player 2 belongs to K^*, because $x_1 \geq 0$. Let L^* be the set of the associates of the members of K^*. Again, $|K^*| = |L^*|$.

If $|K^*| \geq 2$, then the players of K^*, with the exception of at most one of them, can form an objection as follows: One player joins player n and the others form pairs, and they all have a quota split. Denote the objecting players by K; then their associates, who form the set L, cannot counter object unless $|L| = 1$. In this case, however, $|K^*| = 2$, and K^* can object against L^* by having its members form a coalition with a quota split. If $x_1 = 0$, and $\omega_n > 0$, a player who received his quota* in (2.21) is added to the objection, by joining player n and offering him some positive amount. The objection cannot be countered. This contradicts the assumption (2.21), and therefore $|K^*| = 1$.

If now $x_1 > 0$, then player 2 can object against player 1 by joining player n with a quota split, and there exists no counter objection, because player 1 has to "defend" more than his quota. Thus we have proved that $(x; \mathcal{B}^{(2)})$ is the p.c. (2.19).

REMARKS. In the case $n = 3$, with player 1 being weak, one gets a continuum for the partition $\mathcal{B}^{(1)}$. In fact, it is easy to see that

$$(0, x_2, x_3; 1, 23) \in \mathfrak{M}_o, \tag{2.22}$$

if and only if

$$x_2 \geq \omega_1 + \omega_2, \quad x_3 \geq \omega_1 + \omega_3, \quad x_2 + x_3 = \omega_2 + \omega_3. \tag{2.23}$$

---

* This is the only case in which the restriction $n \geq 7$ is used.

In this case we are in the core of the game.*

To be sure, there are other partitions which are represented in $\mathfrak{M}_o$. Take, for example, the case $n = 4$, where no player is weak; it is easy to see that if, e.g., $\omega_3 \geq \omega_1, \omega_2$ and $\omega_4 \geq \omega_1, \omega_2$, then

$$(x_1, x_2, 0, 0; 12, 3, 4) \in \mathfrak{M}_o, \tag{2.24}$$

whenever

$$x_1, x_2 \geq 0, \quad x_1 + x_2 = \omega_1 + \omega_2.$$

In [1] another bargaining set was considered, which differs from $\mathfrak{M}_o$ only by the additional requirement that K and L will always belong to <u>the same coalition</u>, among $B_1, B_2, \ldots, B_m$, of the c.r.p.c. $(x; \mathcal{B})$. The resulting bargaining set $\mathfrak{M}$ always contains $\mathfrak{M}_o$. The following example will show that this inclusion may be strict for quota games.

EXAMPLE 2.1. Let $n = 6$, and let the quotas be 5 for each player; then, e.g., the p.c. $(5 - x, 5 + x, 5 - x, 5 + x, 5 - x, 5 + x; 12, 34, 56) \in \mathfrak{M}$, $0 \leq x \leq 5$. Indeed, if, for example, player 1 objects against player 2 by joining, say, player 3 in any possible split, then player 2 can counter-object by joining player 5 with the $5 + x$, $5 - x$ split.

In practical situations, coalitions do not form simultaneously. In many cases, people negotiate and coalitions start to form one after the other. Once a coalition is formed, it is relatively hard to break it, and one can consider the situation as a new game for the rest of the players, whose characteristic function is unchanged for subsets of the remaining players and has no value for other coalitions. Assume that the participants of a certain quota game desire the stability as demanded by the definition of $\mathfrak{M}_o$. If a coalition is formed, there still remains a quota game for the rest of the players, and the quotas are unchanged. Therefore, as long as

---

* A counter example for $n = 5$ is, e.g., $(0, 5, 8, 12, 0; 12, 34, 5)$, which is stable if the quota is $(-5, 10, 10, 10, 10)$.

more than one or two players* remain in the game, each one of the members of a forming coalition will accept his quota. This situation will continue until one or two players* will be left. Having reached this stage, the remaining players will no longer be subjected to the outside "social pressure" to remain in a quota split, and they can end up anywhere in the core. This situation resembles (and also explains, perhaps) the feature of the benefactor and the beneficiaries which appears in the solutions to Shapley's quota games given in [5].

It is interesting to note that exactly this situation has been observed in some experiments on n-person quota games (without a weak player). See M. Maschler [2].

§ 3. SHAPLEY'S QUOTA GAMES

In [5], L. S. Shapley defines <u>quota games</u> to be games in which <u>all</u> the subsets of the players are permissible coalitions and the characteristic function is <u>superadditive</u>, i.e.,

$$v(B \cup C) \geq v(B) + v(C) \quad \text{if} \quad B \cap C = \emptyset, \tag{3.1}$$

and satisfies**

$$v(i) = 0, \quad v(ik) = \omega_i + \omega_k, \quad i, k = 1, 2, \ldots, n, \quad i \neq k, \tag{3.2}$$

$$v(N) = \omega_1 + \omega_2 + \ldots + \omega_n. \tag{3.3}$$

Here again, the <u>quota</u> ω_i of player i, $i = 1, 2, \ldots, n$, is a fixed real number. At most one player can be <u>weak</u>, i.e., can have a negative quota.

In this section we shall study some stable payoff configurations of these games. Again, we shall mainly be interested in the coalition structures in which at most one player acts as a 1-person coalition.

---

* Or three players, if there is a weak player among them. One assumes that the weak player will be the last one to join a coalition, or that he will remain alone.

** The normalizing condition $v(i) = 0$ is not assumed in [5].

DEFINITION 3.1. A coalition in a game Γ is called _effective_, if it is possible to share its value among its members in such a way that no sub-coalition is _alone_ able to provide more for its members.*

By (2.1), all 1- and 2-person coalitions are effective. It follows from the coalitional rationality demand that only effective coalitions occur in a c.r.p.c. A non-effective coalition cannot enter any stable payoff configuration, or any objection or any counter objection, and therefore no change in the bargaining set will result if we declare such a coalition non-permissible.

DEFINITION 3.2. Let Γ be a Shapley's quota game. A coalition B will be called _superior_, _regular_, or _inferior_, if $v(B)$ is greater than, equal to, or smaller than $\sum_{i \in B} \omega_i$, respectively.

Certainly, the 2- and n-person coalitions are regular. It will turn out later that the existence of superior coalitions may change considerably the bargaining set of the game. The following lemma will furnish some of the basic connections between the various kinds of coalitions.

LEMMA 3.1. Let B be a coalition in a Shapley's n-person quota game.

_a_. If $|B|$ is even, then B is not an inferior coalition.

_b_. If B is a superior coalition, then any coalition obtained by adding two new members to B is also a superior coalition.

_c_. If B is a superior coalition, then $N - B$ is an inferior coalition.

_d_. If n is even, and if $|B|$ is even, then B is a regular coalition (Shapley [5]); therefore B

* Such a share of the value will be called an _effective share_.

is superior or inferior only if $|B|$ is odd.

e. If n is even, then any two superior coalitions, if such exist, must intersect each other.

f. If n is odd, then B is a superior coalition only if $|B|$ is even. (Therefore, there can be no weak player in this case (Shapley [5]).)

The proofs follow immediately from (3.1), (3.2) and (3.3).

LEMMA 3.2. Let Γ be a Shapley's quota game. An inferior coalition B, where $|B| > 1$, cannot be effective.

PROOF. Let B be an effective coalition in Γ. If $|B|$ is even, it is certainly not inferior (Lemma 3.1, a.). If $|B|$ is odd, let T_j be a subcoalition $T_j \equiv B - \{j\}$, $j \in B$. T_j contains an even number of players. Let $\{x_i\}$, $i \in B$, be an effective share of $v(B)$ among the members of B. Certainly,

$$\sum_{i \in T_j} x_i \geq \sum_{i \in T_j} \omega_i, \quad j \in B, \tag{3.4}$$

because the players in T_j can command at least $\sum_{i \in T_j} \omega_j$, by forming 2-person coalitions. Summing up these inequalities and dividing by $|B| - 1$, we find that B is a superior or a regular coalition.

LEMMA 3.3. Let Γ be a Shapley's quota game, and let B be a regular coalition, which contains at least three players. If B is effective, then the effective share must assign each player exactly his quota.

PROOF. If B is regular and effective, any two members of B must, in an effective share, get a total sum equal to the sum of their quotas. This is obvious if $|B|$ is even. If $|B|$ is odd, this follows from the fact that equalities must hold in (3.4). Therefore, since $|B| \geq 3$, any three members of B must get a total sum equal to the sum of their quotas, in any effective share. This implies that any three players in B must share their total sum in a quota split.

Clearly, if B is regular, it need not be effective. It is never effective if B is regular and contains a weak player, and $|B| \geq 3$. Examples 3.1, 3.2, and 3.3 below will furnish other cases.

COROLLARY 3.1. If B and N − B are effective coalitions, $|B|$, $|N - B| \geq 2$, then they must be regular coalitions also.

THEOREM 3.1. Let $(x; \mathcal{B})$ be a c.r.p.c. in a Shapley's quota game, and suppose that each coalition in $\mathcal{B}$ contains at least two players; then the members of any k-person coalition in $\mathcal{B}$, where $k \geq 3$, get exactly their quotas.

PROOF. The coalitions in $\mathcal{B}$ are effective; hence, by Lemma 3.2, they are not inferior. If one of them was superior, then, by (3.1), (3.3) would be violated. Therefore, the coalitions in $\mathcal{B}$ are all regular. Lemma 3.3 now completes the proof, because x induces an effective share in each coalition of $\mathcal{B}$.

COROLLARY 3.2. Note that a weak player cannot belong to a k-person coalition, $k \geq 3$, if $(x; \mathcal{B})$ is subject to the conditions of Theorem 3.1.

THEOREM 3.2. Let Γ be an n-person Shapley's quota game, $n \geq 3$, in which no coalition is superior.* Let $\mathcal{B}$ be a partition of the players into coalitions in which no two 1-person coalitions occur. In order that a p.c. $(x; \mathcal{B})$ be stable, it is necessary and sufficient that:

---

* Therefore, the game does not contain a weak player.

(i) Each coalition in $\mathcal{B}$ which contains more than one player is regular.

(ii) Each player gets his quota, except the "exceptional player" who belongs to a 1-person coalition (if any). If there is an "exceptional player," he gets 0.

PROOF. Let $(x; \mathcal{B})$ satisfy (i) and (ii). It is certainly a p.c. It is also coalitionally rational, because the game has no superior coalitions. The only possible objection can arise if there exists a 1-person coalition, which consists of a player with a positive quota. In this case the objecting K is one player, who joins the "exceptional player," offering him some amount less than this player's quota.* If $|L|$ is even, the players in L can counter object by forming pairs with quota splits. Otherwise, one player of L joins the "exceptional player," and the others form pairs, again with quota splits.

Suppose now that $(x; \mathcal{B})$ is stable. Certainly (i) is satisfied. By Lemma 3.3, the players who perhaps do not satisfy (ii) must belong to 2-person coalitions. To prove that such players do not exist, we apply similar arguments to those used in Theorem 2.1 (starting after (2.8)), if no 1-person coalition occurs in $\mathcal{B}$, and to those used in Theorem 2.2 (starting after (2.12)), otherwise. This is evident, because all the objections were made there by forming two-person coalitions, and so they can similarly be made here; and no additional counter objections are now possible, because the players cannot use k-person coalitions, $k \geq 3$, in their counter objections since they got more than their quotas in $(x; \mathcal{B})$.**

If superior coalitions exist in a Shapley's quota game, an analogue of the previous theorem no longer holds. It is instructive to study the following examples for which each player has a quota equal to 5.

EXAMPLE 3.1. A 6-person Shapley's quota game:

---

* In any effective coalition which contains more than two players, each player must get his quota in an effective share, because there are no superior coalitions in the game (see Lemma 3.3).

** See the first footnote in this proof.

$v(B) = 0, 10, 12, 20, 25, 30$ if $|B| = 1, 2, 3, 4, 5, 6$, respectively, except that $v(135) = 18$, $v(135ij) = 28$ for $1, 3, 5, i, j$ distinct.

Certainly, the payoff configuration

$$(5, 5, 5, 5, 5, 5; \; 123456) \tag{3.5}$$

is not stable because it is not coalitionally rational. The same holds true for the c.r.p.c.

$$(5, 5, 5, 5, 5, 5; 12, 34, 56), \tag{3.6}$$

because the objection

$$(6, 0, 6, 0, 6, 0; 135, 2, 4, 6) \tag{3.7}$$

of $K = \{1, 3, 5\}$ against $L = \{2, 4, 6\}$ cannot be countered. On the other hand,

$$(5, 5, 5, 5, 5, 5; 13, 24, 56) \tag{3.8}$$

is stable, because the only* possible objections are for $K = 5$ against $L = 6$, in which the set of partners of K is $\{1, 3, 5\}$ or $\{1, 3, 5, 2, 4\}$. In the latter case, players 2 and 4 must get exactly 5 each, because of the coalitional rationality of the objection. Therefore, player 6 can always counter object by

$$(0, 0, 0, 5, 0, 5; 46, 1, 2, 3, 5). \tag{3.9}$$

This example shows that the same imputation may be stable for one coalition-structure, and unstable for another.

---

* By (1.4) and (1.5), K cannot contain either player 1 or player 3.

EXAMPLE 3.2. A 4-person Shapley's quota game:*

$v(B) = 0, 10, 10, 20$ for $|B| = 1, 2, 3, 4$, respectively, except that $v(123) = 18$.

The bargaining set** $\mathfrak{M}_o$, which in this game is identical with the bargaining set $\mathfrak{M}$ (see section preceding Example 2.1), is given, up to obvious permutations of the players 1, 2, 3, by:

$$(3.10)\qquad \begin{cases} (0, 0, 0, 0; 1, 2, 3, 4); & \\ (x_1, 10 - x_1, 0, 0; 12, 3, 4), & 0 \le x_1 \le 10; \\ (x_1, 0, 0, 10 - x_1; 14, 2, 3), & 6 \le x_1 \le 10; \\ (5, 5, x_3, 10 - x_3; 12, 34), & 6 \le x_3 \le 6\tfrac{1}{2}. \end{cases}$$

Note that no 3- or 4-person coalition is stable, and no quota imputation appears in $\mathfrak{M}_o$.

EXAMPLE 3.3. An 8-person Shapley's quota game:

$v(B) = 0, 10, 10, 20, 20, 30, 30, 40$ for $|B| =$ 1, 2, 3, 4, 5, 6, 7, 8, respectively, except that $v(123) = 18$; $v(123ij) = 28$, for $1, 2, 3, i, j$, distinct, and $v(N-k) = 38$, for $k \neq 1, 2, 3$.

It is easy to verify that

$$(3.11)\qquad (6, 6, 6, 5, 4, 4, 4, 5; 15, 26, 37, 48)$$

belongs to $\mathfrak{M}_o$. The imputation in this p.c. differs from any imputation in Shapley's "quota-solution" [5].

We shall now turn to the study of Shapley's quota games which contain

* A similar example was suggested in a different context by Dr. H. Kesten (see [2]).

** A method for computing the bargaining set is described in [1].

a weak player. This can happen only if n is even (see [5]; also Lemma 3.1, f.).

LEMMA 3.4. Let Γ be a Shapley's quota game, which contains a weak player; then

(i) a coalition B is superior if and only if $|B|$ is odd and contains the weak player;

(ii) a coalition B is inferior if and only if $|B|$ is odd and does not contain the weak player.

The proof follows from Lemma 3.1, b., c., d. and e., in view of the fact that the weak player forms a 1-person superior coalition.

As in the case of games with no weak players, one can construct examples in which some non-weak players in 2-person coalitions do not get their quotas in stable p.c.'s. This is not the case if the superior coalitions are not "too superior." This will be clarified in Theorem 3.3.

DEFINITION 3.3. Let B be a coalition in a Shapley's quota game. The quantity

$$c(B) \equiv v(B) - \sum_{i\in B} \omega_i \tag{3.12}$$

will be called the excess of B.

LEMMA 3.5. Let B be a superior coalition in a Shapley's quota game; then, the excess of B is not greater than the smallest among the quotas of the players in $N - B$.

PROOF. By Lemma 3.1, the coalition $N - B$ is inferior, hence it contains an odd number of players. We delete from $N - B$ that player whose quota is minimum and let the others form 2-person coalitions. The result now follows from (3.1), (3.2) and (3.3).

LEMMA 3.6. Let B_1 and B_2 be two superior coalitions in a Shapley's quota game. If $B_1 \cap B_2 = \emptyset$, then $c(B_1) + c(B_2)$ is not greater than the smallest among the quotas of the players in $N - (B_1 \cup B_2)$. If $B_1 \subset B_2$, then $c(B_1) \leq c(B_2)$.

PROOF. If $B_1 \subset B_2$, then $|B_2 - B_1|$ is even, because, by Lemma 3.1, $|B_1|$ and $|B_2|$ are both even or both odd. $c(B_1) \leq c(B_2)$ now follows from the superadditivity of the characteristic function. Similarly, if $B_1 \cap B_2 = \emptyset$, then $B_1 \cup B_2$ is also a superior coalition with an excess not smaller than $c(B_1) + c(B_2)$, and the result follows from Lemma 3.5.

LEMMA 3.7. Let B be a superior coalition in a Shapley's quota game, which contains a weak player; then the excess of B is not smaller than the negative of the quota of the weak player.

The proof is an immediate consequence of Lemma 3.4, (3.1), (3.2), and (3.3).

DEFINITION 3.4. A superior coalition in a Shapley's quota game which contains a weak player will be called a <u>superior</u>-<u>regular</u> coalition, if its excess is equal to minus the quota of the weak player.

THEOREM 3.3. Let Γ be an n-person Shapley's quota game, $n \geq 4$, which contains a weak player. Suppose, also, that all the superior coalitions in Γ are superior-regular. Let $\mathcal{B}$ be any partition of the players into coalitions in which no two 1-person coalitions occur. In order that a p.c. $(x; \mathcal{B})$ be stable, it is necessary and sufficient that:

(i) If $B \in \mathcal{B}$, and B does not contain the weak player, then $|B|$ is even or $|B| = 1$.

(ii) If $B \in \mathcal{B}$, and B contains the weak player, then either $|B|$ is odd and ≥ 3, or $|B| = 2$.

(iii) The weak player gets 0. If he has only one associate, then the associate gets the value of their coalition.

(iv) A player in a 1-person coalition (if such exists) gets 0.

(v) All other players get their quotas.

PROOF. We shall make use of the fact that if a superior-regular coalition occurs in Γ, then any effective share of its value among its members must assign the value 0 to the weak player, and if, in addition, this coalition contains more than three players, each non-weak player must get his quota. Similarly, an effective share of the value of a regular coalition, which contains more than two players, must assign each player of this coalition his quota (Lemma 3.3).

Let $(x; \mathcal{B})$ satisfy (i) - (v). It is easy to check that it is a c.r.p.c. The proof that $(x; \mathcal{B})$ is stable is similar to the proofs given in Theorem 3.2 and Theorem 2.3,* taking into account that if the weak player is in L, he counter objects by playing alone.

Suppose now that $(x; \mathcal{B})$ is stable. Certainly (iv) is satisfied. $\mathcal{B}$ must contain only regular, superior, and 1-person coalitions (Lemma 3.2), and therefore (i) is satisfied (Lemma 3.4). Let the weak player belong to a coalition B^o, $B^o \in \mathcal{B}$. If $|B^o|$ is even and greater than 2, then B^o is regular (Lemma 3.4), and by the remark at the beginning of the proof, each player of B^o, including the weak player, must get his quota. This is impossible because the weak player can alone make more. Therefore, $|B^o|$ must be equal to 2. If $|B^o| = 1$, there will exist, as a result of (i), another 1-person coalition in $\mathcal{B}$, contrary to our assumption. We have thus proved that (ii) is satisfied. Also, if $|B^o|$ is odd, the weak player must get 0. By the remark at the beginning of this proof, the players other than the player in a 1-person coalition (if any), who get in $(x; \mathcal{B})$ less than their quotas, must belong to different coalitions in $\mathcal{B}$, all of which are 2-person

* If a 1-person coalition exists in $\mathcal{B}$, or not, respectively.

coalitions, except, perhaps, a 3-person coalition, which contains the weak player. The set of these players will be denoted by K^*. We shall denote L^* the set of their associates, except that the weak player will not be included in L^*, if he belongs to a 3-person coalition. Evidently, $|K^*| = |L^*|$. The rest of the proof is similar to the proof given in Theorem 2.3 (starting after (2.15)), if the weak player belongs to a 2-person coalition, and to Theorem 2.2 (starting after (2.12)), otherwise.

BIBLIOGRAPHY

[1] AUMANN, R. J. and MASCHLER, M., "The bargaining set for cooperative games," (this Study).

[2] MASCHLER, M., "An Experiment on n-person Cooperative Games." Recent Advances in Game Theory, Proceedings of the Princeton University Conference of October 1961, privately printed for members of the Conference (1962), pp. 49-56.

[3] MASCHLER, M., "n-person games with only 1, n—1 and n-person permissable coalitions," J. Math. Analysis and Applications, Vol. 6 (1963), pp. 230-256.

[4] PELEG, B., "On the bargaining set $\mathfrak{M}_0$ of m-quota games," (this Study).

[5] SHAPLEY, L. S., "Quota solutions of n-person games," Contributions to the Theory of Games, Vol. II, Annals of Mathematics Studies, No. 28, Princeton University Press, Princeton, New Jersey (1953), pp. 343-359.

Michael Maschler

Princeton University
The Hebrew University of Jerusalem

ON THE BARGAINING SET $\mathfrak{M}_0$ OF m-QUOTA GAMES

Bezalel Peleg

This paper continues the study of bargaining sets that began in [1]. In the following sections we investigate the bargaining sets $\mathfrak{M}_0$ of n-person m-quota games. Our definition of m-quota games is similar to Maschler's definition ([4]), which is based in turn on Kalisch's definition in [2], and our results generalize some of Maschler's results for quota games (in [3]), to the case of m-quota games.

Specifically, we obtain the following results (all relating to $\mathfrak{M}_0$): If there are no weak players, if the players are partitioned according to a coalition structure that contains the maximum number of m-person coalitions, and if m divides n, then a coalitionally rational payoff configuration is stable (i.e., is an element of $\mathfrak{M}_0$) if and only if every player gets his quota (Theorem 2.1). More generally, if $n = qm + r$, $0 < r < m$, then a coalitionally rational payoff configuration is stable if and only if the r isolated players get zero and the remaining players get their quotas (Theorem 2.2). If there are weak players and $m(m + 1) \leqq n$ then a necessary and sufficient condition for stability is that the weak and the isolated players get zero and the strong (i.e., non-weak) get no more than their quotas (Theorem 3.1). The case $m(m + 1) > n$ remains open (when there are weak players), except when $m = n - 1$ (see [4]).

This paper is entirely self contained; no formal use is made of the material in the references.

§ 1. DEFINITIONS

In this section we give only the definitions that we need in this paper. Further introductory material on bargaining sets and discussions about the definitions are to be found in [1] and [3].

An n-person cooperative game is a pair (N, v) where $N = \{1, \ldots, n\}$ is a set with n members and v is a real valued function defined on the power set of N. We always assume that v is normalized such that $v(\{i\}) = 0$, $i = 1, \ldots, n$ and we assume that $v(B) \geqq 0$ for all B, $B \subset N$.

A subset $\mathcal{B}$ of the power set of N is a coalition structure (c.s.) if $B_1, B_2 \in \mathcal{B}$ implies that $B_1 \cap B_2 = \emptyset$, and $\bigcup_{B \in \mathcal{B}} B = N$; i.e., if $\mathcal{B}$ is a partition of N.

A coalitionally rational payoff configuration (c.r.p.c.) is a pair $(x; \mathcal{B})$, where $\mathcal{B}$ is a c.s. and $x = (x_1, \ldots, x_n)$ is an n-tuple of real numbers that satisfies:

$$\Sigma_B x_i = v(B) \quad \text{for all } B,\ B \in \mathcal{B},$$

and

$$\Sigma_S x_i \geqq v(S) \quad \text{for all } S,\ S \subset B \in \mathcal{B}.$$

If $(x; \mathcal{B})$ is a c.r.p.c. and $S \subset N$ we define

$$P[S; (x; \mathcal{B})] = \cup\{B\colon B \in \mathcal{B} \text{ and } B \cap S \neq \emptyset\}.$$

$P[S; (x; \mathcal{B})]$ is the set of the partners of S in $(x; \mathcal{B})$.

Let $(x; \mathcal{B})$ be a c.r.p.c. and K and L be non-empty disjoint subsets of N such that $P[K; (x; \mathcal{B})] = P[L; (x; \mathcal{B})]$. An objection of K against L in $(x; \mathcal{B})$ is a c.r.p.c. $(y; \mathcal{C})$ that satisfies:

$$P[K; (y; \mathcal{C})] \cap L = \emptyset,$$

$$y_i > x_i, \text{ for all } i,\ i \in K,$$

and

$$y_i \geqq x_i, \text{ for all } i,\ i \in P[K; (y; \mathcal{C})].$$

Let $(x; \mathcal{B})$ be a c.r.p.c. and $(y; \mathcal{C})$ an objection of a K against an L in $(x; \mathcal{B})$. A counter objection of L against K is a c.r.p.c. $(z; \mathcal{D})$ that satisfies:

$P[L; (z; \mathcal{d})] \not\supset K,$

$z_i \geqq x_i$ for all $i, i \in P[L; (z; \mathcal{d})]$,

and

$z_i \geqq y_i$ for all $i, i \in P[L; (z; \mathcal{d})] \cap P[K; (y; c)]$.

We remark that in order to specify an objection $(y; c)$ of a K against an L in a c.r.p.c. $(x; \mathcal{b})$, or a counter objection $(z; \mathcal{d})$ of L against K, it is sufficient to give the set $P[K; (y; c)]$, or $P[L; (z; \mathcal{d})]$, and the payments for their players.

A c.r.p.c. $(x; \mathcal{b})$ is called <u>stable</u> if for each objection of a K against an L in $(x; \mathcal{b})$ there is a counter objection of L against K. The set of all stable c.r.p.c's is denoted by $\mathfrak{M}_0$.

If $S \subset N$ then $|S|$ denotes the number of members of S. If $t = |S|$ then S is called a <u>t-coalition</u>.

An n-person game (N, v) is an <u>m-quota game</u>, $2 \leqq m < n$, if there exist n real numbers $w_1, \ldots, w_n$ such that $v(S) = \Sigma_S w_i$ when $|S| = m$, and $v(S) = 0$ when $|S| \neq m$. w_i is called the <u>quota</u> of player i. The quota vector, if it exists, is unique. (See [4], Lemma 4.2.) We shall use the symbol $[N, m, w]$ to denote the m-quota n-person game with the player set N and the quota vector w.

Let $[N, m, w]$ be an m-quota n-person game. If $w_i < 0$ then player i is called <u>weak</u>. A c.s. $\mathcal{b}$ is a <u>maximal</u> <u>c.s.</u> (m.c.s.) if it contains the maximum number of m-coalitions.

Let $(x; \mathcal{b})$ be a c.r.p.c., $S \subset N$ and $0 < t \leqq |S|$. T is a <u>set of t first (last) members of S with respect to</u> $(x; \mathcal{b})$ if $T \subset S$, $|T| = t$ and $w_i - x_i \geqq w_j - x_j$ $(w_i - x_i \leqq w_j - x_j)$ for all i and j, $i \in T$ and $j \in S - T$.

LEMMA 1.1. Let $(x; \mathcal{b})$ be a c.r.p.c., $K = \{i: x_i < w_i\}$ and $L = \{i: x_i > \max(0, w_i)\}$. If a set $K_1 \subset K$ has an objection $(y; c)$ against a set S such that $\Sigma_{K_2}(w_i - y_i) + \Sigma_{K_3}(w_i - x_i) < \Sigma_{S \cap L}(x_i - w_i)$, where $K_2 = \{i: i \in K \cap P[K_1: (y; c)], w_i > y_i\}$ and

$K_3 = K - P[K_1; (y; c)]$, then S cannot counter object.

The proof is straightforward and is omitted.

§ 2. GAMES WITHOUT WEAK PLAYERS

THEOREM 2.1. Let $[N, m, w]$ be an m-quota n-person game. If $n = qm$, $w_i \geqq 0$ for all i, $i \in N$, and $\mathcal{b}$ is a m.c.s., then a c.r.p.c. $(x; \mathcal{b}) \in \mathfrak{M}_0$ if and only if $x_i = w_i$, $i = 1, \ldots, n$.

PROOF. If $x = w$ then there are no objections in $(x; \mathcal{b})$, so that $(w; \mathcal{b}) \in \mathfrak{M}_0$. Suppose now that $(y; \mathcal{b})$ is a c.r.p.c. Denote $K = \{i: y_i < w_i\}$, $I = \{i: y_i = w_i\}$ and $L = \{i: y_i > w_i\}$. We show that if $L \neq \emptyset$ then $(y; \mathcal{b}) \notin \mathfrak{M}_0$. This will be done by proving the existence of sets U and V such that U has an objection against V in $(y; \mathcal{b})$ and V has no counter objection. We distinguish the following possibilities:

Case (a)

$$|K| + |I| < m.$$

Let $T \subset L$ be a set with $m - |K| - |I|$ members. Since $q \geqq 2$ we have that $\Sigma_{K \cup T}(w_i - y_i) > 0$. It is also true that $P[K; (y; \mathcal{b})] = P[L - T; (y; \mathcal{b})]$. K can object against $L - T$ by forming an m-coalition together with $I \cup T$. The payments will be $y_i + \frac{1}{|K|} \Sigma_{K \cup T}(w_j - y_j)$ for $i \in K$, and y_i for $i \in I \cup T$. If $(z; c)$ is a c.r.p.c. such that $z_i \geqq y_i$ for $i \in L - T$ then $|P[L - T; (z; c)]| = n$ (since $|L - T| = (q - 1)m$ and $\Sigma_{L-T}(y_i - w_i) > 0$), so that $L - T$ cannot counter object.

Case (b)

$$|K| < m \text{ and } |K| + |I| \geqq m.$$

Here K can object against L by forming, together with $m - |K|$ members of I, an m-coalition with a quota split. L has no counter objection.

Case (c)

$$|K| \geq m.$$

We have that $|K| = tm + s$, $t \geq 1$ and $0 \leq s < m$. If $s = 0$ then K can form t m-coalitions with a quota split and object against L. If $s + |I| \geq m$ then K, together with $m - s$ members of I, can form $t + 1$ m-coalitions with a quota split and object against L. In both cases L has no counter objection. So we may assume that $s > 0$ and $s + |I| < m$. Let S be a set of s last members of K and T be a set of $m - s - |I|$ first members of L with respect to $(y; \mathcal{B})$. We have to consider the following sub-cases:

Case (c.1)

$$\Sigma_{S \cup T}(w_i - y_i) < 0.$$

From (c.1) it follows that $P[K - S; (y; \mathcal{B})] = P[L; (y; \mathcal{B})]$. $K - S$ can object against L by forming t m-coalitions with a quota split. We have that $\Sigma_L(y_i - w_i) = \Sigma_K(w_i - y_i) > \Sigma_S(w_i - y_i)$. By Lemma 1.1 L has no counter objection.

Case (c.2)

$$\Sigma_{S \cup T}(w_i - y_i) = 0.$$

If $P[K - S; (y; \mathcal{B})] = P[L; (y; \mathcal{B})]$ then $K - S$ objects against L as in the previous case. If $P[K - S; (y; \mathcal{B})] \neq P[L; (y; \mathcal{B})]$ then, since $P[L; (y; \mathcal{B})] \supset P[K - S; (y; \mathcal{B})]$, there is a $B \in \mathcal{B}$ such that $B \cap K \neq \emptyset$ and $B \cap K \subset S$. From (c.2) it follows that $B \cap K = S$. Hence $P[K - S; (y; \mathcal{B})] = P[L - B; (y; \mathcal{B})]$. $K - S$ can object against $L - B$ by forming t m-coalitions with a quota split. We have that $\Sigma_{L-B}(y_i - w_i) = \Sigma_{K-S}(w_i - y_i) > \Sigma_S(w_i - y_i)$ and so, by Lemma 1.1, $L - B$ has no counter objection.

Case (c.3)

$$\Sigma_{S \cup T}(w_i - y_i) > 0.$$

We have that $P[K; (y; \mathcal{b})] = P[L - T; (y; \mathcal{b})]$. K has the following objection against $L - T$: $K - S$ forms t m-coalitions with a quota split while S joins $I \cup T$ to form an additional m-coalition F. The payments for the members of F are $z_i = y_i + \frac{1}{s} \Sigma_{S \cup T}(w_j - y_j)$ for $i \in S$, and y_i for $i \in I \cup T$. Since $\Sigma_S(w_i - y_i) > \Sigma_T(y_i - w_i)$ and $\Sigma_L(y_i - w_i) = \Sigma_K(w_i - y_i)$ we have that $\Sigma_{L-T}(y_i - w_i) > \Sigma_{K-S}(w_i - y_i) > \Sigma_S(w_i - y_i) > \Sigma_{S_1}(w_i - z_i)$, where $S_1 = \{i: i \in S, z_i < w_i\}$. By Lemma 1.1 $L - T$ has no counter objection.

THEOREM 2.2. Let $[N, m, w]$ be an m-quota n-person game. If $n = qm + r$ where $0 < r < m$, $w_i \geqq 0$ for all i, $i \in N$, and $\mathcal{b}$ is a m.c.s. then a c.r.p.c. $(x; \mathcal{b}) \in \mathfrak{M}_0$ if and only if $x_i = w_i$ for all i, $i \in B \in \mathcal{b}$ and $|B| = m$.

PROOF. Without loss of generality, $\mathcal{b} = \{B_1, \ldots, B_q, B_{q+1}\}$ where $|B_{q+1}| = r$. We show firstly that if $x_i = w_i$ for all i, $i \in B_j$, $1 \leqq j \leqq q$, then $(x; \mathcal{b}) \in \mathfrak{M}_0$. If $(y; c)$ is an objection of a K against an L in $(x; \mathcal{b})$ then we have to show that L has a counter objection. If $K \subset B_{q+1}$ then L can counter object by forming $|L|$ 1-coalitions. If $K \not\subset B_{q+1}$ then there is a $k \in K$ such that $y_k > w_k$. There is an m-coalition $V \in c$ that contains k. Let $|L| = tm + s$, $0 \leqq s < m$. If $s = 0$ then L counter objects by forming t m-coalitions with a quota split. If $s > 0$ then L again forms t m-coalitions with a quota split, and the remaining s players of L form an additional m-coalition U together with a set Q of $m - s$ first members of $V - \{k\}$ with respect to $(y; c)$. The payments for the players in U are w_i for $i \in U \cap L$ and $y_i + \frac{1}{m-s} \Sigma_Q(w_j - y_j)$ for $i \in Q$. (Observe that $\Sigma_Q(w_j - y_j) > 0$.)

Let now $(z; \mathcal{b})$ be a c.r.p.c. We denote $K = \{i: i \in B_j, j \leqq q$ and $z_i < w_i\}$, $I = \{i: i \in B_j, j \leqq q$ and $z_i = w_i\}$ and $L = \{i: z_i > w_i\}$. We shall prove that if $L \neq \emptyset$ then $(z; \mathcal{b}) \notin \mathfrak{M}_0$. The idea of the proof is the same as of the proof of Theorem 2.1. We distinguish the following possibilities:

Case (a)

$$|K| + r + |I| < m.$$

Let T be a set of $m - |K| - r - |I|$ first members of L. Since $r > 0$, $\Sigma_{K \cup T}(w_i - z_i) > 0$ and therefore $P[K; (z; \mathcal{b})] = P[L - T; (z; \mathcal{b})]$. K can object against $L - T$ by forming an m-coalition together with $B_{q+1} \cup I \cup T$ and paying its members the amount $z_i + e_i$, where $0 < e_i < w_i - z_i$ and $\Sigma_K e_i = \Sigma_{K \cup T}(w_i - z_i)$, the members of B_{q+1} the amount w_i, and the members of $I \cup T$ the amount z_i. We show that $L - T$ has no counter objection to the above objection. Suppose, per absurdum, that $(u; \mathcal{d})$ is a counter objection of $L - T$. There must be a $Q \in \mathcal{d}$ that satisfies $|Q| = m$ and $Q \subset P[L - T; (u; \mathcal{d})]$. Since $|Q \cap K| < |K|$ we have that $|Q \cap L| > |T|$ and therefore $\Sigma_Q(w_i - u_i) < \Sigma_{Q \cap K}(w_i - u_i) + \Sigma_T(w_i - z_i) < \Sigma_T(w_i - z_i) + \Sigma_K(w_i - z_i - e_i) = 0$, which is impossible.

Case (b)

$$|K| + r < m \quad \text{and} \quad |K| + r + |I| \geqq m.$$

Here K objects against L by forming, together with B_{q+1} and $m - |K| - r$ members of I, an m-coalition with a quota split. L has no counter objection.

Case (c)

$$|K| + r \geqq m.$$

Let $|K| + r = tm + s$, $t \geqq 1$. If $s = 0$ then K objects against L by forming, together with B_{q+1}, t m-coalitions with a quota split. If $s > 0$ and $s + |I| \geqq m$, let $Q \subset K$ be an s-coalition. Then $K - Q$ forms, together with B_{q+1}, t m-coalitions with a quota split, while Q joins $m - s$ members of I also with a quota split, and thus K obtains an objection against L. In both cases L has no counter objection. So we may assume that $s > 0$ and $s + |I| < m$. If $|L| < m - |I| - s$ let $Q \subset K$ be an s-coalition. We have $P[K - Q; (z; \mathcal{b})] = P[L; (z; \mathcal{b})]$. $K - Q$ can object against L by forming, together with B_{q+1}, t m-coalitions with a quota split. Since $\Sigma_L(z_i - w_i) = \Sigma_K(w_i - z_i) > \Sigma_Q(w_i - z_i)$, L has no counter objection. So we may suppose also that $|L| \geqq m - s - |I|$. Observe that in this case $|P[L; (z; \mathcal{B})]| = |P[K; (z; \mathcal{B})]| \geqq 2m$. Now let S be a set of s last members of K and T a set of $m - s - |I|$ first members of L with respect to $(z; \mathcal{B})$. We have to consider the following sub-cases:

Case (c.1)

$$\Sigma_{S \cup T}(w_i - z_i) < 0.$$

It follows from (c.1) that $P[K - S; (z; \mathcal{B})] = P[L; (z; \mathcal{B})]$. $K - S$ can object against L by forming, together with B_{q+1}, t m-coalitions with a quota split. $\Sigma_L(z_i - w_i) = \Sigma_K(w_i - z_i) > \Sigma_S(w_i - z_i)$. By Lemma 1.1, L cannot counter object.

Case (c.2)

$$\Sigma_{S \cup T}(w_i - z_i) = 0.$$

If $P[K - S; (z; \mathcal{B})] = P[L; (z; \mathcal{B})]$ then $K - S$ objects against L as in the previous case. If $P[K - S; (z; \mathcal{B})] \neq P[L; (z; \mathcal{B})]$ then there is a $B_j \in \mathcal{B}$, $j \leqq q$, such that $B_j \cap K \subset S$. By (c.2) we must have that $B_j \cap K = S$ and that $B_j \cap L = T_1$ is a set of $|T|$ first members of L with respect to $(z; \mathcal{B})$. Since $P[L; (z; \mathcal{B})]$ includes at least two m-coalitions of $\mathcal{B}$ we have that $\Sigma_L(z_i - w_i) \geqq 2\, \Sigma_{T_1}(z_i - w_i)$. If $\Sigma_L(z_i - w_i) > 2\, \Sigma_{T_1}(z_i - w_i)$ then $K - S$ objects against $L - T_1$ by forming, together with B_{q+1}, t m-coalitions with a quota split. Now $\Sigma_{L-T_1}(z_i - w_i) > \Sigma_{T_1}(z_i - w_i) = \Sigma_S(w_i - z_i)$, and so, by Lemma 1.1, $L - T_1$ has no counter objection. If $\Sigma_L(z_i - w_i) = 2\, \Sigma_{T_1}(z_i - w_i)$ then there is a $k \neq j$ such that $|B_k \cap K| = s$. Also $|I| = 0$. To see this observe that if $|I| > 0$ then for every $k \neq j$ either $|B_k \cap K| > s$ or $|B_k \cap L| > |T|$, and the equality $\Sigma_L(z_i - w_i) = 2\, \Sigma_{T_1}(z_i - w_i)$ fails. Let $h_1 \in S$, $h_2 \in B_k \cap K$ and $h_3 \in B_{q+1}$. If $s > 1$ then $\{h_1\} \cup (K - S - \{h_2\})$ can object against L by forming, together with B_{q+1}, t m-coalitions with a quota split, and L cannot counter object. If $s = 1$ then K can object against L by forming, together with $B_{q+1} - \{h_3\}$, t m-coalitions with a quota split. Since in this case $|L| \geqq 2(m - 1)$ and there is only one player, namely h_3, that may agree to accept less than his quota when joining L as a partner in a counter objection, L cannot counter object.

Case (c.3)

$$\Sigma_{S \cup T}(w_i - z_i) > 0.$$

In this case $P[K; (z; \mathscr{b})] = P[L - T; (z; \mathscr{b})]$; also if $j \leqq q$ then $\Sigma_{B_j \cap L}(z_i - w_i) > \Sigma_T(z_i - w_i)$ and therefore $\Sigma_L(z_i - w_i) > 2\ \Sigma_T(z_i - w_i)$. K has the following objection against $L - T$: $K - S$, together with B_{q+1}, forms t m-coalitions with a quota split, while S joins $I \cup T$ to form an additional m-coalition F. The payments for the members of F are z_i for $i \in I \cup T$, and $z_i + e_i$, where $0 < e_i < w_i - z_i$ and $\Sigma_S e_i = \Sigma_{S \cup T}(w_i - z_i)$, for $i \in S$. Since $\Sigma_{L-T}(z_i - w_i) > \Sigma_T(z_i - w_i) = \Sigma_S(w_i - z_i - e_i)$, by Lemma 1.1 $L - T$ has no counter objection.

§ 3. GAMES WITH WEAK PLAYERS

THEOREM 3.1. Let $[N, m, w]$ be an m-quota n-person game and let $A = \{i : w_i < 0\}$ If $n = qm + r$, $0 \leqq r < m$, $q \geqq m + 1$ and $\mathscr{b}$ is a m.c.s., then a c.r.p.c. $(x; \mathscr{b}) \in \mathfrak{M}_0$ if and only if $x_i = 0$ for $i \in A$ and $x_i \leqq w_i$ for $i \in N - A$.

PROOF Without loss of generality, $\mathscr{b} = \{B_1, \ldots, B_q, B_{q+1}\}$ where $|B_{q+1}| = r$. Since $|A| \leqq m - 1$ we may assume that $(B_{q-1} \cup B_q) \cap A = \emptyset$. We show firstly that if $x_i = 0$ for $i \in A$ and $x_i \leqq w_i$ for $i \in N - A$ then $(x; \mathscr{b}) \in \mathfrak{M}_0$. Suppose that $(y; c)$ is an objection of a K against an L in $(x; \mathscr{b})$. We shall prove that L can counter object. Let $L_1 = L \cap A$, with $|L - L_1| = tm + s$, $0 \leqq s < m$, and let $L_2 \subset L - L_1$ be a tm-coalition. We let L_1 form $|L_1|$ 1-coalitions and L_2 form t m-coalitions with a quota split. If $s = 0$ we have already a counter objection. If $s > 0$ we distinguish two possibilities. The first is that there is a $k \in K$ such that $y_k > \max(0, w_k)$. There is an m-coalition $V \in c$ that contains k. Let V_1 be a set of $m - s$ first members of $V - \{k\}$ with respect to $(y; c)$. $L - L_1 - L_2$ forms an m-coalition V_2 together with V_1. The payments in V_2 will be w_i for $i \in L - L_1 - L_2$ and $y_i + \frac{1}{m-s}\Sigma_{V_1}(w_j - y_j)$

for $i \in V_1$. (Observe that $\Sigma_{V_1}(w_j - y_j) > 0$.) So the above counter objection of L is completed. The remaining possibility is that $y_i \leqq \max(0, w_i)$ for all $i \in K$. In this case if $|P[K; (y; \mathcal{C})] \geqq 2m$ we choose $V \subset P[K; (y; \mathcal{C})]$, $V \in \mathcal{C}$ and $h \in V$ with $y^h \geqq w^h$ and proceed as before. If $|P[K; (y; \mathcal{C})]| = m$ then $L - L_1 - L_2$ can find $m - s$ members of $B_{q-1} \cup B_q$ to complete the counter objection.

Let now $(z; \mathcal{B})$ be a c.r.p.c. We denote $K = \{i: i \in B_j, j \leqq q$ and $z_i < w_i\}$, $I = \{i: i \in B_j, j \leqq q$ and $z_i = w_i\}$ and $L = \{i: z_i > \max(0, w_i)\}$. We shall prove that if $L \neq \emptyset$ then $(z; \mathcal{B}) \notin \mathfrak{M}_0$. This will be done by showing the existence of sets U and V such that U has an objection against V in $(z; \mathcal{B})$ and V has no counter objection. Let $A_1 = A - B_{q+1}$, $R = B_{q+1} - A$ and $r_1 = |R|$. We have that $|K| + |I| \geqq m$. If $|K| + r_1 < m$ then K objects against $L \cup A_1$ by forming, together with R and $m - |K| - r_1$ members of I, an m-coalition with a quota split. $L \cup A_1$ has no counter objection. So we may suppose that $|K| + r_1 = tm + s$ where $t \geqq 1$. If $s = 0$ then K objects against $L \cup A_1$ by forming, together with R, t m-coalitions with a quota split. If $s > 0$ and $s + |I| \geqq m$ let $Q \subset K$ be an s-coalition. Then $K - Q$ forms, together with R, t m-coalitions with a quota split, while Q joins $m - s$ members of I also with a quota split, and thus K obtains an objection against $L \cup A_1$. In both cases $L \cup A_1$ has no counter objection. So we may assume that $s > 0$ and $s + |I| < m$. Since $q \geqq m + 1$ we have also that $|L \cup A_1| \geqq m$. Now let S be a set of s last members of K and T a set of $m - s - |I|$ first members of $L \cup A_1$ with respect to $(z; \mathcal{B})$. We distinguish the following possibilities:

Case (a)

$$\Sigma_{S \cup T}(w_i - z_i) < 0.$$

In this case we have that $P[K - S; (z; \mathcal{B})] = P[L \cup A_1; (z \mathcal{B})]$ and that $\Sigma_L(z_i - w_i) \geqq \Sigma_{L \cap B_q}(z_i - w_i) > \Sigma_S(w_i - z_i)$. $K - S$ can object against $L \cup A_1$ by forming, together with R, t m-coalitions with a quota split. By Lemma 1.1, $L \cup A_1$ has no counter objection.

Case (b)

$$\Sigma_{S \cup T}(w_i - z_i) = 0.$$

If $P[K - S; (z; \mathcal{B})] = P[L \cup A_1; (z; \mathcal{B})]$ then, since $\Sigma_L(z_i - w_i) \geqq \Sigma_{L\cap(B_{q-1}\cup B_q)}(z_i - w_i) \geqq 2\,\Sigma_S(w_i - z_i) > \Sigma_S(w_i - z_i)$, $K - S$ can object against $L \cup A_1$ as in the previous case. If $P[K - S; (z; \mathcal{B})] \neq P[L \cup A_1; (z; \mathcal{B})]$ then there is a $B_j \in \mathcal{B}$, $j \leqq q$, such that $B_j \cap K \subset S$. By (b) we must have that $B_j \cap K = S$ and that $B_j \cap (L \cup A_1) = T_1$ is a set of $|T|$ first members of $L \cup A_1$. If $\Sigma_L(z_i - w_i) > 2\,\Sigma_S(w_i - z_i)$ then $\Sigma_{L-T_1}(z_i - w_i) > \Sigma_S(w_i - z_i)$, and $(L \cup A_1) - T_1$ has no counter objection when $K - S$ forms, together with R, t m-coalitions with a quota split and objects against it. If $\Sigma_L(z_i - w_i) = 2\,\Sigma_S(w_i - z_i)$ then we may assume that $j = q$. Also we have that $|B_{q-1} \cap K| = s$ and $|I| = 0$. Let $h_1 \in S$, $h_2 \in B_{q-1} \cap K$ and $h_3 \in K - (B_q \cup B_{q-1})$. Since $q \geqq m + 1 \geqq 3$ and $|I| < m$, $K - (B_q \cup B_{q-1}) \neq \emptyset$. If $s > 1$ then $\{h_1\} \cup (K - S - \{h_2\})$ can object against $L \cup A_1$ by forming, together with R, t m-coalitions with a quota split, and $L \cup A_1$ cannot counter object. There is a B_k, $k \leqq q - 2$, such that $h_3 \in B_k$. If $s = 1$ and $|K \cap B_k| = 1$ then $K - \{h_3\}$ can object against $(L \cup A_1) - B_k$ by forming, together with R, t m-coalitions with a quota split. Since in this case $|L - B_k| \geqq 2(m - 1)$ and there is only one player, namely h_3, that may agree to accept less than his quota when joining $(L \cup A_1) - B_k$ as a partner in a counter objection, $(L \cup A_1) - B_k$ cannot counter object. If $|K \cap B_k| > 1$ then $K - \{h_3\}$ objects against $L \cup A_1$ by forming, together with R, t m-coalitions with a quota split and $L \cup A_1$ cannot counter object since $|L| \geqq 2(m - 1)$ and there is at most one player that may receive less than his quota when joining it as a partner in a counter objection.

Case (c)

$$\Sigma_{S\cup T}(w_i - z_i) > 0.$$

In this case $P[K; (z; \mathcal{B})] = P\,[(L \cup A_1) - T; (z; \mathcal{B})]$. Also $\Sigma_L(z_i - w_i) > 2\,\Sigma_T(z_i - w_i)$. To see this, observe that if $|K \cap (B_q \cup B_{q-1})| \geqq 2s$ then $\Sigma_L(z_i - w_i) \geqq 2\,\Sigma_S(w_i - z_i) > 2\,\Sigma_T(z_i - w_i)$, and if $|K \cap (B_q \cup B_{q-1})| < 2s$ then $|L \cap (B_q \cup B_{q-1})| > 2|T|$ and the inequality follows. K has the following objection against $(L \cup A_1) - T$: $K - S$, together with R, forms t m-coalitions with a quota split, while S joins $I \cup T$ to form an additional m-coalition F. The payments for the members of F are z_i for $i \in I \cup T$, and $z_i + e_i$, where $0 < e_i < w_i - z_i$ and $\Sigma_S e_i = \Sigma_{S\cup T}(w_i - z_i)$, for $i \in S$.

Since $\Sigma_{L-T}(z_i - w_i) > \Sigma_T(z_i - w_i) = \Sigma_S(w_i - z_i - e_i)$, by Lemma 1.1 $(L \cup A_1) - T$ has no counter objection.

This work was carried out under the supervision of Dr. R. J. Aumann as a part of a doctoral thesis to be submitted at the Hebrew University.

BIBLIOGRAPHY

[1] AUMANN, R. J., and MASCHLER, M., "The bargaining set for cooperative games." This study.

[2] KALISCH, G. K., "Generalized quota solutions of n-person games," Annals of Mathematics Study No. 40 (Princeton 1959) pp. 163-177.

[3] MASCHLER, M., "Stable payoff configurations for quota games." This study.

[4] MASCHLER, M., "n-Person games with only 1, n − 1 and n-person permissible coalitions," Research Memorandum No. 37, Econometric Research Program, Princeton University, New Jersey (1962).

Bezalel Peleg

The Hebrew University
Jerusalem, Israel

A PROPERTY OF STABILITY POSSESSED BY CERTAIN IMPUTATIONS[1]

Hans Radström

§ 1. INTRODUCTION

In this paper we consider constant-sum n-person games in normalized form only. For our purposes, therefore, a game is a constant-sum characteristic function, that is, a real-valued set function, v, defined for all subsets of the set I of n players and satisfying:

1. $v(S \cup T) \geq v(S) + v(T)$ if $S \cap T = \theta$ (= the void set)
2. $v(S) + v(I - S) = v(I)$.

Since the intuitive meaning of $v(S)$ is the value of S, that is the amount the set S of players can secure for itself in opposition to $I - S$, it might seem reasonable to require that $v(I)$ be divided among the players so that every subset S of I gets at least $v(S)$. However, it is well known and easy to see that this is possible only if the game is inessential. Namely, it follows from 2 above that such a division would give every coalition S exactly the amount $v(S)$, in particular every single player would get his own value, so that for each S the value $v(S)$ would have to be equal to the sum of the values of the players in S. If the function v has this property the game is called inessential. The really interesting games are of course the essential ones.

Therefore, it is of interest to study families of subsets of I with the following "rationality" property: There exists a distribution of the amount $v(I)$ to the players such that each subset S in the family gets at least $v(S)$. (An example of such a family is the family of all one-element subsets. A distribution of $v(I)$ among the players such that each player gets at least his own value is called an imputation.) In this paper we

1 This work was done under contract with the Swedish Research Institute of National Defense.

consider certain families of subsets of I which we call coalition structures and which turn out to have the rationality property.

The heuristic background is the following. We assume that in the bargaining which accompanies the game, each coalition which succeeds in securing a certain amount for itself, has a tendency to split into two competing parties when the question arises of how to divide that amount between the members of the coalition. In this situation the parties are assumed to contemplate breaking out of the coalition and joining forces with outside groups. These assumptions lead quite naturally to the considerations below.

Accordingly we begin by defining the concept of a coalition structure (Section 2), as a hierarchical arrangement of coalitions obtained as a result of the following process. First the set I of all players is divided into two disjoint parties. Then each of these is divided into two subparties and these parties again divided, and so on until non-divisible parties, e.g., single players, are reached. A coalition structure is the family of all parties formed during such a process. We calculate the number of different coalition structures (Theorem 1).

Next we formulate a distribution principle (Section 3) which generalizes the way the imputations in the non-discriminative solution of a three-person game are determined. By applying (Section 4) this principle step by step to the successive subdivisions in the process by which a given coalition structure is formed we calculate the amount which the principle gives to each coalition (Theorem 2). In particular we obtain a distribution of the amount $v(I)$ among the single players. This turns out to be an imputation which assigns to each coalition in the given structure at least the value of that coalition (Theorem 3). These imputations are called basic. The basic imputations therefore satisfy a requirement intermediate between the impossible requirement that every coalition get at least its value and the requirement which defines imputations, that every one-player coalition get at least its value. This shows that the coalition structures have the rationality property as defined above.

Sections 5 and 6 are devoted to a purely technical matter. The formulas for the basic imputations are a little unmaneuverable but can be much simplified if written in a symbolic way (Section 5). This method is given a rigorous treatment (Section 6) by the introduction of a certain commutative algebra.

In Section 7 it is shown that the basic imputations possess a certain stability property with respect to a restricted class of changes of coalition structure.

In the heuristic arguments about stability which follow the leading idea is that a coalition structure is considered as given and we try to find an imputation such that no coalition in the structure has much hope of gaining by a slight change of the structure. The term "slight" is given an exact definition in Section 2.

§ 2. COALITION STRUCTURES

The precise definition of a coalition structure is the following.

DEFINITION. Let I denote the set of players. Then a family F of subsets of I is a coalition structure if

1. $I \in F$, $\theta \notin F$.

2. To every $S \in F$ such that S contains more than one player there exists precisely one (unordered) pair of disjoint, non-void sets S_1 and S_2 in F with $S_1 \cup S_2 = S$. The sets S_1 and S_2 will be called the parties of S.

3. Every set S with $I \neq S \in F$ is a party of precisely one set in F.

There are various ways of representing a coalition structure in print or graphically. By a suitable ordering of the players it is always possible to describe the structure by brackets as is shown by the following example for nine players:

$$\left[\left[\left[\left[[1\ 2]\ 3\right]\ 4\right]\ 5\right]\ \left[[6\ 7]\ [8\ 9]\right]\right]$$

The same structure is represented as a tree in the figure below.

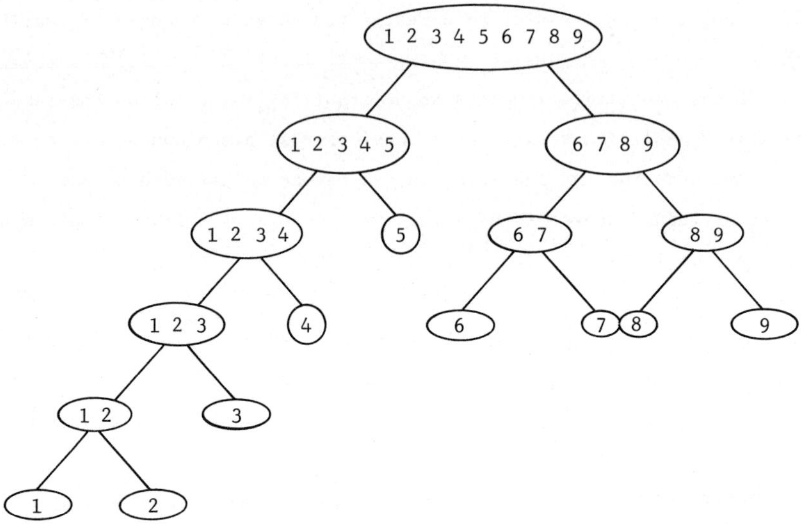

It is obvious that beginning with an arbitrary coalition in a structure one can reach I by successively adjoining other coalitions in the structure and that this can be done in only one way. In our example, beginning with the coalition [1 2 3] we have to adjoin [4], [5] and [6 7 8 9] in that order. We shall call the sequence [1 2 3], [4], [5], [6 7 8 9] the <u>chain</u> of [1 2 3].

In the general case let F be a coalition structure and let $I \neq S_1 \in F$. By 3 the set S_1 is a party of a unique larger set S in F and by 2 this set is of the form $S = S_1 \cup S_2$ where S_2 is also a party of S. Thus the set S_2 is uniquely determined by S_1. We shall call it the coparty of S_1. By the <u>chain of</u> S_1 we mean the longest sequence $S_1, S_2, \ldots, S_k$ of sets in F which satisfies

$$S_2 = \text{coparty of } S_1$$
$$S_3 = \text{coparty of } S_1 \cup S_2$$
$$\cdots$$
$$S_k = \text{coparty of } S_1 \cup S_2 \cup \cdots \cup S_{k-1}.$$

The chain of S_1 is of course finite and $S_1 \cup S_2 \cup \ldots \cup S_k = I$ since the only set in F which has no coparty is I.

It is clear that if $S_1, S_2 \ldots, S_k$ is the chain of S_1 for a coalition structure F then

(a) $S_1 \cup S_2 \in F$

(b) The chain of $S_1 \cup S_2$, is $S_1 \cup S_2, S_3, \ldots, S_k$.

(c) The family F' obtained from F by replacing $S_1 \cup S_2$ by $S_2 \cup S_3$ with no other changes made is a coalition structure and the chain of $S_2 \cup S_3$ in F' is $S_2 \cup S_3, S_1, S_4, \ldots, S_k$.

The change from F to F' as described in (c) will be called a <u>slight</u> change.

If the set of players consists of one or two elements there is of course precisely one coalition structure. For three players, 1, 2, 3, there are the three structures: $\left[[1\ 2]\ 3\right]$, $\left[[1\ 3]\ 2\right]$ and $\left[[2\ 3]\ 1\right]$.

Let us calculate the number of different coalition structures on a set of n players. Denote that number by a_n. There are $\binom{n}{k}$ different ways of choosing a non-void subset $A \neq I$ of $k \neq n$ elements from I. For each such choice we obtain a coalition structure on I by choosing one coalition structure on A and another on its complement $I - A$. For a given choice of A there are therefore $a_k\, a_{n-k}$ different coalition structures on I in which A is one of the two sets into which I is divided. Now, as A varies, we obtain each coalition structure on I exactly twice. Thus

$$2\, a_n = \sum_{k=1}^{n-1} \binom{n}{k} a_k\, a_{n-k}\, .$$

This formula gives a_n recursively.

However, it is easy to obtain an explicit expression for a_n. Putting $b_n = \frac{a_n}{n!}$ and introducing the generating power series

$$f(x) = b_1 x + b_2 x^2 + \ldots$$

we see that the recursion formula becomes

$$2\,b_n = \sum_{k=1}^{n-1} b_k\, b_{n-k}\,.$$

This shows that, formally, $2f(x) - 2b_1x = f^2(x)$. Since $b_1 = a_1 = 1$ we have, still formally,

$$f(x) = 1 - \sqrt{1-2x} = -\sum_{n=1}^{\infty} (-2x)^n \binom{\frac{1}{2}}{n}.$$

This gives

$$b_n = (-1)^{n+1}\, 2^n \binom{\frac{1}{2}}{n} = \frac{1\cdot 1\cdot 3\cdot 5\cdot 7\cdot \ldots\ (2n-3)}{n!}$$

for $n \geq 2$.

Justification of the formal calculations can (for example) be taken from the fact that the series $f(x)$ turns out to converge absolutely to $1 - \sqrt{1-2x}$ for small x. The result is

THEOREM 1. There are $1\cdot 3\cdot 5\cdot \ldots (2n-3)$ different coalition structures on a set of $n \geq 2$ players.

§ 3. THE DISTRIBUTION PRINCIPLE

Consider a given coalition structure on I. Is there a natural way of distributing the amount $v(I)$ among the players without causing instability? To be precise this question should be reformulated in more well-defined terms. (See Section 7.)

Since we are dealing with a constant sum game we have $v(S) + v(I - S) = v(I)$. This shows that it is natural to require that the parties S and $I - S$ of I get $v(S)$ and $v(I - S)$ respectively. Next denote $I - S$ by S_3 and the parties of S by S_1 and S_2. By the usual heuristic arguments forwarded (for example in von Neumann-Morgenstern) in favour of the non-discriminative solution of the constant sum three-person game it follows that it is necessary for the stability of our coalition structure that S_1, S_2 and S_3 get the amounts which they would get if they were three single players involved in a three-person game and with S_1 and S_2 in coalition. These amounts are determined by consideration of the effects of deviations from the given structure and the heuristic arguments lead to the following <u>principle for distribution</u> of $v(S_1 \cup S_2)$ to S_1 and S_2:

> The amounts distributed to S_1 and S_2 when they are in coalition should be so chosen that if either S_1 or S_2 tries to gain by breaking the original coalition and instead joining forces with S_3, then, without losing anything the excluded party can offer S_3 a better coalition.

In the literature the heuristic arguments behind this principle are generally regarded as very strong. Of course, the principle only applies to cases when changes of coalition structure are really comtemplated, that is in non-discriminative situations.

Summing up, we have seen that there are strong heuristic reasons for assigning certain well determined amounts to the parties of I and also to the parties of these parties. Suppose now that somehow a coalition structure and a rule for assigning definite amounts to all coalitions in the structure which have chains of length $\leq k-1$ have been agreed on by the players. Are there any strong heuristic arguments which will cause the players to accept an extension of this rule to coalitions of chain length $\leq k$? What extension will then be accepted, the coalition structure being considered as given but slight changes contemplated and used as threats?

Let S be a coalition of chain length $k-1$, let S, S_3, $S_4, \ldots, S_k$ be its chain and denote its parties by S_1 and S_2. By the rule for chain length $k-1$ it is known how much the coalitions S_3 and $S_1 \cup S_2 = S$ get. It is also known how much the coalitions S_2 and $S_1 \cup S_3$ would get if the coalition structure was changed slightly (see Section 2) by the deviation of S_1 from S_2 to S_3, and it is known what S_1 and $S_2 \cup S_3$ would get if S_2 deviates. The situation is very analogous to that in the three-person game. Essentially the same heuristic arguments can be taken over and they show that the distribution principle as formulated above should be used. However, the precise sense in which application of the distribution principle implies stability of the coalition structure requires further discussion. We postpone that discussion to Section 7 and proceed now to calculate the result of applying the distribution principle.

Denote by w_{12} the amount given to $S_1 \cup S_2$ by the $k-1$ order rule in the original coalition structure in which $S_1 \cup S_2$ has the chain $S_1 \cup S_2, S_3, S_4, S_5, \ldots, S_k$. Define w_{13} and w_{23} similarly by considering the slightly changed structures corresponding to the permutations of the Figures 1, 2, 3. Denote the amounts which the distribution principle assigns to S_1 and S_2 in the coalition $S_1 \cup S_2$ by α and β, respectively. Then $\alpha + \beta = w_{12}$. Moreover, if S_1 deviates from S_2 to S_3 hoping to get $\alpha + \epsilon$ with $\epsilon > 0$ then S_3 gets $w_{13} - \alpha - \epsilon$. According to the principle we shall have

$$w_{23} - \beta > w_{13} - \alpha - \epsilon \quad \text{for all} \quad \epsilon > 0$$

that is $w_{23} - \beta \geq w_{13} - \alpha$. By reversing the roles of 1 and 2 we obtain equality which (in combination with $\alpha + \beta = w_{12}$) shows that

$$\begin{cases} \alpha = \frac{1}{2}(w_{12} + w_{13} - w_{23}) \\ \beta = \frac{1}{2}(w_{12} + w_{23} - w_{13}) \end{cases}$$

Denote by w_1 the amount given to S_1 by the $k-1$ order rule in the coalition structure where the chain of S_1 is $S_1, S_2 \cup S_3, S_4, \ldots, S_k$ and by w_2 the amount given to S_2 when its chain is $S_2, S_1 \cup S_3, S_4, \ldots, S_k$. Since the amount given to $S_1 \cup S_2 \cup S_3$ is the same in both cases we have $w_{13} + w_2 = w_{23} + w_1$ or

$$w_{13} - w_{23} = w_1 - w_2.$$

Combining this with the previous result we get

$$\begin{cases} \alpha = \frac{1}{2}(w_1 - w_2 + w_{12}) \\ \beta = \frac{1}{2}(w_2 - w_1 + w_{12}) \end{cases}$$

which is the desired result expressing the division of w_{12} between S_1 and S_2 in terms of quantities given by the $k-1$ order rule.

§ 4. THE BASIC IMPUTATIONS

Consider now a given coalition structure as formed by the process of successive divisions into parties described in the introduction. Each time a coalition is divided into two parties we can calculate how the gain of the coalition is to be distributed to the parties by use of the distribution principle. This amounts to using the formulas at the end of Section 3. The amount which a given coalition in the structure gets as a result of this process is easily calculated. It is clear from the structure of the formulas in Section 3 that the amount given to a certain coalition S_1 will be determined uniquely by the chain of S_1.

Denote by $v(S_1|S_2, \ldots, S_k)$ the amount given to S_1 by successive applications of the distribution rule to a coalition structure in which the chain of S_1 is $S_1, S_2, \ldots, S_k$. Denote by $v_1, v_2, \ldots, v_k, v_{12}, v_{13}, \ldots, v_{k-1,k}, \ldots$ the values $v(S_1), v(S_2), \ldots, v(S_k), v(S_1 \cup S_2), v(S_1 \cup S_3), \ldots, v(S_{k-1} \cup S_k), \ldots$.

THEOREM 2. $2^{k-1} \cdot v(S_1|S_2, S_3, \ldots, S_k) =$
$\pm v_1 \pm v_2 \pm \quad \pm v_k \mp v_{12} \mp v_{13} \pm \cdots \pm v_{k-1,k} \cdots v_{123\ldots k}$
where $+$ is chosen if 1 is among the indices of the term and $-$ otherwise.

PROOF. There is only one coalition which has a chain of length 1, namely I. This makes the theorem trivially true for $k = 1$. If the chain of S_1 is S_1, S_2 then

$$2\, v(S_1|S_2) = 2\, v(S_1) = v_1 + v_1 = v_1 - v_2 + v_{12}.$$

If $k \geq 3$ the distribution principle applies. Suppose that the theorem is proved for all coalitions with chains of length $\leq k-1$ where $k > 3$. In the notation of Section 3 we have

$$\alpha = v(S_1 | S_2, S_3, \ldots, S_k),$$

$$w_1 = v(S_1 | S_2 \cup S_3, S_4, \ldots, S_k),$$

$$w_2 = v(S_2 | S_1 \cup S_3, S_4, \ldots, S_k),$$

$$w_{12} = v(S_1 \cup S_2 | S_3, S_4, \ldots, S_k) .$$

By the induction assumption we have

$$\begin{aligned} 2^{k-2}(w_1 - w_2 + w_{12}) = &\Big[(v_1 - v_{23} - v_4 - \cdots - v_k) + (v_{123} + v_{14} + \cdots + v_{1k} - \\ &- v_{234} - \cdots - v_{23k} - v_{45} - \cdots) + (\ldots) + \cdots \Big] \\ &- \Big[(v_2 - v_{13} - v_4 - \ldots) + (v_{123} + \ldots) + \ldots\Big] \\ &+ \Big[(v_{12} - v_3 - v_4 - \ldots) + (v_{123} + \ldots) + \ldots\Big] . \end{aligned}$$

Careful inspection (a more elegant method will be given in Section 5) of this expression shows that it is equal to the right member of the equation in the theorem. On the other hand, Section 3 shows that $2^{k-2}(w_1 - w_2 + w_{12}) = 2^{k-1}\alpha = 2^{k-1}v(S_1 | S_2, \ldots, S_k)$. This proves the theorem.

It is clear from the way the distribution of $v(I)$ to the coalitions in a given coalition structure is done according to the distribution principle that the sum of the amounts given to the single players is $v(I)$. We shall see (Theorem 3) that each player (in fact, each coalition in the structure) gets at least his value. Therefore, the amounts distributed to the players form an imputation, which is thus uniquely determined by the coalition structure. This imputation will be called the basic imputation of the structure.

THEOREM 3. The basic imputation of a coalition structure assigns to every coalition in the structure at least the value of that coalition.

PROOF. We want to prove that

$$v(S_1 | S_2, S_3, \ldots, S_k) \geq v(S_1).$$

Consider

$$(v_1 - v_2 - v_3 - \ldots - v_k) + (v_{12} + \ldots + v_{1k} - v_{23} - \ldots - v_{k-1,k}) + \ldots = 2^{k-1} v(S_1 | S_2, \ldots, S_k) .$$

Observe that there are exactly as many terms with a minus sign in one parenthesis as there are terms with a plus in the next. A typical plus term is of the form $v_{1\chi}$ where χ denotes a collection of indices $\neq 1$. If $v_{1\chi}$ is in one parenthesis then $-v_\chi$ is in the preceding parenthesis. By the superadditivity of the characteristic function (1 in Section 1) it follows that $v_{1\chi} - v_\chi \geq v_1$. Therefore, by grouping together corresponding plus terms and minus terms we obtain a sum which is $\geq$ a sum containing 2^{k-1} terms v_1 so that $2^{k-1} v(S_1 | S_2, \ldots, S_k) \geq 2^{k-1} v_1 = 2^{k-1} v(S_1)$.

§ 5. A SYMBOLIC METHOD

If in the expressions in Theorem 2 we write symbolically $v_{12} = v_1 v_2$, $v_{23} = v_2 v_3$, $v_{123} = v_1 v_2 v_3$ and so on, the statement in the theorem can be written in the following elegant form

$$1 - 2^{k-1} v(S_1 | S_2, \ldots, S_k) = (1 - v_1)(1 + v_2)(1 + v_3) \ldots (1 + v_k).$$

This formula is of course pure nonsense, but we shall see in the next section that as long as we only consider linear identities between the linear combinations of the v_i, v_{ij}, $\ldots$, all calculations based on this symbolic notation will lead to correct results.

As a first example, let us consider the identity which was needed in the proof of Theorem 1. We use the notation of that proof. By the induction assumption in the proof we have

$$1 - 2^{k-2}w_1 = (1 - v_1)(1 + v_2v_3)(1 + v_4) \cdots (1 + v_k)$$

$$1 - 2^{k-2}w_2 = (1 - v_2)(1 + v_1v_3)(1 + v_4) \cdots (1 + v_k)$$

$$1 - 2^{k-2}w_{12} = (1 - v_1v_2)(1 + v_3)(1 + v_4) \cdots (1 + v_k).$$

Thus

$$1 - 2^{k-2}(w_1 - w_2 + w_{12})$$

$$= \left[(1 - v_1)(1 + v_2v_3) - (1 - v_2)(1 + v_1v_3) + (1 - v_1v_2)(1 + v_3)\right]$$

$$(1 + v_4) \cdots (1 + v_k) = (1 - v_1)(1 + v_2)(1 + v_3)(1 + v_4) \cdots (1 + v_k).$$

This is the required identity in symbolic notation.

§ 6. THE GAME ALGEBRA

We proceed now to give a rigorous formulation of the symbolic method. We are interested in linear identities between linear combinations of values of the characteristic function v for different sets of players. The word "identity" of course refers to relations satisfied for an arbitrary choice of the characteristic function. It is therefore natural to consider linear relations valid for arbitrary real-valued set functions on the set of players.

If there are n players these set functions form a real vector space A of dimension 2^n (one dimension for each subset of I). If S is a given subset of I denote by $e(S)$ the linear functional on A which assigns to each $a \in A$ the value of the set function a on the set S i.e.,

$$e(S) : a \to a(S).$$

The number of these functionals $e(S)$ is 2^n and they form a basis of the dual space A' of A. Therefore, the identities we want to consider are the linear relations between the elements of A'.

The characteristic function v of a particular game is an element of A and therefore canonically embedded as an element of A''. This linear functional, f, on A' will be called the <u>characteristic functional</u> of the game. We have $v(S) = f(e(S))$.

We are free to turn the vector space A' into an algebra by introducing a multiplication between elements of A'. For this purpose it is enough to define the multiplication for the basic vectors $e(S)$. This can be done in many ways. We adopt the following definition:

$$e(S_1) \cdot e(S_2) = e(S_1 \cup S_2).$$

Putting $e(\theta) = e$ it is easy to see that this definition makes A' a commutative and associative real algebra with e as identity element. This will be called the <u>n-person game algebra</u>.

Since $v(\theta) = 0$ we have $f(e) = 0$. The condition that v be super-additive:

$$v(S_1 \cup S_2) \geq v(S_1) + v(S_2) \quad \text{for} \quad S_1 \cap S_2 = \theta$$

takes the form

$$f\left[(e(S_1) - e)(e(S_2) - e)\right] \geq 0 \quad \text{for} \quad S_1 \cap S_2 = \theta.$$

The statement in Theorem 2 is the following

$$-2^{k-1} v(S_1 | S_2, \ldots, S_3) = f\left[(e - e(S_1))(e + e(S_2)) \ldots (e + e(S_k))\right].$$

Comparing this with the example in Section 5 the reader should have no difficulty in turning the symbolic formulas in that example into meaningful calculations in the game algebra.

As further examples and for use in the next section, we treat two more problems.

I: Suppose S_1 has the chain $S_1, S_2, \ldots, S_k$ in a given coalition structure. Suppose that S_2 deserts S_1 and allies with S_3 so that the new chain of S_1 becomes $S_1, S_2 \cup S_3, S_4, \ldots, S_k$. In what direction does this change influence the amount that S_1 gets in the game?

In other words, is it profitable for S_1 to be in coalition with S_2?

We want to find the sign of the expression $\Delta = v(S_1 | S_2, S_3, \ldots, S_k) - v(S_1 | S_2 \cup S_3, S_4, \ldots, S_k)$. Put $e(S_i) = u_i$. Then

$$- 2^{k-1}\Delta = f\Big[(e - u_1)(e + u_2)(e + u_3) \ldots (e + u_k) - 2(e - u_1)(e + u_2u_3)(e + u_4) \ldots (e + u_k)\Big].$$

This gives

$$2^{k-1}\Delta = f\Big[(e - u_1)(e - u_2)(e - u_3)(e + u_4) \ldots (e + u_k)\Big].$$

Multiplying out, a linear combination of $v_1, v_2, \ldots, v_{12} \ldots$ is obtained. It is easy to verify that there are games in which coalition structures can be found which give a negative Δ. (Such a game is the simple four person game corresponding to the main corner of the fundamental cube. Corner I in von Neumann-Morgenstern. We omit the details.)

There are also games in which this phenomenon does not occur so that coalitions are always profitable. (Such games are the four person games corresponding to points sufficiently close to the center of the fundamental cube.)

The fact that Δ can be negative in some situations complicates the discussion of stability in the next section. We remark that Δ is symmetric in u_1 u_2 and u_3 so that all three coalitions S_1, S_2, S_3 are in the same position with respect to the profit they may get from a coalition.

Our next problem is the following.

II: Suppose S_1 and S_2 are two coalitions in a given structure. Let w_1, w_2 and w_{12} be defined as in Section 3. Is w super-additive? In other words, what is the sign of

$$w_{12} - w_1 - w_2?$$

We have (putting $e(S_i) = u_i$)

$$- 2^{k-2}(w_{12} - w_1 - w_2) = f\Big[((e - u_1u_2)(e + u_3) - (e - u_1)(e + u_2u_3) - (e - u_2)(e + u_1u_3))(e + u_4) \cdots (e + u_k)\Big].$$

It follows that

$$2^{k-2}(w_{12} - w_1 - w_2) = f\Big[(e - u_1)(e - u_2)(e - u_3)(e + u_4) \cdots (e + u_k)\Big] = 2^{k-1}\Delta$$

where Δ is the same quantity as was considered in problem I. Thus

$$w_{12} - w_1 - w_2 = 2\Delta .$$

§ 7. STABILITY WITH RESPECT TO SLIGHT CHANGES

Let a coalition structure be given and suppose that a bracket symbol for that structure contains the part $[[S_1\ S_2]\ S_3]$. In motivating the distribution principle in Section 3 we considered the effects of the transitions to $[[S_1\ S_3]\ S_2]$ and to $[[S_2\ S_3]\ S_1]$ when there is already given a rule, the $k - 1$ order rule of Section 3, which permits determination of the amounts to be given to the two parties of $S_1 \cup S_2 \cup S_3$ in each of the three cases considered. We assume now that the $k - 1$ order rule is obtained by successive applications of the distribution principle, so that a basic imputation is associated to each structure.

We observe that the changes of coalition structure, which may be contemplated by S_1 and S_2, when they try to come to an agreement of how to divide the amount that the $k - 1$ order rule will give to $S_1 \cup S_2$, are not only those listed above but also more revolutionary changes involving other coalitions than S_1, S_2 and S_3. However, we shall not consider such revolutionary changes in the discussion of stability. We recall that a change of coalition structure is <u>slight</u> if there are three coalitions $S_1\ S_2\ S_3$ in the original structure which are allied in the order $[[S_1\ S_2]\ S_3]$ in the bracket notation and so that in the new structure this is changed to $[S_1\ [S_2\ S_3]]$ with no other changes made. (In this case S_2 deviates from S_1 to S_3.) In the three-person game all possible changes are slight, but in the general case this is of course not true. We prove now that *there is stability with respect to slight changes*. *This means that given any slight change we shall be able to exhibit a coalition in the original structure which opposes the change or at least is indifferent to it and which is in the position to prevent the change*.

Let S_1, S_2 and S_3 be the parties involved in the slight change. As in Section 3, let w_i and w_{jk} denote the amounts distributed to S_i and $S_j \cup S_k$ in case the arrangement is $[S_i\ [S_j\ S_k]]$. Let w_{123} denote the amount given to $S_1 \cup S_2 \cup S_3$. By w we denote the set function on the set $\{1, 2, 3\}$ which has the values $0, w_1, w_2, w_3, w_{12}, w_{13}, w_{23}$ and w_{123} for the sets forming the indices. We now refer back to the results of problems I and II in Section 6. There are two cases.

First suppose $\Delta \geq 0$. Then II shows that w is super-additive and hence the characteristic function of a constant sum three-person game. Therefore, the full force of the usual stability arguments in such games is available and we refer to the standard discussion of that case.

Secondly, suppose $\Delta < 0$. If the proposed change is for example

$$[[S_1\ S_2]\ S_3] \rightarrow [S_1\ [S_2\ S_3]]$$

then I shows that S_3 who is the excluded party in the original structure is also the favoured party. Therefore S_3 will oppose accepting S_2 as a coalition partner since otherwise the gain of S_3 would be reduced. This gives the required stability in this case.

In the second case there does not seem to be very strong heuristic reasons for accepting the use of the distribution principle to the distribution between S_1 and S_2 of what $S_1 \cup S_2$ gets, at least not when only slight changes are considered. This shows that it is desirable to investigate problem 1 below. It should be noticed that if $\Delta < 0$ there is a certain tendency for the coalition $S_1 \cup S_2 \cup S_3$ to fall apart since neither S_1 nor S_2 is really interested in forming the coalition $S_1 \cup S_2$. Instead, they will probably try to make coalitions with parties outside of $S_1 \cup S_2 \cup S_3$. This shows the need for a discussion of results of other changes than the slight ones.

We conclude by listing two unsolved problems.

1. For which games does there exist a coalition structure such that $\Delta \geq 0$ for any three S_1, S_2 and S_3 occurring in the order $[[S_1\ S_2]\ S_3]$ in a bracket symbol?

2. For which games is the set of basic imputations a solution?

Hans Rådström

Royal Institute of Technology, Stockholm

COALITION BARGAINING IN n-PERSON GAMES

Evar D. Nering

§ 1. INTRODUCTION

Since von Neumann and Morgenstern first advanced a theory of n-person games there have been many suggestions that their theory is not satisfactory in all respects, and many alternatives have been proposed. [1, 2, 3, 4, 6, 7, 9, 10, 11, 12, 13, 14, 17, 18.] It is evident that no one theory is going to meet all demands. The theory we wish to advance here also makes no claim to universal applicability. It deals with a restricted type of coalition formation in n-person games for relatively small values of n. It is a normative, rather than a descriptive, theory in that it provides a rational estimate of the bargaining position of each player in terms of his position in the structure of the game.

We shall consider n-person zero-sum games in which the sole strategic maneuvers are the formations of coalitions. When a coalition A is formed, the rules shall specify a payment $v(A)$ to the coalition as a whole. Nothing is specified about the payments to individuals within a coalition. That is left for the bargaining process. How strong is each person's bargaining position? What demands can he reasonably make?

§ 2. THE BARGAINING PROCESS AND THE CONTRACT

We shall assume the following bargaining and coalition-formation structure. A coalition already formed can carry on negotiations with another disjoint coalition to form a larger coalition. The coalition functions as a unit in such negotiations, and an individual within a coalition may not negotiate separately. Since an individual cannot engage in further bargaining once he has joined a coalition, he must negotiate at the time he enters any coalition for his share of any future gains of the coalition he joins. Similarly, a coalition which joins with another coalition must bargain at that time for its share of all future gains of the enlarged coalition. Thus, at each stage of coalition formation the potential gains of all subsequent steps must be anticipated and the division of these net gains must be agreed to.

Let A and B be two disjoint coalitions that are negotiating with each other. If they form a larger coalition (A, B) a binding contract must be signed at that time. This contract will specify the shares going to A as a whole and to B as a whole as a result of future gains of the coalition (A, B). This contract need not specify how these gains are distributed within each of the two coalitions, since this was decided in the earlier contracts as the coalitions A and B were built up.

We shall assume that the process of building up the coalitions proceeds in steps. At each step only two coalitions can unite into a larger coalition. Thus the number of coalitions is reduced by exactly one at each step. Three or more coalitions cannot unite into a larger coalition at the same time. These conditions are not unduly unrealistic because negotiations among three or more parties are difficult to carry out. However, other assumptions are clearly possible.

The rules specify a payment of $v(A)$ to the coalition A regardless of how the complementary set is constituted. If the complementary set is not united into a single coalition, further possibilities for the formation of coalitions exist. Unless the coalition game at that particular stage is flat (in the sense of von Neumann and Morgenstern), further coalition formation will be attractive to the participants. If the coalition game is flat, further coalition formation will not change the shares going to any coalition. Thus we shall assume that the formation of larger coalitions proceeds until only two opposing coalitions remain. Under these conditions $v(A)$ can be interpreted to be a characteristic function in the sense of von Neuman and Morgenstern.

§ 3. THE INEQUALITIES

Let $N = \{1, 2, \ldots, n\}$ be the set of players. Let $\pi_k = \{A_1, A_2, \ldots, A_k\}$ be a partition of N into k disjoint coalitions. If this is the set of coalitions that has been achieved by bargaining up to a certain point, A_i and A_j can consider the possibility of uniting to form a larger coalition, (A_i, A_j). If they unite, they can get the amount $v(A_i, A_j)$ even if they do not extend their coalition any further. At any rate they will find themselves in a coalition game involving $k - 1$ coalitions. We shall assume for the moment that we can resolve the shares for the players in a coalition game with fewer then k players and that the share that (A_i, A_j) can expect for extending its coalition in any conceivable way can be evaluated. Let $u(A_i, A_j)$ be the share of (A_i, A_j) at the termination of a particular sequence of coalition formations. The contract between A_i and A_j will have to specify how this amount will be split between them, if it should ever be realized.

Let x_{ij} be A_i's share of $u(A_i, A_j)$ and let x_{ji} be A_j's share. Then

$$x_{ij} + x_{ji} = u(A_i, A_j). \tag{1}$$

Each of A_i and A_j can consider other coalition possibilities. Suppose that A_i considers forming a coalition with each of the other coalitions in π_k, and suppose that $u(A_i, A_1), u(A_i, A_2), \ldots$ are the payoffs to other coalitions including A_i at the end of sequences of coalition formations in some sense comparable to each other. We shall go into detail later on what we mean by comparable sequences of coalition formation. Let $\{x_{i1}, x_{i2}, \ldots, x_{ik}\}$ be A_i's shares of the payoffs to all these other possible coalitions. Let

$$x_i = \max \{x_{i1}, x_{i2}, \ldots, x_{ik}\}. \tag{2}$$

We shall assume that A_i will consider only those coalitions in which its share is x_i. It may well be argued that A_i might be willing to join in a coalition in which its share is less than x_i in anticipation of a larger potential gain. This consideration has merit, but for this discussion we shall take the simpler assumption.

Combining (1) and (2) we obtain the system of inequalities

$$x_i + x_j \geq u(A_i, A_j) \quad \text{for all} \quad i, j. \tag{3}$$

This is the fundamental system of inequalities with which we have to deal.

Certainly, all the inequalities would be satisfied if the x_i are taken large enough. However, the parties A_i and A_j can realize the amounts x_i and x_j by joining in a coalition if and only if

$$x_i + x_j \leq u(A_i, A_j). \tag{4}$$

This additional condition means that at least some of the inequalities in the system (3) must be satisfied as equalities. Let $X_0 = \{x_1, x_2, \ldots, x_k\}$ be a solution of the system (3) of inequalities. Those inequalities that are satisfied as equalities by X_0 are said to be _effective for the solution_ X_0, and the coalitions involved in the effective inequalities are said to be _effective for that solution_.

Let E be the set of effective inequalities for a particular solution X_0, and let r be the rank of E. The rank of all the left-hand sides of the inequalities in (3) is k so that $r \leq k$. Suppose $r < k$. In this case E has a family of solutions of which X_0 is only one. The other inequalities in (3) are satisfied as strict inequalities so that there is an open range of values of the parameters of the solutions of E which includes X_0 and still satisfies all the inequalities of the system (3). Thus the shares of the players could be adjusted so that another inequality becomes effective while the inequalities which are already effective remain so. We interpret this as meaning that the full bargaining possibilities of the situation are not being realized. We therefore define an _effective solution_ of the system (3) as a solution in which the set of effective inequalities has rank k. We shall consider only effective solutions.

§ 4. THE 3-PERSON COALITION GAME

Let $N = \{1, 2, 3\}$. In the first round of coalition negotiation we have the following system of inequalities:

$$\begin{aligned} x_1 + x_2 \phantom{{}+ x_3} &\geq v(1, 2) \\ x_1 \phantom{{}+ x_2} + x_3 &\geq v(1, 3) \\ x_2 + x_3 &\geq v(2, 3). \end{aligned} \tag{5}$$

The additional condition that the system of effective inequalities be of rank 3 implies that all three inequalities are equalities. The solution is then

$$\begin{aligned} x_1 &= \tfrac{1}{2}\{v(1, 2) + v(1, 3) - v(2, 3)\} \\ x_2 &= \tfrac{1}{2}\{v(1, 2) - v(1, 3) + v(2, 3)\} \\ x_3 &= \tfrac{1}{2}\{-v(1, 2) + v(1, 3) + v(2, 3)\}. \end{aligned} \tag{6}$$

No further coalitions are profitable because of the assumption that we are dealing with zero-sum games. Since these are precisely the conditions derived by von Neumann and Morgenstern in their analysis of the 3-person game [15, pp. 220-231] it is no surprise that the results obtained here are the same. The three possible contracts are the three imputations of the finite solution which they obtained.

§ 5. THE 4-PERSON COALITION GAME

Let $N = \{1, 2, 3, 4\}$. In the first round of coalition negotiation we have the following system of inequalities:

$$\begin{aligned} x_1 + x_2 &\geq v(1, 2) \\ x_1 + x_3 &\geq v(1, 3) \\ x_1 + x_4 &\geq v(1, 4) \\ x_2 + x_3 &\geq v(2, 3) \\ x_2 + x_4 &\geq v(2, 4) \\ x_3 + x_4 &\geq v(3, 4). \end{aligned} \tag{7}$$

In this case the condition that the solution be effective implies that all inequalities are effective because of the zero-sum condition. Thus the solution is

$$\begin{aligned} x_1 &= \tfrac{1}{2}\{v(1, 2) + v(1, 3) + v(1, 4)\} \\ x_2 &= \tfrac{1}{2}\{v(1, 2) + v(2, 3) + v(2, 4)\} \\ x_3 &= \tfrac{1}{2}\{v(1, 3) + v(2, 3) + v(3, 4)\} \\ x_4 &= \tfrac{1}{2}\{v(1, 4) + v(2, 4) + v(3, 4)\}. \end{aligned} \tag{8}$$

After a coalition of two persons has been formed the three parties find themselves in a 3-person game. By an application of the previous arguments we can find the reasonable shares for each of these parties. But one of these parties is a coalition and their share must be divided between them.

Suppose, to be particular, that $A = (1, 2)$ is the coalition formed in the first round. If they do not succeed in extending their coalition, their joint share is $v(1, 2)$. If they do succeed in extending their coalition, their joint share is

$$\tfrac{1}{2}\{v(1, 2, 3) + v(1, 2, 4) - v(3, 4)\} = u(1, 2). \tag{9}$$

We have denoted this joint share by $u(1, 2)$. There are five other first-round coalitions that could have been formed. If one of these other coalitions had been formed and extended it would also have earned a second-round share similar to (9). Since the first-round shares for each player do not depend upon the identity of the other partner in the coalition, the second-round shares will also have to compete or else the coalition that proposes too low a second-round share will not be formed. Thus, if $\{z_1, z_2, z_3, z_4\}$ represent the maximum share to each player in the second step of coalition formation, we have

$$
\begin{array}{lllll}
(10) & z_1 + z_2 & & & \geq u(1, 2) \\
& z_1 & + z_3 & & \geq u(1, 3) \\
& z_1 & & + z_4 & \geq u(1, 4) \\
& & z_2 + z_3 & & \geq u(2, 3) \\
& & z_2 & + z_4 & \geq u(2, 4) \\
& & & z_3 + z_4 & \geq u(3, 4).
\end{array}
$$

The system of inequalities (10) resembles the system of inequalities (7), but there is a slight difference. The function $u(A)$ does not satisfy the zero-sum condition. However, we have

$$(11)\quad u(1, 2) + u(3, 4) = \tfrac{1}{2}\{v(1, 2, 3) + v(1, 2, 4) + v(1, 3, 4) + v(2, 3, 4)\}.$$

Since the expression on the right is fully symmetric in the four players, we have $u(1, 2) + u(3, 4) = u(1, 3) + u(2, 4) = u(1, 4) + u(2, 3)$. This is enough to guarantee that if a subsystem of rank 4 is effective, then all six inequalities are effective. Thus

$$
\begin{aligned}
(12)\quad z_1 &= \tfrac{1}{2}\{u(1, 2) + u(1, 3) - u(2, 3)\} \\
&= \tfrac{1}{4}\{v(1, 2, 3) + v(1, 2, 4) - v(3, 4) + v(1, 2, 3) + v(1, 3, 4) \\
&\qquad - v(2, 4) - v(1, 2, 3) - v(2, 3, 4) + v(1, 4)\} \\
&= \tfrac{1}{4}\{v(1, 2, 3) + v(1, 2, 4) + v(1, 3, 4) + v(1, 2) + v(1, 3) \\
&\qquad + v(1, 4) + v(1)\}.
\end{aligned}
$$

The fact that this expression for z_1 is fully symmetric in the players $\{2, 3, 4\}$ shows that 1's second-round share is the same for any first-round coalition he should choose to join. Furthermore, we can write down immediately corresponding expressions for the second-round share of each of the other three players.

§ 6. A NUMERICAL EXAMPLE

Consider the 4-person game with characteristic function

$v(1) = -20$	$v(1, 2) = 30$	$v(3, 4) = -30$	$v(2, 3, 4) = 20$
$v(2) = -40$	$v(1, 3) = 0$	$v(2, 4) = 0$	$v(1, 3, 4) = 40$
$v(3) = -40$	$v(1, 4) = -10$	$v(2, 3) = 10$	$v(1, 2, 4) = 40$
$v(4) = -20$			$v(1, 2, 3) = 20.$

According to formula (8) we have

$$x_1 = 10, \ x_2 = 20, \ x_3 = -10, \ x_4 = -20.$$

According to formula (12) we have

$$z_1 = 25, \ z_2 = 20, \ z_3 = 5, \ z_4 = 10.$$

In addition, there remains the calculations of the shares for those players that do not succeed in joining a coalition on the first round but do succeed on the second round. All the calculations will be summarized in the following table. In this table we shall use a vinculum to denote the first coalition and parenthesis to denote the second. Thus $\overline{12}(34)$ means that a coalition of 1 and 2 was formed in the first round and in the subsequent 3-person game 3 and 4 formed a coalition.

Several interesting and important observations can be made from an examination of the table. Player 2 finds himself in a strong position for the first round. He gains 60 units by joining any coalition. But he cannot share in the future gains of any of these coalitions. On the other hand, player 4 gains nothing from the first round in joining a coalition. All his gains would come if the coalition is extended. For either player to demand more would have the effect of making him an undesirable member of a proposed coalition.

This game was used in an experiment performed by Kalisch, Milnor, Nash, and Nering [5], [8], [10]. The subjects used in the experiment were intelligent but untutored in any aspects of game theory. Furthermore, the stakes in the outcome were so small that motivation for a careful analysis was slight. Under the circumstances, it is not surprising that the outcomes did not resemble the outcomes proposed here. All the players looked to the 30 units awarded to the coalition (1, 2) as dominating the game. Players

Coalition Structure	Share for player 1	2	3	4
$\overline{12}$(34)	10	20	−25	—5
($\overline{123}$)4	25	20	−25	−20
($\overline{124}$)3	25	20	−40	−5
$\overline{13}$(24)	10	−10	−10	10
($\overline{132}$)4	25	−10	5	−20
($\overline{134}$)2	25	−40	5	10
$\overline{14}$(23)	10	5	5	−20
($\overline{142}$)3	25	5	−40	10
(143)2	25	−40	5	10
$\overline{23}$(14)	−5	20	−10	−5
($\overline{231}$)4	−5	20	5	−20
($\overline{234}$)1	−20	20	5	−5
$\overline{24}$(13)	10	20	−10	−20
($\overline{241}$)3	10	20	−40	10
($\overline{243}$)1	−20	20	−10	10
$\overline{34}$(12)	25	5	−10	−20
($\overline{341}$)2	25	−40	5	10
($\overline{342}$)1	−20	5	5	10

3 and 4 considered their positions so hopeless that they did not try to engage in a coalition with 1 or 2. Players 1 and 2 split not their net gains but the absolute reward evenly, i.e., each took 15 units. When the first-round coalitions were formed there were no agreements about the division of future gains. But all future gains were divided evenly according to a tacit understanding of fair play.

Our analysis shows that players 3 and 4 grossly underestimated their bargaining positions. As a result, 1 was assured a coalition with player 2 and at a favorable price. Actually, although the various players differ in bargaining strengths and the demands they can make, each makes just as good a coalition partner as anyone else.

There are six 3-person games that result after the first coalition is formed, and they are quite different from each other. Player 4 finds himself in a better position if players 1 and 3 form a coalition than he does if 1 and 2 or 2 and 3 form a coalition. This raises an interesting point that will not be pursued at this moment. Player 4 cannot force two other players to form a coalition. But in this case he might find it advisable to start negotiations with player 2. If he succeeds in forming a coalition, it doesn't really matter whom it is with. But perhaps, by negotiating with player 2, he can at the same time impede formation of the coalitions (1, 2) and (2, 3). Exactly the same kind of reasoning would lead player 2 to initiate negotiation with player 4, player 1 with player 2, and player 3 with player 2.

§ 7. GENERAL CONSIDERATIONS

Let G be an n-person coalition game, and let $N = \{1, 2, \ldots, n\}$ be the set of players. Let $\pi_k = \{A_1, A_2, \ldots, A_k\}$ be a partition of N into k disjoint coalitions. For each π_k there is a derived coalition game $G(\pi_k)$ in which each A_i is regarded as a single player and in which the payoff to a coalition in $G(\pi_k)$ is the same as it is for the same coalition in the original game G.

We say that π_{k-1} follows π_k immediately if π_{k-1} is obtained from π_k by uniting two coalitions in π_k. The coalitions which unite are said to be active. The sequence $P = \{\pi_n, \pi_{n-1}, \ldots, \pi_2\}$ is a play of the coalition game G if each π_{k-1} follows π_k immediately.

For each player i and each play P, let $I_i(P) = (a_n, a_{n-1}, \ldots, a_3)$ be a sequence of 0's and 1's, where $a_k = 1$ if i is in one of the active coalitions in forming π_{k-1} from π_k, and $a_k = 0$ otherwise. We call $I_i(P)$ i's play index for the play P. As far as i is concerned, two plays with the same play index represent comparable strategic maneuvers. Thus, if two plays with the same play index offer player i different shares, he will prefer the play offering the larger share. He can exercise this preference if these plays branch at points where he is in an active coalition, but he cannot exercise this preference if they branch at a point where he is in an inactive coalition.

We have shown in Sections 4 and 5 how an effective solution of the inequalities (3) leads to the determination of a reasonable bargaining position for each player. In the 4-person game, consider the reasonable shares demanded by each player for the various plays. We see that his share is given by (8) for every play with play index (1, 0), and it is given by (12) for every play with play index (1, 1). Player i's share is v(i) if he joins no coalition, that is, if the play index is (0, 0). The one situation in which a player's reasonable share does not depend on the play index alone is the case in which the play index is (0, 1). This does not present a difficulty because plays with this play index offer different shares only if they branch at a point where the player is inactive, that is, where he has no choice.

These considerations motivate the following definition. A play index in which no 1 follows a 0 is said to be simple.

THEOREM. A fair bargaining position can be determined for each player which will specify a fair share for each play with a simple play index, such that the share will depend only on the play index.

For the 3-person zero-sum game, the analysis of Section 4 and the discussion above apply. Thus the assertions of the theorem are established for such games.

After the first coalition is formed, the players find themselves in the coalition game $G(\pi_{n-1})$. If a player is inactive in the first round he becomes an individual in the resulting (n–1)-person game. Whatever can be said about his bargaining position is dependent on our ability to analyse (n–1)-person games. We assume inductively that this has been done. Our main problem concerns the shares due to a player who is active in the first round.

First we must consider all plays with play index (1, 0, ..., 0). Since the coalition active in such a play is not extended, each coalition (i, j) must divide the amount v(i, j). This leads to a system of inequalities like (3) in which the right sides are the v(i, j). There may be several different effective solutions to this system of inequalities, and several different sets of effective coalitions. Let one effective solution

X_1 be chosen.

Then, consider all plays with play index (1, 1, 0, ..., 0) for which the coalitions active in the first step are effective for X_1. If the coalition (i, j) is active in the first step it finds itself in a coalition game with $n - 1$ persons and it extends its coalition one more step. Let $u(i, j)$ be the share to (i, j) in this shorter game. We have another system of inequalities like (3) in which the right sides are these $u(i, j)$, but only those inequalities can be considered which were effective for the solution X_1. All other coalitions are ignored in the first step of coalition formation and have no future shares to consider. The remaining system of inequalities is still of rank n and they have an effective solution. Let X_2 be the effective solution chosen at this step.

We next consider all plays with play index (1, 1, 1, 0, ..., 0), etc. At each stage only those inequalities which are effective for all previously chosen solutions will be considered. A coalition which eventually becomes ineffective in this way will not be considered as a feasible coalition in the first stage of coalition formation. If we drop these coalitions and the corresponding inequalities from consideration from the first, the solutions will not be changed because the remaining inequalities are of rank n. The coalitions which remain will depend upon the entire sequence of solutions $\mathfrak{X} = X_1, X_2, \ldots$, and these coalitions will be called <u>effective</u> for $\mathfrak{X}$.

For a given $\mathfrak{X}$, each X_k specifies a fair bargaining share for every player for every play with a simple play index in which he is active through k coalition formations. $\mathfrak{X}$ will determine which coalitions are sought, and which plays will occur.

In a previous paragraph when we considered all plays with play index (1, 1, 0, ..., 0) we assumed the share $u(i, j)$ was known for the shorter game in which the coalition (i, j) is a player. In determining this $u(i, j)$ we also determined the fair shares for all other players, who are individuals in this game. For these individuals we have determined fair shares for all plays with play index (0, 1, 0, ..., 0) in which the first coalition formed is (i, j). In a similar way we determine fair share for all other plays with the same play index, and, in general, for all plays for which the play index contains an unbroken sequence of 1's.

In case the sequence of 1's is broken the question remains

unresolved. To see the problem in this case, consider the example given in Section 6 as representing the terminal play of a 5-person game in which the player designated as "4" is the coalition (45) which was active in the first step. For players 4 and 5 there are three plays with play index (1, 0, 1). They are $\overline{12}$(3 (45)), $\overline{13}$(2 (45)), and $\overline{23}$(1 (45)). The shares going to (45) in these cases are −5, 10, and −5. There is no question of what they would prefer, but they cannot decide which coalition will form at a step where they are inactive. For each of 4 and 5 there would be other available partners, and for each of them there would be three possibilities to consider. The question is how to compare each or all of these three possibilities with the possibilities available with other partners.

One possible resolution of this ambiguity is to order the payoffs with each partner and compare payoffs with the same relative rankings. Let E be the set of coalitions effective for $\mathfrak{X}$, and let (i, j) be a coalition in E. For a given broken play index they can order the plays with that play index according to their preference for the payoffs. Other coalitions in E can do the same for all plays with that same play index for them. If u(i, j) is the value of the first preference for (i, j), etc., then we must find an effective solution of the system

$$x_i + x_j \geq u(i, j) \quad \text{for} \quad (i, j) \in E. \tag{13}$$

We then solve the corresponding system for the second preferences, etc.

§ 8. SUMMARY AND CONCLUSIONS

The theory of coalition negotiation advanced here is normative. As such it has an ad hoc flavor, which is to be expected. It is not intended to describe how people do bargain. Nor is there any guarantee that a person who bargains this way will necessarily be successful in bargaining with people who do not. If the game involves a large number of players the contracts would necessarily be so involved that a sophisticated player could not induce an unsophisticated player to consider it. However, in games where the contracts would not have to be complicated it should be expected that a person who can evaluate his bargaining position accurately will fare better, on the average.

The spirit of the analysis is a direct extension of the analysis given the zero-sum 3-person game by von Neumann and Morgenstern. The extension is obtained by emphasizing the balance of bargaining forces rather than the stability of the payoff. In this way we are concerned more with the dynamics of coalition formation than with the static stability of the payoffs. This difference is clear in the numerical example of Section 6. The fair shares obtained there are not contained in any solution in the sense of von Neumann and Morgenstern. However, the Shapley quota point [16] is one of the distributions of shares.

The theory advanced here differs from the von Neumann-Morgenstern theory, and most alternatives proposed, in another important respect. In solving the system of inequalities (3) there is nothing to guarantee that a player's share will be at least as much as he could get alone, i.e., it might happen that $x_i < v(i)$. The assumption that $x_i \geq v(i)$ has been one of the least challenged assumptions of n-person game theory.

Actually, the condition $x_i \geq v(i)$ can be included without difficulty. Just add these inequalities to the system (3) and proceed with the analysis. If they have any effect at all, it will be to exclude some players from any coalition. We can call these additional conditions _conservative conditions_, and the lack of such conditions as we have been assuming can be called _risky conditions_. A position in between is also possible, with some players playing conservatively and some playing with risk. In any case the same principles apply, and only the actual calculations are affected.

BIBLIOGRAPHY

[1] AUMANN, Robert, "Acceptable points in general cooperative n-person games," Contributions to the Theory of Games, IV, Annals of Mathematics Studies, 40, Princeton University Press, 1959, pp. 287-324.

[2] GALE, David, "A theory of games with perfect information," Proc. Nat. Acad. Sci. 39, 1953, pp. 496-501.

[3] HARSANYI, John C., "A bargaining model for the cooperative n-person game," Contributions to the Theory of Games, IV, Annals of Mathematics Studies, 40, Princeton University Press, 1959, pp. 325-355.

[4] ISBELL, John R., "Absolute games," Contributions to the Theory of Games, IV, Annals of Mathematics Studies, 40, Princeton University Press, 1959, pp. 357-396.

[5] KALISCH, G. K., MILNOR, J. W., NASH, J. F., and NERING, E. D., "Some experimental n-person games," Decision Processes, Edited by Thrall, Coombs, and Davis. John Wiley and Sons, Inc., 1954, pp. 301-327.

[6] LUCE, R. Duncan, "A definition of stability for n-person games," Annals of Mathematics, 59, 1954, pp. 357-366.

[7] LUCE, R. Duncan, "K-stability of symmetric and of quota games," Annals of Mathematics 62, 1955, pp. 517-527.

[8] LUCE, R. Duncan, "A note on the article 'Some experimental n-person games'," Contributions to the Theory of Games, IV, Annals of Mathematics Studies 40, Princeton University Press, 1959, pp. 279-285.

[9] LUCE, R. Duncan, "Ψ-stability: A new equilibrium concept for n-person game theory," Proc. of a Symposium on Mathematics Models of Human Behavior, 1954. Dunlap and Associates, Inc., and the Commission of Accidental Trauma. Armed Forces Epidemiological Board, 1955, pp. 1-21.

[10] LUCE, R. Duncan and RAIFFA, Howard, Games and Decision. Introduction and Critical Survey, New York, John Wiley and Sons, Inc., 1957.

[11] LUCE, R. Duncan and TUCKER, Albert W., eds., Contributions to the Theory of Games, IV, Annals of Mathematics Studies 40, Princeton University Press, 1959.

[12] MILNOR, John W., "Reasonable outcomes for n-person games," Research Memorandum RM-916, The RAND Corporation, Santa Monica, 1952.

[13] NASH, John F., "Equilibrium points in n-person games," Proc. Nat. Acad. Sci. 36, 1950, pp. 48-49.

[14] NASH, John F., "Non-cooperative games," Annals of Mathematics 54, 1951, pp. 286-295.

[15] VON NEUMANN, John and MORGENSTERN, Oskar, Theory of Games and Economic Behavior, Princeton, Princeton University Press, 1947.

[16] SHAPLEY, Lloyd S., "Quota solutions of n-person games," Contributions o to the Theory of Games, II, Annals of Mathematics Studies 24, Princeton University Press, 1953, pp. 343-359.

[17] SHAPLEY, Lloyd S., "A solution containing an arbitrary closed component," Contributions to the Theory of Games, IV, Annals of Mathematics Studies 40, Princeton University Press, 1959, pp. 87-93.

[18] SHAPLEY, Lloyd S., "A value for n-person games," Contributions to the Theory of Games, IV, Annals of Mathematics Studies 24, Princeton University Press, 1953, pp. 307-317.

Evar D. Nering

Arizona State University,
Mathematics Research Center,
University of Wisconsin

THE n-PERSON BARGAINING GAME*

Koichi Miyasawa

§ 1. INTRODUCTION

When we consider a bargaining situation among persons, many complicated factors will be found which affect the final outcome to each participant of that bargaining. Even in the case of bargaining between two players, many theories have been proposed [1, 6, 7, 8] which have given rise to much controversy [2, 9].

In this paper we should like to generalize the idea of the Nash solution for a two-person bargaining game to the n-person case. Along this line of generalization, Harsanyi [3] has developed an interesting theory. We received recently his revised paper [4] on this problem, in which he replies to criticisms raised by Isbell [5]. The purpose of this paper is to develop a more generalized approach to the structure of the n-person bargaining game.

§ 2. SUMMARY

In our n-person bargaining model, we take into consideration the preliminary meetings among the members of each subset — that is, of each coalition S — of the all-player set N, as well as the plenary meeting among all the players. Even though the final payoff to each player may be determined at the plenary meeting, we shall assume that the final payoff u_i is constructed by adding up incremental payoffs Δ_i^S which could be attributed to the effect of player i advancing from one coalition to a larger one. On the basis of this final payoff structure, we define the "equilibrium strategy

* I wish to express my gratitude to Professor Oskar Morgenstern for giving me the opportunity to study at the Econometric Research Program of which he is the director and for his valuable advice and suggestions. I am greatly indebted to Dr. Robert J. Aumann for many valuable comments.

set" as the one which satisfies an "internal balance" condition among the incremental payoffs (Definition 1).

We take into consideration explicitly the dependence of the outcome space which a coalition S could attain on the strategy chosen by the complementary coalition $\overline{S} = N - S$.

Our first task then is to prove the existence of such an "equilibrium strategy set" under certain assumptions (Theorem 1). The second task is to choose an "optimal strategy set" from the equilibrium strategy sets. For that we define the deflated final payoff v_i, which can be rewritten as the weighted sum of the quantities which one coalition wishes to maximize and its complementary coalition wishes to minimize. Looking at this expression of the deflated final payoff, we define the optimal strategy set by requiring the realization of the maximin property between each pair of complementary coalitions. However we make clear that the optimal strategy set if it exists actually depends only on the maximin property between the one-player coalition and its complementary (n–1)-person coalition and not on the maximin property among the other pairs of complementary coalitions (Theorem 2).

§ 3. THE STRUCTURE OF THE n-PERSON BARGAINING GAME

We consider the structure of the n-person bargaining game Γ as follows.

Let N be the set of n players $1, 2, \ldots, n$ of the game Γ. We call every subset S of N — including one-member subsets and the all-member subset N, but not the empty set — a <u>coalition</u>. In playing the game Γ, each coalition $S \subseteq N$ announces its joint strategy z^S. This strategy z^S may be cooperative or non-cooperative, but at any rate, it must and may be any element of the given strategy space Θ^S of the coalition S. Let Θ be the product space of all Θ^S, $S \subseteq N$.

Each element Z of Θ consists of $(2^n - 1)$ strategies z^S, that is

$$Z = \{z^S;\ S \subseteq N\}, \text{ where } z^S \in \Theta^S.$$

If such a <u>strategy set</u> $Z \in \Theta$ is announced by all coalitions $S \subseteq N$, then a final payoff u_i to each player i, $i = 1, 2, \ldots, n$, is determined by the all player strategy $Z^N \in Z$ according to the rule of the game Γ, that is

$$u_i = f_i(Z^N), \quad i = 1, 2, \ldots, n.$$

This situation may be considered to represent the fact that a final payoff to each member of the game will be determined at the plenary meeting among all members $1, 2, \ldots, n$. It would appear, therefore, that the strategies announced by the coalitions — in other words the strategies decided at the various preliminary meetings among the members of the coalitions $S \subseteq N$ — have no effect on the final payoffs u_i, $i = 1, 2, \ldots, n$. But, we assume that the final payoffs u_i, $i = 1, 2, \ldots, n$, are constructed by adding up many components which depend on the strategies chosen by the coalitions S in the following way:

(i) Corresponding to a pair of strategies $Z^{(i)}$ and $Z^{N-(i)}$, chosen by a one-person coalition (i) and its complementary coalition $N-(i)$, respectively, the rules of the game guarantee a payoff $\sigma_i^{(i)} \equiv \Delta_i^{(i)}$ to player i. This will be written:

$$\sigma_i^{(i)} \equiv \Delta_i^{(i)} = f_i(Z^{(i)}, Z^{N-(i)}), \quad i = 1, 2, \ldots, n. \tag{1}$$

This payoff $\Delta_i^{(i)}$ may be interpreted as the amount which player i can guarantee by himself.

(ii) Corresponding to a pair of strategies $Z^{(ij)}$ and $Z^{N-(ij)}$, the rules of the game guarantee a payoff $\sigma_\mu^{(ij)}$ to each player μ in the coalition (i, j). This will be called a <u>dividend</u> to player μ from the coalition (i, j) and can be written:

$$\sigma_\mu^{(ij)} = f_\mu(Z^{(ij)}, Z^{N-(ij)}), \quad \mu = i, j. \tag{2}$$

In this case, we define the quantities $\Delta_i^{(ij)}$ and $\Delta_j^{(ij)}$ by the following equations:

$$\Delta_i^{(ij)} = \sigma_i^{(ij)} - \Delta_i^{(i)}, \quad \Delta_j^{(ij)} = \sigma_j^{(ij)} - \Delta_j^{(j)}, \tag{3}$$

that is

$$\sigma_i^{(ij)} = \Delta_i^{(i)} + \Delta_i^{(ij)}, \quad \sigma_j^{(ij)} = \Delta_j^{(j)} + \Delta_j^{(ij)}. \tag{4}$$

Then, $\Delta_i^{(ij)}$ [or $\Delta_j^{(ij)}$] may be interpreted as a net incremental payoff to player i [or j] which is attributable to the formation of a coalition (i, j).

(iii) Proceeding in this way, if we take any proper coalition $S \subset N$, a pair of strategies Z^S and $Z^{\overline{S}}$ announced by a coalition S and its complementary coalition $\overline{S}$ guarantees a payoff, that is a dividend, σ_i^S to each member i of S according to the rule of the game. This will be written:

$$\sigma_i^S = f_i(Z^S, Z^{\overline{S}}), \qquad i \in S. \tag{5}$$

Of course, in this case, the rule of the game also guarantees a payoff, that is a dividend, $\sigma_j^{\overline{S}}$ to each member j of $\overline{S}$. This will be written:

$$\sigma_j^{\overline{S}} = f_j(Z^S, Z^{\overline{S}}), \qquad j \in \overline{S}. \tag{6}$$

For simplicity, we will sometimes express (5) and (6) by

$$\sigma^S = f_S(Z^S, Z^{\overline{S}}) \quad \text{and} \quad \sigma^{\overline{S}} = f_{\overline{S}}(Z^S, Z^{\overline{S}}),$$

respectively.

At this stage, assuming that quantities Δ_i^R, that is the net incremental payoffs to player i from the coalitions R, are defined for all proper subsets R of S, we define Δ_i^S by the following equation:

$$\Delta_i^S = \sigma_i^S - \sum_{\substack{R \ni i \\ R \subset S}} \Delta_i^R, \tag{7}$$

that is

(8) $$\sigma_i^S = \sum_{\substack{R \ni i \\ R \subseteq S}} \Delta_i^R, \quad i \in S.$$

Then, this quantity Δ_i^S may be interpreted as a net incremental payoff to player i, which results from the fact that player i advanced from $(s-1)$-person coalitions to an s-person coalition S, where s is the number of the members in the coalition $S \subset N$.

(iv) As mentioned above, a strategy Z^N chosen by the all-player coalition N determines a final payoff $u_i \equiv \sigma_i^N$ to each player i, that is

(9) $$u_i \equiv \sigma_i^N = f_i(Z^N), \qquad i = 1, 2, \ldots, n.$$

But, by analogy, we assume that this final payoff σ_i^N, $i = 1, 2, \ldots, n$, which results from a strategy set $Z = \{Z^S, S \subseteq N\}$, is made up from the net incremental payoffs Δ_i^S from all coalitions $S \subset N$ of which i is a member, as well as a net incremental payoff Δ_i^N to player i from the all-player coalition N, where Δ_i^N is defined through the equation

(10) $$\sigma_i^N = \sum_{\substack{R \ni i \\ R \subseteq N}} \Delta_i^R, \qquad i = 1, 2, \ldots, n.$$

We shall call a final payoff $u = (u_1, u_2, \ldots, u_n)$ an <u>equilibrium</u> <u>payoff</u> if all net incremental payoffs Δ_i^R constituting the final payoffs u_i possess a kind of "internal balance" which will be defined in the next section.

§ 4. THE EQUILIBRIUM STRATEGY SET AND THE EQUILIBRIUM PAYOFF

Let $\xi = \{\xi^S, S \subseteq N\}$, $\xi^S \in \Theta^S$, be a given strategy set in the game Γ. Then for any coalition S and a given fixed strategy $\xi^{\overline{S}}$, where $\overline{S}$ is the complementary coalition of S, we shall represent the set of all dividends to the members i in S resulting from the choice of a strategy $Z^S \in \Theta^S$, that is

(11) $$f_S(Z^S, \xi^{\overline{S}}) \equiv \{f_i(Z^S, \xi^{\overline{S}}), \quad i \in S\},$$

as a point in s-dimensional Euclidean space E^s, where s is the number of members of S.

Let $P^S(\xi^{\overline{S}})$ be the set of all these points, that is

$$P^S(\xi^{\overline{S}}) = \{f_S(Z^S, \xi^{\overline{S}});\ Z^S \in \Theta^S\}. \tag{12}$$

We should like to remark here that this outcome space $P^S(\xi^{\overline{S}})$ will depend on the strategy $\xi^{\overline{S}}$ chosen by the complementary coalition $\overline{S}$ of S. We assume that $P^S(\xi^{\overline{S}})$ is a compact and convex subset of E^s, and that the upper-right boundary of $P^S(\xi^{\overline{S}})$ is represented by the equation

$$H(x_i, x_j, \dots, x_k;\ \xi^{\overline{S}}) = 0, \tag{13}$$

where $S = (i, j, \dots, k)$. We will denote this upper-right boundary by $H^S(\xi^{\overline{S}})$. For simplicity, we assume the differentiability of the function H with respect to each variable x_μ in it.

Now, in defining the equilibrium strategy set, we will discuss successively the two-person subgame, ..., the s-person subgame, ..., and the final n-person game in the following way:

(i) At first, we consider the two-person subgame $\Gamma^{(ij)}(\xi^{N-(ij)})$, for all two-person coalitions (i, j) in the game Γ, under the condition that the strategy chosen by the complementary coalition $N-(ij)$ is $\xi^{N-(ij)}$.

In this subgame $\Gamma^{(ij)}(\xi^{N-(ij)})$ between the two players i and j, the players may choose any strategy $Z^{(ij)}$ in $\Theta^{(ij)}$. If they choose a strategy $Z^{(ij)}$, then the resulting payoffs to each player i and j are given by (11) with $S = (i, j)$. Accordingly, the outcome space of this subgame is given by (12) with $S = (i, j)$.

In this subgame $\Gamma^{(i,j)}(\xi^{N-(ij)})$, we require that the dividends resulting from $\xi^{(ij)} \in \xi$, that is

$$\sigma_i^{(ij)} = f_i(\xi^{(ij)}, \xi^{N-(ij)}),\quad \sigma_j^{(ij)} = f_j(\xi^{(ij)}, \xi^{N-(ij)}), \tag{14}$$

constitute the Nash bargaining solution, when we take the point

$$
(15) \qquad d^{(ij)} \begin{cases} d_i^{(ij)} \equiv \Delta_i^{(i)} = f_i(\xi^{(i)}, \xi^{N-(i)}), \\ d_j^{(ij)} \equiv \Delta_j^{(j)} = f_j(\xi^{(j)}, \xi^{N-(j)}), \end{cases}
$$

as the claim point of this bargaining subgame. I should like to remark that here and in the following, we will use the definition of the Nash solution in the generalized sense which is given and the existence of which is proved in Harsanyi's paper [3].

Let

$$
(16) \quad a_\mu^{(ij)}(\xi^{N-(ij)}) = \left. \frac{\partial H(x_i, x_j; \xi^{N-(ij)})}{\partial x_\mu} \right|_{x = \sigma^{(ij)}}, \qquad \mu = i, j.
$$

Then the above condition will be expressed by

$$
(17) \qquad H(\sigma_i^{(ij)}, \sigma_j^{(ij)}; \xi^{N-(ij)}) = 0,
$$

and

$$
(18) \qquad a_i^{(ij)}(\xi^{N-(ij)})\Delta_i^{(ij)} = a_j^{(ij)}(\xi^{N-(ij)})\Delta_j^{(ij)},
$$

for all two-person coalitions (i, j), where $\Delta_i^{(ij)}$ and $\Delta_j^{(ij)}$ are defined by (14), (15) and the relation (3).

The situation may occur in which the Nash solution can not be expressed by conditions such as (18) and the ones which will appear later. However, I should like to remark that in proving the existence of an equilibrium strategy set, we will not rely upon the conditions expressed in (18), but just upon the fact that there exists at least one Nash solution in the subgame, as proved in Harsanyi's paper [3]. But in Section 6 below, in our treatment of the optimal strategy, we shall consider only the cases in which the Nash solution of the subgame can be characterized by the analytical expressions, like (17) and (18).

(ii) Next, we consider three-person subgames $\Gamma^{(ijk)}(\xi^{N-(ijk)})$, for all three-person coalitions (ijk) in the game Γ, under the condition that the strategy chosen by the complementary coalition $N-(ijk)$ is $\xi^{N-(ijk)}$.

In this subgame among players i, j and k, the domain of joint strategies $Z^{(ijk)}$ is $\Theta^{(ijk)}$, the resulting payoffs to each player i, j and k are given by (11), and the outcome space is given by (12), with $S = (ijk)$.

In this subgame $\Gamma^{(ijk)}(\xi^{N-(ijk)})$, we require that the dividends resulting from $\xi^{(ijk)} \in \xi$, that is

$$\sigma_i^S = f_i(\xi^S, \xi^{\overline{S}}),\ \sigma_j^S = f_j(\xi^S, \xi^{\overline{S}}),\ \sigma_k^S = f_k(\xi^S, \xi^{\overline{S}}), \tag{19}$$

where $S = (ijk)$, constitute the Nash solution, when we take the point:

$$d^{(ijk)} \begin{cases} d_i^{(ijk)} \equiv \Delta_i^{(i)} + \Delta_i^{(ij)} + \Delta_i^{(ik)}, \\ d_j^{(ijk)} \equiv \Delta_j^{(j)} + \Delta_j^{(ij)} + \Delta_j^{(jk)}, \\ d_k^{(ijk)} \equiv \Delta_k^{(k)} + \Delta_k^{(ik)} + \Delta_k^{(jk)}, \end{cases} \tag{20}$$

as the claim point of this bargaining subgame, where $\Delta_i^{(i)}$, $\Delta_i^{(ij)}$ etc., are defined in stage (i).

Let

$$a_\mu^{(ijk)}(\xi^{N-(ijk)}) = \left.\frac{\partial H(x_i, x_j, x_k; \xi^{N-(ijk)})}{\partial x_\mu}\right|_{x = \sigma^{(ijk)}}, \quad \mu = i, j, k. \tag{21}$$

Then the above condition will be expressed by

$$H(\sigma_i^S, \sigma_j^S, \sigma_k^S; \xi^{\overline{S}}) = 0, \tag{22}$$

and

$$a_i^S(\xi^{\overline{S}})\Delta_i^S = a_j^S(\xi^{\overline{S}})\Delta_j^S = a_k^S(\xi^{\overline{S}})\Delta_k^S, \tag{23}$$

for all three-person coalitions $S = (ijk)$, where Δ_μ^S, $\mu = i, j, k$ are defined by (19), (20) and the relation (7).

(iii) In this way, assuming that Δ_i^R, $i \in R$, which satisfy conditions (17), (18), (22), (23), and so on, are defined through the given strategy ξ for all proper subsets R of $S \subseteq N$, we consider the s-person subgame

$\Gamma^S(\xi^{\overline{S}})$, under the condition that the strategy chosen by the complementary coalition $\overline{S}$ is $\xi^{\overline{S}}$. In this subgame $\Gamma^S(\xi^{\overline{S}})$ played by the members of the coalition S, the domain of the joint strategies Z^S is Θ^S, the payoff to each member $i \in S$ is given by (11), and the outcome space $P^S(\xi^{\overline{S}})$ is given by (12). We require that the dividends of this subgame to each member μ of the coalition S through the strategy $\xi^S \in \xi$, that is

$$\sigma^S_\mu = f_\mu(\xi^S, \xi^{\overline{S}}), \ \mu = i, j, \ldots, k; \ S = (i, j, \ldots, k), \tag{24}$$

constitute the Nash bargaining solution in this subgame $\Gamma^S(\xi^{\overline{S}})$, when we take the point

$$d^S : d^S_\mu = \sum_{\substack{R \ni \mu \\ R \subset S}} \Delta^R_\mu, \quad \mu = i, j, \ldots, k; \quad S = (i, j, \ldots, k), \tag{25}$$

as the claim point of this game.

Let

$$a^S_\mu(\xi^{\overline{S}}) = \left.\frac{\partial H(x_i, x_j, \ldots, x_k; \xi^{\overline{S}})}{\partial x_\mu}\right|_{x = \sigma^S}, \ \mu = i, j, \ldots, k; \ S = (i, j, \ldots, k), \tag{26}$$

and define Δ^S_μ by the equation

$$\Delta^S_\mu = \sigma^S_\mu - \sum_{\substack{R \ni \mu \\ R \subset S}} \Delta^R_\mu. \tag{27}$$

That is,

$$\sigma^S_\mu = f_\mu(\xi^S, \xi^{\overline{S}}) = \sum_{\substack{R \ni \mu \\ R \subseteq S}} \Delta^R_\mu, \tag{28}$$

using the net incremental payoffs Δ^R_μ, for the proper subsets R of S, which are defined in the preceding subgames through the Nash solutions. Then the condition for $\sigma^S = \{\sigma^S_\mu\}$ to be the Nash solution of the subgame $\Gamma^S(\xi^{\overline{S}})$ will be expressed as follows:

$$H(\sigma^S_i, \sigma^S_j, \ldots, \sigma^S_k; \xi^{\overline{S}}) = 0, \tag{29}$$

(30) $$a_i^S(\xi^{\overline{S}})\Delta_i^S = a_j^S(\xi^{\overline{S}})\Delta_j^S = \ldots = a_k^S(\xi^{\overline{S}})\Delta_k^S,$$

where $S = (i, j, \ldots, k) \subseteq N$.

DEFINITION 1. If in a strategy set $\xi = \{\xi^S, S \subseteq N\}$, $\xi^S \in \Theta^S$, each strategy $\xi^S \in \xi$, $s \geq 2$, $S \subseteq N$, where s is the number of members of S, gives the Nash solution for each subgame $\Gamma^S(\xi^{\overline{S}})$ among the members of the coalition S, taking the point d^S as defined by (25) as the claim point of this game, then we call ξ an <u>equilibrium strategy set</u>, and we call the set of the net incremental payoffs $\Delta = \{\Delta_\mu^S;\ \mu \in S, S \subseteq N\}$ an <u>equilibrium incremental payoff set</u>, and the resulting set of the final payoffs $u = (u_1, u_2, \ldots, u_n)$, where

(31) $$u_i = \sum_{\substack{R \ni i \\ R \subseteq N}} \Delta_i^R,$$

an <u>equilibrium payoff</u>.

§ 5. THE EXISTENCE OF AN EQUILIBRIUM STRATEGY SET

In order to prove the existence of an equilibrium strategy set, we assume that our n-person game Γ satisfies the following additional conditions:

ASSUMPTION 1. In each subgame $\Gamma^S(Z^{\overline{S}})$, $S \subseteq N$, for each given $Z^{\overline{S}} \in \Theta^{\overline{S}}$ the outcome space $P^S(Z^{\overline{S}})$ is a compact and convex subset of s-dimensional Euclidean space E^s.

DEFINITION 2. If $f_S(Z^S, Z^{\overline{S}}) \in H^S(Z^{\overline{S}})$, then we call Z^S <u>good</u> for $Z^{\overline{S}}$. If Z^S is good for $Z^{\overline{S}}$, and $Z^{\overline{S}}$ is good Z^S, then Z^S and $Z^{\overline{S}}$ are <u>mutually good</u>.

ASSUMPTION 2. If Z^S is good for some $Z^{\overline{S}} \in \Theta^{\overline{S}}$, then

z^S is good for all $z'^{\overline{S}} \in \Theta^{\overline{S}}$. (We should like to remark that in some cases we can do without this strong assumption as will be shown later on.)

ASSUMPTION 3. If z^S and z'^S are both good for some $z^{\overline{S}}$, then

$$H^{\overline{S}}(z^S) = H^{\overline{S}}(z'^S).$$

ASSUMPTION 4. If z^S is good for some $z^{\overline{S}}$, and $H^S(z^{\overline{S}}) = H^S(z'^{\overline{S}})$ for some $z^{\overline{S}}$ and $z'^{\overline{S}} \neq z^{\overline{S}}$, then

$$f_S(z^S, z^{\overline{S}}) = f_S(z^S, z'^{\overline{S}}).$$

THEOREM 1. Under the assumptions 1 - 4, there exists at least one equilibrium strategy set $\xi = \{\xi^S, S \subseteq N\}$ for the n-person bargaining game Γ.

PROOF. (i) At first we shall prove, using only Assumption 2, that, for any two complementary coalitions S and $\overline{S}$, there exists a pair of mutually good strategies $y^S \in \Theta^S$ and $y^{\overline{S}} \in \Theta^{\overline{S}}$. We choose arbitrarily a pair of strategies $z^S \in \Theta^S$ and $z^{\overline{S}} \in \Theta^{\overline{S}}$. Then there exists $y^S \in \Theta^S$ such that

$$f_S(y^S, z^{\overline{S}}) \in H^S(z^{\overline{S}}), \tag{32}$$

and there exists $y^{\overline{S}} \in \Theta^{\overline{S}}$ such that

$$f_{\overline{S}}(y^S, y^{\overline{S}}) \in H^{\overline{S}}(y^S), \tag{33}$$

by the definition of $H^S(z^{\overline{S}})$ and $H^{\overline{S}}(y^S)$. Since (32) means that y^S is good for $z^{\overline{S}}$, by Assumption 2 y^S is also good for $y^{\overline{S}}$, that is

(34) $$f_S(y^S, y^{\overline{S}}) \in H^S(y^{\overline{S}}).$$

Accordingly, (33) and (34) show that y^S and $y^{\overline{S}}$ are mutually good strategies (see Fig. 1). In this way, we obtain a strategy set $y = \{y^S, S \subseteq N\}$

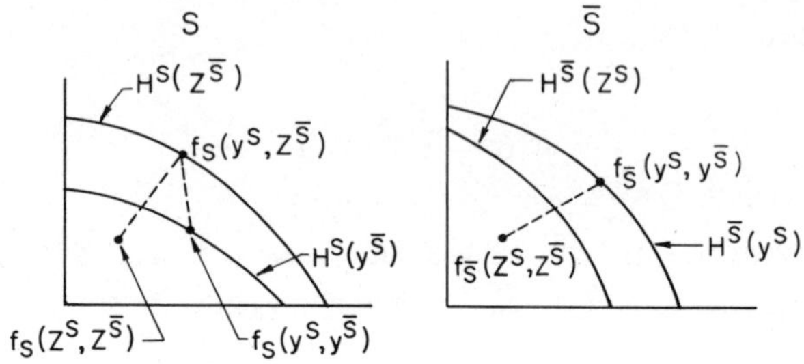

Figure 1.

in which each pair y^S and $y^{\overline{S}}$ for two complementary coalitions S and $\overline{S}$ form a pair of mutually good strategies. In this strategy set y, y^N is an arbitrary element in Θ^N.

Note. In the following case we can prove the existence of a pair of mutually good strategies y^S and $y^{\overline{S}}$ without relying on Assumption 2. We assume Θ^S and $\Theta^{\overline{S}}$ are compact and convex subsets in E^s and E^{n-s} respectively. For any $Z^S \in \Theta^S$, let $\varphi_1(Z^S)$ be the set of all elements $Z^{\overline{S}} \in \Theta^{\overline{S}}$ such that $Z^{\overline{S}}$ is good for Z^S, that is

$$\varphi_1(Z^S) = \{Z^{\overline{S}} \mid f_{\overline{S}}(Z^S, Z^{\overline{S}}) \in H^{\overline{S}}(Z^S)\},$$

and for any $Z^{\overline{S}} \in \Theta^{\overline{S}}$, let $\varphi_2(Z^{\overline{S}})$ be the set of all $Z^S \in \Theta^S$ such that Z^S is good for $Z^{\overline{S}}$, that is

$$\varphi_2(Z^{\overline{S}}) = \{Z^S \mid f_S(Z^S, Z^{\overline{S}}) \in H^S(Z^{\overline{S}})\}.$$

Now we define a mapping ψ from points in $\Theta^S \times \Theta^{\overline{S}}$ to subsets of $\Theta^S \times \Theta^{\overline{S}}$ by

$$\psi : (Z^S, Z^{\overline{S}}) \to (\varphi_2(Z^{\overline{S}}), \varphi_1(Z^S)).$$

If φ_1 and φ_2 are upper semicontinuous, and $\varphi_1(Z^S)$ and $\varphi_2(Z^{\overline{S}})$ are convex, then by Kakutani's fixed point theorem, there exist y^S and $y^{\overline{S}}$ such that

$$(y^S, y^{\overline{S}}) \in (\varphi_2(y^{\overline{S}}), \varphi_1(y^S)).$$

That is, there exists a pair of mutually good strategies y^S and $y^{\overline{S}}$.

(ii) Starting from the strategy set $y = \{y^S, S \subseteq N\}$ constructed in stage (i), we will attain an equilibrium strategy set ξ through several steps of the modification of strategies, as follows:

(1) We consider all two-person bargaining subgames $\Gamma^{(ij)}(y^{N-(ij)})$, for all two-person coalitions $(ij) \subset N$. In such a subgame, the strategy space is $\Theta^{(ij)}$ and the outcome space is $P^{(ij)}(y^{N-(ij)})$ with the upper-boundary $H^{(ij)}(y^{N-(ij)})$. We take the point

$$d^{(ij)} = (f_i(y^{(i)}, y^{N-(i)}), f_j(y^{(j)}, y^{N-(j)})) \tag{35}$$

as the claim point of this bargaining subgame, and we define

$$\begin{cases} \sigma_i^{(i)} \equiv \Delta_i^{(i)} = f_i(y^{(i)}, y^{N-(i)}), \\ \sigma_j^{(j)} \equiv \Delta_j^{(j)} = f_j(y^{(j)}, y^{N-(j)}). \end{cases} \tag{36}$$

Let $\sigma^{(ij)} = (\sigma_i^{(ij)}, \sigma_j^{(ij)})$ be the Nash solution of this bargaining subgame. Then, by the definition of the Nash solution,

$$\sigma^{(ij)} \in H^{(ij)}(y^{N-(ij)}), \tag{37}$$

and there exists $\xi^{(ij)} \in \Theta^{(ij)}$ such that

$$f_{(ij)}(\xi^{(ij)}, y^{N-(ij)}) = \sigma^{(ij)}. \tag{38}$$

Then we define $\Delta_i^{(ij)}$ and $\Delta_j^{(ij)}$ through the equations

$$\sigma_i^{(ij)} = \Delta_i^{(i)} + \Delta^{(ij)},$$

(39) and

$$\sigma_j^{(ij)} = \Delta_j^{(j)} + \Delta_j^{(ij)},$$

where $\Delta_i^{(i)}$ and $\Delta_j^{(j)}$ are defined by (36). Moreover, we define $\xi^{(i)}$ as

(40) $$\xi^{(i)} = y^{(i)},\ i \quad 1,\ 2,\ \ldots,\ n.$$

Now we construct a new strategy set $y' = \{y'^S;\ S \subseteq N\}$ from the given strategy set $y = \{y^S;\ S \subseteq N\}$, by defining

$$y'^{(ij)} = \xi^{(ij)}, \text{ for all two-person coalitions } (ij),$$
$$y'^S = y^S, \text{ for all other coalitions } S,$$

where $\xi^{(i)}$ and $\xi^{(ij)}$ are defined by (40) and (38) respectively.

Now, all the terms which may be affected by this modification from y to y' are $f_{(ij)}(y^{(ij)},\ y^{N-(ij)})$, $f_{N-(ij)}(y^{(ij)},\ y^{N-(ij)})$ and $H^{N-(ij)}(y^{(ij)})$. But, $f_{(ij)}(y^{(ij)},\ y^{N-(ij)})$ is changed so as to give the Nash solution to the subgame $\Gamma^{(ij)}(y'^{N-(ij)})$, by our definition of y'. From Assumption 3 and the definition of $y'^{(ij)}$, it is clear that

(41) $$H^{N-(ij)}(y^{(ij)}) = H^{N-(ij)}(y'^{(ij)}).$$

Also we have

(42) $$f_{N-(ij)}(y^{(ij)},\ y^{N-(ij)}) \in H^{N-(ij)}(y^{(ij)})$$

by the property of y. That is, $y^{N-(ij)}$ is good for $y^{(ij)}$. Accordingly, by Assumption 2, $y^{N-(ij)}$ is also good for $y'^{(ij)}$, and we have (41). Accordingly, from Assumption 4, we have

(43) $$f_{N-(ij)}(y'^{(ij)},\ y^{N-(ij)}) = f_{N-(ij)}(y^{(ij)},\ y^{N-(ij)}).$$

Consequently, from (41) and (43), we see that, by changing from y to y', we just obtain the Nash solution of the subgame $\Gamma^{(ij)}(y^{N-(ij)})$, without giving any variation to other terms.

(2) Now, with respect to the new strategy set y', we consider three-person subgames $\Gamma^{(ijk)}(y'^{N-(ijk)})$ for all three-person coalitions (ijk). In this subgame $\Gamma^{(ijk)}(y'^{N-(ijk)})$, the strategy space is $\Theta^{(ijk)}$, and the outcome space is $P^{(ijk)}(y'^{N-(ijk)})$ with the upper-boundary $H^{(ijk)}(y'^{N-(ijk)})$. Taking the point $d^{(ijk)}$ defined by the relation (20) as the claim point of this game $\Gamma^{(ijk)}(y'^{N-(ijk)})$, we can find a strategy $\xi^{(ijk)}$ which gives the Nash solution $\sigma^{(ijk)}$ to this subgame. Then, we construct a new strategy set $y'' = \{y''^S;\ S \subseteq N\}$ by the following rule:

$$y''^{(ijk)} = \xi^{(ijk)}, \text{ for all three-person coalitions } (ijk),$$
$$y''^S = y'^S, \text{ for all other coalitions } S \subseteq N.$$

It can be easily seen, as in step (1), that by changing from y' to y'', we just obtain the Nash solution of the subgames $\Gamma^{(ijk)}$, without giving any variation to other terms.

(3) Proceeding in this way, we can attain a strategy set $\xi = \{\xi^S;\ S \subseteq N\}$ in which each strategy ξ^S gives the Nash solution to the subgame $\Gamma^S(\xi^{\overline{S}})$, when we take the point $d^S = \{d_i^S,\ i \in S\}$ defined by

$$d_i^S = \sum_{\substack{R \ni i \\ R \subset S}} \Delta_i^R, \quad i \in S$$

as the claim point of this subgame. Accordingly, the resulting strategy set $\xi = \{\xi^S;\ S \subseteq N\}$ is an equilibrium strategy set of the n-person bargaining game Γ. This proves the existence of an equilibrium strategy set ξ as we intended.

§ 6. THE OPTIMAL STRATEGY SET AND THE SOLUTION OF THE GAME

In the preceding section, we defined the equilibrium strategy set of the n-person bargaining game Γ and proved the existence of such strategy sets. But, usually, there may be many such strategy sets, as will be seen later in Example 3. So, in this section we will try to define the optimal strategy set among the equilibrium strategy sets and to prove its existence.

Let $\xi = \{\xi^S;\ S \subseteq N\}$ be an equilibrium strategy set of the n-person bargaining game Γ. We define the quantities W^S, U^S and v_i as follows:

(44) $$W^S = \sum_{\mu\in S} \sum_{\substack{R\ni\mu \\ R\subseteq S}} a_\mu^R(\xi^{\overline{R}})\Delta_\mu^R,$$

(45) $$U^S = \sum_{\mu\in S} a_\mu^S(\xi^{\overline{S}})\Delta_\mu^S,$$

(46) $$v_i = \sum_{\substack{R\ni i \\ R\subseteq N}} a_i^R(\xi^{\overline{R}})\Delta_i^R, \quad i = 1, 2, \ldots, n,$$

where $a_\mu^R(\xi^{\overline{R}})$ and Δ_μ^R are defined by (26) and (27), respectively, and in the case where R is a one-person coalition $R = (i)$, then we define $a_i^{(i)}(\xi^{N-(i)}) = 1$ for $i = 1, 2, \ldots, n$. Then, from these definitions, it is clear that

(47) $$W^S = \sum_{R\subseteq S} U^R, \quad \text{for each } S \subseteq N.$$

From these equations, solving for U^R, we have

(48) $$U^R = \sum_{S\subseteq R} (-1)^{r-s} W^S,$$

where r and s are the numbers of members of R and S respectively.

In this section we treat the case where the analytical characterization of the Nash solution of each subgame considered in Section 3 is given by conditions (29) and (30) (see Harsanyi [3] and [4]). Then, from (30) we have

(49) $$a_\mu^S(\xi^{\overline{S}})\Delta_\mu^S = \frac{1}{s}U^S, \quad \text{for each } \mu \in S, \text{ and for each } S \subseteq N.$$

Accordingly, from (46), (48) and (49), we have

(50) $$v_i = \sum_{\substack{R\ni i\\ R\subseteq N}} \frac{1}{r}U^R = \sum_{\substack{R\ni i\\ R\subseteq N}} \sum_{S\subseteq R} \frac{1}{r}(-1)^{r-s}W^S .$$

So, following the same reasoning as Harsanyi [3], we have

(51) $$v_i = \sum_{\substack{S\ni i\\ S\subseteq N}} \frac{(s-1)!\,(n-s)!}{n!} (W^S - W^{\bar{S}}), \quad i = 1, 2, \ldots, n.$$

Now, as was shown in (31), the final payoff u_i to each player i, resulting from the strategy set $\xi = \{\xi^S;\ S \subseteq N\}$, has the structure of the sum of the net incremental payoffs Δ_i^S, that is

(52) $$u_i = \sigma_i^N = \sum_{\substack{S\ni i\\ S\subseteq N}} \Delta_i^S, \quad i = 1, 2, \ldots, n.$$

Accordingly, the quantity v_i defined by (46) as the weighted sum of Δ_i^S is not the final payoff itself, but we can give v_i the following interpretation. As far as we follow the idea of the Nash solution in each subgame $\Gamma^S(\xi^{\bar{S}})$, we do not consider the incremental payoffs Δ_i^S themselves, but rather the modified incremental payoffs $a_i^S(\xi^{\bar{S}})\Delta_i^S$, for $i \in S$; and we try to equate these modified payoffs among the members of the coalition S. Therefore, $a_i^S(\xi^{\bar{S}})$ has the meaning of a deflator in comparing incremental payoffs among the members of the coalition S.

Accordingly, for each player i who wishes to obtain the relative advantage of his final payoff among the players of the game Γ, it is natural to search for a strategy set which maximizes the sum of the modified incremental payoffs, v_i. In this sense, v_i may be called a deflated final payoff to player i, $i = 1, 2, \ldots, n$.

Now, from the expression of v_i in (51), it is natural to assume that each member of a coalition S will try to maximize $(W^S - W^{\bar{S}})$, and each member of a complementary coalition $\bar{S}$ will try to maximize $(W^{\bar{S}} - W^S)$, that is, to minimize $(W^S - W^{\bar{S}})$. So that, roughly speaking, an equilibrium strategy set which gives the maximin value of $(W^S - W^{\bar{S}})$ for each pair of the coalitions S and $\bar{S}$ may be defined as the optimal strategy set of the n-person bargaining game Γ.

Now let $\xi = \{\xi^S;\ S \subseteq N\}$ be an arbitrary equilibrium strategy set of the game Γ, and let us consider any coalition S, and the difference $(W^S - W^{\overline{S}})$. If all strategies $\xi^R \in \xi$ other than ξ^S, $\xi^{\overline{S}}$, and ξ^N are fixed, then it can be easily seen from (28) and (44) that $(W^S - W^{\overline{S}})$ is a function only of ξ^S and $\xi^{\overline{S}}$. Let this function be written

$$(53)\quad W^S - W^{\overline{S}} = g_{S,\overline{S}}(\xi^S, \xi^{\overline{S}};\ \text{all other } \xi^R,\ R \neq S, \overline{S},\ R \subset N \text{ being fixed}).$$

Then, in this case, each member of the coalition S will wish to use a joint strategy ξ^S which maximizes $g_{S,\overline{S}}$ and each member of the complementary coalition $\overline{S}$ will wish to use a joint strategy $\xi^{\overline{S}}$ which minimizes $g_{S,\overline{S}}$. For this function $g_{S,\overline{S}}$, we consider just the following kind of variations of ξ^S and $\xi^{\overline{S}}$. That is, ξ^S and $\xi^{\overline{S}}$ are allowed to change to ξ'^S and $\xi'^{\overline{S}}$, such that, by choosing the appropriate ξ'^N, the whole set of strategies ξ'^S, $\xi'^{\overline{S}}$; ξ^R, $R \neq S, \overline{S}$, $R \subset N$ and ξ'^N constitute an equilibrium strategy set.

DEFINITION 3. If an equilibrium strategy set $\xi_o = \{\xi_o^S;\ S \subseteq N\}$ satisfies the following conditions (54) for all $S \subset N$, then ξ_o is called an <u>optimal strategy set</u> of the n-person bargaining game Γ, and the resulting final payoff set $u_o = (u_1^o, u_2^o, \ldots, u_n^o)$ is called a <u>solution</u> of the game Γ.

$$(54)\quad g_{S,\overline{S}}(\xi_o^S, \xi_o^{\overline{S}};\ \text{all other } \xi_o^R,\ R \neq S, \overline{S},\ R \subset N \text{ being fixed})$$
$$= \max_{\xi^S} \min_{\xi^{\overline{S}}} g_{S,\overline{S}}(\xi^S, \xi^{\overline{S}};\ \text{all other } \xi_o^R,\ R \neq S, \overline{S},\ R \subset N \text{ being fixed}),$$

where the function $g_{S,\overline{S}}$ is defined by (53) and the maximin is taken with respect to those ξ^S and $\xi^{\overline{S}}$ which will constitute an equilibrium strategy set when taken together with all other strategies ξ_o^R, $R \neq S, \overline{S}$, $R \subset N$, and with an appropriate strategy ξ^N.

Now, for the optimal strategy set ξ_o as defined by Definition 3, let us look at the pair of strategies ξ_o^S and $\xi_o^{\overline{S}}$, where S is neither a one-player coalition nor an (n–1)-person coalition. Then, ξ_o satisfies the condition (54) with respect to the coalitions S and $\overline{S}$. Let ξ^S and $\xi^{\overline{S}}$ be any strategies of the coalitions S and $\overline{S}$, respectively, which constitute an equilibrium strategy set when taken together with ξ_o^R, $R \neq S, \overline{S}$, $R \subset N$ and with an appropriate strategy ξ^N. Let this equilibrium strategy set be ξ^*. Then, applying the definition of an equilibrium strategy set to ξ^*, we know that ξ^S gives the Nash solution to the subgame $\Gamma^S(\xi^{\overline{S}})$. Let this solution be

$$\sigma^{*S} = f_S(\xi^S, \xi^{\overline{S}}), \tag{55}$$

where

$$\sigma_i^{*S} = \sum_{\substack{R \ni i \\ R \subset S}} \Delta_i^R + \Delta_i^{*S}, \quad \text{for} \quad i \in S; \tag{56}$$

since Δ_i^R, for $R \subset S$, is determined by ξ_o^R and $\xi_o^{\overline{R}}$, they are not affected by ξ^S and $\xi^{\overline{S}}$.

Now, let k be any player who is not a member of the coalition S. In the subgame $\Gamma^{S+(k)}(\xi_o^{\overline{S+(k)}})$, $\xi_o^{S+(k)}$ gives the Nash solution

$$\sigma^{S+(k)} = f_{S+(k)}(\xi_o^{S+(k)}, \xi_o^{\overline{S+(k)}}), \tag{57}$$

since $\xi_o^{S+(k)}$ and $\xi_o^{\overline{S+(k)}}$ are elements of the equilibrium strategy set ξ_o. Of course, $\xi_o^{S+(k)}$ and $\xi_o^{\overline{S+(k)}}$ may also be considered elements of the equilibrium strategy set ξ^*.

If we consider $\xi_o^{S+(k)}$ as an element of ξ_o, then the claim point of the subgame $\Gamma^{S+(k)}(\xi_o^{\overline{S+(k)}})$ is the point $d^{S+(k)}$, given by

$$d_i^{S+(k)} = \sum_{\substack{R \ni i \\ R \subset [S+(k)] \\ R \neq S}} \Delta_i^R + \Delta_i^S, \quad \text{for} \quad i \in S, \tag{58}$$

and

$$d_k^{S+(k)} = \sum_{\substack{R \ni k \\ R \subset [S+(k)]}} \Delta_k^R. \tag{59}$$

On the other hand, if we consider $\xi_o^{S+(k)}$ as an element of ξ^*, then the claim point of the subgame $\Gamma^{S+(k)}(\xi_o^{\overline{S+(k)}})$ is the point $d^{*S+(k)}$, given by

$$(60) \qquad d_i^{*S+(k)} = \sum_{\substack{R \ni i \\ R \subset [S+(k)] \\ R \neq S}} \Delta_i^R + \Delta^{*S}_i, \quad \text{for } i \in S,$$

and

$$(61) \qquad d_k^{*S+(k)} = \sum_{\substack{R \ni k \\ R \subset [S+(k)]}} \Delta_k^R,$$

where Δ_i^{*S} is defined by (56). That is, in the subgame $\Gamma^{S+(k)}(\xi_o^{\overline{S+(k)}})$, $\sigma^{S+(k)}$ as defined by (57) is the Nash solution whether we take $d^{S+(k)}$, defined by (58) and (59), as the claim point, or $d^{*S+(k)}$, defined by (60) and (61), as the claim point of the game.

Accordingly, from the conditions for the Nash solution, in the (s+1)-dimensional Euclidean space, the three points $\sigma^{S+(k)}$, $d^{S+(k)}$ and $d^{*S+(k)}$ must lie on the same straight line. But, as can be seen from (59) and (61), the two points $d^{S+(k)}$ and $d^{*S+(k)}$ have the same $(s+1)^{st}$ coordinate

$$(62) \qquad d_k^{S+(k)} = d^{*S+(k)}_k.$$

Therefore, in order to meet the condition mentioned above, the two points $d^{S+(k)}$ and $d^{*S+(k)}$ must coincide, except in the case where $\sigma^{S+(k)}$ has the same $(s+1)^{st}$ coordinate (59) as $d^{S+(k)}$ and $d^{*S+(k)}$, that is, the case where $\Delta_k^{S+(k)} = 0$.

We assume that in the case we are treating here, this kind of singularity never occurs.

Accordingly, we have

$$(63) \qquad d_j^{S+(k)} = d^{*S+(k)}_j, \quad \text{for all } j \in S+(k).$$

From (58), and (60) and (63), we have

$$(64) \qquad \Delta_i^S = \Delta^{*S}_i, \quad \text{for } i \in S.$$

So that we have

(65) $$\sigma_i^S = \sigma*_i^S, \quad \text{for} \quad i \in S,$$

where $\sigma*_i^S$ is defined by (55) and

(66) $$\sigma_i^S = f_i(\xi_o^S, \xi_o^{\overline{S}}), \quad \text{for} \quad i \in S.$$

Therefore

(67) $$f_S(\xi_o^S, \xi_o^{\overline{S}}) = f_S(\xi^S, \xi^{\overline{S}}),$$

and, similarly

(68) $$f_{\overline{S}}(\xi_o^S, \xi_o^{\overline{S}}) = f_{\overline{S}}(\xi^S, \xi^{\overline{S}}).$$

Accordingly, we have shown that, for a coalition S, which is different from a one-person coalition and an (n–1)-person coalition, in order for ξ^S to be an element of the equilibrium strategy set $\xi*$, it must be equivalent to ξ_o^S as a strategy. So that for any equilibrium strategy set ξ_o, its components ξ_o^S and $\xi_o^{\overline{S}}$ satisfy the maximin condition of (54). This is true, because any ξ^S and $\xi^{\overline{S}}$ which satisfy the conditions for taking the maximin value of the function $g_{S,\overline{S}}$ in (54) must be equivalent to ξ_o^S and $\xi_o^{\overline{S}}$ respectively.

Next, we consider the case where S is a one-person coalition (i) and $\overline{S}$ is an (n–1)-person coalition $N-(i)$, i.e., the case where an equilibrium strategy set $\xi*$ is given as follows:

(69) $$\xi* = \{\xi^{(i)}, \xi^{N-(i)}, \xi_o^R \text{ for all other } R \subset N, \text{ and } \xi^N\},$$

where ξ^N is to be chosen so that $\xi*$ constitutes an equilibrium strategy set.

Let

(70) $$\Delta*_i^{(i)} = f_i(\xi^{(i)}, \xi^{N-(i)}).$$

Then, in the subgame $\Gamma^{(ik)}(\xi_o^{N-(ik)})$, the strategy $\xi_o^{(ik)}$ gives the Nash solution $\sigma^{(ij)}$, whether we take the point $(\Delta_i^{(i)}, \Delta_k^{(k)})$, where

$$(71)\qquad \begin{aligned} \Delta_i^{(i)} &= f_i(\xi_o^{(i)}, \xi_o^{N-(i)}), \\ \Delta_k^{(k)} &= f_k(\xi_o^{(k)}, \xi_o^{N-(k)}), \end{aligned}$$

as the claim point, or the point $(\Delta*_i^{(i)}, \Delta*_k^{(k)})$ as the claim point, where $\Delta*_i^{(i)}$ is given by (70) and

$$(72)\qquad \Delta*_k^{(k)} = f_k(\xi_o^{(k)}, \xi_o^{N-(k)}) = \Delta_k^{(k)}.$$

Following the same reasoning as above, we conclude that

$$(73)\qquad \Delta*_i^{(i)} = \Delta_i^{(i)}.$$

Now, it is clear that in the subgames $\Gamma^{N-(i)}(\xi_o^{(i)})$ which result from ξ_o and $\Gamma^{N-(i)}(\xi^{(i)})$ which result from ξ^*, as defined by (69), the claim points are the same point in both cases. However, the outcome spaces $P^{N-(i)}(\xi_o^{(i)})$ and $P^{N-(i)}(\xi^{(i)})$ of these games may differ, and consequently, the Nash solutions of $\Gamma^{N-(i)}(\xi_o^{(i)})$ and $\Gamma^{N-(i)}(\xi^{(i)})$ may differ. This difference gives rise to the difference in the value of $(W^{(i)} - W^{N-(i)})$ according to whether ξ_o is used or ξ^* is used.

From the above reasoning we can state the following theorem:

THEOREM 2. An optimal strategy set ξ_o which is defined by Definition 3 is equivalent to an equilibrium strategy set ξ_o which satisfies the following conditions for $i = 1, 2, \ldots, n$:

$$(74)\qquad \begin{aligned} &g_{(i),N-(i)}(\xi_o^{(i)}, \xi_o^{N-(i)};\ \text{all other } \xi_o^R \text{ being fixed}) \\ &= \max_{\xi^{(i)}} \min_{\xi^{N-(i)}} g_{(i),N-(i)}(\xi^{(i)}, \xi^{N-(i)};\ \text{all other } \xi_o^R \text{ being fixed}), \end{aligned}$$

where the maximin is taken with respect to those $\xi^{(i)}$ and

$\xi^{N-(i)}$ which will constitute an equilibrium strategy set when taken together with all other strategies ξ_o^R, $R \neq (i)$, $N-(i)$, $R \subset N$, and with an appropriate strategy ξ^N.

Example 1. If each hypersurface $H^S(\xi^{\overline{S}})$ is an (s−1)-dimensional hyperplane, with direction numbers (1, 1, ..., 1), for all $S \subseteq N$, then it is clear that

$$a_\mu^S(\xi^{\overline{S}}) = 1, \text{ for all } \mu \in S \text{ and all } S \subseteq N. \tag{75}$$

Accordingly, in this case, we have

$$W^S = \sum_{\mu \in S} \sum_{\substack{R \ni \mu \\ R \subseteq S}} \Delta_\mu^R = \sum_{\mu \in S} \sigma_\mu^S = \sum_{\mu \in S} f_\mu(\xi^S, \xi^{\overline{S}}). \tag{76}$$

Thus, W^S is a function only of ξ^S and $\xi^{\overline{S}}$; that is, we can write

$$W^S = h_S(\xi^S, \xi^{\overline{S}}).$$

Similarly we have

$$W^{\overline{S}} = \sum_{\nu \in \overline{S}} \sigma_\nu^{\overline{S}} = \sum_{\nu \in \overline{S}} f_\nu(\xi^S, \xi^{\overline{S}}), \tag{77}$$

that is, $W^{\overline{S}}$ is also a function only of ξ^S and $\xi^{\overline{S}}$, so we can write

$$W^{\overline{S}} = h_{\overline{S}}(\xi^S, \xi^{\overline{S}}).$$

Accordingly, we have

$$\begin{aligned} W^S - W^{\overline{S}} &= h_S(\xi^S, \xi^{\overline{S}}) - h_{\overline{S}}(\xi^S, \xi^{\overline{S}}) \\ &= g_{S,\overline{S}}(\xi^S, \xi^{\overline{S}}), \end{aligned} \tag{78}$$

that is, $W^S - W^{\overline{S}}$ is a function only of ξ^S and $\xi^{\overline{S}}$.

Consequently, in this case, there will exist — under the relevant conditions on the strategy spaces Θ^S — the maximin strategies ξ_o^S and $\xi_o^{\overline{S}}$

of the function $g_{S,\overline{S}}(\xi^S, \xi^{\overline{S}})$, for all $S \subseteq N$. That is, there will exist an optimal strategy ξ_o of the game. We note that in this case we have

$$v_i = \sum_{\substack{S \ni i \\ S \subseteq N}} \Delta_i^S = u_i, \quad i = 1, 2, \ldots, n, \tag{79}$$

that is, v_i is equal to the final payoff u_i.

Example 2. If we have

$$a_\mu^S(\xi^{\overline{S}}) = a_\mu, \text{ for all } S \ni \mu, \quad S \subseteq N,$$

then, as in Example 1, it will readily be seen that the difference $(W^S - W^{\overline{S}})$ is a function only of ξ^S and $\xi^{\overline{S}}$. Thus, there will exist a maximin strategy ξ_o in this case. Moreover, we remark that, in this case, we have

$$v_i = \sum_{\substack{S \ni i \\ S \subseteq N}} a_i \Delta_i^S = a_i u_i, \tag{80}$$

and this corresponds to the case Harsanyi treated in his paper [4].

Example 3. Let us consider the three-person bargaining game Γ with $N = (1, 2, 3)$ defined as follows:

The strategy sets $\Theta^{(i)}$, $i = 1, 2, 3$, of each player i is the set of real numbers such that

$$\Theta^{(1)} = \{z^1;\ 0 \le z^1 \le 5\},$$
$$\Theta^{(2)} = \{z^2;\ 0 \le z^2 \le 9\},$$
$$\Theta^{(3)} = \{z^3;\ 0 \le z^3 \le 10\}.$$

The strategy set $\Theta^{(ij)}$ of joint strategies $z^{(ij)}$ of a two-person coalition (ij) is the set of points in the (i,j)-plane with coordinates $(z_i^{(ij)}, z_j^{(ij)})$ such that

$$\Theta^{(12)} = \{z^{(12)};\ 0 \le z_1^{(12)} + z_2^{(12)} \le 15,\ z_1^{(12)} \ge 0,\ z_2^{(12)} \ge 0\},$$
$$\Theta^{(13)} = \{z^{(13)};\ 0 \le z_1^{(13)} + z_3^{(13)} \le 18,\ z_1^{(13)} \ge 0,\ z_3^{(13)} \ge 0\},$$
$$\Theta^{(23)} = \{z^{(23)};\ 0 \le z_2^{(23)} + z_3^{(23)} \le 20,\ z_2^{(23)} \ge 0,\ z_3^{(23)} \ge 0\}.$$

The strategy set $\Theta^{(123)}$ of all-player joint strategies $Z^{(123)}$ is the set of points in the three-dimensional Euclidean space E^3 with coordinates $(Z_1^{(123)}, Z_2^{(123)}, Z_3^{(123)})$ such that

$$\Theta^{(123)} = \{Z^{(123)};\ 0 \leq Z_1^{(123)} + Z_2^{(123)} + Z_3^{(123)} \leq 30,\ Z_i^{(123)} \geq 0,\ i = 1, 2, 3\}.$$

The payoff functions f are defined as follows (see Fig. 2):

$$(81)\qquad \begin{cases} f_1(Z^{(1)}, Z^{(23)}) = Z^1, \\ f_2(Z^{(1)}, Z^{(23)}) = (1 - \frac{Z^1}{20})\, Z_2^{(23)}, \\ f_3(Z^{(1)}, Z^{(23)}) = (1 - \frac{Z^1}{20})\, Z_3^{(23)}, \end{cases}$$

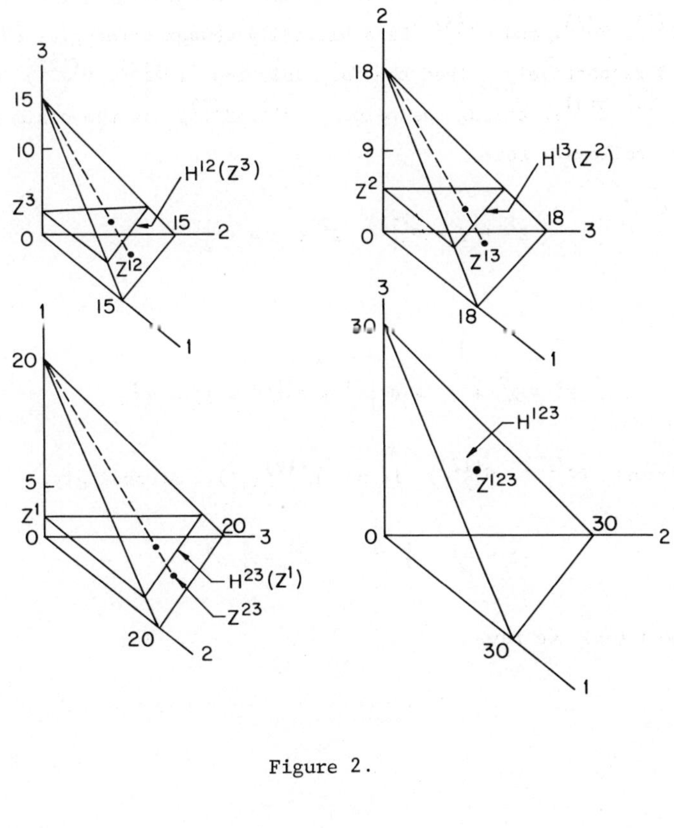

Figure 2.

(82)
$$\begin{cases} f_2(z^{(2)}, z^{(13)}) = z^2, \\ f_1(z^{(2)}, z^{(13)}) = (1 - \frac{z^2}{18})\, z_1^{(13)}, \\ f_3(z^{(2)}, z^{(13)}) = (1 - \frac{z^2}{18})\, z_3^{(13)}, \end{cases}$$

(83)
$$\begin{cases} f_3(z^{(3)}, z^{(12)}) = z^3, \\ f_1(z^{(3)}, z^{(12)}) = (1 - \frac{z^3}{15})\, z_1^{(12)}, \\ f_2(z^{(3)}, z^{(12)}) = (1 - \frac{z^3}{15})\, z_2^{(12)}, \end{cases}$$

(84)
$$f_i(z^{(123)}) = z_i^{(123)}, \quad i = 1, 2, 3.$$

In this game Γ, it is clear that all Assumptions 1 through 4 hold.

At first we will find out an equilibrium strategy set of this game Γ. Let $z^{(1)}$, $z^{(2)}$, and $z^{(3)}$ be arbitrarily chosen strategies of players 1, 2, and 3 respectively. Then the Nash solution $(\sigma_1^{(12)}, \sigma_2^{(12)})$ of the subgame $\Gamma^{(12)}(z^{(3)})$, taking the point $(z^{(1)}, z^{(2)})$ as the claim point, is obtained as follows. Let

(85)
$$z^1 + \ell = \sigma_1^{(12)}, \quad z^2 + \ell = \sigma_2^{(12)}.$$

Then

$$z^1 + z^2 + 2\ell = \sigma_1^{(12)} + \sigma_2^{(12)} = 15 - z^3,$$

since the point $(\sigma_1^{(12)}, \sigma_2^{(12)})$ is on $H^{(12)}(z^3)$. Accordingly,

(86)
$$\ell = \frac{15}{2} - \frac{1}{2} z^1 - \frac{1}{2} z^2 - \frac{1}{2} z^3.$$

From (85) and (86), we have

$$(87) \qquad \begin{cases} \sigma_1^{(12)} = \frac{15}{2} + \frac{1}{2} z^1 - \frac{1}{2} z^2 - \frac{1}{2} z^3, \\ \sigma_2^{(12)} = \frac{15}{2} - \frac{1}{2} z^1 + \frac{1}{2} z^2 - \frac{1}{2} z^3 . \end{cases}$$

Similarly, the Nash solution $(\sigma_1^{(13)}, \sigma_3^{(13)})$ of the subgame $\Gamma^{(13)}(z^{(2)})$, taking $(z^{(1)}, z^{(3)})$ as the claim point, is given by

$$(88) \qquad \begin{cases} \sigma_1^{(13)} = 9 + \frac{1}{2} z^1 - \frac{1}{2} z^2 - \frac{1}{2} z^3, \\ \sigma_3^{(13)} = 9 - \frac{1}{2} z^1 - \frac{1}{2} z^2 + \frac{1}{2} z^3 . \end{cases}$$

The Nash solution $(\sigma_2^{(23)}, \sigma_3^{(23)})$ of the subgame $\Gamma^{(23)}(z^{(1)})$, taking $(z^{(2)}, z^{(3)})$ as the claim point, is given by

$$(89) \qquad \begin{cases} \sigma_2^{(23)} = 10 - \frac{1}{2} z^1 + \frac{1}{2} z^2 - \frac{1}{2} z^3, \\ \sigma_3^{(23)} = 10 - \frac{1}{2} z^1 - \frac{1}{2} z^2 + \frac{1}{2} z^3 . \end{cases}$$

Now, from (81), (82), and (83), we have

$$(90) \qquad \begin{cases} \sigma_1^1 = \Delta_1^1 = z^1, \\ \sigma_2^2 = \Delta_2^2 = z^2, \\ \sigma_3^3 = \Delta_3^3 = z^3. \end{cases}$$

Since $\Delta_1^{(12)} = \sigma_1^{(12)} - \Delta_1^{(1)}$, etc., we have, from (87), (88), (89), and (90)

$$(91) \qquad \begin{cases} \Delta_1^{(12)} = \Delta_2^{(12)} = \frac{15}{2} - \frac{1}{2} z^1 - \frac{1}{2} z^2 - \frac{1}{2} z^3, \\ \Delta_1^{(13)} = \Delta_3^{(13)} = 9 - \frac{1}{2} z^1 - \frac{1}{2} z^2 - \frac{1}{2} z^3, \\ \Delta_2^{(23)} = \Delta_3^{(23)} = 10 - \frac{1}{2} z^1 - \frac{1}{2} z^2 - \frac{1}{2} z^3. \end{cases}$$

Accordingly, in the final all-player bargaining game, the claim point d^N is given by

$$(92)\quad \begin{cases} d_1^N = \Delta_1^{(1)} + \Delta_1^{(12)} + \Delta_1^{(13)} = \frac{33}{2} - z^2 - z^3, \\ d_2^N = \Delta_2^{(2)} + \Delta_2^{(12)} + \Delta_2^{(23)} = \frac{35}{2} - z^1 - z^3, \\ d_3^N = \Delta_3^{(3)} + \Delta_3^{(13)} + \Delta_3^{(23)} = 19 - z^1 - z^2, \end{cases}$$

and the upper-right boundary of the outcome space of this game is given by

$$u_1 + u_2 + u_3 = 30.$$

Thus, it can be easily seen that the Nash solution $u = (u_1, u_2, u_3)$ of this final all-player bargaining game is given by

$$(93)\quad \begin{cases} u_1 = \frac{53}{6} + \frac{2}{3} z^1 - \frac{1}{3} z^2 - \frac{1}{3} z^3, \\ u_2 = \frac{59}{6} - \frac{1}{3} z^1 + \frac{2}{3} z^2 - \frac{1}{3} z^3, \\ u_3 = \frac{68}{6} - \frac{1}{3} z^1 - \frac{1}{3} z^2 + \frac{2}{3} z^3. \end{cases}$$

This is an equilibrium solution of the game Γ. Moreover, in this game we have

$$(94)\quad \begin{cases} W^{(1)} = \Delta_1^{(1)} = z^1,\ W_2^{(2)} = \Delta_2^{(2)} = z^2,\ W^{(3)} = \Delta_3^{(3)} = z^3, \\ W^{(12)} = \sigma_1^{(12)} + \sigma_2^{(12)} = 15 - z^3, \\ W^{(13)} = \sigma_1^{(13)} + \sigma_3^{(13)} = 18 - z^2, \\ W^{(23)} = \sigma_2^{(23)} + \sigma_3^{(23)} = 20 - z^1, \\ W^{(123)} = 30. \end{cases}$$

From (93) and (94), we can check easily that, for example, u_1 is expressed as follows:

$$u_1 = \frac{2!}{3!}(W^{(1)} - W^{(23)}) + \frac{1!}{3!}[(W^{(12)} - W^{(3)}) + (W^{(13)} - W^{(2)})] + \frac{2!}{3!} W^{(123)}, \text{ etc.}$$

The optimal solution $u^o = (u_1^o, u_2^o, u_3^o)$ of this game is given by

$$u_1^o = \frac{53}{6} + \frac{2}{3} \times 5 - \frac{1}{3} \times 9 - \frac{1}{3} \times 10 = \frac{35}{6} \approx 5.83,$$

$$u_2^o = \frac{59}{6} - \frac{1}{3} \times 5 + \frac{2}{3} \times 9 - \frac{1}{3} \times 10 = \frac{65}{6} \approx 10.83, \tag{95}$$

$$u_3^o = \frac{68}{6} - \frac{1}{3} \times 5 - \frac{1}{3} \times 9 + \frac{2}{3} \times 10 = \frac{80}{6} \approx 13.33.$$

BIBLIOGRAPHY

[1] HARSANYI, J. C., "Approach to the bargaining problem before and after the theory of games: A critical discussion of Zeuthen's, Hicks', and Nash's theories," Econometrica, Vol. 24, April, 1956, pp. 144-157.

[2] HARSANYI, J. C., "Notes on the bargaining problem," The Southern Economic Journal, Vol. XXIV, April, 1958, pp. 471-476.

[3] HARSANYI, J. C., "A bargaining model for the cooperative n-person game," Annals of Mathematics Studies No. 40 (Princeton 1959), pp. 325-355.

[4] HARSANYI, J. C., "A simplified bargaining model for the n-person cooperative games," (Mimeographed Paper), January, 1961.

[5] ISBELL, J. R., "A modification of Harsanyi's bargaining model," Bulletin of the American Mathematical Society, Vol. 66, No. 2, March, 1960, pp. 70-73.

[6] NASH, John, "The bargaining problem," Econometrica, Vol. 18, April, 1950, pp. 155-162.

[7] NASH, John, "Two-person cooperative games," Econometrica, Vol. 21, January, 1953, pp. 128-140.

[8] WAGNER, H. M., "A unified treatment of bargaining theory," The Southern Economic Journal, Vol. XXIII, April, 1957, pp. 380-397.

[9] WAGNER, H. M., "Rejoinder on the bargaining problem," The Southern Economic Journal, Vol. XXIV, April, 1958, pp. 476-482.

Koichi Miyasawa

Faculty of Economics
University of Tokyo

VALUATION OF n-PERSON GAMES*

Reinhard Selten

INTRODUCTION

The problem of valuation in game theory can be loosely formulated in this way: What is the value of an n-person game for a player taking part in it? The von Neumann value [12], which is the natural consequence of a theory of rational behavior, is defined only for 2-person constant-sum games. Up to now there has been no satisfying theory of rational behavior for general n-person games on which a value concept could be based.

As far as the value problem is concerned, however, the difficult question of rational behavior can be avoided if another approach is taken, namely, axiomatic characterization of value functions. The first result in this direction, the characterization of the Shapley value [11], is not wholly satisfactory because it presupposes that general n-person games are adequately represented by their characteristic functions. This objection cannot be raised against the characterizations given here.

The main results of this paper are two characterization theorems. In §7 a theorem is proved characterizing the Shapley value as a value function for n-person constant-sum games. The value function characterized in §8 coincides with the Shapley value only for the region of n-person constant-sum games. This value function is the modified Shapley value introduced by John C. Harsanyi [2], [3]. Harsanyi has constructed bargaining models for which this value can be derived as an equilibrium solution with reference to a certain arbitration scheme.** Bargaining models and arbitration schemes offer

* This paper is essentially a translation of the author's doctoral dissertation at the Johann Wolfgang Goethe University, Frankfurt am Main, Germany. I am indebted to Professor Dr. E. Burger who gave valuable advice.

** Harsanyi's results can be considered as generalizations of the Nash theory of cooperative 2-person games; cf. [7].

an approach to the value problem basically different from the characterization by inner properties.

Most of the postulates used here for characterization have been intuitively justified already in [10]. The newly-introduced postulates (IXa) and (XI) are discussed in §2. As in [10], value functions are understood as valuations for finite cooperative games with transferable utility and unlimited side-payments; furthermore, it is implicitly assumed that the players want to maximize their expected payoff in the sense of the probability calculus. In [10] it is discussed at length why some of the postulates are reasonable only on these general assumptions.*

Although it is an immediate consequence of postulate (I) that the value depends on the normal form only, all the other postulates are formulated in terms of the extensive form. Probably it would have been possible to use the normal form exclusively, but the extensive form seemed to be more advantageous with respect to the interpretation of the postulates and the simplicity of the proofs. Postulate (I) can presumably be replaced by several simpler postulates concerning the extensive form. In this way, a system of postulates uniformly dealing with the extensive form could be obtained.

The characterization theorems proved in §7 and §8 can be considered generalizations of similar theorems on 2-person value functions already proved in [10], which are restated together with other results from [10] in §1. (There they appear as E 6 and E 7.) The 2-person theorems, however, are not contained in the general theorems as special cases, because fewer characterizing postulates are needed in the 2-person case.

The mutual independence of the characterizing postulates will not be discussed here. In [10], however, two theorems were proved which give a partial answer to this question for the characterization theorems in §7 and §8 (E 9 and E 10 in §1).

There are some open questions** with regard to the class of η-values described in §5. The η-values are value functions for general n-person games which satisfy all the postulates occurring in the characterization theorems of §7 and §8 with exception of postulate (X). The Shapley value and the modified Shapley value characterized in §8 are contained in this class

* See the additional comment at the end of this paper.

** Cf. [10] p. 240.

as extreme special cases. Presumably it is possible to characterize the Shapley value, too, and even other η-values for the region of general n-person games, by using appropriate other postulates instead of postulate (X).

§ 1. FORMER RESULTS

In this paragraph some results of [10] are restated. As far as well-known game-theoretic concepts are concerned, we only explain our notation. Definitions in full detail are given in [10].

GAMES AND THEIR CONSTITUENTS. For games and their constituents, the following letters are used: Γ, and K, $\mathcal{P}$, $\mathfrak{U}$, $\mathfrak{A}$, p, h. Equal indices mean relationship to the same game. K with origin o is the game tree* of the game Γ. A vertex y comes "after" x, if x is between o and y, x comes "after" x. The set X(K) of vertices, other than endpoints, is partitioned by the player partition $\mathcal{P}$ into player sets P_i $(i = 0, \ldots, n)$, and by the information partition $\mathfrak{U}$ into information sets U. The move partition $\mathfrak{A}$ partitions the alternative set A(K) into moves u.** (Alternatives are edges in K.) An alternative a is "at" x, if it connects x and a vertex after x. A move u is "at" U, and U "belongs" to u, if the alternatives of u are at vertices in U. The probability assignment p fixes the probabilities of the chance moves. The payoff function h defines a payoff vector $h(z) = (h_1(z), \ldots, h_n(z))$ for each z in the endpoint set Z(K). The symbol (x, b) stands for the "one-point game" with a single vertex x and payoff vector b. A player with $P_j = \emptyset$ and $h_j(z) = a$ for all $z \in Z(K)$ is called a "dummy with a".

GRAPHICAL REPRESENTATION. The graphical representation describes information sets by dashed lines and payoff vectors by column vectors. The symbol i to the left of a vertex x shows that x is in P_i.

STRATEGIES. Strategies of the player i are marked by the lower index i. A pure strategy π_i assigns moves $\pi_i(U)$ to information sets.

* We consider only games with finite trees.

** The use of the word "move" is not the same as in [5].

A mixed strategy fixes probabilities $q_i(\pi_i)$ for pure strategies. We identify a pure strategy π_i with the mixed strategy q_i which assigns 1 to π_i. A "combination" $\pi = (\pi_1, \ldots, \pi_n)$ of pure strategies contains a pure strategy for each of the players $1, \ldots, n$; if a component π_j in π is replaced by ς_j the combination π/ς_j results. The same notation is used for combinations $q = (q_1, \ldots, q_n)$ of mixed strategies. The strategy-set vector $\Pi = (\Pi_1, \ldots, \Pi_n)$ contains the sets Π_i of the pure strategies of a player i. $Q = (Q_1, \ldots, Q_n)$ is the corresponding strategy-set vector for mixed strategies.

EXPECTED PAYOFF. $H(\pi) = (H_1(\pi), \ldots, H_n(\pi))$ is the expected payoff for pure strategies. The maximum of $H_1(\pi) + \ldots + H_n(\pi)$ is called "maximum component sum $M(\Gamma)$". The game Γ is a "constant-sum game", if $H_1(\pi) + \ldots + H_n(\pi)$ is constant. $E(q) = (E_1(q), \ldots, E_n(q))$ is the expected payoff for mixed strategies. The "security level" $s(q_j)$ is the minimum of $E_j(q)$ for q_j in q.

THE NORMAL FORM. A strategy-set vector $\Pi = (\Pi_1, \ldots, \Pi_n)$ together with an expected payoff function H forms an n-person "normal form" $G = (\Pi, H)$. The constituents do not have to originate from a game. Two normal forms $G = (\Pi, H)$ and $G' = (\Pi', H')$ are called isomorphic if there are one-to-one payoff-preserving mappings $f_1, \ldots, f_n$ of the Π_i onto the Π_i'. $N(\Gamma)$ is the "normal form of Γ" with the constituents Π and H from Γ. To each normal form G there is a "related normal game" Γ, which can be described intuitively in this way: The players $i_1, \ldots, i_k$ $(i_1 < \ldots < i_k)$ with more than one strategy in G successively select a strategy each, without being informed about previous moves of the others. The payoff is the expected payoff $H(\pi)$ corresponding to this selection.

OPERATIONS. The "sum" $\Gamma'' = \Gamma + \Gamma'$ is formed from Γ by substituting subgames Γ^z for the payoff vectors $h(z)$; apart from the payoff function h^z the Γ^z do not differ from Γ'; h^z is defined by $h^z(z') = h(z) + h'(z')$. The game $\lambda \cdot \Gamma$ results if all payoffs in Γ are multiplied by λ. "k dummies with 0 are added" by appending k 0-components to all payoff vectors. If a move u in Γ is "forbidden", all parts of K only realizable through u are taken away; apart from this, the rules of Γ are transferred to the resulting game Γ'. Two players are "interchanged" by interchanging their player sets and their payoffs. "The payoff to player j is reduced" in Γ by substituting for the payoff function h a new payoff function h' with the following properties: (1) $h_j'(z) \leq h_j(z)$; (2) $h_i'(z) \geq h_i(z)$ for $i \neq j$; (3) for the resulting game Γ' we have $M(\Gamma') = M(\Gamma)$. From a 2-person game Γ, the "η-replenished" game $F_\eta(\Gamma)$ results by substituting for the payoff function h the following h_η:

$$\begin{aligned} h_{\eta,1}(z) &= h_1(z) + \eta(M(\Gamma) - h_1(z) - h_2(z)) \\ h_{\eta,2}(z) &= h_2(z) + (1-\eta)(M(\Gamma) - h_1(z) - h_2(z)) \end{aligned} \qquad (0 \leq \eta \leq 1) .$$

If the "indexed* mixed strategies $q_{j_1}^1, \ldots, q_{j_m}^m$ are added" to a normal form G, a normal form G' results which, besides the pure strategies of G, contains the $q_{j_r}^r$ $(r = 1, \ldots, m)$ as pure strategies of the player j_r. The payoff for G' is given by $H'(\pi') = E(\pi')$.

LINEAR EQUIVALENCE. The normal forms G and G' are "linearly equivalent" if there is a chain of normal forms $G, G_1, \ldots, G_m, G'$, in which two neighboring links always satisfy one of the following conditions: (a) they are isomorphic, (b) one of them results from adding a mixed strategy to the other. Γ and Γ' are linearly equivalent if $N(\Gamma)$ and $N(\Gamma')$ are linearly equivalent. The notation $G \sim G'$ and $\Gamma \sim \Gamma'$ is used for linear equivalence.

VALUE FUNCTIONS. Value functions w assign values $w(\Gamma) = (w_1(\Gamma), \ldots, w_n(\Gamma))$ to games. The specifications "for general n-person games" and "for n-person constant-sum games" characterize the region on which w is defined. The "von Neumann value function" v is a value for 2-person constant-sum games; $v_i(\Gamma)$ in $v(\Gamma) = (v_1(\Gamma), v_2(\Gamma))$ is the maximum security level of player i. The "η-stable agreements" V_η $(0 \leq \eta \leq 1)$ are value functions for general 2-person games defined by $V_\eta(\Gamma) = v(F_\eta(\Gamma))$. The "2-person η-values" w_η $(0 \leq \eta \leq 1/2)$ are defined by $w_\eta(\Gamma) = \frac{1}{2} V_\eta(\Gamma) + \frac{1}{2} V_{1-\eta}(\Gamma)$.

* The same strategy may appear several times with different indices.

POSTULATES. Our postulates concern value functions. The postulates (IXa) and (XI), which did not occur in [10], are related to operations which will be described in §2. The postulates are listed below.

(I) LINEAR EQUIVALENCE. If $\Gamma \sim \Gamma'$ then $w(\Gamma) = w(\Gamma')$.

(II) SYMMETRY. If Γ' results from Γ by interchanging two of the players, then $w(\Gamma')$ is formed from $w(\Gamma)$ by interchanging the corresponding components.

(III) MAXIMUM JOINT PAYOFF. $w_1(\Gamma) + w_2(\Gamma) + \ldots + w_n(\Gamma) = M(\Gamma)$.

(IV) DUMMY. If player j in Γ is a dummy with a, then $w_j(\Gamma) = a$.

(V) STRATEGIC MONOTONICITY. If Γ' results from Γ by forbidding a move u of player j, then $w_j(\Gamma') \leq w_j(\Gamma)$.

(VI) HOMOGENITY. If $\Gamma' = \lambda \cdot \Gamma$ with $\lambda > 0$, then $w(\Gamma') = \lambda w(\Gamma)$.

(VII) ADDITIVITY. If $\Gamma'' = \Gamma + \Gamma'$ then $w(\Gamma'') = w(\Gamma) + w(\Gamma')$.

(VIIa) WEAK ADDITIVITY. If Γ' is a one-point game and $\Gamma'' = \Gamma + \Gamma'$, then $w(\Gamma'') = w(\Gamma) + w(\Gamma')$.

(VIII) TRUNCATION. If Γ' results from Γ by substituting for a subgame Γ_y its value $w(\Gamma_y)$, then $w(\Gamma') = w(\Gamma)$.

(IX) PAYOFF MONOTONICITY. If Γ' results from Γ by reducing the payoff of player j, then $w_j(\Gamma') \leq w_j(\Gamma)$.

(IXa) WEAK PAYOFF MONOTONICITY. If Γ' results from Γ by transferring payoff from player j to other players, then $w_j(\Gamma') \leq w_j(\Gamma)$.

(X) NORMALIZATION. $w_1({}^n\Gamma^o) = w_2({}^n\Gamma^o) = 1$. Here ${}^n\Gamma^o$ is the normalization game. It results from ${}^2\Gamma^o$ (Figure 1) by adding $n-2$ dummies with 0.

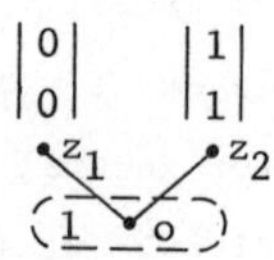

Figure 1: ${}^2\Gamma^o$

(XI) INFORMATION MONOTONICITY. If Γ' results from Γ by informing player k at U^1 about the move u of player j ($j \neq k$), then $w_j(\Gamma') \leq w_j(\Gamma)$.

RESULTS. Results from [10] are listed below by numbers of the form E ... .

E 1: If G' results from G by adding the indexed mixed strategies $q_{j_1}^{1}, \ldots, q_{j_m}^{m}$, then $G' \sim G$ (§2, p. 247).

E 2: If the subgame Γ_y of a game Γ is replaced by a game Γ'_y with $\Gamma_y \sim \Gamma'_y$, a game Γ' results with $\Gamma \sim \Gamma'$ (§2, p. 248).

E 3: The von Neumann value function v satisfies postulates (I) - (IX) (§5, p. 256).

E 4: The η-stable agreements V_η satisfy postulates (I) and (III) - (IX) (§6, p. 261).

E 5: The 2-person η-values w_η satisfy postulates (I) - (VII) and (IX) (§6, p. 263).

E 6: There is one and only one value function for 2-person constant-sum games which satisfies postulates (I), (III), (IV), (V), and (IX), namely the von Neumann value function v (§5, p. 259).

E 7: There is one and only one value function for general 2-person games which satisfies postulates (I) - (V) and (VIII), namely $w_{1/2}$ (§7, p. 264).

E 8: There is one and only one value function for general 2-person games, which satisfies (I) - (V), (VII), and (X), namely $w_{1/2}$ (§7, p. 264).

E 9: (I), (III), (IV), (V) as postulates for value functions for 2-person constant-sum games are mutually independent (§8, p. 277).

E 10: (I) - (V), (VII), and (X) as postulates for value functions for general 2-person games are mutually independent (§8, p. 277).

E 11: There is no value function for general n-person games $(n \geq 3)$ satisfying postulates (I) – (V), and (VIII) (§9, p. 281).

§ 2. POSTULATES (IXa) AND (XI)

PAYOFF TRANSFER. Γ' results from Γ "by transferring payoff from player j to other players," if only the payoff functions h' and h'' differ in both n-person games and if, furthermore, the following conditions are satisfied:

(1) $h'_j(z) \leq h_j(z)$;

(2) $h'_i(z) \geq h_i(z)$ for $i \neq j$;

(3) $\sum_{i=1}^{n} h'_i(z) = \sum_{i=1}^{n} h_i(z)$.

If Γ' results from Γ by transferring payoff from player j to other players, then obviously Γ' results from Γ also by reducing the payoff of player j. Therefore postulate (IXa) already stated in §1 is weaker than postulate (IX).

As a postulate for value functions for general n-person games, postulate (IX) is too strong. If the payoff to player j is reduced, there may be an endpoint z where the payoff to another player k is increased to such an extent that both players together receive more than before; this could be an improvement of player j's position if he, together with player k, can enforce realization of z. Obviously this objection cannot be raised against (IXa). It will be shown in §5 that the value functions for general n-person games described here satisfy (IXa) but not (IX).

INFORMATION TRANSFER. Let U^2 be an information set of player j in an n-person game Γ and u a move at U^2. We define a vertex set $Y(u)$ "realizable with u". This set contains all vertices coming after vertices to which alternatives of u lead. An information set U of player k is called "partitionable by u", if it contains only vertices which come after vertices of U^2. Thus partitionability by u depends only on the information set belonging to u. Let U^1 be an information set of player k partitionable by u (we don't exclude $j = k$). U^1 can be partitioned into two sets $B = U^1 \cap Y(u)$ and $U^1 - B$. Each set is, if not empty, a set "into which u partitions U^1". A move "into which u partitions a move u^1 at U^1" contains all the alternatives of u^1 which are at vertices of a certain one of the sets into which u partitions U^1. We define the game Γ' which u results from Γ if "player k at U^1 is informed about the move u of player j". This game differs from Γ only with regard to the information partition $\mathfrak{U}'$ and the move partition $\mathfrak{A}'$. $\mathfrak{U}'$ differs from $\mathfrak{U}$ by not containing U^1 and by containing the sets into which u partitions U^1, instead. Correspondingly, the move partition $\mathfrak{A}'$ does not contain the moves at U^1, but contains the moves into which u partitions the moves at U^1; otherwise $\mathfrak{A}'$ is not different from $\mathfrak{A}$. Speaking of player k at U^1 being informed about the move u of player j, we always presuppose that U^1 is partitionable by u.

The operation "information transfer" can be interpreted in the following way: The rules of the game are changed by an additional rule. Before the play begins, player j has to inform the "game director" whether his agent at U^2 will or will not select move u in case he will have to decide. After the beginning of the play, but before the decision of the agent at U^1, the game director gives this information to the agent at U^1. Thus the agent at U^1 always knows to which one of the two subsets B and $U^1 - B$ the play has advanced when and if he has to decide. If U^1 were not partitionable by u he would also always know that only one of two subsets of U^1 can be decisive, but the two subsets would have in common those vertices which do not follow vertices of U^2. Therefore we must always have partitionability of U^1 by u.

It is intuitively clear that the position of player j is not improved if another player is informed about one of his moves. Admittedly it could be advantageous for player j if player k receives information about u ; in this case, however, he can also give this information to player k before the beginning of the play. Therefore postulate (XI) (information monotonicity) can be considered as reasonable with regard to our intuitive assumptions, because we deal only with fully cooperative games, in which the players can make arbitrary contracts before the beginning of the play which can be thought of as enforcible by law, once they are agreed upon.

THEOREM ON INFORMATION TRANSFER. Let Γ' be the game which results from Γ if player k at U^1 is informed about move u of player j. Then Γ' has the following properties: (a) $M(\Gamma') = M(\Gamma)$; (b) if $j = k$, then $\Gamma' \sim \Gamma$. (c) For any value function w, for general n-person games or for n-person constant-sum games, which satisfies (I), (III), and (V), we have

$$w_k(\Gamma') \geq w_k(\Gamma).$$

Furthermore, if $n = 2$, the value function w satisfies (XI).

COROLLARY. From (c), and E 3, E 4, E 5 it is clear that the 2-person value functions v, V_η, and w_η satisfy (XI).

PROOF. We can assume without loss of generality that both $B = U^1 \cap Y(u)$ and $U^1 - B$ are not empty, because otherwise Γ and Γ' are the same game and therefore our propositions are trivially true.

In Γ' all players with the exception of player k have the same strategies as in Γ. For every strategy π_k in Γ, player k has a strategy $\pi'_k = f(\pi_k)$ with $\pi'_k(U) = \pi_k(U)$ for the information sets with $U \neq U^1$ and $\pi'_k(B) = \pi'_k(U^1 - B) = \pi_k(U^1)$. For all combinations π in Γ we have

$$H(\pi/\pi_k) = H'(\pi/f(\pi_k)). \tag{1}$$

The combinations π/π_k and $\pi/f(\pi_k)$ realize the same endpoints with the same probabilities, indeed, because for vertices x of player k we always have $\pi_k(x) = \pi'_k(x)$, if $\pi'_k = f(\pi_k)$. For every strategy π'_k in Γ' player k has two strategies $\pi^1_k = g_1(\pi'_k)$ and $\pi^2_k = g_2(\pi'_k)$ with $\pi^1_k(U) = \pi^2_k(U) = \pi'_k(U)$ for the sets U with $U \neq U^1$ and $\pi^1_k(U^1) = \pi'_k(B)$ and $\pi^2_k(U^1) = \pi'_k(U^1 - B)$. If $\pi'_j(U^2) = u$ is valid, a combination π' containing π'_j can lead only to plays which intersect with U^1 in B or not at all. For

$\pi'_j(U^2) \neq u$ we have the same situation with regard to $U^1 - B$. Therefore for combinations π' in Γ' we have

$$(2) \qquad H(\pi'/\pi'_j/g_1(\pi'_k)) = H'(\pi'/\pi'_j/\pi'_k) \qquad \text{for } \pi'_j(U^2) = u$$

$$(3) \qquad H(\pi'/\pi'_j/g_2(\pi'_k)) = H'(\pi'/\pi'_j/\pi'_k) \qquad \text{for } \pi'_j(U^2) \neq u.$$

If π is a combination with maximum component sum for $H(\pi)$, and if π_k is player k's component in π, then because of (1), the component sum for $H'(\pi/f(\pi_k))$ is $M(\Gamma)$, too. Therefore we have $M(\Gamma') \geq M(\Gamma)$. If π' is a combination with the maximum component sum for $H'(\pi')$ and if π'_k is player k's component in π', then because of (2) and (3) one of the two combinations $\pi'/g_1(\pi'_k)$ and $\pi'/g_2(\pi'_k)$ in Γ also results in the component sum $M(\Gamma')$. Consequently we have $M(\Gamma) \geq M(\Gamma')$. This proves (a). We now proceed to show (b). Thus we have $j = k$. Let $g(\pi'_k) = g_1(\pi'_k)$ for $\pi'_k(U^2) = u$ and $g(\pi'_k) = g_2(\pi'_k)$ for $\pi'_k(U^2) \neq u$. Because of (2) and (3) we have for all π'

$$(4) \qquad H(\pi'/g(\pi'_k)) = H'(\pi'/\pi'_k).$$

It can also easily be seen that we always have $\pi_k = g(f(\pi_k))$. We extend the region, on which f is defined: For every π'_k which cannot be expressed by $\pi'_k = f(\pi_k)$, we construct an indexed mixed strategy q_k which is $g(\pi'_k)$ completed by an index and then we define $\pi'_k = f(q_k)$. To $N(\Gamma)$ we add the set of all these q_k and get a normal form G^1. $N(\Gamma) \sim G^1$ follows from E 1 (cf. §1). Because of (4) we have

$$(5) \qquad H'(\pi'/f(q_k)) = H(\pi'/g(f(q_k))) = H^1(\pi'/q_k);$$

in addition, because of (1) we have

$$(6) \qquad H'(\pi'/f(\pi_k)) = H(\pi'/\pi_k) = H^1(\pi'/\pi_k).$$

It can be seen from (5) and (6) that f together with the identical mappings of the strategy sets π_i with $i \neq k$ onto themselves gives the isomorphism of G^1 and $N(\Gamma')$. Because of $N(\Gamma) \sim G^1$ we have $\Gamma \sim \Gamma'$. This proves (b).

We now proceed to show (c). Here we assume $j \neq k$ and a value function w satisfying (I), (III), and (V). Let G^2 be the normal form which is different from $N(\Gamma')$ only in that player k has fewer strategies in G^2, that is, he has only those which can be expressed as $f(\pi_k)$; the payoff function is taken from $N(\Gamma')$ without change. Obviously we can get a game isomorphic to the normal game Γ^2 of G^2 by successively forbidding moves in the normal game of $N(\Gamma')$ where the forbidden moves correspond to the strategies π_k' which cannot be expressed as $\pi_k' = f(\pi_k)$. Because of (I) and (V) we have

$$w_k(\Gamma^2) \leq w_k(\Gamma').$$

It can be seen from (1) that f together with the identical mappings of the π_i with $i \neq k$ gives the isomorphism of $N(\Gamma)$ and G^2. Therefore we have

$$w_k(\Gamma') \geq w_k(\Gamma).$$

This is the first part of (c). Because of (a) and (III) we have $n = 2$

$$w_j(\Gamma') \leq w_j(\Gamma).$$

This proves the rest of the theorem.

§ 3. COALITION GAMES

COALITIONS. A coalition in an n-person game is a subset C of the set $N = (1, \ldots, n)$. The players i with $i \in C$ are the "members" of C. The coalition C' which results from C by "interchanging players i and j" contains one of the two players if and only if C contains the other; otherwise C' has the same members as C.

COALITION GAMES. We define for each coalition C in an n-person game Γ a coalition game ${}_C\Gamma$. Only with regard to the player partition and to the payoff function is this game different from Γ. ${}_C\Gamma$ is a 2-person game. The player-set ${}_CP_1$ of player 1 in ${}_C\Gamma$ is the union of all player sets of members of C in Γ. The player set ${}_CP_2$ is the union of all player sets of members of the complementary coalition $N - C$ in Γ. At all endpoints z, the payoff ${}_Ch_1(z)$ is the sum of the payoffs to the members of C in Γ. Similarly, ${}_Ch_2(z)$ is the sum of the payoffs to the members of $N - C$ in Γ. If we have $C = N$, this is to be understood as ${}_Ch_2(z) = 0$ everywhere; similarly we have ${}_Ch_1(z) = 0$ if $C \neq \emptyset$.

COALITION STRATEGIES AND COALITION FORMS. We use $|M|$ for the number of elements in a set M. Let $|C| = k$. A "coalition strategy" for C is a k-tuple π_C which contains a pure strategy for Γ for each member of C. By its components π_C assigns moves $\pi_C(U)$ to the information sets U in the player-set ${}_CP_1$, and therefore can be considered as a pure strategy of player 1 for ${}_C\Gamma$. Thus the coalition strategies for C are the pure strategies for player 1 in ${}_C\Gamma$ and the coalition strategies for $N - C$ are the pure strategies for player 2 in ${}_C\Gamma$.

In the same way we can define coalition strategies for normal forms, and for any normal form G we can form 2-person normal forms ${}_CG$, in which player 1 has the coalition strategies for C and player 2 has the coalition strategies for $N - C$. The payoff function ${}_CH$ is derived in the same way from H as ${}_Ch$ from h. Obviously, the "coalition form" ${}_CG$ of a normal form G and the normal form $N({}_C\Gamma)$ of the coalition game ${}_C\Gamma$ derived from the normal game Γ related to G are isomorphic.

NOTATION. We always use the coalition as the left lower index for coalition games and coalition forms; otherwise the same indices are used as in in the original game or form.

THEOREM ON COALITION GAMES.

(a) If $\Gamma \sim \Gamma'$, then ${}_C\Gamma \sim {}_C\Gamma'$.

(b) ${}_{N-C}\Gamma$ results from ${}_C\Gamma$ if the players are interchanged.

(c) If Γ' and C' result from Γ and C respectively by interchanging players j and k, then ${}_C\Gamma$ and ${}_{C'}\Gamma'$ are the same game.

(d) $M({}_C\Gamma) = M(\Gamma)$.

(e) If in Γ player j in C is a dummy with a, then ${}_C\Gamma = {}_{C-(j)}\Gamma + (x, (-a, 0))$.

(f) If Γ' results from Γ by forbidding the move u of player j, then ${}_C\Gamma'$ results from ${}_C\Gamma$ by forbidding u; in ${}_C\Gamma'$, u is a move of player 1 for $j \in C$ and u is a move of player 2 for $j \in N - C$.

(g) If Γ' results from Γ by transferring payoff from a member of C to other players, then ${}_C\Gamma'$ results from ${}_C\Gamma$ by transferring payoff from player 1 to other players.

(h) If $\Gamma' = \lambda\Gamma$, then ${}_C\Gamma' = \lambda {}_C\Gamma$.

(i) If $\Gamma'' = \Gamma + \Gamma'$, then ${}_C\Gamma'' = {}_C\Gamma + {}_C\Gamma'$.

(j) If Γ' results from Γ by informing player k at U^1 about the move u of player j and if $j \in C$ and $k \in N-C$, then ${}_C\Gamma'$ results from ${}_C\Gamma$ by informing player 2 at U^1 about the move u of player 1.

(k) If Γ' results from Γ by informing player k at U^1 about move u of player j and if $j \in C$ and $k \in C$, then we have ${}_C\Gamma' \sim {}_C\Gamma$.

PROOF. Statements (b) – (j) are immediate consequences of the definition of coalition game, and (k) follows from the theorem on information transfer in §1 (statement (b)). We now proceed to prove (a). As $\Gamma \sim \Gamma'$ and $N(\Gamma) \sim N(\Gamma')$ are equivalent, we have only to show that ${}_CG \sim {}_CG'$ is a consequence of $G \sim G'$. It is clear from the definition of linear equivalence that we have to consider only the special case in which G' is formed by adding a mixed strategy q_j of player j to G. Without loss of generality we can assume $j \in C$, because, if the theorem has been proved already for $j \in C$, we would have for $j \in N-C$ that ${}_{N-C}\Gamma \sim {}_{N-C}\Gamma'$ is a consequence of $\Gamma \sim \Gamma'$, then by interchanging both players in ${}_{N-C}\Gamma$ and ${}_{N-C}\Gamma'$ we do not disturb the linear equivalence, and we get ${}_C\Gamma \sim {}_C\Gamma'$ because of (b).

In ${}_CG'$ player 2 has the same strategies as in ${}_CG$; player 1 retains his strategies from ${}_CG$, too, but he has additional strategies π_C'

containing q_j. For each additional strategy π'_C we construct a related mixed strategy $q_C = f(\pi'_C)$ for ${}_C G$; this is uniquely defined by $q_C(\pi'_C/\pi_j) = q_j(\pi_j)$. Obviously, $q_C = f(\pi'_C)$ assigns the probability 0 to all coalition strategies which do not agree with π'_C about the behavior of the players $i \neq j$. We consider the normal form ${}_C G''$, which results from ${}_C G$ by adding the indexed set of all mixed strategies $q_C = f(\pi'_C)$. We now define mappings which give the isomorphism of ${}_C G'$ and ${}_C G''$: the strategies of ${}_C G'$ which also occur in ${}_C G$ are mapped onto themselves, and the additional strategies of ${}_C G'$ are mapped by f onto the added strategies of ${}_C G''$. It can easily be seen that these mappings do not disturb the payoff function. Hence ${}_C G'$ and ${}_C G''$ are isomorphic. ${}_C G \sim {}_C G''$ follows from E 1. Therefore we have ${}_C G \sim {}_C G'$. This proves the theorem.

§ 4. CHARACTERISTICS

CHARACTERISTICS. An "n-person characteristic" is a function c which assigns real numbers $c(C)$ to the coalitions $C \subseteq N$ and satisfies

$$c(C) + c(N - C) = c(N). \tag{1}$$

A player j is a "dummy" in c, if for $j \in C$ we always have

$$c(C - (j)) = c(C) - c((j)). \tag{2}$$

If all players are dummies, c is called "inessential". "Superadditivity" is defined in the usual way. The coalition function c' results from c by "interchanging players j and k", if for coalitions C and C', where C' results from C by interchanging players j and k, we always have $c'(C') = c(C)$. A characteristic c is "symmetric", if there is a coalition R called a "symmetry carrier" with the property that c depends only on $|C \cap R|$; clearly, because of $c(\emptyset) = 0$, all players outside R are dummies in c with $c((j)) = 0$. If c has N as symmetry carrier, c is "fully symmetric".

LEMMA ON SYMMETRY. A nontrivial symmetric characteristic has only one symmetry carrier R. (The "trivial" characteristic assigns 0 to all coalitions.)

PROOF. The proof is indirect. If c has several symmetry carriers, then there is at least one player j who belongs to one of the carriers R, but not to another, R'. Player j is a dummy in c and we have $c((j)) = 0$. As $c(C)$ depends only on $|C \cap R|$, it follows that we have $c(C) = 0$ for $|C| = 1$. If $c(C) = 0$ is true for $|C \cap R| = k < |R|$, then a coalition C with $|C \cap R| = k$, which does not contain j, can be extended by adding the dummy j; thus we get a coalition C_1 with $c(C_1) = 0$ and $|C_1 \cap R| = k + 1$; hence we have $c(C) = 0$ also for $|C \cap R| = k + 1$. By induction $c(C) = 0$ is always true. This proves the lemma.

COMBINATION THEOREM: All characteristics can be linearly combined from superadditive symmetric characteristics.

This theorem has already been proved by Isbell in [4], p. 393-394; Isbell has shown that all characteristics are linear combinations of the characteristic functions of direct majority games. Obviously these characteristic functions are superadditive symmetric characteristics.

η-CHARACTERISTICS. The "η-characteristic" $c_\eta (0 \leq \eta \leq 1/2)$ of an n-person game Γ is the n-person characteristic given by $c_\eta(C) = w_{\eta,1}({}_C\Gamma)$.

We must justify this definition by showing that c_η really is a characteristic. From E 5 we know that w_η satisfies (I) – (IV). The theorem in §3 states under (b) that ${}_{N-C}\Gamma$ results from ${}_C\Gamma$ by interchanging both players, and under (d) we find $M({}_C\Gamma) = M(\Gamma)$. By (I) and (b) we get $c_\eta(N-C) = w_{\eta,2}({}_C\Gamma)$. Therefore, because of (III) and (d), we have $c_\eta(C) + c_\eta(N-C) = M(\Gamma)$; this is condition (1) from the definition of the characteristic. $c_\eta(\emptyset) = 0$ is an immediate consequence of (IV).

The symbol η is always used as lower index for η-characteristics; otherwise the same indices are used as in the original game.

CHARACTERISTIC FUNCTION. From our definitions it can easily be seen that the characteristic function introduced by von Neumann and Morgenstern is given by

$$V(C) = V_{0,1}({}_C\Gamma) \qquad \text{for all} \quad C \subseteq N.$$

THEOREM ON CONSTANT-SUM GAMES. All the η-characteristics $c_\eta(0 \leq \eta \leq 1/2)$ of an n-person constant-sum game Γ are the same, namely the characteristic function V of Γ.

PROOF. For 2-person constant-sum games the value functions V_η, w_η, and the von Neumann value function v coincide, because here η-replenishment does not change the expected payoff. From this we get the theorem as an immediate consequence.

THEOREM ON η-CHARACTERISTICS.

(a) If $\Gamma \sim \Gamma'$, then $c_\eta = c'_\eta$.

(b) If Γ' results from Γ by interchanging players j and k, then c'_η results from c_η by interchanging players j and k.

(c) $c_\eta(N) = M(\Gamma)$.

(d) If player j in Γ is a dummy with a, then he is a dummy in c_η, too, and we have $c((j)) = a$.

(e) If Γ' results from Γ by forbidding move u of player j, then for $j \in C$ we always have

$$c'_\eta(C) \leq c_\eta(C) + \frac{1}{\eta}\,(M(\Gamma') - M(\Gamma)).$$

(f) If

(1) $$\Gamma = \sum_{s=1}^{r} \lambda_s \Gamma^s \quad \text{where} \quad \lambda_s \geq 0 \quad \text{for} \quad s = 1, \ldots, r,$$

then we have

(2) $$c_\eta = \sum_{s=1}^{r} \lambda_s c_\eta{}^s \ .$$

(g) If Γ' results from Γ by transferring payoff from player j to other players, then for $j \in C$ we always have $c'_\eta(C) \leq c_\eta(C)$.

(h) If Γ' results from Γ by informing player k at U^1 about the move u of player $j (j = k$ is not excluded), then for $j \in C$ we always have $c'_\eta(C) \leq c_\eta(C)$.

PROOF. From §1, E 5, and §2 we know that the 2-person η-values satisfy (I) – (VII), (IX), and (XI). Considering this, we get (a) – (d) and (f) – (h) as immediate consequences of the theorem on coalition games in §3. We now proceed to prove (e).

${}_C\Gamma'$ results from ${}_C\Gamma$ by forbidding move u of player 1 (statement (f) in the theorem of §3). The η-replenished game $F_\eta({}_C\Gamma')$ results from $F_\eta({}_C\Gamma) + (x, u)$ by forbidding move u, where u is the vector $(\eta, 1 - \eta)$ multiplied by $M(\Gamma') - M(\Gamma)$ (cf. [10], §6, p. 260*). Using $M({}_C\Gamma) = (M(\Gamma)$, (statement (d) in the theorem of §3) and considering the fact that v satisfies (V) and (VIIa), we conclude:

$$v_{\eta,1}({}_C\Gamma') \leq v_{\eta,1}({}_C\Gamma) + \eta(M(\Gamma') - M(\Gamma))$$

$$v_{1-\eta,1}({}_C\Gamma') \leq v_{1-\eta,1}({}_C\Gamma) + (1-\eta)(M(\Gamma') - M(\Gamma)).$$

By the definition of the 2-person η-values, this yields

$$c'_\eta(C) = w_{\eta,1}({}_C\Gamma') \leq c_\eta(C) + \frac{1}{2}\,(M(\Gamma') - M(\Gamma)).$$

§ 5. THE n-PERSON η-VALUES

In this paragraph a class of value functions for general n-person games is described. These value functions are generalizations of the 2-person η-values (§1).

In the following, $\mathfrak{C}_i$ stands for the set of all coalitions $C \subseteq N$ containing i; $\mathfrak{C}$ is the set of all coalitions $C \subseteq N$.

n-PERSON η-VALUES. An η-value $(0 \leq \eta \leq 1/2)$ is a value function w_η for general n-person games; the components $w_{\eta,i}$ are given by

* There is an error in the third line of the theorem "Satz über das Verbot eines Zuges". The first letter should have been Γ'', not Γ'.

(A) $$w_{\eta,i}(\Gamma) = \sum_{C\in\mathfrak{C}_i} \gamma_n(C)(c_\eta(C) - c_\eta(N-C))$$

where

$$\gamma_n(C) = (n - |C|)!(|C| - 1)!/n!$$

It can easily be seen that this definition for $n = 2$ agrees with the definition of w given in §1.

LEMMA ON γ_n. γ_n has the following properties:

(1) $$\sum_{C\in\mathfrak{C}_i} \gamma_n(C) = 1$$

(2) $$\gamma_n(C) = \gamma_n((N - C) \cup (i)) \qquad \text{for all } C \in \mathfrak{C}_i.$$

PROOF. It can easily be seen that (2) is true. There are exactly $\binom{n-1}{k-1}$ k-element coalitions in $\mathfrak{C}_i$. Therefore we have

$$\sum_{C\in\mathfrak{C}_i} \gamma_n(C) = \sum_{k=1}^{n} \binom{n-1}{k-1} \cdot \frac{(n-k)!(k-1)!}{n!} = 1.$$

This proves (1).

OTHER EXPRESSIONS FOR THE η-VALUES.

(B) $$w_{\eta,i}(\Gamma) = \sum_{C\in\mathfrak{C}_i} \gamma_n(C)(c_\eta(C) - c_\eta(C - (i)))$$

(C) $$w_{\eta,i}(\Gamma) = -M(\Gamma) + 2 \sum_{C\in\mathfrak{C}_i} \gamma_n(C)c_\eta(C)$$

for $i = 1, \ldots, n$.

PROOF. Expression (A) can be split into two sums:

$$\sum_{C\in\mathfrak{C}_i} \gamma_n(C)c_\eta(C) - \sum_{C\in\mathfrak{C}_i} \gamma_n(C)c_\eta(N - C).$$

If $(N-C') \cup (i)$ is substituted for C in the second sum, expression (B) is seen to result with the aid of (2). Also, because of $c_\eta(N) = M(\Gamma)$, we have

$$c_\eta(C) - c_\eta(N-C) = 2c_\eta(C) - M(\Gamma).$$

Using this, (A) is easily transformed into (C) with the aid of (1).

THEOREM ON THE n-PERSON η-VALUES. The n-person η-values satisfy postulates (I) – (VII), (IXa), and (XI).

PROOF. We use (a) – (h) from the theorem on η-characteristics (§4). (I) is satisfied because of (a). With the aid of (C), (b) gives (II). From (d) and property (1) of γ_n it can be seen with the aid of (B) that (IV) is satisfied. (V) follows from (e) and (C). Statement (f) gives (VI) and (VII) with the aid of (A). Postulates (IXa) and (XI) are given by (g) and (h) with the aid of (C), where property (a) from the theorem on information transfer (§2) is used with regard to (XI). Expression (A) is formed in the same way from c_η as the components of the Shapley value are formed from the characteristic function (cf. [11], p. 312, (13), or formula (S) in the next section). Therefore Shapley's axiom 2 ([11], p. 309), which states that the value-components sum up to the characteristic function value of N, together with (c) gives (III). (Shapley's axioms 1 and 3 could have been used in the same way in order to show that (II) and (VII) are satisfied.)

THE SHAPLEY VALUE. Let V be the characteristic function of Γ. The value function s with components

$$\text{(S)} \qquad s_i(\Gamma) = \sum_{C \in \mathfrak{C}_i} \gamma_n(C)(V(C) - V(C - (i))) \qquad \text{for } i = 1, \ldots, n$$

is called the "Shapley value" (cf. [11]).

THEOREM ON THE SHAPLEY VALUE. The Shapley value s and the η-value w_0 are the same value function.

PROOF. If the 2-person game Γ' results from Γ by interchanging both players, then $F_{1-\eta}(\Gamma')$ results from $F_\eta(\Gamma)$ by interchanging both players (cf. [10], p. 260). This, together with (b) from the theorem on coalition games (§3), gives

$$v_{1,1}({}_C\Gamma) = v_{0,2}({}_{N-C}\Gamma).$$

Hence, because (III) is satisfied by v_0 (E 4, §1), we have

$$v_{1,1}({}_C\Gamma) = M({}_C\Gamma) - v_{0,1}({}_{N-C}\Gamma).$$

Therefore the definition of w_0 with the aid of $M({}_C\Gamma) = M(\Gamma)$ and $v(C) = v_{0,1}(C)$ yields

$$c_0(C) = w_{0,1}({}_C\Gamma) = \tfrac{1}{2}v(C) - \tfrac{1}{2}v(N-C) + M(\Gamma)/2.$$

Therefore the right side of (B) can be transformed into the following expression:

$$\frac{1}{2}\sum_{C\in\mathfrak{C}_i}\gamma_n(C)(v(C) - v(C-(i)) + \frac{1}{2}\sum_{C\in\mathfrak{C}_i}\gamma_n(C)(v((N-C)\cup(i)) - v(N-C)).$$

In the second sum the coefficient $\gamma_n(C)$ can be replaced by $\gamma_n((N-C)\cup(i))$ with the aid of property (2) of γ_n; obviously $\mathfrak{C}_i$ is also the set of all coalitions $(N-C)\cup(i)$ with $C\in\mathfrak{C}_i$. Therefore the second sum equals the first one. Consequently the expression equals the right side of equation (S). This proves the theorem.

THEOREM ON CONSTANT-SUM GAMES. For n-person constant-sum games all η-values are equal to the Shapley value.

PROOF. It follows from the theorem on constant-sum games in §4 that all w_η are equal for constant-sum games. Therefore the proposition is a consequence of the above theorem on the Shapley value.

FURTHER PROPERTIES OF THE η-VALUES. It follows from §1, E 11, that the n-person η-values do not satisfy (VIII) for $n > 2$. The behavior of the w_η with regard to (IX) and (X) will be described by two theorems.

THEOREM. Postulate (IX) is not satisfied by the n-person η-values for $n > 2$.

PROOF. In all games occurring in this proof the players $4, \ldots, n$ are dummies with 0. Only the 3-person case is graphically represented in Figures 2 and 3. For $n > 3$ all payoff vectors must be completed by adding vanishing components for the players $4, \ldots, n$. The game Γ^2 (Figure 3) results from the game Γ^1 (Figure 2) by reducing the payoff of player 1. We have $M(\Gamma^1) = M(\Gamma^2) = 1$. We investigate the η-characteristics c_η^1 and c_η^2 of Γ^1 and Γ^2 by comparing $c_\eta^1(C)$ and $c_\eta^2(C)$ for $C \in \mathfrak{C}_1$. We consider four cases: (1) $2 \in C$, $3 \in C$. ${}_C\Gamma^1$ and ${}_C\Gamma^2$ are linearly equivalent; hence we have $c_\eta^1(C) = c_\eta^2(C)$. (2) $2 \in C$, $3 \notin C$. Player 1 can secure 1 for himself in $F_\eta({}_C\Gamma^1)$ and $F_\eta({}_C\Gamma^2)$; player 2 always receives at least 0. Therefore we have $c_\eta^1(C) = c_\eta^2(C) = 1$. (3) $2 \notin C$, $3 \notin C$. In $F_\eta({}_C\Gamma^1)$ and $F_\eta({}_C\Gamma^2)$ player 1 can secure η for himself and player 2 always receives at least $1 - \eta$. Therefore we have $V_\eta({}_C\Gamma^1) = V_\eta({}_C\Gamma^2) = (\eta, 1 - \eta)$. Consequently we have $c_\eta^1(C) = c_\eta^2(C) = 1/2$. (4) $2 \notin C$, $3 \in C$. In $F_\eta({}_C\Gamma^1)$ player 1 can secure η and player 2 always receives at least $1 - \eta$. As in (3) we have $c_\eta^1(C) = 1/2$. But player 1 in $F_\eta({}_C\Gamma^2)$ can secure 1 for himself while player 2 always receives at least 0. Therefore we have $c_\eta^2(C) = 1$. Since in the cases (1), (2), (3) we have $c_\eta^1(C) = c_\eta^2(C)$, but $c_\eta^1(C) < c_\eta^2(C)$ holds in case (4), we see from the formula (C) for $w_{\eta,1}$ that $w_{\eta,1}(\Gamma^1) < w_{\eta,1}(\Gamma^2)$ is true. This contradicts (IX).

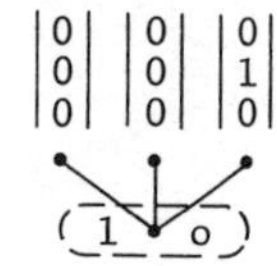

Figure 2: Γ^1

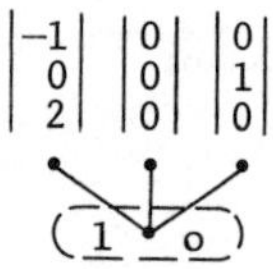

Figure 3: Γ^2

THEOREM. The n-person value function $w_{1/2}$ satisfies (X).

PROOF. Let $\mathfrak{C}_{1,2}$ be the set of all coalitions containing players 1 and 2. If C is not in $\mathfrak{C}_{1,2}$ but in $\mathfrak{C}_1$ or $\mathfrak{C}_2$, then we have

${}^{n}c^{o}_{1/2}(C) - {}^{n}c^{o}_{1/2}(N-C) = 0$, because then the coalition games ${}_{C}{}^{n}\Gamma^{o}$ and ${}_{N-C}{}^{n}\Gamma^{o}$ of the normalization game from (X) and the 2-person normalization game ${}^{2}\Gamma^{o}$ with value (1, 1) are the same game. Therefore, applying formula (A) to $w_{1/2,i}({}^{n}\Gamma^{o})$ for $i = 1, 2$, we may change the regions of summation from $\mathfrak{C}_1$ and $\mathfrak{C}_2$ to $\mathfrak{C}_{1,2}$. This shows that both value-components are equal; they are equal to 1 since we have $M({}^{n}\Gamma^{o}) = 2$, and (III) and (IV) are satisfied by $w_{1/2}$. This proves the theorem.

COROLLARY. $w_{1/2}$ is the only n-person η-value satisfying (X).

This is an immediate consequence of the characterization theorem which will be proved in §8.

§ 6. LEMMATA

In this paragraph eight lemmata are proved which prepare for the characterization theorems in §7 and §8.

LEMMA 1. For every n-person constant-sum game Γ there is an n-person constant-sum game Γ' with the following properties: (1) Γ' is the normal game of a normal form G'; (2) each coalition $C \in \mathfrak{C}_1$ has an optimal pure strategy $\pi_{C'}$ for Γ'; (3) if w is a value function for n-person constant-sum games which satisfies (I) and (XI), we have $w_1(\Gamma') \leq w_1(\Gamma)$; (4) the characteristic functions V and V' of Γ and Γ' are equal to each other.

PROOF. Let Γ^1 be the normal game of Γ. We construct Γ^2: The origin o of Γ^2 belongs to the player set of player 1. There is a one-to-one correspondence between the alternatives a_C at o and the coalitions $C \in \mathfrak{C}_1$. The alternative a_C leads to a vertex X_C. Each vertex x_C forms a one-element information set U_C of player 1. The alternative $a(\pi_C)$ leads from x_C to a vertex $x(\pi_C)$. There is a one-to-one correspondence between the pure coalition strategies π_C in Γ and the alternatives $a(\pi_C)$. Each vertex $x(\pi_C)$ is origin of a subtree of Γ^2 isomorphic to the tree K^1 of Γ^1. The vertices in these subtrees which correspond to vertices in the same information set U_i^1 of player i in Γ^1, together form an information set U_i^2 of player i $(i = 1, \ldots, n)$. The alternatives on vertices from U_i^2, which correspond to alternatives in the same move u_i^1, together form a move u_i^2 belonging to U_i^2. In this way exactly one information set U_i^2 or move u_i^2 respectively corresponds to each information set U_i^1 and to each move u_i^1 of Γ^1. We write $U_i^1 \to U_i^2$ or $u_i^1 \to u_i^2$, respectively, if this correspondence holds. We assign the same payoffs to the endpoints of the subtree on the $x(\pi_C)$ as to the corresponding endpoints of Γ^1. This completes the construction of Γ^2.

We now proceed to prove the linear equivalence of Γ and Γ^2. It is sufficient to show $\Gamma^1 \sim \Gamma^2$. We write $\pi_i^2 \to \pi_i^1$ if $\pi_i^1(U^1) \to \pi_i^2(U^2)$ is true for $U_i^1 \to U_i^2$, and we write $\pi^2 \to \pi^1$ if the components of both combinations in this way correspond to each other; we use the same notation for coalition strategies. For $i = 2, \ldots, n$ this correspondence evidently is one-to-one, but to each π_1^1 correspond several π_1^2. For all these π_1^2 we have $H^2(\pi^2/\pi_1^2) = H^1(\pi^1/\pi_1^1)$ if $\pi^2 \to \pi^1$, since the payoff is only influenced by the decisions after $x(\pi_C)$. Therefore it is clear that a normal form isomorphic to $N(\Gamma^2)$ can be formed by adding a set of indexed mixed strategies to $N(\Gamma^1)$. Therefore we have $\Gamma^1 \sim \Gamma^2$ (§1, E 1).

Let U_o be the information set containing the origin of Γ^2. The game Γ^2 is changed into Γ^3 by transferring information about all moves u at U_o in the following way: All players are informed about u successively at all information sets partitionable by u.

The game Γ^3 is changed into Γ^4 by transferring information about all moves u at the U_C in the following way: All members of C successively are informed about u at all their information sets partitionable by u. The information sets U^4 of a coalition $C \in \mathfrak{C}_1$ in Γ^4 evidently have the following property: If U^4 contains vertices coming after $x(\pi_C)$, then U^4 contains only such vertices. Therefore in Γ^4 we can define special strategies for all members of C called "C-strategies": The alternatives assigned to vertices after any $x(\pi_C)$ by a C-strategy are indirectly indicated by π_C through the isomorphism of the subtree at $x(\pi_C)$ and K^1; the decisions correspond to those of the strategy taking the position of π_C in Γ^1. Furthermore, a C-strategy of player 1 assigns a_C to the origin.

For each π_C we arbitrarily select one of player 1's C-strategies assigning $a(\pi_C)$ to x_C; this strategy is called "the π_C-strategy" of player 1. Let q_C be a mixed coalition strategy for $C \in \mathfrak{C}_1$ (i.e., a mixed strategy of player 1 in ${}_C\Gamma$). We construct a mixed strategy q_1^4 "related" to q_C in the following way: if π_1^4 is the π_C-strategy of player 1, then we define $q_1^4(\pi_1^4) = q_C(\pi_C)$; consequently, if π_1^4 is not a π_C-strategy, we have $q_1^4(\pi_1^4) = 0$. Let $\overline{Q}_1^4$ be a set of indexed mixed strategies which for each $C \in \mathfrak{C}_1$ contains exactly one mixed strategy $\overline{q}_1^4$ related to a fixed optimal strategy $\overline{q}_C$ of player 1 in ${}_C\Gamma$; let only these strategies be in $\overline{Q}_1^4$. We add $\overline{Q}_1^4$ to $N(\Gamma^4)$ and thereby obtain the normal form G' with the normal game Γ'. We have $\Gamma' \sim \Gamma^4$ (§1, E 1).

We have to show that Γ' has properties (1) – (4). Obviously (1) is true. Γ^4 results from Γ^2 by successively transferring information about moves of player 1. Hence, if w satisfies (I) and (XI), we have $w_1(\Gamma^4) \leq w_1(\Gamma^2)$. Here, beside (XI), we use (I) and (b) from the theorem on information transfer in §1. Because of $\Gamma \sim \Gamma^2$ and $\Gamma' \sim \Gamma^4$ it follows with the aid of (I) that Γ' has the property (3). For each $C \in \mathfrak{C}_1$ let $\overline{q}_C^4$ be a mixed coalition strategy which is used if player 1 uses the strategy $\overline{q}_1^4 \in \overline{Q}_1^4$ related to $\overline{q}_C$ and all other members of C use arbitrary but fixed C-strategies. A strategy π_{N-C}^4 is in the "π_{N-C}-class" $B(\pi_{N-C})$ if in the subgame at x_C the decisions prescribed by π_{N-C}^4 are indirectly indicated by π_{N-C} in the following way: Let π_{N-C}^1 correspond to π_{N-C} in Γ^1 and let π_{N-C}^2 correspond to π_{N-C}^1 by $\pi_{N-C}^2 \to \pi_{N-C}^1$; in the subgame at x_C the decisions of strategies in $B(\pi_{N-C})$ coincide with those of π_{N-C}^2. This definition makes use of the fact that in the subgame at x_C every information set U^4 of a member of $N-C$ is a subset of a unique information set U^2 for Γ^2. It can be seen easily that the payoff which results, if a strategy π_{N-C}^4 is played against $\overline{q}_C^4$, depends only on the π_{N-C}-class of π_{N-C}^4, and, moreover, this payoff is equal to the payoff which results if the corresponding π_{N-C} is used against $\overline{q}_C$. Therefore $\overline{q}_C^4$ must have the same security level in ${}_C\Gamma^4$ as $\overline{q}_C$ in ${}_C\Gamma$. Hence we have $V^4(C) \geq V(C)$ for $C \in \mathfrak{C}_1$. On the other hand, as Γ^4 results from Γ^2 by transferring information about moves of player 1, with the aid of $V = c_\eta$ (theorem on constant-sum games, §4) it follows from (h) in the theorem on η-characteristics (§4) that we have $V^4(C) \leq V^2(C)$ for $C \in \mathfrak{C}_1$. But V and V^2 coincide because of $\Gamma \sim \Gamma^2$ (§4, (a) in the theorem on η-characteristics). Hence we have $V^4(C) = V(C)$ for $C \in \mathfrak{C}_1$. Since V is a characteristic, V and V^4 coincide. Therefore $\bar{q}_C^4$ has the security level $V^4(C)$ in ${}_C\Gamma^4$. It follows that $\bar{q}_C^4$ is optimal. V' and V^4 coincide because of $\Gamma' \sim \Gamma^4$. Therefore Γ' has the property (4). Obviously the pure coalition strategy $\bar{\pi}_C'$ corresponding to $\bar{q}_C^4$ in Γ' is optimal, too. This proves (2).

SUPERFLUOUS ALTERNATIVES. A "superfluous" alternative a_0 at a vertex x of player j is characterized by the following properties: (1) x belongs to a one-element information set; (2) a_0 leads to an endpoint z_0; (3) there is another alternative a_1 at x which leads to an endpoint z_1 with $h_j(z_1) \geq h_j(z_0)$ and $h_i(z_1) \leq h_i(z_0)$ for $i \neq j$; besides we have

$$\sum_{i=1}^{n} h_i(z_0) = \sum_{i=1}^{n} h_i(z_1)$$

A move is called "superfluous", if it contains a superfluous alternative.

LEMMA 2. Let w be a value function for general n-person games or for n-person constant-sum games which satisfies (I), (III), (IV), (V), (VII), and (IXa). Let u_0 be a superfluous move in an n-person game Γ. For the game Γ' which results from Γ by forbidding u_0, we have $w(\Gamma') = w(\Gamma)$.

PROOF. We consider the game Γ^k (Figure 4; k = 1, ..., n); the dots stand for vanishing payoff components). By forbidding the move from o to y to player k in Γ^k, a game with value (0, ..., 0) results; this is because of (III) and (IV). Therefore because of (V) we have $w_k(\Gamma^k) \geq 0$. Thus $w_j(\Gamma^k) \geq 0$ holds for $j = k$. By forbidding in Γ^k the move from y to

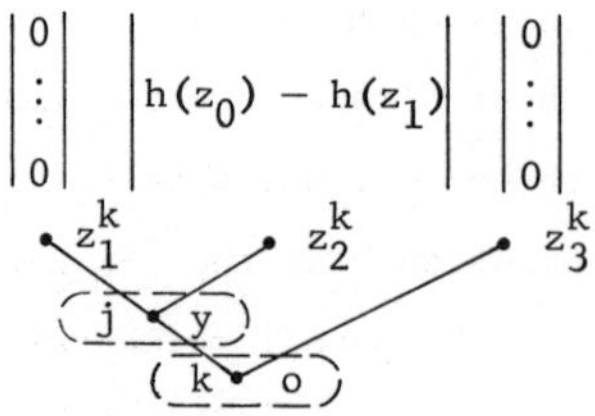

Figure 4: Γ^k

z_2^k a game is formed which is linearly equivalent to a one-point game with vanishing payoffs. Therefore, because of (I), (IV), and (V), we have $w_j(\Gamma^k) \geq 0$ also for $j \neq k$. For $i \neq j$ and $i \neq k$ we have $h_i(z_0) - h_j(z_1) \geq 0$; therefore by substituting 0 for all payoffs at z_2^k in Γ^k we form a game which results from Γ^k by transferring payoff from player i to other players. Since player i in this game is a dummy with 0, it follows with the aid of (IV) and (IXa) that we have $w_i(\Gamma^k) \geq 0$. Hence, because of (III) and $M(\Gamma^k) = 0$, the value of Γ^k is $(0, \ldots, 0)$. We form the games $\Gamma^{n+k} = \Gamma' + \Gamma^k$. Let u_1 be the move of Γ^{n+k} at the endpoint z_1 (from property (3) of a_0) which corresponds to the move from o to z_3^k in Γ^k. By forbidding in Γ^{n+k} first u_1 and then all the moves at the endpoints $z \neq z_1$ which correspond to the move from o to y in Γ^k, a game Γ^{2n+k} results. The subgame of Γ^{2n+k} at the endpoints $z \neq z_1$ and at the vertex x_1 coming immediately before z_1 are linearly equivalent to the subgames of Γ at the same points. It follows with the aid of §1, E 2, that Γ^{2n+k} and Γ are linearly equivalent. Because of (I) we have $w(\Gamma^{2n+k}) = w(\Gamma)$. Because of (V) we have $w_k(\Gamma^{n+k}) \geq w_k(\Gamma^{2n+k})$. Postulate (VII) gives $w(\Gamma^{n+k}) = w(\Gamma')$. With the aid of $M(\Gamma) = M(\Gamma')$ it follows by (III) that $w(\Gamma) = w(\Gamma')$ is true. This proves the lemma.

THE EQUALIZING GAME. For $n = 2, 3, \ldots$ we now construct an n-person game ${}^n\Gamma$, the "equalizing game". The intuitive motive for this name will be made clear by later definitions. ${}^n\Gamma$ is a constant-sum game. The origin o is a vertex of player 1. There are $n - 1$ alternatives at o leading to the endpoints $z_2, \ldots, z_n$.

The payoff is given by

$$ {}^n h_i(z_k) = \begin{cases} 0 & \text{for } i = k \text{ or } i = 1 \\ 1 & \text{for } i \neq k \text{ and } i \neq 1 \end{cases} $$

for $k = 2, \ldots, n$. Since at each endpoint z_k at most one of the members of a nonempty coalition $C \notin \mathcal{C}_1$ receives 0, the common payoff of C is always at least $|C| - 1$. Player 1 can secure $|C| - 1$ for any coalition $C \in \mathcal{C}_1 - (N)$ by realizing a z_k with $k \in N-C$. Therefore the characteristic function ${}^{n}V$ of ${}^{n}\Gamma$ can be described as follows: (1) ${}^{n}V(\emptyset) = 0$; (2) ${}^{n}V(N) = n - 2$; (3) ${}^{n}V(C) = |C| - 1$ for $0 < |C| < n$. Evidently ${}^{n}V$ is fully symmetric.

LEMMA 3. If a value function w for n-person constant-sum games satisfies (I) — (V), (VII), and (IXa), then we have $w_i(a\cdot{}^{n}\Gamma) = a(n - 2)/n$ for $a \geq 0$ and $i = 1, \ldots, n$.

PROOF. Let Γ^1 be the game $a\cdot{}^{n}\Gamma$. We construct an n-person game Γ^2: the trees of Γ^1 and Γ^2 are isomorphic; the endpoints $z_2, \ldots, z_n$ of Γ^1 correspond to the endpoints $z_2^2, \ldots, z_n^2$ of Γ^2. In Γ^2 all players, with exception of players 1 and 2, are dummies with 0; we have $h_1^2(z_k) = +a$ and $h_2^2(z_k) = -a$ for $k \neq 2$ and $h_1^2(z_2) = h_2^2(z_2) = 0$; player 2 decides at the origin. Then $w_i(\Gamma^2) = 0$ holds for $i = 3, \ldots, n$ because of (IV). By transferring payoff from player 1 to other players, Γ^2 can be changed into a game in which all players receive 0 everywhere. Because of (III) and (IV) the value of this game is $(0, \ldots, 0)$. Hence by (IXa) we have $w_1(\Gamma^2) \geq 0$. By forbidding in Γ^2 all moves not leading to z_2^2 a game is formed the value of which is $(0, \ldots, 0)$ because of (III) and (IV). Therefore by (V) we have $w_2(\Gamma^2) \geq 0$. With the aid of (III) it follows from what has already been said that $(0, \ldots, 0)$ is the value of Γ^2.

Next we form $\Gamma^3 = \Gamma^2 + \Gamma^1$. Because of (VII) we have $w(\Gamma^3) = w(\Gamma^1)$. By forbidding for $k = 2, \ldots, n$ all moves at z_k^2 which do not correspond to the move leading to z_k in Γ^1, the game Γ^4 results from Γ^3. By substituting linearly equivalent one-point games for the subgames at the z_k^2, the game Γ^4 is changed into Γ^5. Clearly $\Gamma^4 \sim \Gamma^5$ (§1, E 2). By interchanging the players 1 and 2, a game Γ^6 results from Γ^1. Obviously Γ^5 and Γ^6 are isomorphic. Thus by (I) and (II) it follows that $w_1(\Gamma^5) = w_2(\Gamma^1)$. By (I) and (V) we have $w_1(\Gamma^5) \leq w_1(\Gamma^3)$. Since $w(\Gamma^3) = w(\Gamma^1)$, we see that $w_2(\Gamma^1) \leq w_1(\Gamma^1)$. It will be shown that $w_1(\Gamma^1) \leq w_2(\Gamma^1)$ holds, too. From $w_1(\Gamma^1) = w_2(\Gamma^1)$ it can be seen by (II) that $w_1(\Gamma^1) = w_i(\Gamma^1)$ for $i = 2, \ldots, n$; indeed, Γ^1 is not changed if two of the players $2, \ldots, n$ are interchanged. Since $M(\Gamma^1) = a(n-2)$ the proposition of the theorem is implied by this. Therefore, only $w_1(\Gamma^1) \leq w_2(\Gamma^1)$ remains to be shown.

By substituting 0 for all payoffs we change Γ^2 into Γ^7, which by (III) and (IV) has $(0, \ldots, 0)$ as value. We form $\Gamma^8 = \Gamma^7 + \Gamma^1$. We construct a game Γ^9 which differs from Γ^8 only by a different payoff function: At the endpoint z_k^2 which corresponds to z_k in Γ^1 we substitute $h^6(z_k)$ for $h^1(z_k)$; we do this for $k = 2, \ldots, n$; the payoff at all other endpoints of Γ^9 is $h^1(z_2) = h^6(z_2)$. These changes nowhere increase the payoff of player 2, since player 2 in Γ^6 receives 0 everywhere. The vector $h^6(z_k)$ differs from $h^1(z_k)$ only with regard to the first two components and to the advantage of player 1. Similarly $h^1(z_2)$ is at least as good as $h^1(z_k)$ for all players with exception of player 2. Hence Γ^9 results from Γ^8 by transferring payoff from player 2 to other players. Therefore (IXa) implies $w_2(\Gamma^9) \leq w_2(\Gamma^8)$. By (VII) we have $w(\Gamma^8) = w(\Gamma^1)$. Clearly, $w_2(\Gamma^9) \leq w_2(\Gamma^1)$. It can easily be seen that in Γ^9 all the moves at z_k^2, $k = 3, \ldots, n$, which do not lead to the endpoint corresponding to z_k of Γ^1 are superfluous in the sense of Lemma 2. A game Γ^{10} results from Γ^9 by forbidding all these superfluous moves. By Lemma 2 we have $w(\Gamma^{10}) = w(\Gamma^9)$. Hence $w_2(\Gamma^{10}) \leq w_2(\Gamma^1)$. In Γ^{10} we can substitute linearly equivalent subgames for the subgames at the z_k^2, $k = 2, \ldots, n$. A game Γ^{11} results. By §1 , E 2, we have $\Gamma^{11} \sim \Gamma^{10}$. It follows from (I) that $w(\Gamma^{11}) = w(\Gamma^{10})$. Hence $w_2(\Gamma^{11}) \leq w_2(\Gamma^1)$. Evidently Γ^{11} and Γ^6 are isomorphic. This yields $w_2(\Gamma^6) \leq w_2(\Gamma^1)$. By (II) we have $w_2(\Gamma^6) = w_1(\Gamma^1)$. Hence $w_1(\Gamma^1) \leq w_2(\Gamma^1)$. This proves the lemma.

EQUALIZING. Since in $\Gamma = a \cdot {}^n\Gamma$ only player 1 has more than one pure strategy, the expected payoff for mixed strategies can be described as a function $E(q_1) = (E_1(q_1), \ldots, E_n(q_1))$ of player 1's mixed strategy q_1 alone. Let C be a nonempty coalition not containing player 1. The mixed strategy q_1 "equalizes" the vector $b = (b_1, \ldots, b_n)$ with regard to C, if the following conditions are satisfied:

1) $b_i + E_i(q_1)$ has the same value for all $i \in C$

2) $E_i(q_1) = a$ for all $i \in N-C-(1)$.

Because of 2) the behavior prescribed by q_1 is optimal for $N-C$.

LEMMA 4. For every vector $b = (b_1, \ldots, b_n)$ there is a number $a_0 \geq 0$ such that for $a \geq a_0$ there is a mixed strategy q_1 in $\Gamma = a \cdot {}^n\Gamma$ which equalizes b with regard to a nonempty coalition $C \subseteq N - (1)$.

PROOF. The vectors $E(q_1) = (E_1(q_1), \ldots, E_n(q_1))$ can be considered as points in an n-dimensional "payoff space". $E(q_1)$ satisfies 2) if and only if q_1 assigns positive probabilities only to those $|C|$ pure strategies leading to the payoff 0 for one of C's members. All these $E(q_1)$ form a $(|C| - 1)$-dimensional simplex S in the payoff space. The center of gravity of S is $m = (m_1, \ldots, m_n)$, where $m_1 = 0$ and $m_i = a$ for $i \in N{-}C - (1)$; for $i \in C$ we have $m_i = a(|C| - 1)/|C|$. Let b_C be the sum of all b_i with $i \in C$. Evidently q_1 satisfies 1) from the definition of "equalizing", if we have

$$(1) \qquad b_i + E_i(q_1) = m_i + b_C/|C| \qquad \text{for } i \in C.$$

By substituting $b_C/|C|$ for all b_i with $i \in C$, the vector b is changed into a vector $\overline{b}$. (1) is implied by

$$(2) \qquad b + E(q_1) = m + \overline{b}.$$

(2) in turn is equivalent to

$$(3) \qquad E(q_1) = m + \overline{b} - b.$$

(3) implies not only 1) but also 2), because $\overline{b}_i = b_i$ holds for $i \in N{-}C$. Let r be the radius of the biggest $(|C| - 1)$-dimensional sphere which has m as center and is wholly contained in S. Clearly, if $(\overline{b} - b)^2 < r^2$, then a q_1 can be found which satisfies (3). Since r proportionally depends on a, the inequality for r can be satisfied by choosing a_0 sufficiently great. This proves the lemma.

THE NORMAL GAME OF A SYMMETRIC SUPERADDITIVE CHARACTERISTIC: For every symmetric superadditive characteristic c we construct a normal form

G which has c as characteristic function. If c is the trivial characteristic with $c(C) = 0$ for all $C \in \mathfrak{C}$, then G is the normal form of the one-point game with vanishing payoffs. If c is different from this trivial characteristic, then c has a unique symmetry carrier R (cf. the lemma in §4). G is constructed as follows: Each member of $N - R$ receives 0 everywhere and has exactly one pure strategy. The strategies of a player $i \in R$ are the coalitions $C \in \mathfrak{C}_i$ contained in R. A combination π of pure strategies "realizes" C, if it contains C as component for all members of C. Let $C_1, \ldots, C_s$ be all coalitions realized by π; define

$$u(\pi) = \frac{1}{|R|}\left(c(R) - \sum_{j=1}^{s} c(C_j)\right).$$

If player $i \in R$ is in the realized coalition C_j, then

$$H_i(\pi) = c(C_j)/|C_j| + u(\pi);$$

if player $i \in R$ is in none of the realized coalitions, then

$$H_i(\pi) = u(\pi).$$

This completes the construction of G.

Since u is nonnegative, because of the superadditivity of c, the common payoff $c(C)$ is enforced for C if all members of $C \cap R$ select $C \cap R$ as their strategy. Therefore c is indeed the characteristic of G.

We call the normal game Γ of G the "normal game of c". Let w be a value function satisfying (I) – (IV). We prove that w assigns the Shapley value $s(\Gamma)$ to Γ. If in Γ two members of R are interchanged, a linearly equivalent game results. Therefore by (I) and (II) all members of R have the same value. Since all players in $N - R$ are dummies with 0, their value is 0 by (IV). It follows by (III) that $w_i(\Gamma) = c(R)/|R|$ for $i \in R$. Thus the value of Γ is fixed by (I) – (IV). Since the Shapley value satisfies these postulates, we have $w(\Gamma) = s(\Gamma)$.

The results of this section are stated in the following Lemma 5.

LEMMA 5. For every symmetric superadditive characteristic c there is a normal game Γ which has c as characteristic. Every value function satisfying (I) – (IV) assigns the Shapley value $s(\Gamma)$ to Γ.

THE LEVELLING GAME. For $n = 2, 3, \ldots$ we construct an n-person game ${}^n\Gamma'$, the "levelling game": n alternatives $a_1, \ldots, a_n$ lead from the origin o to the endpoints $z_1, \ldots, z_n$. Player 1 decides at o. At the endpoint z_1 all players receive 0. At $z_k (k = 2, \ldots, n)$ the players 1 and k receive -1, while all other players receive 0.

LEMMA 6. Let w be a value function for general n-person games; if w satisfies (I) – (V), (VII), and (X), then $w_i(a \cdot {}^n\Gamma') = 0$ for $a \geq 0$ and $i = 1, \ldots, n$.

PROOF. A game isomorphic to $\Gamma^1 = a \cdot {}^n\Gamma'$ results from Γ^1 by interchanging any two members of $N - (1)$. Therefore by (I) and (II) all members of $N - (1)$ have the same value. Thus because of (III) it is sufficient to show that $w_1(\Gamma^1) = 0$. By forbidding all moves not leading to z_1, a game with value $(0, \ldots, 0)$ results from Γ^1; this is because of (III) and (IV). Hence by (V) we have $w_1(\Gamma^1) \geq 0$. In the following way a value function w^2 for general 2-person games is induced by w: if Γ' is the n-person game resulting from the 2-person game Γ by adding $n - 2$ dummies with 0, then $w_i^2(\Gamma) = w_i(\Gamma')$ for $i = 1, 2$. It is easily seen that (I) – (V), (VII), and (X) carry over from w to w^2. It has been proved in [10], §7, that, with regard to 2-person value functions, (I) – (V), (VII), and (X) imply the following property (VIa).

$$\text{(VIa)} \qquad w(\lambda\,{}^2\Gamma^o) = \lambda w({}^2\Gamma^o) \qquad \text{for } \lambda > 0.$$

Hence w^2 satisfies (VIa). (E 8, §1, could be used instead of (VIa), but we don't need this stronger result.) Since $a \cdot {}^n\Gamma^o$ results from $a \cdot {}^2\Gamma^o$ by adding $n - 2$ dummies with 0, by (IV) we have $w(a \cdot {}^n\Gamma^o) = aw({}^n\Gamma^o)$. By adding the one-point game $(x, (-a, -a, 0, \ldots, 0))$ to $a \cdot {}^n\Gamma^o$, a game Γ_1 results. Since by (IV) the value of a one-point game is its payoff vector, the value of Γ_1 is $(0, \ldots, 0)$ by (VII) and (X). By interchanging players 1 and i, a game Γ_i results from Γ_1. We form

$$\Gamma^2 = \sum_{i=2}^{n} \Gamma_i .$$

By (II) we have $w(\Gamma_i) = (0, \ldots, 0)$. Hence $w(\Gamma^2) = (0, \ldots, 0)$, by (VII). In Γ^2 we designate n "special" strategies of player 1: If x is one of the vertices corresponding to the origin of Γ_k, for $k \neq i$ the move corresponding to the move of Γ_k which leads to the payoff $(0, \ldots, 0)$ is assigned to x by the ith special strategy; for $k = i$ the opposite decision is prescribed $(i = 2, \ldots, n)$. Only moves corresponding to moves leading to $(0, \ldots, 0)$ are selected by the first special strategy. A game Γ^3 results from the normal game of Γ^2 by forbidding all moves not corresponding to special strategies. Since the kth special strategy leads to the same payoff as the move to z_k in Γ^1, the games Γ^1 and Γ^3 are isomorphic. Thus by (V) we have $w_1(\Gamma^1) \leq w_1(\Gamma^2)$. Hence $w_1(\Gamma^1) \leq 0$. We have already shown that $w_1(\Gamma^1) \geq 0$; hence $w_1(\Gamma^1) = 0$. This proves the lemma.

LEVELLING. Since in $\Gamma = a \cdot {}^n\Gamma'$ only player 1 has more than one pure strategy, the expected payoff for mixed strategies can be described as a function $E(q_1) = (E_1(q_1), \ldots, E_n(q_1))$ of player 1's mixed strategy q_1 alone. A strategy q_1 is "levelling against j", if it assigns positive probabilities only to moves leading to z_1 and to z_j $(j = 2, \ldots, n)$; if moreover,

$$\sum_{i=1}^{n} b_i + E_i(q_1) = g,$$

then q_1 "levels the vector $b = (b_1, \ldots, b_n)$ to g". For $C \in \mathfrak{C}_1$ the payoff in $F_{1/2}({}_C\Gamma)$ is $(0, 0)$ at the endpoints z_k with $k = 1$ or $k \in N{-}C$, and $(-a, +a)$ at the other endpoints. Therefore the moves to the endpoints z_1 and z_j constitute 1/2-optimal* coalition strategies for C if $j \notin C$. Hence, q_1 is 1/2-optimal for each coalition $C \in \mathfrak{C}_1$ not containing j, if q_1 is levelling against j.

* A strategy is called "η-optimal" for Γ, if it is optimal for $F_\eta(\Gamma)$.

LEMMA 7. For every vector $b = (b_1, \ldots, b_n)$ and to every $g \leq b_1 + \ldots + b_n$ there is an $a_0 \geq 0$, so that for $a \geq a_0$ mixed strategies q_1^j $(j = 2, \ldots, n)$ exist in $\Gamma = a \cdot {}^n\Gamma'$ which level b against j to g.

PROOF. It is easily seen that it is sufficient to take

$$a_0 = \frac{1}{2}\left(-g + \sum_{i=1}^{n} b_i\right)$$

LEMMA 8. The Shapley value s is the only value function for n-person constant-sum games which satisfies (I) – (IV), (V), and (VII), and which furthermore assigns $s(\Gamma)$ to all Γ with fully symmetric characteristic functions.

PROOF. Let Γ be an arbitrary n-person constant-sum game. Its characteristic function V can be linearly combined from symmetric superadditive characteristics (§4, combination theorem). Therefore we can assume that

$$V = \sum_{k=1}^{m} \lambda_k V_k, \tag{1}$$

where the λ_k are real numbers and the V_k are symmetric superadditive characteristics. In the following Equation (2) the negative terms of (1) are collected on the left side:

$$V + \sum_{k=1}^{m} \lambda_k' V_k = \sum_{k=1}^{m} \lambda_k'' V_k, \quad (\lambda_k' \geq 0,\ \lambda_k'' \geq 0, \qquad k = 1, \ldots, m). \tag{2}$$

The $\lambda_k' V_k$ are symmetric and superadditive, since these properties are not lost by nonnegative multiplication. Therefore a normal game Γ_k exists for each $\lambda_k' V_k$. We form

$$\Gamma^1 = \Gamma + \sum_{k=1}^{m} \Gamma_k. \tag{3}$$

It follows from Lemma 5 with the aid of (VII) that

(4) $$w(\Gamma^1) = w(\Gamma) + \sum_{k=1}^{m} s(\Gamma_k).$$

On the other hand we have

(5) $$s(\Gamma^1) = s(\Gamma) + \sum_{k=1}^{m} s(\Gamma_k).$$

Comparison of (4) and (5) shows that $w(\Gamma^1) = s(\Gamma^1)$ implies $w(\Gamma) = s(\Gamma)$. Therefore it is sufficient to prove $w(\Gamma^1) = s(\Gamma^1)$. Let Γ_k' be the normal game of $\lambda_k'' V_k$, $k = 1, \ldots, m$. We form

(6) $$\Gamma_1'' = \sum_{k=1}^{m} \Gamma_k'.$$

There are exactly $n!$ games $\Gamma_1'', \ldots, \Gamma_{n!}''$ which can result if, starting from Γ_1'', several interchanges of players take place successively. There is one such game for every permutation of the players $1, \ldots, n$. It follows from Lemma 5 with the aid of (VII) that $w(\Gamma_1'') = s(\Gamma_1'')$. Since w and s both satisfy (II), this yields

(7) $$w(\Gamma_r'') = s(\Gamma_r'') \qquad \text{for } r = 1, \ldots, n!\,.$$

We form

(8) $$\Gamma^2 = \Gamma^1 + \sum_{r=2}^{n!} \Gamma_r''\,.$$

Since Γ_1'' and Γ^1 have the same characteristic function, because of (2), (3) and (6), the characteristic function of Γ^2 is given by

(9) $$V^2 = \sum_{r=1}^{n!} V_r''.$$

Evidently V^2 is fully symmetric. Hence $w(\Gamma^2) = s(\Gamma^2)$. It follows from (8) by (VII) and Lemma 5 that

(10) $$s(\Gamma^2) = w(\Gamma^1) + \sum_{r=2}^{n!} s(\Gamma_r'').$$

Since (10) holds for $w = s$, we must have $w(\Gamma^1) = s(\Gamma^1)$. This proves the lemma.

§ 7. CHARACTERIZATION OF THE SHAPLEY VALUE FOR n-PERSON CONSTANT-SUM GAMES

THEOREM. There is one and only one value function for n-person constant-sum games which satisfies (I) – (V), (VII), (IXa), and (XI), namely the Shapley value s.

PROOF It has already been proved in §5 that s satisfies the postulates of the theorem. Therefore we only have to show that $w(\Gamma) = s(\Gamma)$ holds for any w satisfying these postulates. Because of Lemma 8, we can assume that the characteristic function V of Γ is fully symmetric. Since s and w both satisfy (III) it is sufficient to show

$$s_i(\Gamma) \leq w_i(\Gamma) \qquad \text{for } i = 1, \ldots, n. \tag{1}$$

Since (II) is satisfied by s and w, the inequalities (1) are implied by

$$s_1(\Gamma) \leq w_1(\Gamma), \tag{2}$$

because (2) can be applied to the game which results from Γ by interchanging players 1 and i. Let Γ' be related to Γ as in Lemma 1. It follows from (3) in Lemma 1 that $w_1(\Gamma') \leq w_1(\Gamma)$, and because of (4) in Lemma 1, we have $s_1(\Gamma') = s_1(\Gamma)$. Therefore (2) is implied by

$$s_1(\Gamma') \leq w_1(\Gamma'). \tag{3}$$

Because of Lemma 4, the parameter a in $\Gamma^1 = a \cdot {}^n\Gamma$ can be selected sufficiently great to ensure the existence of a set Q_1^1 containing exactly one indexed strategy q_1^1 for every pair $h'(z')$, C, where $h'(z')$ is a payoff vector of Γ' and C is a nonempty coalition not in $\mathfrak{C}_1$, and q_1^1 equalizes $h'(z')$ with regard to C. By adding such a set Q_1^1 to $N(\Gamma^1)$, a normal form G^2 results. Let Γ^2 be the normal game of G^2. Since s and w satisfy the postulates of Lemma 3, we have $s(\Gamma^1) = w(\Gamma^1)$. Clearly, $\Gamma^1 \sim \Gamma^2$

by §1, E 1. Since (I) is satisfied by s and w, it follows that

$$s(\Gamma^2) = w(\Gamma)^2. \tag{4}$$

We now construct an n-person normal form G^3. In this construction a parameter $b \geq 0$ appears which will be fixed later. For $k = 1, \ldots, n$, the strategies of player k in G^3 are the numbers $j \in N - (k)$. If a player $k \neq 1$ chooses 1 as his strategy, he recieves 0; his payoff is $-b$ if he selects another strategy. Player 1 has the following payoff:

$$H_1^3(\pi^3) = - \sum_{i=2}^{n} H_i^3(\pi^3). \tag{5}$$

Let Γ^3 be the normal game of G^3. Evidently the characteristic function v^3 of Γ^3 assigns 0 to all coalitions. By transferring payoff from player 1 to other players, we can change Γ^3 into a game Γ^4 where all players receive 0 everywhere. Since Γ^4 is linearly equivalent to a one-point game with vanishing payoffs, the value of Γ^4 is $(0, \ldots, 0)$ by (I) and (IV). Thus by (IXa) we have $w_1(\Gamma^3) \geq 0$. By forbidding all the moves of player $j \neq 1$ in Γ^3 which do not correspond to the strategy 1 in G^3, a game Γ^5 results; player j is a dummy with 0 in the normal game of Γ^5. Hence $w_j(\Gamma^3) \geq 0$ because of (I), (IV), and (V). By (III) we have

$$w(\Gamma^3) = s(\Gamma^3) = (0, \ldots, 0). \tag{6}$$

We form $\Gamma^6 = \Gamma^3 + \Gamma' + \Gamma^2$. The application of (VII) with regard to s_1 and w_1 shows with the aid of (4) and (6) that the difference between $w_1(\Gamma^6)$ and $w_1(\Gamma')$ is the same as that between $s_1(\Gamma^6)$ and $s_1(\Gamma')$. Therefore (3) is implied by

$$s_1(\Gamma^6) \leq w_1(\Gamma^6). \tag{7}$$

Since (7) implies (3) which implies (2) which implies (1), only (7) remains to be shown.

The proof of (7) will use the following sequence of games: Γ^6, Γ^8,

$\Gamma^9, \Gamma^{10}, \Gamma^{11}, \Gamma^{12}$. The construction of this sequence will ensure that w_1 is never increased by the transition from one of these games to the next. The last game Γ^{12} will have strong symmetry properties which will be sufficient to prove $w(\Gamma^{12}) = s(\Gamma^6)$. This result will imply (7).

We now construct Γ^8: After every endpoint z^7 of $\Gamma^7 = \Gamma^3 + \Gamma'$ we change Γ^6 in the following way: Let z^3 be the endpoint leading to z^7, and let π^3 be the strategy combination corresponding to z^3 in G^3. In the future $C_k(z^3)$ will always denote the coalition formed by the player having k as strategy in π^3. Let z^7 correspond to z' in Γ'. If z^7 comes after a z^3 with $C_1(z^3) \neq \emptyset$, we forbid at z^7 all those moves of player 1 which do not correspond to the equalizing strategy $q_1^1 \in Q_1^1$ for $h'(z')$ with regard to $C_1(z^3)$. If, however, z^7 comes after a z^3 with $C_1(z^3) = \emptyset$, we forbid at z^7 all moves with exception of one which can be selected arbitrarily. This completes the construction of Γ^8. We have $w_1(\Gamma^8) \leq w_1(\Gamma^6)$ because of (V). By substituting linearly equivalent one-point games for the subgames at the endpoints of Γ^7, a game Γ^9 results from Γ^8. Because of §1, E 2, we have $\Gamma^8 \sim \Gamma^9$. It follows by (I) that $w_1(\Gamma^9) = w_1(\Gamma^8)$.

We now construct Γ^{10} by changing the payoff function of Γ^9. Later it will be shown that Γ^{10} results from Γ^9 by transferring payoff from player 1 to other players. We define h^{10} by the following description of the payoff in the subgame $\Gamma^{10}_{z^3}$ after any endpoint z^3 of Γ^3. The subgame has at most $|N-C_1(z^3)| + 1$ different payoff vectors. One of those is the vector $u(z^3)$ with components given by

$$(8)\quad u_i(z^3) = \frac{1}{|C_k(z^3)|} v^6(C_k(z^3)) + \frac{1}{n}v^6(N) - \frac{1}{n}\sum_{r=1}^{n} v^6(C_r(z^3)), \quad i = 1, \ldots, n,$$

where $C_k(z^3)$ is that coalition among the $C_1(z^3), \ldots, C_n(z^3)$, which contains i. This means that k is player i's component in the π^3 corresponding to z^3. v^6 is the characteristic function of Γ^6. For each player $j \notin C_1(z^3)$, there is another payoff vector $u^j(z^3)$ which may appear. The components are given by

$$(9)\quad u_i^j(z^3) = \begin{cases} u_i(z^3) & \text{for } i \in N-C_1(z^3) \text{ and } i \neq j \\ u_j(z^3) - \bar{g}|C_1(z^3)| + \sum_{s\in C_1(z^3)} u_s(z^3) & \text{for } i = j \\ \bar{g} & \text{for } i \in C_1(z^3) \end{cases}$$

where the parameter $\bar{g}$ must be chosen in accordance with three inequalities (10), (11), and (12), but may be fixed arbitrarily otherwise. (11) and (12) ensure that $u^j(z^3)$ is worse than $u(z^3)$ for player j, but not for the other players.

$$\bar{g} > \max_{z^9 \in Z(K^9)} \max_{i \in N-(1)} h_i^9(z^9) \tag{10}$$

$$u_j^j(z^3) < u_j(z^3) \tag{11}$$

$$u_i^j(z^3) > u_i(z^3) \qquad \text{for } i \in C_1(z^3). \tag{12}$$

V^3 assigns 0 to all coalitions, independently of b. Therefore V^6 does not depend on b either. The inequalities (10), (11), and (12) only require that $\bar{g}$ be sufficiently great; $\bar{g}$ can be selected independently of b.

Let $\varsigma(z^3)$ be a pure coalition strategy for $N-C_1(z^3)$ in the subgame $\Gamma_{z^3}^{10}$ which corresponds to an optimal pure coalition strategy in Γ'. Because of (2) in Lemma 1 such a strategy exists in Γ'. Let $Z(z^3)$ be the set of all endpoints of $\Gamma_{z^3}^{10}$ which can be realized, if $\varsigma(z^3)$ is used. If an endpoint z^9 of the subgame is not in $Z(z^3)$, then the way from z^3 to z^9 contains at least one alternative which cannot be realized, if $\varsigma(z^3)$ is used. Among those players who have a move containing such an alternative, let player j be the lowest in the natural order; we call player j "first deviator" for z^9. Evidently the first deviator is always a member of $N-C_1(z^3)$. Let $Z_j(z^3)$ be the set of all endpoints z^9, for which player j is the first deviator. h^{10} is given by

$$h^{10}(z^9) = \begin{cases} u(z^3) & \text{for } z^9 \in Z(z^3) \\ u^j(z^3) & \text{for } z^9 \in Z_j(z^3) \end{cases} . \tag{13}$$

This completes the construction of Γ^{10}.

Γ^{10} results from Γ^9 by transferring payoff from player 1 to other players, provided that the following conditions (14) – (18) are satisfied:

$$\left.\begin{array}{lll} (14) & u_1(z^3) \leq h_1^9(z^9) & \\ (15) & u_i(z^3) \geq h_i^9(z^9) & \text{for } i = 2, \ldots, n \end{array}\right\} \text{ for } z^9 \in Z(z^3)$$

$$\left.\begin{array}{lll} (16) & u_1^j(z^3) \leq h_1^9(z^9) & \\ (17) & u_i^j(z^3) \geq h_i^9(z^9) & \text{for } i = 2, \ldots, n \end{array}\right\} \begin{array}{l} \text{for } z^9 \in Z_j(z^3), \\ j \in N\text{—}C_1(z^3) \end{array}$$

$$(18) \qquad \sum_{i=1}^{n} h_i^{10}(z^9) = \sum_{i=1}^{n} h_i^9(z^9) = v^6(N) .$$

We shall show that (14) — (18) can be ensured by an appropriate selection of the parameter b in G^3; it is only required that b be sufficiently great.

We consider the subgames $\Gamma_{z^3}^6$ of Γ^6 at the endpoints z^3 of Γ^3. Evidently the game $\Gamma^{z^3} = \Gamma' + \Gamma^2 + (x, h^3(z^3))$ is isomorphic to $\Gamma_{z^3}^6$. A strategy which is optimal for $\Gamma' + \Gamma^2$ is optimal for Γ^{z^3}, too; the addition of a one-point game does not destroy the optimality. Since the members of a coalition $C \in \mathfrak{C}_1$ always have a nonnegative common payoff in Γ^3, the characteristic function v^{z^3} of Γ^{z^3} satisfies $v^{z^3}(C) \geq v^6(C)$ for $C \in \mathfrak{C}_1$. The strategy $\varsigma(z^3)$ corresponds to an optimal strategy for $N\text{—}C_1(z^3)$ in Γ'. When Γ^8 was constructed, only one move at every endpoint z^7 of $\Gamma^3 + \Gamma'$ was not forbidden. For $C_1(z^3) \neq \emptyset$ this move corresponds to a strategy q_1^1, which equalizes $h'(z')$ with regard to $C_1(z^3)$; for $C_1(z^3) = \emptyset$ an arbitrary move was left over. In both cases, the only move remaining at z^7 in Γ^8 corresponds to a coalition strategy in Γ^2 which is optimal for $N\text{—}C_1(z^3)$. Therefore $\varsigma(z^3)$, together with the only possible decisions at z^7, forms an optimal strategy for the subgame $\Gamma_{z^3}^8$ in Γ^8. It follows that $\varsigma(z^3)$ is optimal for the subgame $\Gamma_{z^3}^9$ in Γ^9. Clearly, $\varsigma(z^3)$ secures $v^{z^3}(N\text{—}C_1(z^3))$ for $N\text{—}C_1(z^3)$ in $\Gamma_{z^3}^9$. Therefore the common payoff for $N\text{—}C_1(z^3)$ is at least $v^6(N\text{—}C_1(z^3))$ at endpoints $z^9 \in Z(z^3)$.

We consider the special case $N\text{—}C_1(z^3) = (1)$: Condition (14) is satisfied, since $u_1(z^3) = v^6((1))$. In (16) we must have $j = 1$. Because of (10) the components of $u^1(z^3)$ are better for the members of $N - (1)$ than their best payoffs in Γ^9. Therefore player 1's component is worse than his least payoff in Γ^9; this implies (16).

If player 1 is not the only member of $N\text{—}C_1(z^3)$, we have

$$(19) \qquad h_1^9(z^9) \geq h_1'(z') + b$$

where z' is the endpoint of Γ' which corresponds to z^9. (19) implies (14) and (16), if b is sufficiently great. Thus we have seen that (14) and (16) can be ensured.

We now proceed to show the same for (15) and (17). If b is increased, $h_i^9(z^9)$ is decreased for $i \in N - C_1(z^3) - (1)$. Therefore we can ensure (15) and (17) for these players. In the following we can assume $C_1(z^3) \neq \emptyset$, because otherwise nothing remains to be shown about (15) and (17). Let z^7 be an endpoint of $\Gamma^7 = \Gamma^3 + \Gamma'$ which comes after z^3. Since all members of $C_1(z^3)$ receive 0 at z^3 in Γ^3, the vector $h^7(z^7)$ is equalized with regard to $C_1(z^3)$ by the strategy q_1^1 which corresponds to the only move left over at z^7 in Γ^8. Therefore in $\Gamma_{z^3}^9$ the common payoff of $C_1(z^3)$ is always divided equally among the members of $C_1(z^3)$. Since the common payoff of $N - C_1(z^3)$ is at least $v^6(N - C_1(z^3))$ for $z^9 \in Z(z^3)$, the common payoff of $C_1(z^3)$ is at most $v^6(C_1(z^3))$. Since v^6 is a characteristic function, it follows with the aid of (8) that

$$(20) \quad h_i^9(z^9) \leq v^6(C_1(z^3))/|C_1(z^3)| \leq u_i(z^3) \quad \text{for} \quad i \in C_1(z^3) \quad \text{and} \quad z^9 \in Z(z^3).$$

(20) implies (15). Condition (17) for $i \in C_1(z^3)$ and (18) are immediate consequences of (8) and (9).

Now, having shown the possibility of doing so, we fix b in a way which ensures that Γ^{10} results from Γ^9 by transferring payoff from player 1 to other players. It follows by (IXa) that $w_1(\Gamma^{10}) \leq w_1(\Gamma^9)$.

For every endpoint z^3 of Γ^3 we construct a game $\Gamma_{z^3}^{11}$. Let $S(z^3)$ be the set of all members of $N - C_1(z^3)$, which have more than one strategy for the subgame $\Gamma_{z^3}^{10}$ at z^3 in Γ^{10}. If $S(z^3)$ is empty, $\Gamma_{z^3}^{11}$ is the one-point game $(x_1, u(z^3))$. In the following we assume $S(z^3) \neq \emptyset$. Let $i_1, \ldots, i_r$ with $i_1 < i_2 < \cdots < i_r$ be the members of $S(z^3)$. In $\Gamma_{z^3}^{11}$ the player i_k has exactly one vertex x_k in his player set, $k = 1, \ldots, r$. The vertex x_1 is the origin of $\Gamma_{z^3}^{11}$. The alternative a_k connects x_k and x_{k+1}, $k = 1, \ldots, r-1$. The alternative a_r at x_r leads to the endpoint z_0. There are only two alternatives at each x_k; the second one, a_k', leads to the endpoint z_k, $k = 1, \ldots, r$. The payoff is $u(z^3)$ at z_0 and $u^{i_k}(z^3)$ at z_k. This completes the construction of $\Gamma_{z^3}^{11}$.

Now we proceed to show that $\Gamma_{z^3}^{10} \sim \Gamma_{z^3}^{11}$. Since Γ^{10} and

$\Gamma^7 = \Gamma^3 + \Gamma'$ differ only with regard to the payoff function and, since Γ' is a normal game (cf. (1) in Lemma 1, §6), the subgame $\Gamma_{z^3}^{10}$ can be considered to be the normal game of a normal form $G_{z^3}^{10}$. Therefore each player in $\Gamma_{z^3}^{10}$ has to decide only once, and the order of these decisions is given by the natural order of the players. Clearly, there must be a first deviator if $\zeta(z^3)$ is not used; he is in $S(z^3)$ since the other members of $N-C_1(z^3)$ have only one strategy which must be a component of $\zeta(z^3)$. Among those members of $S(z^3)$ who do not use their components of $\zeta(z^3)$, the first deviator is the lowest in the natural order. The payoff in $\Gamma_{z^3}^{10}$ is $u^j(z^3)$, if player j is the first deviator and it is $u(z^3)$ if there is no first deviator. The payoff in $\Gamma_{z^3}^{11}$ depends in the same way on the first deviator, if player i_k is called "first deviator" for z_k, $k = 1, \ldots, r$. It does not matter which one of his strategies a member of $C_1(z^3)$ is using in $\Gamma_{z^3}^{10}$, and it does not matter which one of the possibly many strategies deviating from $\zeta(z^3)$ a player of $S(z^3)$ is using. From what has already been said, it is easily seen that $N(\Gamma_{z^3}^{11})$ can be changed into a normal form isomorphic to $G_{z^3}^{10}$ by adding a set Q of indexed mixed strategies which contains sufficiently many copies of the strategies for the players of $C_1(z^3)$ and sufficiently many copies of the strategies which prescribe alternatives a_k'. Therefore we have $\Gamma_{z^3}^{10} \sim \Gamma_{z^3}^{11}$ by E 1, §1. By substituting the $\Gamma_{z^3}^{11}$ for the corresponding $\Gamma_{z^3}^{10}$, a game Γ^{11} results from Γ^{10}. We have $\Gamma^{11} \sim \Gamma^{10}$ by E 2, §1. Hence $w_1(\Gamma^{11}) = w_1(\Gamma^{10})$ because of (I).

By substituting $u(z^3)$ for $h^3(z^3)$ everywhere, a game Γ^{12} results from Γ^3. This game Γ^{12} will also result from Γ^{11} by a number of successive changes which will be made alternately in the following ways: (a) superfluous moves are forbidden; (b) linearly equivalent one-point games are substituted for subgames. The changes will take place successively within all subgames $\Gamma_{w^3}^{11}$ with $S(z^3) \neq \emptyset$. Since the required postulates are satisfied by w, it will be clear by Lemma 2 and by E 2, §1, that $w_1(\Gamma^{12}) = w_1(\Gamma^{11})$.

Let Γ_k be the game which results from $\Gamma_{z^3}^{11}$ by substituting the one-point game $(x_k, u(z^3))$ for the subgame at x_k. Let Γ_{r+1} be the game $\Gamma_{z^3}^{11}$. The changes mentioned above are the following: (a) the move containing a_k' is forbidden in Γ_{k+1}; (b) the resulting game Γ_{k+1}' is changed into Γ_k by substituting the one-point game $(x_k, u(z^3))$ for the subgame at x_k. This is done for $k = r, \ldots, 1$. It can be seen with the aid of (11) and (12) that the forbidden moves are indeed superfluous. Since Γ_1 is the one-point game $(x_1, u(z^3))$, the final result is indeed Γ^{12}.

Let Γ^{13} be the game which results from Γ^{12} by interchanging players j and k. The strategies of player i in G^3 are the numbers from $N - (i)$. Thereby an enumeration of player i's strategies for Γ^3 is given. We use this enumeration also for Γ^{13}, but with the following slight change: k is substituted for j and j is substituted for k. Thus the numbers of $N - (i)$ correspond to the strategies of player i in Γ^{13}, too. Every strategy combination for Γ^{12} or Γ^{13} partitions N into coalitions C_r of players whose strategies have the same number r. Since Γ', Γ^2, and Γ^3 have fully symmetric characteristic functions, V^6 is fully symmetric, too. Therefore the payoffs in Γ^{12} and Γ^{13} depend in the same way on the coalitions C_r. It follows that the normal forms of Γ^{12} and Γ^{13} are isomorphic. Consequently, we have $w(\Gamma^{12}) = w(\Gamma^{13})$ because of (I). It follows with the aid of (II) that $w_j(\Gamma^{12}) = w_k(\Gamma^{12})$. Since j and k can be chosen arbitrarily, all players must have the same value in Γ^{12}. Since V^6 is fully symmetric, it follows with the aid of (III) that

$$(21) \qquad w_i(\Gamma^{12}) = M(\Gamma^{12})/n = V^6(N)/n = s_i(\Gamma^6), \qquad i = 1, \ldots, n.$$

Γ^{12} is the last game in the sequence $\Gamma^6, \Gamma^8, \Gamma^9, \Gamma^{10}, \Gamma^{11}, \Gamma^{12}$. We have seen that the value of player 1 is not increased by the transition from one of these games to the next. Therefore we have (7). This proves the theorem.

§ 8. CHARACTERIZATION OF THE VALUE FUNCTION $w_{1/2}$ FOR GENERAL n-PERSON GAMES

THEOREM. There is one and only one value function for general n-person games which satisfies (I) – (V), (VII), (IXa), (X), and (XI), that is $w_{1/2}$.

PROOF. It has been already proved in §5 that $w_{1/2}$ satisfies the postulates of the theorem. Let w be a value function for general n-person games which satisfies these postulates. We must have $w(\Gamma) = s(\Gamma)$ for n-person constant-sum games, since otherwise we would have a contradiction to the characterization theorem of §7. We know from §5 that $w_{1/2}$ and s coincide for constant-sum games. Therefore we have $w(\Gamma) = w_{1/2}(\Gamma)$ for constant-sum games. It remains to be shown that the same is true for non-constant-sum games. Since $w_{1/2}$ and w both satisfy (III), it is sufficient to prove

$$(1) \qquad w_{1/2,i}(\Gamma) \leq w_i(\Gamma) \qquad \text{for } i = 1, \ldots, n.$$

Since (II) is satisfied by $w_{1/2}$ and w, the inequalities (1) are implied by

$$(2) \qquad w_{1/2,1}(\Gamma) \leq w_1(\Gamma)$$

because (2) can be applied to the game which results from Γ by interchanging players 1 and i. Thus only (2) remains to be shown.

Let Γ' be the normal game of Γ. Because of (I) we have

$$(3) \qquad w(\Gamma') = w(\Gamma).$$

Define

$$(4) \qquad g = \min_{z' \in Z(K')} \sum_{i=1}^{n} h'_i(z').$$

Because of Lemma 7, the parameter a in $\Gamma^1 = a \cdot {}^n\Gamma'$ can be selected sufficiently great to ensure the existence of a set Q_1^1 containing exactly one indexed mixed strategy q_1^1 for every pair $h'(z')$, j, where $h'(z')$ is a payoff vector of Γ' and j is a member of $N - (1)$, and q_1^1 levels $h'(z')$ against j to g. By adding such a set Q_1^1 to $N(\Gamma^1)$, a normal form G^2 results. Let Γ^2 be the normal game of G^2. Since $w_{1/2}$ and w satisfy the postulates of Lemma 6, we have $w(\Gamma^1) = w_{1/2}(\Gamma^1) = (0, \ldots, 0)$. Clearly $\Gamma^1 \sim \Gamma^2$ by E 1, §1. Since (I) is satisfied by w and $w_{1/2}$, it follows that $w(\Gamma^2) = w_{1/2}(\Gamma^2) = (0, \ldots, 0)$. We form $\Gamma^3 = \Gamma' + \Gamma^2$. Because of (VII), we have $w(\Gamma^3) = w(\Gamma')$.

Let z_0' be an endpoint of Γ' which satisfies

$$\sum_{i=1}^{n} h_i'(z_0') = M(\Gamma'). \tag{5}$$

In Γ^2 player 1 can enforce the payoff vector $(0, \ldots, 0)$ by a certain move u_0^2. Let u_0^3 be the move at z_0' in Γ^3 which corresponds to u_0^2. After every endpoint z' of Γ' we change Γ^3 in the following way: We forbid all the moves different from u_0^3 which do not correspond to a strategy q_1^1 levelling $h'(z')$ against a $j \in N - (1)$ to g. The result is Γ^4. Because of (5) the move u_0^3 leads to an endpoint z_0^4 of Γ^4 with

$$\sum_{i=1}^{n} h_i^4(z_0^4) = M(\Gamma'). \tag{6}$$

Since all the other moves at the z' in Γ^4 correspond to strategies which level $h(z')$ to g, we have

$$\sum_{i=1}^{n} h_i^4(z^4) = g \tag{7}$$

for all endpoints $z^4 \neq z_0^4$. Because of (V) we have $w_1(\Gamma^4) \leq w_1(\Gamma^3)$. It follows with the aid of $w(\Gamma^3) = w(\Gamma')$ and (3) that

$$w_1(\Gamma^4) \leq w_1(\Gamma). \tag{8}$$

By forbidding u_0^3, a game Γ^5 results from Γ^4. It follows by (5) that Γ^5 is a constant-sum game with $M(\Gamma^5) = g$. We now proceed to show that the characteristic function V^5 of Γ^5 and the 1/2-characteristic $c_{1/2}$ are related as follows

$$V^5(C) = c_{1/2}(C) - \tfrac{1}{2}\,(M(\Gamma) - g) \qquad \text{for all} \quad C \in \mathfrak{C} - (N) - (\emptyset). \tag{9}$$

It is sufficient to prove (9) for $C \in \mathfrak{C}_1 - (N)$; this can be seen in the following way: If C is not in $\mathfrak{C}_1 - (N)$, then $N-C$ is in $\mathfrak{C}_1 - (N)$, and (9) can be applied to $N-C$. Since V^5 and $c_{1/2}$ are characteristic, we can substitute $g - V^5(C)$ for $V^5(N-C)$ and $M(\Gamma) - c_{1/2}(C)$ for $c_{1/2}(N-C)$. This yields an equation which is equivalent to (9) applied to C.

Let C be a coalition in $\mathfrak{C}_1 - (N)$. We consider the coalition game ${}_C\Gamma^5$. There are two kinds of moves at every z': (a) moves corresponding to a q_1^1 which levels against a $j \in C$; (b) moves corresponding to a q_1^1 which levels against a $j \in N{-}C$. The moves of the kind (a) are worse for C and better for $N{-}C$; therefore they can be forbidden as superfluous in ${}_C\Gamma^5$. The result is a game ${}_C\Gamma^6$. Since the postulates of Lemma 2 are satisfied by the von Neumann value v, we have

$$(10) \qquad v_1({}_C\Gamma^6) = v_1({}_C\Gamma^5) = V^5(C).$$

We change the coalition game ${}_C\Gamma'$ into a game ${}_C\Gamma^7$ by substituting the following payoff function ${}_Ch^7$ for ${}_Ch'$:

$$(11) \qquad {}_Ch_i^{\,7}(z') = {}_Ch_i{}'(z') - \frac{1}{2}\,({}_Ch_1{}'(z') + {}_Ch_2{}'(z') - g), \qquad i = 1,\,2.$$

In ${}_C\Gamma^6$ all moves at a z' are of the kind (b); the corresponding strategies q_1^1 level to g by taking away equal amounts from the payoff of player 1 and from the payoff of a player $j \in N{-}C$. Therefore we have ${}_Ch^6(z^6) = {}_Ch^7(z')$ for all endpoints z^6 after z' in ${}_C\Gamma^6$. It follows that ${}_C\Gamma^6$ can be changed into ${}_C\Gamma^7$ by substituting all the subgames at the z' by linearly equivalent one-point games. Clearly, ${}_C\Gamma^7 \sim {}_C\Gamma^6$ by §1, E 2. Thus by (I) and (10) we have

$$(12) \qquad v_1({}_C\Gamma^7) = v_1({}_C\Gamma^6) = V^5(C).$$

Define

$$(13) \qquad {}_C\Gamma^8 = {}_C\Gamma^7 + (x,\, \tfrac{1}{2}(M(\Gamma) - g),\, \tfrac{1}{2}(M(\Gamma) - g))).$$

The von Neumann value of a one-point game is its payoff vector. It follows by (VII) and (12) that

$$(14) \qquad v_1({}_C\Gamma^8) = V^5(C) + \tfrac{1}{2}(M(\Gamma) - g).$$

Because of (11) we have

(15) $$ {}_C h_1^{\,7}(z') - {}_C h_1'(z') = {}_C h_2^{\,7}(z') - {}_C h_2'(z'). $$

It follows with the aid of $M(\Gamma) = M(\Gamma')$ that ${}_C\Gamma^8$ is the 1/2-replenished coalition game $F_{1/2}({}_C\Gamma')$. Therefore we have

(16) $$ v_1({}_C\Gamma^8) = w_{1/2}({}_C\Gamma') = c_{1/2}(C). $$

With the aid of (14), this yields (9).

Since V^5 is the 1/2-characteristic of Γ^5 (theorem on constant-sum games, §4), we can use formula (A) from §5 in order to describe $w_{1/2}(\Gamma^5)$ as follows:

(17) $$ w_{1/2,i}(\Gamma^5) = \gamma_n(N)V^5(N) + \sum_{C\in\mathfrak{C}_i-(N)} \gamma_n(C)(V^5(C) - V^5(N-C)),\ i = 1, \ldots, n. $$

Because of (9) we have

(18) $$ V^5(C) - V^5(N-C) = c_{1/2}(C) - c_{1/2}(N-C) $$

for all C which can occur in one of the sets $\mathfrak{C}_i - (N)$. With the aid of $V^5(N) = g$ and $\gamma_n(N) = 1/n$ it follows that

(19) $$ w_{1/2,i}(\Gamma^5) = g/n + \sum_{C\in\mathfrak{C}_i-(N)} \gamma_n(C)(c_{1/2}(C) - c_{1/2}(N-C)),\quad i = 1, \ldots, n. $$

Let $d = (d_1, \ldots, d_n)$ be the vector with the components

(20) $$ d_i = (M(\Gamma) - g)/n. $$

It follows that

(21) $$ d_i + \frac{g}{n} = M(\Gamma)/n = \gamma_n(N)c_{1/2}(N). $$

Therefore, the right side of (19) can be changed into the right side of formula (A) applied to Γ, simply by adding d_i. Consequently we have

(22) $$w_{1/2}(\Gamma^5) + d = w_{1/2}(\Gamma).$$

We now construct Γ^9: Player i has exactly one vertex x_i in his player set $(i = 1, \ldots, n)$. x_1 is the origin of Γ^9. There are two alternatives a_k and a_k' at each vertex x_i $(i = 1, \ldots, n)$. For $i = 1, \ldots, n-1$, the alternative a_k connects x_i with x_{i+1}; but a_n leads to the endpoint z_0^9. The alternative a_k' leads to the endpoint z_k' $(k = 1, \ldots, n)$. The payoff at all endpoints is 0 for all players.

By substituting a one-point game with vanishing payoffs for the subgame at x_k a game Γ_k^9 results from Γ^9. The players $k, \ldots, n$ in Γ^9 are dummies with 0.

By substituting d for the payoff vector at z_0^9, a game Γ^{10} results from Γ^9.

With the aid of (I) – (IV) it can be seen easily that

$$\left.\begin{array}{ll}(23)\ w_i(\Gamma^9) = w_{1/2,i}(\Gamma^9) = 0 & \\ (24)\ w_i(\Gamma_k^9) = w_{1/2,i}(\Gamma_k^9) = 0 & \text{for } k = 1, \ldots, n \\ (25)\ w_i(\Gamma^{10}) = w_{1/2,i}(\Gamma^{10}) = d_i & \end{array}\right\} \text{ for } i = 1, \ldots, n$$

We form $\Gamma^{11} = \Gamma^9 + \Gamma^4$. By (VII) we have $w(\Gamma^{11}) = w(\Gamma^4)$. By (6) we have $M(\Gamma^4) = M(\Gamma')$. Clearly, $M(\Gamma') = M(\Gamma)$ by $\Gamma \sim \Gamma'$. Therefore we have $M(\Gamma^4) = M(\Gamma)$. This yields $M(\Gamma^{11}) = M(\Gamma)$.

We form $\Gamma^{12} = \Gamma^5 + (x, d)$. Since $M(\Gamma^5) = g$, we have $M(\Gamma^{12}) = M(\Gamma)$. We change Γ^{11} into Γ^{13} by substituting Γ^{12} for the subgame at z_0^9. Evidently we have $M(\Gamma^{13}) = M(\Gamma)$. We forbid the move containing a_k in Γ^{13} and then substitute the linearly equivalent game Γ^4 for the subgame at x_k; thus Γ^{13} is changed into $\Gamma_k^{13} = \Gamma_k^9 + \Gamma^4$. By (VII) and (24) we have

(26) $$w(\Gamma_k^{13}) = w(\Gamma^4).$$

By (I) and (V) we have

(27) $$w_k(\Gamma_k^{13}) \leq w_k(\Gamma^{13}).$$

With the aid of (26) this yields

(28) $$w_k(\Gamma^4) \leq w_k(\Gamma^{13}).$$

Since $M(\Gamma^4)$ and $M(\Gamma^{13})$ coincide with $M(\Gamma)$, it follows by (III) that

(29) $$w(\Gamma^4) = w(\Gamma^{13}).$$

Hence, by (8) we have

(30) $$w_1(\Gamma^{13}) \leq w_1(\Gamma).$$

We change Γ^{13} into a game Γ^{14} by forbidding all those moves in the subgames at the endpoints $z_1^9, \ldots, z_n^9$ of Γ^9 which correspond to the move u_0^3 in Γ^4. By (V) we have $w_1(\Gamma^{14}) \leq w_1(\Gamma^{13})$. It follows by (30) that

(31) $$w_1(\Gamma^{14}) \leq w_1(\Gamma).$$

The subgames of Γ^{14} at the $z_1^9, \ldots, z_n^9$ all are isomorphic to Γ^5; the subgame at z_0^9 is isomorphic to $\Gamma^{12} = \Gamma^5 + (x, d)$. This shows that Γ^{14} is isomorphic to the game $\Gamma^{10} + \Gamma^5$. It follows by (VII) that

(32) $$w(\Gamma^{14}) = w(\Gamma^{10}) + w(\Gamma^5).$$

Since Γ^5 is a constant-sum game, we have $w(\Gamma^5) = w_{1/2}(\Gamma^5)$. It follows from (32) with the aid of (22) and (25) that

(33) $$w_1(\Gamma^{14}) = d_1 + w_{1/2,1}(\Gamma^5) = w_{1/2,1}(\Gamma).$$

(33) and (31) imply (2). We have already seen that it is sufficient to show (2). This proves the theorem.

ADDED IN PROOF

In reference to the discussion in the Introduction, there is one important interpretative problem, which was not mentioned in [10] because then I was not aware of it. This is the question of commitment power, which has been raised by Schelling [9]. The monotonicity postulates (V), (IX), (IXa) and (XI) presuppose that nothing can be gained by some forms of commitment. This is reasonable if every player can commit himself to whatever he wants before the beginning of the play, because then the commitment possibilities should be reflected by the value. Therefore we should add the assumption of "full commitment power" to the general assumptions mentioned in the fourth paragraph of the Introduction.

BIBLIOGRAPHY

[1] BURGER, E., Einführung in die Theorie der Spiele, Berlin 1959.

[2] HARSANYI, J. C., "A bargaining model for the cooperative n-person game," Contributions to the Theory of Games, Vol. IV, Princeton 1959, pp. 325-355.

[3] HARSANYI, J. C., "A simplified bargaining model for the cooperative n-person game," (mimeographed), Canberra, about 1960.

[4] ISBELL, J. R., "Absolute games," Contributions to the Theory of Games, Vol. IV, Princeton 1959, pp. 357-396.

[5] KUHN, H. W., "Extensive games and the problem of information," Contributions to the Theory of Games, Vol. II, Princeton 1953, pp. 193-216.

[6] LUCE, R. D., and RAIFFA, H., Games and Decisions, New York, 1957.

[7] NASH, J., "Two-person cooperative games," Econometrica, Vol. 21 (1953), pp. 128-140.

[8] RAIFFA, H., "Arbitration schemes for generalized two-person games," Contributions to the Theory of Games, Vol. II, Princeton 1953, pp. 361-387.

[9] SCHELLING, T. C., The Strategy of Conflict, Cambridge, Massachusetts, 1960.

[10] SELTEN, R., "Bewertung strategischer Spiele," Zeitschrift für die gesamte Staatswissenschaft, 116. Band, 2. Heft (1960), pp. 221-282.

[11] SHAPLEY, L. S., "A value for n-person games," Contributions to the Theory of Games, Vol. II, Princeton 1953, pp. 307-317.

[12] VON NEUMANN, J., "Zur Theorie der Gesellschaftsspiele," Mathematische Annalen, 100 (1928), pp. 295-320.

[13] VON NEUMANN, J. and MORGENSTERN, O., Theory of Games and Economic Behavior, Princeton, 1944.

Johann Wolfgang Goethe-University
Frankfurt am Main, Germany

Reinhard Selten

MIXED AND BEHAVIOR STRATEGIES IN INFINITE EXTENSIVE GAMES

Robert J. Aumann

§ 1. INTRODUCTION

We are concerned with infinite extensive games — not necessarily of perfect information — in which there may be a continuum of alternatives at some or all the moves; the games may also have unbounded or infinite play length. Our object is to define the notion of mixed strategy for such games, and to use this definition to prove the appropriate generalization of Kuhn's theorem on optimal behavior strategies in games of perfect recall [K_1]. Also, our methods give a solution to the conceptual problem raised by McKinsey under the heading "games played over function space" [Mc, pp. 355-357].

By-products are that our proof of Kuhn's theorem makes no use of the rather cumbersome "tree" model for extensive games, that it explicitly uses conditional probabilities (which are implicitly used by Kuhn), and that it explicitly proves that in a game which is of perfect recall for one player, that player can restrict himself to behavior strategies (this also is implicit in Kuhn's proof). Our proof is longer and more complicated than Kuhn's proof, but only because of the problems introduced by the non-denumerably infinite character of the game; the treatment of finite games by our methods would be considerably shorter.

§ 2. EXAMPLES

We give four examples to motivate this study and to illustrate some of the difficulties.

In our first example, there are two players, the "attacker" and the "defender"; for concreteness, one may think of the attacker as a bomber. The attacker starts the play by choosing a course of action (such as a flight course). The defender has some mechanism (such as radar) for determining the course chosen by the attacker, and he decides on <u>his</u> course of action on the basis of the information he gets from this mechanism. But the mechanism is not perfect; it only gives an <u>apparent</u> attacker course x, which is distributed around the true attacker course z according to a known probability distribution (which may vary with z). Thus the defender gets some information about the attacker's course, but not perfect information.

Denote by X the set of all possible apparent attacker courses, i.e., the set of possible information states of the defender. Denote by Y the set of courses of action available to the defender. Clearly a pure defender strategy is a function from X into Y. What about mixed strategies? If X and Y are finite, then there are only finitely many pure strategies, so there is no difficulty about defining mixed strategies. But in many cases the most appropriate model would be one in which X and Y are, say, copies of the unit interval. It is then still possible to define <u>some</u> kinds of mixed strategies; for example, we can mix finitely or denumerably many pure strategies, or we can adopt a fixed continuous distribution over Y regardless of what information we have — i.e., we can mix a continuum of pure strategies, each of which is a constant function from X into Y. But is this the best we can do? Can't we mix a continuum of pure strategies that are <u>not</u> constants?

A mixed strategy can be thought of as a probability distribution, i.e., a measure, on the set of all pure strategies. But before one defines a measure on a non-denumerable space, one must define a <u>measurable structure</u> on the space, i.e., one must define which subsets are measurable. It is by no means clear how this should be done in our case, or even what kind of measurable structure on the pure strategy space should be considered "appropriate" for this purpose.

For our second example we can do no better than quote McKinsey [Mc, p. 356]:

> "A game has four moves: in the first move P_1 (player 1) chooses a real number x_1; in the second move, P_2, knowing x_1, chooses a real number y_1; in the third

> move, P_1, knowing y_1, but having forgotten x_1, chooses a real number x_2; and in the last move, P_2, knowing y_1 and x_2, but not knowing x_1, chooses a real number y_2. (The payoff is then some function of the four variables x_1, x_2, y_1, and y_2.) A pure strategy for P_1 is now an ordered couple $\{a, f\}$, where a is a real number and f is a function of one real variable (it depends on y_1); and a pure strategy for P_2 is an ordered couple $\{g, h\}$, where g is a function of one real variable (it depends on x_1) and h is a function of two real variables (it depends on y_1 and x_2) ...
>
> It is clear that the payoff function for a game of the type just described need not necessarily have a saddle point, and hence it is natural to suppose that the players will make use of mixed strategies ... ."

The difficulties that McKinsey goes on to describe correspond precisely to those we discussed in connection with the first example.

Our third example involves the notion of the supergame of a given game G. This is a game each play of which consists of a number of repeated plays of G; the payoff to the "superplay" usually is defined as some kind of average of the payoffs to the individual plays. The supergame and related notions* have received considerable attention in the literature; this is partly because supergames occur naturally in the applications, and partly because an analysis of a supergame sometimes yields clues as to rational behavior for a single play.**

Supergames are usually analyzed on a step-by-step basis; that is, it is assumed that each player decides on a strategy for each of the component plays separately. These "local" strategies may or may not depend on the outcomes of the previous component plays, and may be pure or mixed; but the possibility of mixing a number of pure "grand strategies" for the whole supergame is usually ignored. Of course this makes analysis of the supergame much easier.

The supergame may be considered a game in extensive form, a move being a choice of a pure strategy for a component play. Obviously it is a game of perfect recall — at each component play each player remembers what

---

* Such as that of stochastic game.

** Cf. [Mc, the discussion at top of p. 134]; also [A_1] and [A_2, § 10].

he knew at previous component plays. What we are doing when we limit analysis of the supergame to consideration of mixed strategies for the component plays is that we are considering only behavior strategies in the supergame. Now we lose no generality by this restriction if Kuhn's theorem on behavior strategies in games of perfect recall* applies, which is the case when the originally given game is finite and is only repeated finitely often. Wolfe [W, p. 15] has pointed out that Kuhn's theorem may be extended to games with infinite play length, and it is easily seen that we can also allow a denumerable infinity of alternatives at some (or all) of the moves. The difficulties enter when there may be a continuum of alternatives at some of the moves; in our case this corresponds to a G with a continuum of strategies.

What is the importance of supergames of games with a continuum of strategies? Suppose we wish to consider the supergame of a cooperative game. To analyze this supergame properly, we must formalize the pre-play bargaining for each component play. Such a formalization must involve a continuum of pure strategies for the bargaining session — for example we already have a continuum in the set of correlated strategies that can be offered by a player for the consideration of a coalition that he wishes to form.** Thus a satisfactory analysis of a cooperative supergame cannot proceed without first proving an analogue of Kuhn's theorem for the continuous case. Indeed it was this problem that originally motivated this study.

For our last example, we start by recalling Ville's theorem [V]: Every 2-person 0-sum game on the unit square with continuous payoff has optimal mixed strategies, and hence a value. Now in a game on the unit square, each player picks a point in the unit interval $[0, 1]$, with no knowledge of the point chosen by the other; the payoff is a function of the two points chosen. Fox [F] asked whether Ville's result could be extended to multimove games of imperfect (but partial) information, where each move consists of the choice of a point in the unit interval. In these games, the players alternatively choose points $y_i \in [0,1]$ for $i = 1, \ldots, 2n$, say. After the ith

---

* Kuhn's theorem asserts that in a game of perfect recall each mixed strategy m has an equivalent behavior strategy, i.e., a behavior strategy which yields the same payoff as m (to all players) no matter what the other players do.

** Cf. [A_1, § 6] or [A_2, § 10].

move the player who is to make the (i+1)st move is informed of the value of $\psi_i(y_1, \ldots, y_i)$, where ψ_i is a real function. The payoff is $f(y_1, \ldots, y_{2n})$, where f is a continuous, real function. Note that the space of pure strategies of, say, player 1 is $\mathfrak{F} = [0,1] \times F_3 \times \ldots \times F_{2n-1}$, where F_i is the set of all functions from the range of ψ_{i-1} to $[0,1]$.

The object of [F] is to show that in general Ville's result does not extend to this situation, and there is no value. Of course, before this can be done the concept of mixed strategy must be defined. When each ψ_i has finite range, then each F_i and therefore also $\mathfrak{F}$ is a product of finitely many copies of $[0,1]$; in that case, therefore, mixed strategies can be defined as multidimensional distribution functions. When the range of ψ_i is a continuum, however, it is by no means clear how to define mixed strategies. One of Fox's examples, in which the ψ_i as well as f are continuous, is precisely of this kind.

Fox defined mixed strategies as finite mixtures of pure strategies. But this is not at all satisfactory, especially when one wants to show that a game does <u>not</u> have a value. Indeed, there is an example of a game on the unit square which has value 0 when distributions are admitted as mixed strategies, but for which the supinf of the payoff is -1 when one is restricted to finite mixtures of pure strategies [K_2, p. 118]. A priori, it could well be that the counterexamples of [F] vanish when a wider, more natural class of mixed strategies is admitted. In fact, it appears that this does not happen, since the proof apparently makes no essential use of the restrictive nature of the mixed strategies. It is the statement, rather than the proof, that is unsatisfactory; and the reason is that up to now there has been no definition of mixed strategy appropriate to such games.

§ 3. MIXED STRATEGIES

Let us take a closer look at the first example in the previous section; take X and Y to be copies of the unit interval. We shall need to consider probability distributions on X and Y, and as we remarked in the previous section, this involves defining measurable structures on them. Any such measurable structure should be rich enough to enable us to define the probability of an interval; this means it would have to contain all Borel sets. Let us denote by I the unit interval on which has been imposed a measurable structure consisting of all the Borel sets, and let us once and for all* take X and Y to be copies of I.

Henceforth we will write "m–" for "measurable."

Suppose the defender has adopted a strategy f, i.e., a function from X into Y, and that the action of chance and the strategy of the attacker have induced a probability distribution on X. The strategy f, acting on this X-distribution, should induce a distribution on Y. Does it? Suppose $B \subset Y$ is a Borel set. The probability that a member of B is chosen by the defender

$$= \text{prob } \{x : f(x) \in B\}$$
$$= \text{prob } \{f^{-1}(B)\}.$$

This expression is meaningless unless $f^{-1}(B)$ is measurable in X! The same holds for all m– subsets B of Y. In order to have an induced distribution on Y, we want the inverse image under f of a measurable set in Y to be measurable in X. In other words, we want f to be a measurable transformation. So we redefine a pure defender strategy; it is not just any function from X into Y, but an m– transformation.** We denote by Y^X the set of all m– transformations from X into Y.

A <u>mixed</u> strategy, then, should be a probability measure on Y^X, the latter having been endowed with an "appropriate" measurable structure R. Let

---

* We have adopted the smallest structure that fills our needs. An overly rich structure is self-defeating. For example, if the structure on X consists of <u>all</u> subsets, then the only measures on X are purely atomic (under the continuum hypothesis [S, p. 107]); if it consists of all Lebesgue measurable sets, then the only measures are sums of absolutely continuous and purely atomic ones (thus excluding all those with a singular non-atomic component). We therefore see that increasing the set of measurable sets beyond a certain point actually reduces the set of available measures. If we want all intervals to be measurable, the largest set of measures is obtained if we let the structure consist of the Borel sets. (In this connection we remark that there is a confusing misprint in [Mc, p. 357, line 7]; here "Lebesgue measurable" should read "Borel.")

** This redefinition of pure strategy is a consequence of the demand that distributions on X induce distributions on Y. Besides being intuitively desirable, this is absolutely necessary for the formal analysis, as the reader will see later. Perhaps the most compelling intuitive argument, though, is that this is needed so that a pair of pure attacker and defender strategies should induce a <u>payoff</u> distribution, for example so that we should be able to assign a probability to the attacker's payoff being positive.

us define a function $\varphi : Y^X \times X \to Y$ by $\varphi(f, x) = f(x)$. Suppose we again start out with a distribution on X, and suppose that the defender has chosen a chosen a mixed strategy; we wish to calculate the induced distribution on Y. For $\mathfrak{m}$- sets $B \subset Y$, the probability that the defender chooses a member of B

$$= \text{prob}\ \{(f, x) : f(x) \in B\}$$
$$= \text{prob}\ \{(f, x) : \varphi(f, x) \in B\}$$
$$= \text{prob}\ \{\varphi^{-1}(B)\}.$$

As before, we conclude that <u>the structure</u> R <u>must be chosen so that</u> φ <u>is an</u> $\mathfrak{m}$<u>- transformation</u>. But as we have shown elsewhere [A_3], there is no structure R for which this is so; <u>no</u> structure on Y^X is "appropriate"!

We have attempted to define mixed strategies as distributions (i.e., probability measures) on Y^X, and have run into difficulties. In fact these difficulties can be overcome, at least to a certain extent (cf. [A_4]). In the current context, though, we find that a completely different approach is both more convenient and more natural. Let us recall the intuitive meaning of a mixed strategy: It is a method for choosing a pure strategy by the use of a random device. Physically, one tosses a coin, and according as to which side comes up chooses a corresponding pure strategy; or, if one wants to randomize over a continuum of pure strategies, one uses a continuous roulette wheel. Mathematically, the random device — the set of sides of the coin or of points on the edge of the roulette wheel — constitutes a probability measure space, sometimes called the <u>sample space</u>; a mixed strategy is a function from this sample space to the set of all pure strategies. In other words <u>what we have here is precisely a random variable whose values are pure strategies</u>. We previously attempted to work with something corresponding to the distribution of this random variable; now we propose to use the random variable itself.

Let us denote by Ω the measure space that results when we impose Lebesgue measure on I. All of our sample spaces will be copies of Ω. The intuitive justification for this is that every "real-life" random device is either "discrete," "continuous," or a combination of the two; that is, the sample space involved must either be finite or denumerable, or it must be a copy of I (with a measure that is not necessarily Lebesgue measure).* All such random devices can be represented by random variables whose sample space is actually a copy of Ω.

In our example, therefore, we should define a mixed strategy to be a function from Ω to the space Y^X of all pure strategies. We can expect that not all such functions will be "eligible" as mixed strategies, because of the by now familiar condition that a mixed strategy and a distribution on X must induce a distribution on Y. Fortunately, the appropriate condition is not that the mixed strategy as defined above be a measurable transformation, because this would again involve defining a measurable structure on Y^X. To state the correct condition, we recall that to every function from Ω to Y^X there is a corresponding function from $\Omega \times X$ to Y; to $f : \Omega \to Y^X$ there corresponds the function $g : \Omega \times X \to Y$ defined by $g(\omega, x) = f(\omega)(x)$. The correct condition on a mixed strategy is that this corresponding function be an m– transformation. Thus we define a mixed strategy to be an m– transformation from $\Omega \times X$ into Y.

As it now stands, this definition of mixed strategy applies only to the highly simplified situation treated in the first example of the introduction. However, it can be extended without difficulty to more complicated, many-move games, as we shall show in Section 5.

§ 4. EXTENSIVE GAMES

In this section we present the formal structure for extensive games that we will use in the sequel. We first give the definitions, then discuss their intuitive meaning and their relation to other definitions in the literature.

An m– space is called standard** either if it is finite or denumerable with the discrete structure (i.e., all subsets are measurable), or if it is isomorphic*** with I. Most m– spaces that one "encounters in practice" are standard; for example, any Borel subset of any Euclidean space or of

---

* Physically, of course, all sample spaces are discrete and even finite; but it is often convenient to use a continuous or a denumerable model.

** This use of the word is due to Mackey [M].

*** An isomorphism is a one-one correspondence that is measurable in both directions.

Hilbert space is standard.

DEFINITION. A _game from an individual player's viewpoint_, or simply a _game_, consists of

(i) A (finite or infinite) sequence $Y_1, Y_2, \ldots$ of standard m- spaces called _action spaces_;

(ii) A corresponding sequence $X_1, X_2, \ldots$ of standard m- spaces called _information spaces_;

(iii) A set Z called the set of _strategies of the opponents_;

(iv) A sequence of functions

$$g_i : Z \times Y_1 \times \ldots \times Y_{i-1} \to X_i,$$

called _information functions_, which for each fixed $z \in Z$, are m- transformations on $Y_1 \times \ldots \times Y_{i-1}$ into X_i;

(v) A standard m- space H called the _payoff space_;

(vi) A function

$$h : Z \times Y_1 \times Y_2 \times \ldots \to H$$

called the _payoff function_. The payoff function is assumed to be an m- transformation for each fixed $z \in Z$.

Intuitively, the game is played as follows: First the "opponents," including chance, each pick a strategy; the composite of these strategies is a member of z of Z. Next, our player is informed of the value of $g_1(z)$; this is a member of X_1, and represents our player's state of information for his first move. Our player then chooses a member y_1 of Y_1. Next, he is informed of the value of $g_2(z, y_1)$; this is a member of X_2, and on the basis of this he must choose a member y_2 of Y_2. Next, he is informed of the value of $g_3(z, y_1, y_2)$; the game continues in this way. The payoff is determined as a function of the strategy z chosen by the opponents, and the actions $y_1, y_2, \ldots$ taken by our player. Usually it will be most convenient to take the payoff space H to be a Euclidean space of dimension equal to the number of players. However this need not always be so,* and since we do not use any particular form for H in the sequel, we have left H as general as possible.

Note that up to the present we have not assumed that our player remembers anything on the occasion of a given choice except what he is told by the value of the function g_i. This can be made plausible if we think of the choices of $y_1, y_2, \ldots$ as being made by distinct agents of our player, who are not allowed to communicate with each other.

The mappings g and h have been assumed to be m– transformations in the variables y_i for the familiar reason, namely to ensure that distributions on the domain spaces induce a distribution on the range space. This has not been required for the variable z in order to avoid the necessity of defining a measurable structure on the strategy space Z, which leads to difficulties, as we have seen. The results should thus be conceived as holding for each z separately. In a particular case it might be possible to integrate over some components of z (e.g., that belonging to chance); this can be done without difficulty after the results have been established for fixed z.

The above definition is a compromise between the normal and extensive forms of a game. The game has been retained in extensive form for our player, but has been normalized for the other players. Even for finite games, the theorem on games of perfect recall is best stated for one player at a time; the process of normalizing the game for the other players enables us to focus attention on the single player and thus simplifies the proofs.

Not all finite extensive games in the sense of Kuhn [K_1] (or in the more general sense of Isbell [I]) are included in the above definition; however all games of perfect recall are included, as are all finite extensive games in the sense of von Neumann and Morgenstern [N-M]. The condition for a Kuhn game to be included is that the game can be "serialized," time-wise, for the player in question. For example the game in Figure 1 does not come under our definition if the information sets A and B belong to the same player,

* For instance, for some purposes it is convenient to consider the payoff to a supergame as being simply the sequence of payoffs to the individual plays, rather than the average (in some sense) of these payoffs.

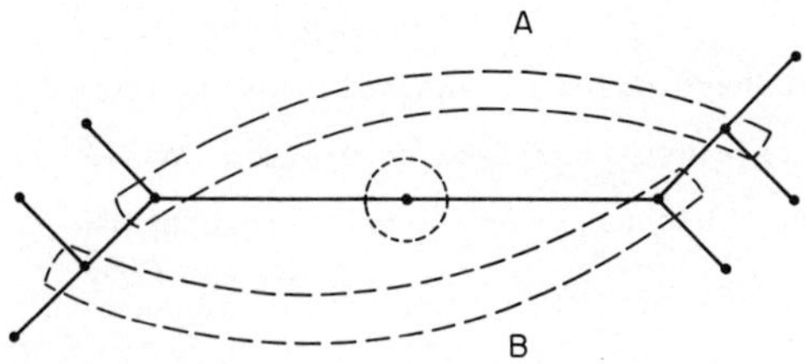

Figure 1

but does come under our definition if they belong to distinct players.* Of course the possibility of serialization is not at all equivalent with perfect recall (but the latter implies the former).

Most extensive-game models used by authors other than those mentioned above are similar to Kuhn's model, and the same remarks apply.

Next, we define games of perfect recall in our model.

DEFINITION. A game is said to be of <u>perfect recall</u> if there are sequences of m— transformations

$$u_j^i : X_i \to Y_j, \quad j < i$$

and

$$t_j^i : X_i \to X_j, \quad j < i$$

such that

$$u_j^i g_i(z, y_1, \ldots, y_{i-1}) = y_j$$

and

$$t_j^i g_i(z, y_1, \ldots, y_{i-1}) = g_j(z, y_1, \ldots, y_{j-1}).$$

Intuitively, u is the function by which a player remembers what he previously did, and t is the function by which he remembers what he

* The example is taken from [K_1].

previously knew.

Note that we have given an analytic definition of games of perfect recall which, while retaining complete generality, avoids the cumbersome geometric tree model. This has been made possible by the device of normalizing the game for all but one player.

§ 5. FORMAL DEFINITIONS OF MIXED AND BEHAVIOR STRATEGIES; KUHN'S THEOREM

We may assume without loss of generality that the X_i and Y_i are all copies of I;* for if one of them is only finite or denumerable, we can always add a continuum of identical copies. The cartesian products $\times_i X_i$ and $\times_i Y_i$ will be denoted $\mathbf{X}$ and $\mathbf{Y}$ respectively,** and their members will be denoted $\mathbf{x} = (x_1, x_2, \ldots)$ and $\mathbf{y} = (y_1, y_2, \ldots)$. We remind the reader that the phrase "sample space" means a copy of Ω. Sample spaces will be denoted by Ω, $\boldsymbol{\Omega}$, Ω_1, etc.; the measures on them by λ, $\boldsymbol{\lambda}$, λ_i, etc., respectively.

In this and the following three sections, the word "subset," when applied to an m– space, will always mean "measurable subset," and $B \subset Y$, when Y is an m– space, will mean "B is an m– subset of Y."

DEFINITION. A <u>mixed strategy</u> is a sequence $\mathbf{m} = (m_1, m_2, \ldots)$ of m– transformations $m_i : \Omega \times X_i \to Y_i$, where Ω is a fixed sample space. A <u>behavior strategy</u> is a mixed strategy $\mathbf{b}$, such that for $i \neq j$, $b_i(\cdot, x_i)$ and $b_j(\cdot, x_j)$ are mutually independent random variables ($x_i \in X_i$ and $x_j \in X_j$ being arbitrary).***

Every triple $(\omega, \mathbf{m}, z)$ consisting of a member of the sample space, a mixed strategy, and a strategy of the opponents uniquely determines a member $\mathbf{v}(\omega; \mathbf{m}, z)$ of $\mathbf{Y}$; $\mathbf{v} = (v_1, v_2, \ldots)$ is defined recursively by

$$v_i = m_i(\omega, g_i(z, v_1, \ldots, v_{i-1})).$$

---

* For the definitions of I and Ω see Section 3.

** The reader should be careful to distinguish between boldface letters $\mathbf{X}$, $\boldsymbol{\Omega}$, $\boldsymbol{\lambda}$, etc., and ordinary letters X, Ω, λ, etc.

*** See Section 9 for a discussion of this definition.

Intuitively, $\mathbf{v}$ is the sequence of choices that actually occur when the game is played. Furthermore every pair $(\mathbf{m}, z)$ uniquely determines a distribution (i.e., measure) μ on $\mathbf{Y}$; this is defined for $\mathbf{B} \subset \mathbf{Y}$ by

$$\mu(\mathbf{B}) = \mu(\mathbf{B}; \mathbf{m}, z) = \lambda\{\omega : \mathbf{v}(\omega; \mathbf{m}, z) \in \mathbf{B}\}.$$

Intuitively, μ is the distribution of the random variable $\mathbf{v}(\cdot; \mathbf{m}, z)$. Two mixed strategies are said to be <u>equivalent</u> if for each $z \in Z$, they determine the same distribution on $\mathbf{Y}$.

We are now ready to state

KUHN'S THEOREM. In a game of perfect recall, every mixed strategy has an equivalent behavior strategy.

§ 6. LEMMAS FOR THE PROOF OF KUHN'S THEOREM

Since we will make extensive use of conditional probabilities where the probability of the condition vanishes, we briefly review the properties of such conditional probabilities. Let Ω be a sample space with measure λ, let Y be a copy of I, and let $v: \Omega \to Y$ be an m– transformation. For arbitrary $\Gamma \subset \Omega$ and $y \in Y$, we are interested in the conditional probability of Γ given that $v(\omega) = y$. Note that the condition $v(\omega) = y$ may well have probability zero, that is, we may have $\lambda\{\omega : v(\omega) = y\} = 0$, possibly even for all y. However, it can be proved [H, pp. 206-209] that for each y there is an essentially unique* probability measure on Ω, denoted** by $\text{cond } \lambda(\cdot \mid v(\omega) = y)$, such that for each $B \subset Y$ and $\Gamma \subset \Omega$, we have

$$\text{(A0)} \qquad \int_B \text{cond } \lambda(\Gamma \mid v(\omega) = y)\, d\lambda v^{-1}(y) = \lambda(v^{-1}(B) \cap \Gamma).$$

---

* Actually cond λ is defined uniquely only up to a set of y which is of (λv^{-1})-measure 0. But all our statements will hold for any particular version of cond λ, so the particular choice can be made arbitrarily.

** We trust that our notation for conditional probabilities, though not standard, is sufficiently transparent as to cause no confusion. There are good reasons for using it rather than one of the standard notations.

Formula (A0) is the analogue of the familiar "partition formula" in elementary probability theory. That $\text{cond}\ \lambda(\cdot \mid v(\omega) = y)$ is a probability measure follows from the fact that Ω is standard [H, p. 210, example 5]; this is the only place where we use standardness.

Let Ω, λ, Y, and v be as above, Y' a copy of I, $g : \Omega \times Y \to Y'$ an m— transformation, $B' \subset Y'$ and $B \subset Y$.

LEMMA A. Under the above conditions

$$\int_B \text{cond}\ \lambda(\{\omega : g(\omega, y) \in B'\} \mid v(\omega) = y) d\lambda\, v^{-1}(y)$$

$$= \lambda\{\omega : g(\omega, v(\omega)) \in B' \text{ and } v(\omega) \in B\}\ .$$

REMARK. The unusual feature of the integral on the left-hand side is that the subset of Ω of which the conditional probability is being taken — the set $\{\omega : g(\omega, y) \in B'\}$ — varies with the condition y. If it were not that $\text{cond}\ \lambda$ is defined essentially uniquely as a probability measure (for example if Ω were not standard), the integral would have no meaning, because $\text{cond}\ \lambda$ could be assigned an arbitrary value for each y. What the lemma says is that since the condition asserts $v(\omega) = y$, we may substitute $v(\omega)$ for y on the left side of the $\mid$ sign, and then obtain the correct answer by using (A0).

PROOF. Let $C \subset \Omega \times Y$ be defined by $C = g^{-1}(B')$. Denoting by C^y the section $\{\omega : (\omega, y) \in C\}$, we obtain that Lemma A is equivalent to

$$\text{(A1)} \qquad \int_B \text{cond}\ \lambda(C^y \mid v(\omega) = y)\, d\lambda\, v^{-1}(y) = \lambda\{\omega : (\omega, v(\omega)) \in C \text{ and } v(\omega) \in B\}.$$

Both sides of (A1), as functions of C, are measures on $\Omega \times Y$ (since $\text{cond}\ \lambda$ is a measure on Ω for each y). Hence it is sufficient to prove (A1) when C is a rectangle $\Gamma \times A$ in $\Omega \times Y$. In this case the left side of (A1) becomes

$$\int_{A\cap B} \text{cond}\ \lambda(\Gamma \mid v(\omega) = y)\, d\lambda v^{-1}(y)$$

$$= \lambda\{\Gamma \cap v^{-1}(A \cap B)\}$$

$$= \lambda(\omega : \omega \in \Gamma \text{ and } v(\omega) \in A \text{ and } v(\omega) \in B\}$$

$$= \lambda\{\omega : (\omega, v(\omega)) \in \Gamma \times A \text{ and } v(\omega) \in B\}$$

$$= \lambda\{\omega : (\omega, v(\omega)) \in C \text{ and } v(\omega) \in B\} .$$

This demonstrates (A1) when C is a rectangle, and (A1) and therefore also Lemma A follows in the general case.

Now let us return to our game. First we introduce some further notation. We write $\mathbf{Y}_i = Y_1 \times \cdots \times Y_i$. Similarly, for $\mathbf{y} \in \mathbf{Y}$, we write $\mathbf{y}_i = (y_1, \ldots, y_i)$. If $B_1 \subset Y_1$, $B_2 \subset Y_2$, ... , then we write $\mathbf{B}_i = B_1 \times \cdots \times B_i$, and $\mathbf{B} = B_1 \times B_2 \times \cdots$. The symbol $\mathbf{B}$ will always be reserved for a rectangle of this kind.

Let us consider a mixed strategy $\mathbf{m}$ with sample space Ω, a strategy z of the opponents, and a sequence $\mathbf{y} \in \mathbf{Y}$. Then for each $i = 1, 2, \ldots$ we may define a sequence $\mathbf{v}^i = (v_1^i, v_2^i, \ldots) = \mathbf{v}^i(\omega, \mathbf{y}; \mathbf{m}, z)$ inductively as follows:

$$v_j^i = \begin{cases} y_j, & \text{for } j < i \\ m_j(\omega, g_j(z, \mathbf{v}_{j-1}^i)), & \text{for } j \geq i. \end{cases}$$

We have $\mathbf{v}^1 = \mathbf{v}$, and $v_i^i = m_i(\omega, g_i(z, \mathbf{y}_{i-1}))$, which is the decision on the ith play if $\mathbf{y}_{i-1}$ has been chosen on the previous plays. Denote $(v_i^i, \ldots, v_k^i)$ by $\hat{v}_k^i$ (for $k \geq i$). Note that v_j^i depends only on $\mathbf{y}_{i-1}$ rather than on all of $\mathbf{y}$, so we may write $v_j^i(\omega, \mathbf{y}_{i-1}; \mathbf{m}, z)$ rather than $v_j^i(\omega, \mathbf{y}; \mathbf{m}, z)$. As $\mathbf{m}$ and z will be fixed throughout most of this discussion, we will usually write $v_j^i(\omega, \mathbf{y}_{i-1})$, and omit explicit mention of $\mathbf{m}$ and z. The expression $v_j^i(\cdot, \mathbf{y}_{i-1})^{-1}(B_j)$ means $\{\omega : v_j^i(\omega, \mathbf{y}_{i-1}) \in B_j\}$. For future reference note that

$$\text{(B1)} \qquad v_j^{i+1}(\omega, (\mathbf{y}_{i-1}, v_i^i(\omega, \mathbf{y}_{i-1}))) = v_j^i(\omega, \mathbf{y}_{i-1}) .$$

Next, remembering that $\mathbf{y}$ is fixed, define a sequence $\lambda, \lambda_{y_1}, \lambda_{y_2}, \ldots$ of measures on Ω as follows:

$$\lambda_{\mathbf{y}_i}(\Gamma) = \text{cond } \lambda_{\mathbf{y}_{i-1}}(\Gamma \mid v_i^i(\omega, \mathbf{y}_{i-1}) = y_i)$$

(where of course $\lambda_{\mathbf{y}_0}$ stands for λ).

LEMMA B. Let $B_i \subset Y_i, \ldots, B_k \subset Y_k$. Then

$$\int_{B_i} \cdots \int_{B_k} d\lambda_{\mathbf{y}_{k-1}} \, v_k^k(\cdot, \mathbf{y}_{k-1})^{-1}(y_k) \ldots d\lambda_{\mathbf{y}_{i-1}} \, v_i^i(\cdot, \mathbf{y}_{i-1})^{-1}(y_i)$$

$$= \lambda_{\mathbf{y}_{i-1}} \, \hat{v}_k^i(\cdot, \mathbf{y}_{i-1})^{-1}(\hat{B}_k^i),$$

where $\hat{B}_k^i = B_i \times \ldots \times B_k$.

PROOF. We use reverse induction on i. The start, at $i = k$, is immediate. For the inductive step $(i + 1$ implies $i)$ we have

$$\int_{B_i} \int_{B_{i+1}} \cdots \int_{B_k} = \int_{B_i} \lambda_{\mathbf{y}_i} \, \hat{v}_k^{i+1}(\cdot, \mathbf{y}_i)^{-1}(\hat{B}_k^{i+1}) \, d\lambda_{\mathbf{y}_{i-1}} \, v_i^i(\cdot, \mathbf{y}_{i-1})^{-1}(y_i)$$

$$= \int_{B_i} \text{cond } \lambda_{\mathbf{y}_{i-1}} (\{\omega : \hat{v}_k^{i+1}(\omega, (\mathbf{y}_{i-1}, y_i)) \in \hat{B}_k^{i+1}\} \mid$$

$$v_i^i(\omega, \mathbf{y}_{i-1}) = y_i) \, d\lambda_{\mathbf{y}_{i-1}} v_i^i(\cdot, \mathbf{y}_{i-1})^{-1}(y_i) .$$

Applying Lemma A with $\lambda_{\mathbf{y}_{i-1}}$ instead of λ, B_i instead of B, Y_i instead of Y, $\hat{B}_k^{i+1}$ instead of B', $\hat{v}_k^{i+1}(\cdot, (\mathbf{y}_{i-1}, \cdot))$ instead of g, $\hat{Y}_k^{i+1}$ instead of Y', and $v_i^i(\cdot, \mathbf{y}_{i-1})$ instead of v, we obtain that the last expression above is equal to

$$\lambda_{\mathbf{y}_{i-1}}\{\omega : \hat{v}_k^{i+1}(\omega, (\mathbf{y}_{i-1}, v_i^i(\omega, \mathbf{y}_{i-1}))) \in \hat{B}_k^{i+1} \text{ and } v_i^i(\omega, \mathbf{y}_{i-1}) \in B_i\},$$

and from (B1) we deduce that this is equal to

$$\lambda_{\mathbf{y}_{i-1}}\{\omega : \hat{v}_k^i(\omega, \mathbf{y}_{i-1}) \in \hat{B}_k^i\} .$$

This completes the induction.

COROLLARY C. Let $B_1 \subset Y_1, \ldots, B_k \subset Y_k$. Then

$$\int_{B_1} \cdots \int_{B_k} d\lambda_{\mathbf{y}_{k-1}} \; v_k^k(\cdot, \mathbf{y}_{k-1})^{-1}(y_k) \ldots d\lambda \; v_1^{-1}(y_1) = \lambda \mathbf{v}_k^{-1}(\mathbf{B}_k).$$

COROLLARY D. Let f be an $m-$ transformation from Y_k to the real numbers. Then

$$\int_{B_1} \cdots \int_{B_k} f(\mathbf{y}_k) \; d\lambda_{\mathbf{y}_{k-1}} \; v_k^k(\cdot, \mathbf{y}_{k-1})^{-1}(y_k) \ldots d\lambda \; v_1^{-1}(y_1)$$

$$= \int_{\mathbf{B}_k} f(\mathbf{y}_k) \; d\lambda \; \mathbf{v}_k^{-1}(\mathbf{y}_k) .$$

PROOF. If f is the characteristic function of a rectangular parallelepiped in $\mathbf{Y}_k$, this follows from Corollary C. The general case follows by the usual methods.

§ 7. FURTHER LEMMAS

The object of this section is to prove that a family of distributions can be "inverted" to yield a family of random variables; the precise statement is Lemma F below. We shall need the following lemma, which says that a single distribution can be inverted to yield a single random variable.

LEMMA E. Let f be a non-decreasing upper semi-continuous function* on I such that $f(0) \geq 0$ and $f(1) = 1$. For $0 \leq y \leq 1$ define

$$f^{-1}(y) = \begin{cases} \sup \{x : f(x) \leq y\}, & \text{if } \{x : f(x) \leq y\}, \text{ is non-empty} \\ 0 & \text{if it is empty.} \end{cases}$$

Then

* I.e., $f(x) = \limsup_{y \to x} f(y) = \lim_{y \to x+} f(y)$.

(1) f^{-1} is non-decreasing

(2) f^{-1} is upper-semi-continuous

(3) $f^{-1}(0) \geq 0$, $f^{-1}(1) = 1$

(4) $(f^{-1})^{-1} = f$.

The proof is straightforward, and will be omitted.

Now let X and Y be copies of I, and let $\mathcal{B}$ be the σ-ring of m— sets in Y. Let $\beta : \mathcal{B} \times X \to \Omega$ be a function which is measurable in X for each fixed $B \in \mathcal{B}$ and a probability in $\mathcal{B}$ for each fixed $x \in X$.

LEMMA F. Under the above conditions, there is a family of random variables whose distributions are given by β; more precisely, there is an m— transformation $b : \Omega \times X \to Y$ such that

$$\lambda\{\omega : b(\omega, x) \in B\} = \beta(B, x)$$

for each $x \in X$ and $B \in \mathcal{B}$.

PROOF. For $y \in Y$, define

$$\pi(x, y) = \beta([0, y], x).$$

Write $\pi_x = \pi(x, \cdot)$; π_x is a non-decreasing upper-semi-continuous function of y, so by Lemma E it has a well-defined inverse, which we denote b_x; set $b(\omega, x) = b_x(\omega)$.

LEMMA F1. $b(\omega, \cdot)$ is Borel measurable in x for each fixed ω.

PROOF. For $B \in \mathcal{B}$ we must show that $\{x : b(\omega, x) \in B\}$ is measurable in X. It is sufficient to show this when B is of the form $[0, y_0)$. Now

$$\begin{aligned}\{x : b(\omega, x) \in [0, y_0)\} &= \{x : \sup\{y : \pi(x,y) \leq \omega\} < y_0\} \\ &= \{x : \exists \text{ rational } r < y_0 \text{ such that } \pi(x,r) > \omega\} \\ &= \bigcup_{r<y_0} \{x : \pi(x,r) > \omega\} \\ &= \bigcup_{r<y_0} \{x : \beta([0, r], x) > \omega\} \\ &= \text{union of Borel sets} = \text{a Borel set.}\end{aligned}$$

This completes the proof of Lemma F1.

Next we show that b is measurable in the two variables simultaneously. It is sufficient to prove that sets of the form $[y_0, 1]$ have measurable inverse images. Indeed,

$$\begin{aligned}b^{-1}[y_0, 1] &= \{(\omega, x) : b(\omega, x) \geq y_0\} \\ &= \{(\omega, x) : (\forall \text{ rational } s), (s > \omega \Rightarrow b(s, x) \geq y_0)\} \\ &\qquad \text{(because of upper semi-continuity of } b) \\ &= \cap_s \{(\omega, x) : (b(s, x) \geq y_0) \text{ or } (s \leq \omega)\} \\ &= \cap_s \big(\{(\omega, x) : b(s, x) \geq y_0)\} \cup \{(\omega, x) : (s \leq \omega)\}\big) \\ &= \cap_s \big((\Omega \times \{x : b(s, x) \geq y_0\}) \cup ([s, 1] \times X)\big),\end{aligned}$$

and this is Borel measurable in $\Omega \times X$ (by Lemma F1)

Finally, we show that

$$\lambda\{\omega : b(\omega, x) \in B\} = \beta(B, x).$$

It is sufficient to demonstrate this when B is of the form $[0, y]$. Then

$$\lambda\{\omega : b(\omega, x) \in [0, y]\} = \lambda\{\omega : b(\omega, x) \leq y\} = \sup\{\omega : b_x(\omega) \leq y\} = b_x^{-1}(y).$$

But $b_x = \pi_x^{-1}$; so $b_x^{-1} = \pi_x$ (by Lemma E). Hence

$$\lambda\{\omega : b(\omega, x) \in [0, y]\} = b_x^{-1}(y) = \pi_x(y) = \pi(x, y) = \beta([0, y], x),$$

and the proof of Lemma F is complete.

§ 8. PROOF OF KUHN'S THEOREM

Fix $\mathfrak{m}$; we wish to find an equivalent behavior strategy, which we will call $\mathfrak{b}$. We first define the distributions β_i of the random variables $b_i(\cdot, x)$, and then only the random variables themselves. For $B \subset Y_i$ and $x \in X_i$, define

$$\beta_i(B, x) = \text{cond}\ \lambda\left(\{\omega : m_i(\omega, x) \in B\} \,|\, m_{i-1}(\omega, t^i_{i-1}(x)) = u^i_{i-1}(x) \,| \right.$$
$$\left. \ldots \,|\, m_1(\omega, t^i_1(x)) = u^i_1(x)\right).$$

The expression on the right is to be interpreted as an iterated conditional probability, similar to the definition of $\lambda_{\mathbf{y}_i}$. To underscore the similarity, note that

$$\text{(K1)} \qquad \beta_i(B, g_i(z, \mathbf{y}_{i-1})) = \lambda_{\mathbf{y}_{i-1}}\ v^i_i(\cdot, \mathbf{y}_{i-1}; \mathfrak{m}, z)^{-1}(B).$$

Let $\Omega_1, \Omega_2, \ldots$ be a sequence of copies of Ω. According to Lemma F we can find $b'_i: \Omega_i \times X_i \to Y_i$ so that the $\beta_i(\cdot, x)$ are the distributions of the $b'_i(\cdot, x)$, i.e., so that

$$\text{(K2)} \qquad \lambda_i\{\omega_i : b'_i(\omega_i, x) \in B\} = \beta_i(B, x).$$

Let $\boldsymbol{\Omega} = \Omega_1 \times \Omega_2 \times \ldots$, and note that $\boldsymbol{\Omega}$ is a copy of Ω; define a behavior strategy $b_i: \boldsymbol{\Omega} \times X_i \to Y_i$ by $b_i(\boldsymbol{\omega}, X_i) = b'_i(\omega_i, X_i)$, where $\boldsymbol{\omega} = (\omega_1, \omega_2, \ldots)$. Let $B_1 \subset Y_1$, $B_2 \subset Y_2$, ... ; for each n write $\mathbf{B}^*_n = \mathbf{B}_n \times Y_{n+1} \times \ldots$. To show that $\mathfrak{m}$ and $\mathfrak{b}$ are equivalent, it is only necessary to show that

$$\text{(K3)} \qquad \mu(\mathbf{B}^*_n; \mathfrak{m}, z) = \mu(\mathbf{B}^*_n; \mathfrak{b}, z)$$

for every $z \in Z$ and every n and arbitrary choice of the B_i.

Let us write $\boldsymbol{\omega}_n$ for $(\omega_1, \ldots, \omega_n)$. We first note that

$\mathbf{v}_n(\boldsymbol{\omega};\ \mathbf{b},\ z)$ depends only on $\boldsymbol{\omega}_n$ rather than on all of $\boldsymbol{\omega}$. In fact, if we define $\mathbf{w}_n(\boldsymbol{\omega}_n)$ recursively by

$$w_n(\boldsymbol{\omega}_n) = b_n'(\omega_n,\ g_n(z,\ \mathbf{w}_{n-1}(\boldsymbol{\omega}_{n-1}))),$$

and $\mathbf{w}_n = (w_1,\ \ldots,\ w_n)$, then

$$\text{(K4)} \qquad w_n(\boldsymbol{\omega}_n) = v_n(\boldsymbol{\omega};\ \mathbf{b},\ z).$$

Henceforth we will use $\mathbf{v}$ for $\mathbf{m}$ exclusively (unless we explicitly indicate otherwise); thus $\mathbf{v}(\omega)$ will mean $\mathbf{v}(\omega;\ \mathbf{m},\ z)$, and similarly for v_j^i, etc.

The proof of (K3) is by induction on n; the induction is easily started (at $n = 1$). For the inductive step (n implies $n + 1$) note that

$$\boldsymbol{\lambda}_n \mathbf{w}_n^{-1}(\mathbf{B}_n) = \mu(\mathbf{B}_n^*;\ \mathbf{b},\ z)$$

because of (K4), where $\boldsymbol{\lambda}_n = \lambda_1 \times \ldots \times \lambda_n$; furthermore

$$\lambda \mathbf{v}_n^{-1}(\mathbf{B}_n) = \mu(\mathbf{B}_n^*;\ \mathbf{m},\ z).$$

Since by induction hypothesis the two right sides are equal, it follows that the left sides also are; but since this holds for all $\mathbf{B}_n \subset \mathbf{Y}_n$, it follows that

$$\text{(K5)} \qquad \boldsymbol{\lambda}_n \mathbf{w}_n^{-1} = \lambda \mathbf{v}_n^{-1}$$

as measures on $\mathbf{Y}_n$. Next, we have

$$\begin{aligned}
\mu(B_{n+1}^*;\ \mathbf{b},\ z) &= \boldsymbol{\lambda}\{\boldsymbol{\omega} : \mathbf{w}_{n+1}(\boldsymbol{\omega}_{n+1}) \in \mathbf{B}_{n+1}\} \\
&= \boldsymbol{\lambda}_{n+1}\{\boldsymbol{\omega}_{n+1} : w_{n+1}(\boldsymbol{\omega}_{n+1}) \in B_{n+1} \text{ and } \mathbf{w}_n(\omega_n) \in \mathbf{B}_n\} \\
&= \boldsymbol{\lambda}_{n+1}\{\boldsymbol{\omega}_{n+1} : b_{n+1}'(\omega_{n+1},\ g_{n+1}(z,\ \mathbf{w}_n(\omega_n))) \in B_{n+1} \text{ and } \mathbf{w}_n(\boldsymbol{\omega}_n) \in \mathbf{B}_n\} \\
&= \int_{\mathbf{w}_n^{-1}(\mathbf{B}_n)} \lambda_{n+1}\{\omega_{n+1} : b_{n+1}'(\omega_{n+1},\ g_{n+1}(z,\ \mathbf{w}_n(\boldsymbol{\omega}_n))) \in B_{n+1}\}\, d\boldsymbol{\lambda}_n(\boldsymbol{\omega}_n)
\end{aligned}$$

$$= \int_{\mathbf{B}_n} \beta_{n+1}(B_{n+1}, g_{n+1}(z, \mathbf{y}_n))\, d\boldsymbol{\lambda}_n \mathbf{w}_n^{-1}(\mathbf{y}_n),$$

(because of (K2) and the change of variables $\mathbf{y}_n = \mathbf{w}_n(\boldsymbol{\omega}_n)$)

$$= \int_{\mathbf{B}_n} \lambda_{\mathbf{y}_n} v_{n+1}^{n+1}(\cdot, \mathbf{y}_n)^{-1}(B_{n+1})\, d\lambda \mathbf{v}_n^{-1}(\mathbf{y}_n)$$

(because of (K1) and (K5))

$$= \int_{B_1} \cdots \int_{B_n} \lambda_{\mathbf{y}_n} v_{n+1}^{n+1}(\cdot, \mathbf{y}_n)^{-1}(B_{n+1}) d\lambda_{\mathbf{y}_{n-1}} v_n^n(\cdot, \mathbf{y}_{n-1})^{-1}(y_n) \cdots d\lambda v_1^{-1}(y_1)$$

(because of Corollary D)

$$= \int_{B_1} \cdots \int_{B_n} \int_{B_{n+1}} d\lambda_{\mathbf{y}_n} v_{n+1}^{n+1}(\cdot, \mathbf{y}_n)^{-1}(y_{n+1}) \cdots d\lambda v_1^{-1}(y_1)$$

$$= \lambda \mathbf{v}_{n+1}^{-1}(\mathbf{B}_{n+1})$$

(because of Corollary C)

$$= \mu(B_{n+1}^*; \mathrm{m}, z).$$

This completes the proof of Kuhn's theorem.

§ 9. REMARKS ON THE DEFINITION OF BEHAVIOR STRATEGY

Intuitively, a player using a behavior strategy "postpones crossing his bridges until he gets to them;" that is, he randomizes independently on each occasion of a choice, rather than letting his choices be governed by a single randomization performed before the start of play. Thus a behavior strategy is a family of independent random procedures for choosing an action, one for each possible information state. The direct translation of this intuitive concept into technical language yields a sequence of functions $b_i : \Omega \times X_i \to Y_i$, where the functions $b_i(\cdot, x)$ are mutually independent random variables for distinct x, even when we are dealing with a single i. This is a little different from the definition of Section 5, in which it was demanded only that the $b_i(\cdot, x)$ be independent for distinct i; that is, the player randomizes independently at each stage, but before the information for that stage is received, rather than afterwards.

To demand that the $b_i(\cdot, x)$ be independent for distinct x — even for a fixed i — would mean that we must have a non-denumerable number of mutually independent bounded random variables on the same sample space;

and except for the degenerate case in which almost all of these are constants, this is in fact impossible (when the phrase "sample space" is used in our restricted sense, which corresponds to the intuitive idea of random device). Indeed, suppose $\{b_x\}$ is a nondenumerable family of bounded nonconstant mutually independent random variables on the sample space Ω, and let $c_x = b_x - \text{mean}\,(b_x)$. Then from independence it follows that the c_x are mutually orthogonal in the Hilbert space $L^2(\Omega)$, and from the fact that the b_x are nonconstant it follows that the c_x do not vanish identically. So we have a nondenumerable number of nonzero mutually orthogonal members of $L^2(\Omega)$, and hence $L^2(\Omega)$ has nondenumerable dimension. But its dimension is known to be denumerable, so our contention is proved.

It may seem that this makes any genuine analogue of Kuhn's theorem in the continuous case impossible. The difference, however, is illusory, and there is no real loss of strength in our theorem. We have seen that the $b_i(\cdot, x)$ must necessarily be correlated as x ranges over X_i, simply because of the cardinality of X_i. However, this correlation is entirely irrelevant to the game, and cannot affect the payoff in any way. In fact, <u>the payoff distribution depends only on the distributions of the individual $b_i(\cdot, x)$, and not on any of the joint distributions</u> (this follows from Section 8). In other words, the $b_i(\cdot, x)$ are correlated (for fixed i and varying x) not because this correlation is necessary to mimic the effect of the given mixed strategy m; in fact, if this were necessary (as it may be when the game is not of perfect recall), this kind of in-stage correlation could not accomplish it,* and we would have to resort to interstage correlation. The correlation is rather in the way of being an irrelevant mathematical accident.

Yet another way of saying this is that as long as they have the proper distributions, the $b_i(\cdot, x)$ can be chosen in any way we please, without any regard to each other, except that in the end b_i must be simultaneously measurable in both ω and x. Though the $b_i(\cdot, x)$ must be correlated, what form the correlation takes is of no concern to us.

Finally, we remark that it would have been possible to define behavior strategies as functions that take members of X_i into distributions

---

* Cf. the underlined statement above.

over Y_i, in a manner directly analogous to that of $[K_1]$. With that approach, the independence assumptions would have been implicit in the formula for the payoff to a behavior strategy, and the question under discussion would not have arisen at all. We prefer our approach because it underscores the natural relation between mixed and behavior strategies.

BIBLIOGRAPHY

$[A_1]$ AUMANN, R. J., "Acceptable points in general cooperative n-person games," Annals of Mathematics Study No. 40 (Princeton 1959) pp. 287-324.

$[A_2]$ AUMANN, R. J., "The core of a cooperative game without side payments," Trans. Amer. Math. Soc. 98 (1961), pp. 539-552.

$[A_3]$ AUMANN, R. J., "Borel structures for function spaces," Illinois J. of Math. 5 (1961), pp. 614-630.

$[A_4]$ AUMANN, R. J., "On choosing a function at random," Ergodic Theory, Academic Press, 1963, pp. 1-20.

[F] FOX, M., "Some zero sum two-person games with moves in the unit interval," Pac. J. Math. 10 (1960), pp. 1235-1242.

[H] HALMOS, P. R., Measure Theory, Van Nostrand, 1950.

[I] ISBELL, J. R., "Finitary games," Annals of Mathematics Study No. 39 (Princeton, 1957), pp. 79-96.

$[K_1]$ KUHN, H. W., "Extensive games and the problem of information," Annals of Mathematics Study No. 28 (Princeton, 1953), pp. 193-216.

$[K_2]$ KUHN, H. W., Lectures on the Theory of Games, Princeton University Lecture Notes, 1952 (mimeographed).

[N-M] VON NEUMAN, J. and MORGENSTERN, O., Theory of Games and Economic Behavior, Princeton University Press, 1944, Third Edition, 1953.

[M] MACKEY, G. W., "Borel structures in groups and their duals," Trans. Amer. Math. Soc. 85 (1957), pp. 283-295.

[Mc] MCKINSEY, J.C.C., Introduction to the Theory of Games, McGraw-Hill, 1952.

[S] SIERPINSKI, W., Hypothèse du continu, Chelsea, 1956.

[V] VILLE, J., "Sur la théorie générale des jeux où intervient l'habilité des joueurs," note in Traité du Calcul des Probabilites et de ses Applications by E. Borel, Tome IV, Fascicule II, "Applications aux jeux de hasard," (1938), pp. 105-113.

[W] WOLFE, P., editor, Report of an informal conference on recent developments in the theory of games, Princeton University, 1955 (mimeographed).

Robert J. Aumann

The Hebrew University of Jerusalem

A GENERAL SOLUTION FOR FINITE NON-COOPERATIVE GAMES, BASED ON RISK-DOMINANCE*

John C. Harsanyi

§ 1. INTRODUCTION

By a non-cooperative game Γ we shall mean a game where the players cannot conclude <u>enforceable agreements</u> so that only equilibrium points in Nash's sense [7] can represent stable situations[1] — irrespective of whether inter-player <u>communication</u> is permitted (<u>vocal</u> games) or is not (<u>tacit</u> games). We shall also distinguish <u>semi-vocal</u> games where the players can tell one another when they are ready to make a <u>concession</u>, but cannot communicate sufficiently to allow <u>coordination</u> of their strategies when this would be desirable.

We shall define a solution concept, called <u>V-solution</u>, which in general can be used only in fully <u>vocal</u> games. We shall also introduce another solution concept, called <u>T-solution</u>, which will be defined in the first instance for <u>semi-vocal</u> games. Once defined, however, it can be used also in <u>tacit</u> games (this is why it is called T-solution). Our concept of semi-vocal games is in fact only an auxiliary device to obtain a solution usable also in <u>tacit</u> games.

An interesting solution concept for non-cooperative games has already been developed by Nash [7]. But it exists only in very special cases. The V and T-solutions, which we shall define, always <u>exist</u>. Like the Nash solution when it exists, they are also always <u>unique</u>. Basically they are defined

---

* The original version of this paper was written while the author was at the Research School of Social Sciences, Australian National University, Canberra, Australia.

1 Except that in some very special cases (which in the space of all finite non-cooperative games represent only a set of <u>measure zero</u>), rational players, as we shall argue, may have to fall back on using <u>maximin</u> strategies even if these are <u>not</u> equilibrium strategies.

by first eliminating from the game all equilibrium points whose payoff vectors are dominated by some other equilibrium-point payoff vector; and then by introducing a new type of dominance relation, called risk-dominance, among the remaining equilibrium points.

§ 2. DEFINITIONS AND PRELIMINARY CONCLUSIONS

In this section we shall suggest a few informal preliminary Conclusions, based on heuristic considerations, about rational behavior in non-cooperative games.

The set of all n players will be denoted by N.

Let $s = (s_1, \ldots, s_n)$ be a strategy n-tuple for players $1, \ldots, n$. Then by s^i we denote the $(n-1)$-tuple that remains if the ith component s_i is omitted from s. We shall write $s = (s_i, s^i)$. If strategy s_i is a component of the n-tuple s then we shall say that s_i is used in s.

U_i denotes the payoff function for player i while U stands for the vector $U = (U_1, \ldots, U_n)$.

The maximin payoff of player i is defined as

$$(2.1) \qquad \check{U}_i = \max_{s_i \in S_i} \min_{t^i \in S^i} U_i(s_i, t^i),$$

where S_i is the set of all strategies available to player i whereas $S^i = S^1 \times \ldots \times S^{i-1} \times S^{i+1} \times \ldots \times S^n$ is the set of all strategy $(n-1)$-tuples available to the other $(n-1)$ players. The strategy $\check{s}_i$ is a maximin strategy if

$$(2.2) \qquad \min_{t^i \in S^i} U_i(\check{s}_i, t^i) = \check{U}_i.$$

If all components $\check{s}_1, \ldots, \check{s}_n$ of a given n-tuple $\check{s}$ are maximin strategies then $\check{s}$ is called a maximin point.

An n-tuple s is an equilibrium point in Nash's sense (to be abbreviated as EP, and in the plural as EPS) if, for every player i, the strategy s_i is a best reply against the other players' strategies in the $(n-1)$-tuple s^i; that is, if

(2.3) $U_i(s_i, s^i) \geqq U_i(t_i, s^i),$ all $t_i \in S_i$, $i \in N$.

Any strategy s_i which is used in some EP $s = (s_i, s^i)$ is called an equilibrium strategy.

CONCLUSION I. Let $\bar{U}_i$ be a payoff higher than player i's maximin payoff $\check{U}_i$. Then, in a non-cooperative game, by using a certain strategy s_i player i can expect to achieve payoff $\bar{U}_i$ only if s_i yields $\bar{U}_i$ at some equilibrium point $s = (s_i, s^i)$ so that $U_i(s_i, s^i) = \bar{U}_i$.

EXPLANATION. If s is not an EP then some player(s) j will always have an incentive to use a strategy t_j not belonging to s, even if all the other players k did use strategies s_k belonging to s. Therefore no player can rationally act on the expectation that all players will simultaneously use strategies belonging to s.[2]

CONCLUSION II. Let $\check{U}_i$ be player i's maximin payoff. Then player i can be sure to obtain at least payoff $\check{U}_i$ if he uses any maximin strategy $\check{s}_i$, even if $\check{s}_i$ is not an equilibrium strategy.

EXPLANATION. This follows from the definition of maximin strategies. To obtain a payoff higher than his maximin payoff, a player needs cooperation by the other players, and in a non-cooperative game stable cooperation among the players is possible only at an EP. But, to obtain merely his maximin payoff, a player does not need cooperation by the other players, and so he need not restrict himself to strategies leading to EPS.

[2] This conclusion retains its validity even in the case where condition (2.3) is satisfied for all players $j \neq i$ and fails to be satisfied only for player i himself. In this case, also, player i cannot assume that these players j will stick to their strategies s_j because they will now act on the expectation that at least player i himself will depart from strategy s_i (even if in fact this is not his intention) since s_i is not a best reply to the strategy (n–1)-tuple s^i.

CONCLUSION III. Suppose that in a given non-cooperative game Γ there is no equilibrium point s where player i's payoff $U_i(s)$ exceeds his maximin payoff $\check{U}_i$. Then it is more rational for him to use a <u>maximin</u> <u>strategy</u> $\check{s}_i$ (even if $\check{s}_i$ is <u>not</u> an equilibrium strategy) than it is to use an <u>equilibrium</u> <u>strategy</u> s_i (if s_i is <u>not</u> a maximin strategy).

EXPLANATION. In game Γ player i cannot rationally expect to obtain a payoff <u>higher</u> than $\check{U}_i$, whether he uses strategy $\check{s}_i$ or s_i or any other strategy. But if he uses $\check{s}_i$ he can be <u>sure</u> to obtain at least $\check{U}_i$ while if he uses s_i he runs the risk of obtaining less than $\check{U}_i$.

For instance, consider the game:

	β_1	β_2
α_1	(3,1)	(0,2)
α_2	(1,4)	(2,0)

EXAMPLE 1.

This game has only one EP $s = (s_1, s_2)$ with the equilibrium strategies $s_1 = \frac{4}{5}\alpha_1 + \frac{1}{5}\alpha_2$ and $s_2 = \frac{1}{2}\beta_1 + \frac{1}{2}\beta_2$ for players 1 and 2. This EP yields the two players only their maximin payoffs $\check{U}_1 = \frac{3}{2}$ and $\check{U}_2 = \frac{8}{5}$, which they can also obtain by using their maximin strategies $\check{s}_1 = \frac{1}{4}\alpha_1 + \frac{3}{4}\alpha_2$ and $\check{s}_2 = \frac{2}{5}\beta_1 + \frac{3}{5}\beta_2$ respectively. But if e.g., player 1 uses strategy $\check{s}_1$ then he can be <u>sure</u> to obtain $\check{U}_1 = \frac{3}{2}$ whereas if he used strategy s_1 then he <u>could</u> possibly obtain as little as $\frac{2}{5}$. Of course his payoff can fall below $\frac{3}{2}$ only if the other player fails to use the equilibrium strategy s_2. (For example, it would fall to $\frac{2}{5}$ only if player 2 used strategy β_2.) However, if player 1 plays strategy s_1 then in fact player 2 will have no incentive to stick to strategy s_2 because any other strategy will also yield him exactly the same payoff, viz. $\frac{8}{5}$.

Let $\overline{\Gamma}$ be the set of all EPS in a given game Γ, and let A be a nonempty subset of $\bar{\Gamma}$ (possibly $\bar{\Gamma}$ itself). Let $A(s_i)$ be the set of all EPS t in A such that $t_i = s_i$, where s_i is some given strategy of player i. Then we define player i's safe payoff within this set A from strategy s_i [or equivalently from any given EP t belonging to $A(s_i)$] as the quantity

$$U_i^*(s_i; A) = U_i^*(t; A) = \min_{t \in A(s_i)} U_i(t). \tag{2.4}$$

The set A of EPS for which this safe payoff is defined will be called the admissible set.

The n-vector whose i^{th} component is $U_i^*(t; A)$ for $i = 1, \ldots, n$ will be denoted by $U^*(t; A)$ and will be called the safe payoff vector associated with EP t within the admissible set A. In contrast to the safe payoff $U_i^*(t; A)$, the payoff $U_i(t)$ that the original payoff function U_i of the game assigns to EP t will be called player i's definitional payoff from t.

CONCLUSION IV. Let A be a set of EPS in a given game Γ. Suppose that all EPS t in A are admissible in the sense that the (n–1) players other than player i have no independent reasons of their own for avoiding any EP t in A. Then player i cannot count on any payoff higher than his safe payoff $U_i^*(s_i; A)$ when he uses strategy s_i.

EXPLANATION. For instance, let both $s = (s_i, s^i)$ and $t = (s_i, t^i)$ be in A, with $U_i(s) > U_i(t)$. But suppose that, apart from i, the other players have no reason to prefer EP s to EP t. Then player i cannot act on the expectation that if he himself chooses strategy s_i then the other players j will choose strategies s_j belonging to s^i, rather than choosing strategies t_j belonging to t^i. If the game were a cooperative game then of course player i might be able to reach a binding agreement with the other players to the effect that if he himself used s_i then the other players j would use the strategies s_j. But in a non-cooperative game player i cannot obtain a binding commitment of this sort from the other players.

Consider, for example, the game:

	β_1	β_2
α_1	(1,1)	(2,1)
α_2	(1,2)	(2,2)

EXAMPLE 2.

Here both players would be better off if they could enter into a binding agreement to use strategies α_2 and β_2 respectively. But in a non-cooperative game such agreement would have no force. Yet, in the absence of such an agreement player 1 will have no incentive to use strategy α_2 rather than α_1 because, whatever strategy his opponent uses, his choice between α_1 and α_2 will make no difference to his own payoff. Likewise, player 2's choice between β_1 and β_2 will make no difference to player 2's payoff. Hence e.g., player 1, whether he uses strategy α_1 or α_2, cannot _count_ on a payoff larger than 1.

A strategy n-tuple r is a _recombination_ of EPS s and t if, for every player i, _either_ $r_i = s_i$ _or_ $r_i = t_i$. Two EPS s and t are _interchangeable_ if every possible recombination r of s and t is also an EP.

A set Ψ of EPS is called by Nash [7] a _sub-solution_ if any pair of EPS s and t in Ψ are interchangeable, provided that Ψ is a maximal set with respect to this property. If all EPS of a game are pairwise interchangeable then the sub-solution Ψ comprising all EPS of the game is called a _Nash solution_ (ibid.).

For any sub-solution or Nash solution Ψ, we define the _upper value_ $V_i^*(\Psi)$ and the _lower value_ $V_i^-(\Psi)$ of sub-solution or Nash solution Ψ to player i as

$$(2.5) \qquad V_i^+(\Psi) = \max_{s \in \Psi} U_i(s), \qquad V_i^-(\Psi) = \min_{s \in \Psi} U_i(s).$$

LEMMA. For any sub-solution Ψ and for any $s = (s_i, s^i) \in \Psi$, player i's _safe payoff_ within Ψ from strategy s_i equals the _lower value_ of Ψ to player i. That is

(2.6) $$U_i^*(s_i;\ \Psi) = V_i^-(\Psi).$$

PROOF. For any pair of EPS s and t in Ψ, in view of (2.3) we must have

(2.7) $$U_i(s_i,\ s^i) = U_i(t_i,\ s^i).$$

Hence, by (2.4)

(2.8) $$U_i^*(s_i;\ \Psi) = U_i^*(t_i;\ \Psi) = V_i^-(\Psi)$$

as desired.

We say that a given strategy s_i of player i is _used_ in sub-solution Ψ if s_i is used in any EP $s = (s_i,\ s^i)$ belonging to Ψ. The set $\Phi(s_i)$ of all sub-solutions Ψ in which a given strategy s_i is used is called the _sub-solution range_ of strategy s_i. If two strategies s_i and t_i of a given player i have the same sub-solution range $\Phi(s_i) = \Phi(t_i)$ then we say that s_i and t_i are _similar_ strategies. Clearly, similarity is reflexive, symmetric and transitive. Hence the relation of similarity will partition the strategy space S_i of each player i into equivalence classes $X_i^1, \ldots, X_i^k$ of similar strategies, called _similarity sets_.

Two EPS s and t are _equivalent_ if $U(s) = U(t)$.

Two EPS s and t are _safely equivalent_ within some admissible set A if they are equivalent in terms of the relevant _safe payoffs_, that is, if $U^*(s;\ A) = U^*(t;\ A)$.

Suppose that in a given game Γ the players have to choose between EPS s and t. We speak of a _coordinated_ choice if _either_ all players choose s (i.e., if all of them use the strategies $s_1, \ldots, s_n$ belonging to s) _or_ if all players choose t (i.e., if all of them use the strategies $t_1, \ldots, t_n$ belonging to t).

We say that a given game Γ involves a _coordination problem_ if it contains at least one pair of EPS s and t such that s and t are _safely equivalent_ within some set $A \subseteq \bar{\Gamma}$ but are _not interchangeable_.

CONCLUSION V. Only if a game Γ involves a coordination problem does it make any difference whether the game is fully vocal on the one hand, or is tacit or semi-vocal on on the other hand.

EXPLANATION. If two EPS s and t yield different safe payoff vectors[3] $U^*(s; A)$ and $U^*(t; A)$ within all sets $A \subseteq \bar{\Gamma}$ then, as we shall see, the choice between s and t can always be uniquely decided by choice criteria based on dominance and on bargaining considerations.[4] If the players follow these criteria then all of them will make the same choice between s and t even without communication and no problem of strategy coordination will arise.

In contrast, if within the relevant admissible set A the EPS s and t yield the same safe payoff vector $U^*(s; A) = U^*(t; A)$ then the players will have no a priori criterion for choosing between s and t. Even if all players were willing to accept the safe payoff vector associated with s and t, in the absence of communication in general they will make uncoordinated choices, some players i choosing strategies s_i belonging to s while other players j choosing strategies t_j belonging to t. Thus the result will be in general neither s nor t but rather some recombination r of s and t.

But if s and t are interchangeable then this fact will pose no problem because any recombination r of s and t will itself be also an

---

3 In view of Conclusion IV, the players will in general compare two EPS s and t, not in terms of the definitional payoffs they yield, but rather in terms of the safe payoffs they yield within the relevant admissible set $A \subseteq \bar{\Gamma}$.

4 If one of the two EPS is preferable to the other for all players (or even if it is preferable for some players and is at least indifferent for the other players) in terms of the relevant safe payoffs then it will dominate the other EP.

On the other hand, if the players' preferences are divided between the two EPS then the choice between them will be a matter of bargaining among the players. As we shall see, the factors deciding the outcome of this bargaining process can be formalized by means of our concept of risk-dominance. The latter will provide an unambiguous criterion to decide when rational behavior will involve acceptance of EP s by all players; when it will involve acceptance of EP t by all players; and when it will involve acceptance of a compromise outcome different from both s and t.

EP, yielding the same safe payoff vector as s and t do.[5] Hence in this case lack of communication will cause no difficulty and we shall say that the safe equivalence of EPS s and t does <u>not</u> give rise to a coordination problem.

On the other hand, if the two safely equivalent EPS s and t are <u>not</u> interchangeable then without communication[6] the players will be in general unable to reach the payoff vector $U^* = U^*(s; A) = U^*(t; A)$ at all. This is so because an uncoordinated choice, resulting in some recombination r of s and t, in general will yield a payoff vector different from U^*; and will also result in an unstable situation because in general r will not be an EP. Thus the players may have to choose some EP less favorable than s and t, or they may have to fall back even on using their maximin strategies. Hence, in such a case it does make a difference whether the game is a <u>vocal</u> game or not, and we shall say that the choice between s and t involves a <u>coordination problem</u>.

Consider, for instance, the following game:

---

5 This can be shown as follows. The problem of choice between s and t arises only if the relevant admissible set A contains both s and t. (Moreover, in view of the possibility of uncoordinated choices, A must also contain all recombinations r of s and t.) If s and t are safely equivalent then

$$U^*(s; A) = U^*(t; A).$$

For any recombination r and for any $i \in N$, either $r_i = s_i$ or $r_i = t_i$. Suppose first that $r_i = s_i$. Then

$$U_i^*(r; A) = U_i^*(s_i, r^i; A) = U_i^*(s_i, s^i; A)$$

$$= U_i^*(s; A) = U_i^*(t; A)$$

where the second equality sign holds in view of (2.4). We can show by similar reasoning that if $r_i = t_i$ then

$$U_i^*(r; A) = U_i^*(t; A) = U_i^*(s; A).$$

But this argument applies to all $i \in N$. Hence

$$U^*(r; A) = U^*(s; A) = U^*(t; A)$$

as desired.

6 To be sure, coordination of the different players' strategies need not require <u>explicit</u> communication even in this case. As Schelling [9, Chapters 3 and 4] has pointed out, intelligent players may be able to <u>guess</u> each other's strategy intentions even without explicit communication. But in such a case, for the purposes of a formal theory, we would classify the game as a <u>vocal</u> game and would speak of <u>implicit</u> communication among the players, even if physically it were a <u>tacit</u> game. In other words, we would analyse the game in terms of its V-solution and not in terms of its T-solution.

	β_1	β_2
α_1	(1,1)	(0,0)
α_2	(0,0)	(1,1)

EXAMPLE 3.

This game has three EPS, viz. $s = (\alpha_1, \beta_1)$; $t = (\alpha_2, \beta_2)$ and $r = (\frac{1}{2}\alpha_1 + \frac{1}{2}\alpha_2, \frac{1}{2}\beta_1 + \frac{1}{2}\beta_2)$. Both s and t yield the payoff vector (1,1) while r yields only $(\frac{1}{2}, \frac{1}{2})$. Obviously, if the game is played as a <u>vocal</u> game then the players will choose s or t (they will not care which) since these both dominate r. But if the game is played as a <u>tacit</u> or as a <u>semi-vocal</u> game then the players will be unable to coordinate their choices between s and t, and this will leave them no other alternative than choosing r. (Though r will yield the players only their maximin payoffs $\check{U}_1 = \check{U}_2 = \frac{1}{2}$, the choice of r in this case is not inconsistent with Conclusion III because the strategies r_1 and r_2 are not only equilibrium strategies but are also maximin strategies.)[7]

CONCLUSION VI. The V-solution for fully <u>vocal</u> games must be defined either as a set M of <u>maximin points</u> or as a set E of <u>equilibrium points safely equivalent</u> with respect to some admissible set $A \supseteq E$.

CONCLUSION VII. The T-solution for <u>tacit</u> and <u>semi-vocal</u> games must be defined either as a set M* of <u>maximin points</u> or as a set E* of <u>interchangeable equilibrium</u> points, such that if both s and t are in E* then

[7] This game happens to have the interesting property that if the players try to choose between s and t so as to leave coordination of their choices to <u>chance</u>, this will automatically lead them to r. This is so because, in the absence of explicit coordination, each player will choose both of his two strategies presumably with 50-50 per cent probability, the two players' choices being statistically independent. But this amounts to the same thing as if player 1 used mixed strategy

$r_1 = \frac{1}{2}\alpha_1 + \frac{1}{2}\alpha_2$ while player 2 used mixed strategy

$r_2 = \frac{1}{2}\beta_1 + \frac{1}{2}\beta_2$.

every <u>recombination</u> r of s and t is also in E*. (Being interchangeable, the equilibrium points in E* will necessarily be also <u>safely equivalent</u> within any admissible set $A \supseteq E^*$.)[8]

EXPLANATION. The purpose of our two solution concepts is to select a unique <u>payoff vector</u> (or at least a unique safe payoff vector) as the outcome of the game, assuming rational behavior by all players; and to indicate the eligible individual strategies and the eligible strategy n-tuples by which this payoff vector can be actually achieved by the players.

As a matter of convenience, formally each solution concept will be defined simply as the set of all eligible strategy n-tuples. In view of Conclusions II and III in certain cases the player have to use maximin strategies, and so one or both solutions will have to be defined as a set of <u>maximin points</u>. In all other cases, in view of Conclusion I, each solution must be defined as a set of <u>equilibrium points</u>.

Finally, the safe-equivalence requirement for the V-solution, and the interchangeability requirement for the T-solution, follow from Conclusions IV and V and the subsequent discussion. In the case of fully <u>vocal</u> games we can use a solution concept based on the assumption that the players can always coordinate their strategy choices when they find this desirable. Thus it is sufficient if the solution specifies a set E of safely equivalent EPS. The choice among these EPS will be a matter of indifference to the players, and since they can communicate they will have no difficulty in achieving a coordinated choice.

In contrast, in the case of <u>tacit</u> and <u>semi-vocal</u> games, we need a solution concept which can be used without the possibility of strategy coordination. Thus the solution must consist of a set E* of EPS such that any recombination r of two EPS s and t in E* should itself be also an EP in E*. Hence E* must be a set of mutually interchangeable EPS closed under interchangeability.

It may be noted that the Nash solution, which is meant to apply only to what we have called <u>tacit</u> non-cooperative games, is defined as the

---

8 Cf. Footnote 5.

set of all EPS in a game in which all EPS are interchangeable, and so satisfies our Conclusion VII.

§ 3. THE REDUCED GAME Γ^* DEFINED BY PAYOFF-DOMINANCE

In this section, by _dominance_ we shall mean _weak_ dominance, denoted by $\succ$.

Accordingly, in the case of two _payoff vectors_ U and $\bar{U}$ (where U or $\bar{U}$ or both may also be safe payoff vectors), $U \succ \bar{U}$ if

$$U_i \geqq \bar{U}_i \tag{3.1}$$

for _all_ $i \in N$, with a $>$ sign for at least _one_ $i \in N$.

In the case of two EPS s and t, we define two dominance relations, _simple dominance_ and _strict dominance_.

EP s _simply dominates_ EP t within some admissible set A, written as

$$s \underset{A}{\succ} t \tag{3.2}$$

if

$$U^*(s;\ A) \succ U^*(t;\ A). \tag{3.3}$$

EP s _strictly dominates_ EP t within some admissible set A, written as

$$s \overset{*}{\underset{A}{\succ}} t \tag{3.4}$$

if

$$U^*(s;\ A) \succ U(t). \tag{3.5}$$

Clearly, in view of (2.4), strict dominance always implies simple dominance but not conversely. Thus strict dominance can be regarded intuitively as a conservative assessment of dominance relations between EPS.

Moreover, if $\bar{A} \subseteq A$ then in view of (2.4)

$$(3.6) \qquad s \overset{*}{\underset{A}{\succ}} t \quad \text{implies} \quad s \overset{*}{\underset{\bar{A}}{\succ}} t$$

but in general

$$(3.7) \qquad s \underset{A}{\succ} t \quad \text{does not imply} \quad s \underset{\bar{A}}{\succ} t.$$

An EP s is called <u>simply undominated</u> (or <u>strictly undominated</u>) within some admissible set A if s is not simply dominated (is not strictly dominated) by some other EP t within the set A.

We shall construct a reduced game Γ^* from the original game Γ by progressively deleting more and more strategies for various players on the basis of certain dominance considerations. At each stage of this construction process we shall define the admissible set A as the set of all original EPS still included in the game, and will define strict dominance and simple dominance relations in terms of this admissible set A. In view of (3.6), our assessment of <u>strict dominance</u> relations will be valid even if we happen to use an admissible set A larger than strictly necessary, i.e., even if we have not yet excluded from the game all EPS that can be excluded on the basis of the information available. In contrast, in view of (3.7), in assessing <u>simple dominance</u> relations we must be sure that we have already excluded from the game, and therefore from the admissible set A, all EPS that can be excluded at all on the basis of the information available at that stage.[9] For

---

[9] For example, consider a three-person game Γ in which one of player 1's strategies, $s_i = t_i$, is used both in EP $s = (s_1, s_2, s_3)$ and in EP $t = (t_1, t_2, t_3)$, with $U_1(s) = 2$ and $U_1(t) = 1$. Let r be a third EP not using the same strategies. Clearly, if both s and t are included in the admissible set A then $U_1^*(s; A) = 1$ while if t is excluded from A then $U_1^*(s; A) = 2$. Thus it may happen that the third EP r will <u>simply dominate</u> s if t is included in A but will <u>not</u> simply dominate s <u>if t is</u> excluded from A.

In such a case, if we want to use the concept of simple dominance, it will be essential to decide in advance whether there is any reason for excluding t from A. (Such a reason may be e.g., that s yields higher payoffs than t does for both players 2 and 3 so that they will always prefer s to t.)

But if we want to decide only whether r <u>strictly dominates</u> s then the status of t with respect to the admissible set A is immaterial because this strict-dominance relation depends only on the definitional payoff $U_1(s) = 2$ and not on the safe payoff $U_1^*(s; A)$.

this reason we shall first eliminate from the game the EPS dominated by other EPS in terms of strict dominance relations and shall only then eliminate the EPS dominated merely in terms of simple dominance relations.

All these dominance relations defined in this section will be called payoff-dominance relations, to be distinguished from the risk-dominance relations we shall define in Section 6 and subsequent sections.

We now construct the reduced game Γ^* by the following steps.

Step (α). We delete all strategies s_i not used in some equilibrium point $s = (s_i, s^i)$.

An EP is called profitable if it yields every player a payoff higher than his maximin payoff, that is, if

$$U_i(s) > \check{U}_i \qquad \text{for all} \quad i \in N. \tag{3.8}$$

Step (β). We delete all strategies s_i not used in some profitable equilibrium point $s = (s_i, s^i)$.

This step is motivated by Conclusion III, which suggests that the players should not use equilibrium strategies associated only with unprofitable EPS (but rather should use other equilibrium strategies, or even maximin strategies instead, if necessary).

Step (γ). We define the set of all equilibrium points still included in the game as the admissible set A. We delete all strategies s_i not used in some equilibrium point $s = (s_i, s^i)$ strictly undominated within A. Then we re-define A as the set of still remaining equilibrium points and repeat this procedure until the set A achieves a stable size.

Step (δ). At each stage again defining A as the set of all equilibrium points still included in the game, we delete all strategies s_i not used in some equilibrium point $s = (s_i, s^i)$ simply undominated within A. We repeat this procedure until A reaches a stable size.

Step (ε). For each remaining equilibrium point s we replace the definitional payoff vector $U(s)$ by the safe payoff vector $U^*(s; A)$. In what follows the symbols U^* and $U_1^*, \ldots, U_n^*$ will refer to this re-defined payoff function. That is

$$U_i^*(s) = U_i^*(s; A) \quad \text{and} \quad U^*(s) = U^*(s; A) \tag{3.9}$$

if s is an EP of the original game Γ whereas

$$U_i^*(s) = U_i(s) \quad \text{and} \quad U^*(s) = U(s) \tag{3.10}$$

if s is not an EP of Γ.

This last step is introduced because, in view of Conclusion IV, we want to define all risk-dominance relations in terms of the relevant safe payoffs, rather than in terms of the corresponding definitional payoffs.

The smaller game obtained by the iterative procedure consisting of Steps (α) to (ϵ) is called the reduced game Γ^*. The set of all equilibrium points in Γ^* will be denoted by $\overline{\Gamma}^*$.

For any game Γ containing only a finite number of sub-solutions Ψ, the iterative procedure just described always comes to an end after a finite number of operations. This is obviously true for Steps (α), (β), and (ϵ). But it is also true for Steps (γ) and (δ), because in implementing these steps whenever we delete a given strategy s_i we also delete all strategies $\overline{s}_i$ in the same similarity set X_i^j; and the number of such similarity sets in Γ is finite if Γ contains only a finite number of sub-solutions.

However, if in a given game Γ the number of sub-solutions is infinite[10] then Steps (γ) and (δ) may give rise to two infinite sequences of admissible sets A, which we shall denote by $A_1, A_2, \ldots$ and $\overline{A}_1, \overline{A}_2 \ldots$ respectively. But both these sequences will converge because $A_1 \supset A_2 \supset \ldots$ and $\overline{A}_1 \supset \overline{A}_2 \supset \ldots$ and so

$$A_o = \lim_{m\to\infty} A_m = \bigcap_{m=1}^{\infty} A_m \tag{3.11}$$

$$\overline{A}_o = \lim_{m\to\infty} \overline{A}_m = \bigcap_{m=1}^{\infty} \overline{A}_m. \tag{3.12}$$

Moreover, if Step (β) has not made already the first admissible set A_1 empty, then both A_o and $\overline{A}_o$ will be nonempty sets.

To show that A_o is nonempty, consider the set B of all definitional payoff vectors $U(s)$ such that $s \in A_1$. Clearly there must be at least one vector $U(s^*)$ undominated by any other vector $U(t)$ in B. The

---

10 I am indebted to Lloyd Shapley for pointing out to me that even a finite game may contain infinitely many, and indeed uncountably many, different sub-solutions.

EP s^* associated with this vector $U(s^*)$ will be a member of every admissible set A_m, $m = 1, 2, \ldots$. Hence $s^* \in A_o$ and so A_o will be nonempty.

We will now show that $\bar{A}_o$ is also nonempty. Suppose, to the contrary, that $\bar{A}_o$ is empty. Then for every EP s in set A_o there exists in the sequence $\bar{A}_1, \bar{A}_2, \ldots$ a <u>last</u> admissible set A_m containing s, so that $\bar{A}_m \ni s$ but $\bar{A}_{m+1} \not\ni s$. Let us write $\bar{U}(s) = U^*(s; \bar{A}_m)$ for the safe payoff vector that EP s yields within this last admissible set $\bar{A}_m$. Now consider the set C of all vectors $\bar{U}(s)$ such that $s \in A_o$. In this set C there must be at least one vector $\bar{U}(s^{**})$ undominated by any other vector $\bar{U}(t)$ in C.

Let $A_{m^{**}}$ be the <u>last</u> admissible set A_m containing this EP s^{**}. Then there must be in $A_{m^{**}}$ some EP t such that

$$U^*(t; A_{m^{**}}) \succ U^*(s^{**}; A_{m^{**}}) = \bar{U}(s^{**}). \tag{3.13}$$

But this implies that

$$\bar{U}(t) \succ \bar{U}(s^{**}) \tag{3.14}$$

because by (3.7)

$$\bar{U}_i(t) \geqq U_i^*(t; A_{m^{**}}) \qquad \text{for all } i \in N. \tag{3.15}$$

However, this is a contradiction because (3.14) is inconsistent with the fact already established that the vector $\bar{U}(s^{**})$ is undominated by any other vector $\bar{U}(t)$ in C. Therefore $\bar{A}_o$ must be nonempty as desired.

When Step (ε) has been completed four cases are possible:

($\alpha\alpha$) The reduced game Γ^* may be <u>empty</u> because of Step (γ). Then in view of Conclusion III, we define both the V-solution and the T-solution as the set M of all <u>maximin points</u> $\check{s}$ in the original game Γ.

($\beta\beta$) All equilibrium points in Γ^* may be <u>interchangeable</u>, i.e., the reduced game Γ^* may have a Nash solution. Then we define both the V-solution and the T-solution as the set $\bar{\Gamma}^*$ of all equilibrium points in Γ^*.

($\gamma\gamma$) All equilibrium points in Γ^* may be equivalent[11] without all of them being interchangeable. Then we define the V-solution as the set $\bar{\Gamma}^*$ of all equilibrium points in Γ^* but define the T-solution in the same way as in Case ($\delta\delta$).

($\delta\delta$) The equilibrium points in Γ^* may be neither equivalent nor interchangeable. Then we define both the V-solution and the T-solution in terms of risk-dominance relations as will be described below.

§ 4. ZEUTHEN'S PRINCIPLE

The concept of risk-dominance we shall use is a generalization of a decision rule, first suggested by Zeuthen [10, Chapter IV] for two-person bargaining. He essentially argues as follows. Let x_1 be party 1's last offer and x_2 be party 2's last offer while let c denote the conflict situation. Then, assuming that each party tries to maximize the mathematical expectation of his payoff, the maximum risk (the largest subjective probability of a conflict) that party i (i, j = 1, 2) would be prepared to face in order to achieve x_i rather than x_j is

$$(4.1) \qquad R_i = R_i(x_i, x_j) = \frac{U_i(x_i) - U_i(x_j)}{U_i(x_i) - U_i(c)} .$$

Now Zeuthen's decision rule says that

(a) Party i should make a concession and party j should not make one if $R_i < R_j$ whereas

(b) Both parties should make some concessions if $R_i = R_j$.

I have shown elsewhere [1, 3] that Zeuthen's principle is the only decision rule consistent with certain very natural rationality postulates. If applied to the two-person cooperative game it yields a solution mathematically equivalent to Nash's solution [6, 8, 1]. (That is, to Nash's cooperative solution for the two-person game, to be distinguished from his non-cooperative solution concept, which we have been using in this paper.) It can also be used to generalize Nash's two-person solution for the n-person

---

11 In view of Step (ε) this means that as equilibrium points of the original game Γ they are safely equivalent within $\bar{\Gamma}^*$ as admissible set.

cooperative game [2, 4]. But in my view the <u>same</u> rationality postulates which motivate its use in cooperative games equally justify its use also in non-cooperative games.

As we shall see (in Section 6), in the case of non-cooperative games Zeuthen's principle and the resulting concept of risk-dominance are closely related to Luce and Raiffa's notion of "psychological dominance" [5, p. 110].

§ 5. THE DIRECT RISK FUNCTION $P_i(s, t)$ AND THE EXTENDED RISK FUNCTION $R_i(s, t)$

Let $s = (s_i, s^i)$ and $t = (t_i, t^i)$ be two EPS in the reduced game Γ^*, with $s_i \neq t_i$, and with $U_i^*(s) > U_i^*(t)$. Suppose that all players, apart from player i, are willing to accept the payoffs belonging to the payoff vector $U^*(t)$ and are therefore willing to cooperate in achieving EP t. Then, if player i tells the other players[12] that he is likewise willing to accept payoff $U_i^*(t)$ and to use strategy t_i, the other players j will use the strategies t_j belonging to t^i and so player i will obtain $U_i^*(t)$ with certainty. But if he insists on using strategy s_i in order to obtain the higher payoff $U_i^*(s)$ then he <u>may</u> in fact obtain $U_i^*(s)$ because the other players j may in the end come round to using the strategies s_j belonging to s^i. Yet he also runs the <u>risk</u> of obtaining only $U_i^*(s_i, t^i)$ because the other players j just as well may stick to their original intention of using the strategies t_j belonging to t^i, even if player i himself uses strategy s_i.

The <u>maximum risk</u> player i would take in order to achieve EP s rather than EP t (i.e., the highest subjective probability he may assign to the possibility that the other players j will use the strategies t_j, without being himself thereby deterred from using strategy s_i) is

$$P_i(s, t) = \frac{U_i^*(s) - U_i^*(t)}{U_i^*(s) - U_i^*(s_i, t^i)}. \tag{5.1}$$

This quantity $P_i(s, t)$ we call player i's <u>direct risk function</u> for s and against t.

[12] By assuming that player i can <u>tell</u> the other players about his willingness to accept payoff $U_i^*(t)$ and to use strategy t_i, our model becomes (at least) a <u>semi-vocal</u> game.

Obviously player i will take no risk whatever for the sake of s if he does not clearly prefer s to t. Hence we write

$$P_i(s, t) = 0 \tag{5.2}$$

whenever $U_i^*(s) \leqq U_i^*(t)$.

Let $W_i(t)$ be the set of all EPS r in the reduced game Γ^*, such that <u>either</u> r = t or $U_i^*(r) < U_i^*(t)$. Thus $W_i(t)$ contains EP t itself and also all EPS r less favorable than t from the standpoint of player i. We shall say that player i is <u>uniformly</u> willing to take the risk R in order to achieve some given EP s rather than <u>any</u> EP r in this set $W_i(t)$, if for <u>every</u> EP r in $W_i(t)$ the maximum risk $P_i(s, r)$ that he is willing to take for the sake of s and against r is uniformly <u>at least</u> as high as R; that is, if

$$P_i(s, r) \geqq R \qquad \text{for all } r \in W_i(t). \tag{5.3}$$

The largest number R satisfying (5.3) is obviously

$$R = R_i(s, t) = \min_{r \in W_i(t)} P_i(s, r). \tag{5.4}$$

Thus $R_i(s, t)$ is the maximum risk that player i would be <u>uniformly</u> willing to take in order to achieve EP s and in order to defeat EP t and all other EPS r <u>even less favorable</u> than t itself. This quantity $R_i(s, t)$ we call player i's <u>extended risk function</u> for s and against t.

Clearly R_i has the monotonity property that

$$R_i(s, t) \leqq R_i(s, t^*) \tag{5.5}$$

if $U_i^*(t) > U_i^*(t^*)$.

We now propose to show that for the application of Zeuthen's principle in non-cooperative games and for the definition of risk-dominance relations, the proper risk function to use is the <u>extended</u> risk function $R_i(s, t)$ rather than the direct risk function $P_i(s, t)$.

In the derivation of Zeuthen's principle as a decision rule for bargaining situations, one of the important rationality postulates used [3, p. 184] is the

> ACCEPTANCE OF HIGHER PAYOFFS POSTULATE. If a given player is willing to accept some particular outcome x then he is also willing to accept any outcome y yielding him an even higher payoff.

A natural generalization of this postulate is the

> RISK MONOTONITY POSTULATE. Suppose that a given player i would take a risk no higher then $P_i = P_i(s, r)$ in order to defeat EP r, for the sake of EP s as his positive objective. Then he will <u>not</u> take a risk <u>higher</u> than P_i in order to defeat some EP t <u>less unfavorable</u> to him than r, for the sake of the same EP s as his positive objective. That is, if player i dislikes t less than he dislikes r then he will not oppose t more strongly than he opposes r.

Thus if $U_i^*(r) < U_i^*(t)$ then the risk $\bar{R}_i(s, t)$ that player i will take in order to defeat r for the sake of s will satisfy $\bar{R}_i(s, t) \leqq P_i(s, r)$. Hence, in view of (5.4) this risk $\bar{R}_i(s, t)$ will be defined by the extended risk function $\bar{R}_i(s, t) = R_i(s, t)$.

Indeed, the Risk Monotonity postulate is not only a natural generalization of the Acceptance of Higher Payoffs postulate, but directly <u>follows</u> from the latter and from Zeuthen's principle taken together. If Zeuthen's principle were applied to a risk function $\bar{R}_i$ not satisfying the Risk Monotonity then we would obtain a decision rule violating the Acceptance of Higher Payoffs postulate. For in that case it could happen that $\bar{R}_i(s, t)$ would have a high value while $\bar{R}_i(s, r)$ would have a low value, even though $U_i^*(t) > U_i^*(r)$. Hence, by Zeuthen's principle player i might be able to refuse accepting EP t even though he might have to accept EP r, which would be inconsistent with the Acceptance of Higher Payoffs postulate.

§ 6. THE CONCEPT OF PRIMARY RISK-DOMINANCE

Let s and t be two EPS in the reduced game Γ^*. We say that s weakly risk-dominates t (at the risk level $\tilde{R}$) if

$$\tilde{R} = \max_i R_i(s, t) \geqq \max_j R_j(t, s). \tag{6.1}$$

If we can replace the $\geqq$ sign by the $>$ sign in (6.1) then we say that s strongly risk-dominates t (at the risk level $\tilde{R}$). These relations we shall call weak and strong primary risk-dominance relations respectively.

If for a given player i^*

$$\tilde{R} = R_{i^*}(s, t) \tag{6.2}$$

then we say that his opposition to t is (strongly or weakly) decisive because his opposition to t at risk level $\tilde{R}$ gives rise to this (strong or weak) risk-dominance relation.

Our concept of strong primary risk-dominance can be regarded as a quantitative re-statement and generalization of Luce and Raiffa's qualitative notion of psychological dominance [5, p. 110].

Consider, for instance, the following game:

	β_1	β_2
α_1	(2,1)	(−1,−100)
α_2	(0,0)	(1,3)

EXAMPLE 4.

In this game player 1 would prefer EP $s = (\alpha_1, \beta_1)$ whereas player 2 would prefer EP $t = (\alpha_2\ \beta_2)$. But, if both players used strategies associated with their own favorite EPS, i.e., if player 1 used strategy α_1 while player 2 used strategy β_2, then player 2 would suffer a much heavier loss (as compared with the payoff difference to him between the two EPS). In such a case Luce and Raiffa would argue that player 2 would be more afraid to use strategy β_2 in defiance of player 1 using strategy α_1, than player 1 would be afraid to use strategy α_1 in defiance of player 2 using strategy β_2. Therefore EP $s = (\alpha_1, \beta_1)$ would psychologically dominate EP $t = (\alpha_2, \beta_2)$.

In our own terminology, EP s would strongly risk-dominate EP t. This would be so because player 1 would be willing to use strategy α_1 even if he assigned the probability $R_1(s, t) = P_1(s, t) = 1/3$ to the possibility that player 2 may use strategy β_2. In contrast, player 2 would be willing to use strategy β_2 only if the probability he would assign to the possibility that player 1 might use strategy α_1 did not exceed $R_2(t, s) = P_2(t, s) = 2/103$, which is much less than $R_1(s, t) = 1/3$.

§ 7. THE CONCEPT OF SECONDARY RISK-DOMINANCE

These primary risk-dominance relations, however, in general do not select a determinate outcome for the game, because in most games there will not be any EP s having strong or even weak primary risk-dominance over all other EPS t. In order for a given EP s to have at least weak primary risk-dominance over all other EPS, it is necessary and sufficient that s itself should not be subject to strong primary risk-dominance by any other EP t. But in most games every EP s will be in fact subject to some strong risk-dominance relations. Basically this is due to the fact that primary risk-dominance is not a transitive relation. Hence even if s strongly risk-dominates t while t strongly risk-dominates u this does not exclude the possibility of s being in turn itself subject to strong risk-dominance by u.

But we do obtain a transitive risk-dominance relation, called secondary risk-dominance, and obtain a determinate outcome for the game, if we analyse the primary risk-dominance relations themselves in terms of Zeuthen's principle.

Let Γ^* be a reduced game in which every EP t is strongly risk-dominated at least by one other EP s at some risk level $\tilde{R} = H(s, t)$. Let $i = i^*(s, t)$ be the player whose opposition to t is decisive in establishing this risk-dominance relation, i.e., the player i for whom $R_i(s, t) = \tilde{R} = H(s, t)$. But EP s itself, which player i favors, will also be strongly risk-dominated by some other EP r, now player $j = i^*(r, s)$ being the player whose opposition to s is decisive, etc. Moreover, any given EP may be strongly risk-dominated by more than one EP. For instance, EP t may be risk-dominated not only by s but also by some other EP s', now perhaps another player $i' = i^*(s', t)$ being the decisive opponent of t, etc.

Hence in this game, on the basis of the primary risk-dominance relations, every EP t is subject to the veto power of one or more players whose active opposition to t would be decisive. Consequently, the players will be able to agree on one particular EP t only if the player or players having veto power against this EP t refrain from making use of it — presumably because they come to the conclusion that they cannot achieve any EP s more favorable than t anyhow since any such EP s would be vetoed by some other players.

It is at this point that we can make use of Zeuthen's principle, now for a second time, to decide the outcome. If it is necessary that _either_ a given player i _or_ a given player j should give up his active opoosition to some EP then by Zeuthen's principle this concession must come from that player who would be _less_ willing to risk a conflict in order to defeat the EP that he opposes, i.e., it must come from that player whose opposition is associated with a _lower risk level_ $\tilde{R}$. (If both players' opposition is associated with exactly the same risk level $\tilde{R}$ then by Zeuthen's principle both of them must make some concessions — we shall come back to this case in Section 8.) In other words, primary risk-dominance relations associated with _higher_ risk levels $\tilde{R} = R^*$ always _take precedence_ over risk-dominance relations associated with lower risk levels $\tilde{R} = R^{**}$.

Thus Zeuthen's principle in the first instance establishes risk-dominance relations, called _primary_ risk-dominance relations, between the EPS of a given reduced game Γ^*. But then it also establishes a precedence order, which may be regarded as a _second-order_ risk-dominance relation, between these primary risk-dominance relations themselves. We shall now show how this precedence order in turn generates a new type of risk-dominance relation, called _secondary_ risk-dominance, between the various EPS.

For any pair of EPS s and t in the reduced game Γ^*, let

(7.1)

$$H(t,\ s) = \tilde{R} \quad \text{if } t \text{ has strong primary risk-dominance over } s \text{ at risk level } \tilde{R}.$$

$$H(t,\ s) = 0 \quad \text{if } t \text{ has no strong primary risk-dominance over } s \text{ at all.}$$

Moreover, let

$$K(s) = \max_t H(t,\ s). \tag{7.2}$$

In other words $K(s)$ is the highest risk level $\tilde{R}$ at which s is subject to strong primary risk-dominance by any EP t in Γ^*.

We also write

$$K_i(s) = \max_t R_i(t,s). \tag{7.3}$$

Thus $K_i(s)$ is the highest risk level $R_i(t,\ s)$ at which a given player i opposes s in favor of some other EP t (including cases where player i's opposition to s for the sake of some EP t is ineffective, i.e., does not lead to primary risk-dominance of t over s).

In view of (6.1) we can also define $K(s)$, equivalently to (7.2), as

$$K(s) = \max_t \max_i R_i(t,\ s) = \max_i \max_t R_i(t,\ s) = \max_i K_i(s). \tag{7.4}$$

Thus $K(s)$ is the highest risk level $R_i(t,\ s)$ at which s is opposed by any player (again including cases of ineffective opposition).

We say that EP s has <u>weak secondary risk-dominance</u> over t if

$$K(s) \leqq K(t). \tag{7.5}$$

We say that s has <u>strong secondary risk-dominance</u> over t if in (7.5) we can replace the $\leqq$ sign by a $<$ sign. Thus intuitively s has strong (or weak) secondary risk-dominance over t if s meets with <u>less intensive</u> (or with no more intensive) opposition among the players that t does.

Clearly secondary risk-dominance, whether weak or strong, is a transitive relation, and in fact weak secondary risk-dominance is a complete quasi-ordering of all EPS in the reduced game Γ^*.

In what follows, the phrase "... strongly (or weakly) risk-dominates ..." will always refer to secondary risk-dominance.

COROLLARY I. In any reduced game Γ^* there exists at least one EP s weakly risk-dominating all other EPS t, viz. the EP s satisfying

$$K(s) = \min_t K(t) \tag{7.6}$$

COROLLARY II. A given EP s strongly risk-dominates all other EPS t if and only if s is the only EP weakly risk-dominating all other EPS.

§ 8. DEFINITION OF THE V AND T-SOLUTIONS

The V-solution is the set E of all EPS s in the reduced game Γ^* such that

1. either s strongly risk-dominates all other EPS t in Γ^*
2. or s weakly risk-dominates all other EPS t and strongly risk-dominates at least those EPS t which are not equivalent to s,

provided that this set E is not empty.

The T-solution is the set E* of all EPS s in the reduced game Γ^* such that

1. either s strongly risk-dominates all other EPS t in Γ^*
2. or s weakly risk-dominates all other EPS t and strongly risk-dominates at least those EPS t which are not interchangeable with s,

provided that

(i) the set E* is not empty, and provided that

(ii) every recombination r of two EPS s and t in E* also belongs to E*.

These definitions are based on Part (a) of Zeuthen's principle and on Conclusions VI and VII.

Let F be the set of all EPS s such that s <u>weakly</u> risk-dominates every other EP t in Γ^*. By Corollary I, the set F is never empty. Moreover, by Corollary II, if F contains only <u>one</u> EP s then the set E = E* = F itself will satisfy the definitions of both the V-solution and the T-solution.

But if F contains two or more EPS then it may happen that neither E = F (or E* = F) itself nor any nonempty subset E (or E*) of F will satisfy the definition of the V-solution (or of the T-solution respectively). Intuitively this means that in such cases our rationality criteria will not enable the players to choose among the alternative EPS in set F. In these cases we shall say that the game is <u>at deadlock</u> among the various EPS in F. Such deadlock situations, however, can occur only in very special cases, which represent a set of <u>measure zero</u> in the space of all possible finite noncooperative games.[13]

In the case of a deadlock situation we apply Part (b) of Zeuthen's principle, which requires that both (or all) players must make concessions.

---

[13] The simplest types of <u>deadlock situations</u> occur in certain <u>symmetric</u> two-person games. Consider, for instance, the two games:

	β_1 β_1	β_2 β_2
α_1	(1,1)	(0,0)
α_2	(0,0)	(1,1)

EXAMPLE A

	β_1	β_2
α_1	(2,1)	(0,0)
α_2	(0,0)	(1,2)

EXAMPLE B

(Example A is identical to Example 3 of Section 2 above.)

We assume that Example A is a <u>tacit</u> or a <u>semi-vocal</u> game. In the case of Example B it does not matter whether it is <u>vocal</u>, or <u>semi-vocal</u>, or <u>tacit</u>. In both examples there is a deadlock between the two EPS s = (α_1, β_1) and t = (α_2, β_2). In both cases, because of the complete symmetry of the game, no <u>conceivable</u> rational criterion could decide between the two EPS.

However, in Example A there is no conflict of interest between the two players because the two EPS are equivalent. If the game allowed free communication between the players they would have no difficulty in making a coordinated choice between s and t, and the deadlock arises only from the lack of free communication. In such cases we speak of a <u>coordination deadlock</u>.

In contrast, in Example B the two players have opposite interests: player 1 would prefer EP s while player 2 would prefer EP t. Here free communication would not help to decide the issue. In such cases we speak of a <u>bargaining deadlock</u>.

Let

$$\widetilde{U}_i = \max_{s \in F} U_i^*(s). \tag{8.1}$$

Then, what Zeuthen's principle requires in a deadlock situation is that every player i should be ready to accept some "compromise solution" u with $U_i^*(u) < \widetilde{U}_i$ and accordingly should give up the use of all equilibrium strategies s_i associated with any EP $s = (s_i, s^i)$ yielding him a payoff $U_i^*(s) \geqq \widetilde{U}_i$. The compromise solution u eventually accepted may be an EP of the reduced game Γ^*; or it may be an EP of the original game Γ, excluded from Γ^* in Step (δ) or (ϵ) because it was strictly dominated or simply dominated by some other EP;[14] or it may be even a maximin point of the original game Γ if no suitable EP can be found.

Formally, in the case of a deadlock situation we can proceed as follows. Let G_i be the set of all strategies s_i of player i such that s_i is used in some EP $s = (s_i, s^i)$ in the reduced game Γ^* such that

$$U_i^*(s) \geqq \widetilde{U}_i. \tag{8.2}$$

Then we go back to the original game Γ and delete all strategies s_i belonging to the sets G_i for each player $i \in N$. Let the resulting smaller game be called Γ_2.

We now construct a new reduced game Γ_2^* from Γ_2, going through Steps (α) to (ϵ). If now Γ_2^* has a V-solution or a T-solution (whichever we want to construct) then this will be regarded as the V-solution or the T-solution of the original game Γ. If however Γ_2^* gives rise again to a deadlock situation, we construct a still smaller new game Γ_3 by excluding from Γ_2 all strategies s_i in the new sets G_i. Then we construct a new reduced game Γ_3^* from Γ_3, etc.

---

14 For instance, in the case of Example A of Footnote 12, the EP $r = (\frac{1}{2}\alpha_1 + \frac{1}{2}\alpha_2, \frac{1}{2}\beta_1 + \frac{1}{2}\beta_2)$ will be excluded from the game in Step (δ) because it is strictly dominated both by s and by t. Nevertheless r will be the only rational "compromise solution" between s and t once the players realize that they cannot achieve either s or t for lack of strategy coordination.

If at any stage of this iterative procedure the game becomes empty then we define the relevant solution as the set of all maximin points of the original game Γ.

Once more, if the original game Γ contains only a finite number of sub-solutions then this iterative procedure will always come to an end, because in any reduced game $\Gamma^* = \Gamma_1^*$, or Γ_2^*, etc., all EPS s using similar strategies s_i yield the same payoffs $U^*(s)$. Therefore all strategies s_i belonging to the same similarity set X_i^j will always be excluded from the game at the same time; and the number of such similarity sets will be finite.

If the original game Γ contains infinitely many sub-solutions then we may obtain an infinite sequence of reduced games $\Gamma_1^*, \Gamma_2^*, \ldots$ with corresponding sequences of reduced strategy spaces $S_i^*(1), S_i^*(2), \ldots$ for each player i. Since $S_i^*(1) \supseteq S_i^*(2) \supseteq \ldots$ we can write

$$S_i^* = \lim_{k\to\infty} S_i^*(k) = \bigcap_{k=1}^{\infty} S_i^*(k). \tag{8.3}$$

Let $S_i^{**} = S_i - S_i^*$ where S_i is player i's strategy space in the original game Γ.

We now define the limiting reduced game Γ_o^* as the game which results if in Γ we delete all strategies s_i belonging to S_i^{**} for every player i. If Γ_o^* contains a V-solution (or T-solution) then this is regarded as the solution of the original game Γ. If Γ_o^* does not contain such a solution (e.g., because it is empty) then again we define the relevant solution as the set of all maximin points in Γ.

§ 9. CONCLUSION

The V and the T-solutions for non-cooperative games, which we have defined in terms of payoff-dominance and risk-dominance relations, try to capture the possibilities of rational cooperation for mutual benefit, and the possibilities for explicit or implicit bargaining, which exist even in what we call non-cooperative games (except in the case of two-person zero-sum games, where the two players have no common interests at all). In our view the essential difference between cooperative and non-cooperative games consists only in the fact that in the former the players are unable to cooperate in achieving a payoff vector U _outside_ the set of _equilibrium points_ — however desirable this may be for all of them[15] — but there is no reason why they should not cooperate _within_ the set of equilibrium points.

15 This fact underlies A. W. Tucker's "prisoner's dilemma" example, cf. Luce and Raiffa [5, p. 95].

BIBLIOGRAPHY

[1] HARSANYI, John C., "Approaches to the bargaining problem before and after the theory of games: a critical discussion of Zeuthen's, Hicks', and Nash's theories," Econometrica, 24 (1956), pp. 144-157.

[2] HARSANYI, John C., "A bargaining model for the cooperative n-person game," Contributions to the Theory of Games, IV, Princeton, 1959, pp. 325-355.

[3] HARSANYI, John C., "On the rationality postulates underlying the theory of cooperative games," J. Conflict Resolution, 5 (June 1961), pp. 179-196.

[4] HARSANYI, John C., "A simplified bargaining model for the n-person cooperative game," to be published in International Economic Review 4, (May 1963).

[5] LUCE, Duncan R. and RAIFFA, Howard, Games and Decisions, New York, John Wiley and Sons, 1957.

[6] NASH, John F., "The bargaining problem," Econometrica, 18 (1950), pp. 155-162.

[7] NASH, John F., "Non-cooperative games," Annals of Mathematics, 54 (1951), pp. 286-295.

[8] NASH, John F., "Two-person cooperative games," Econometrica, 21 (1953), pp. 128-140.

[9] SCHELLING, Thomas C., The Strategy of Conflict, Cambridge, Massachusetts, Harvard University Press, 1960.

[10] ZEUTHEN, Frederic, Problems of Monopoly and Economic Warfare, London, G. Routledge and Sons, 1930.

John C. Harsanyi

Wayne State University
Detroit, Michigan